VOLUME FOUR HUNDRED AND NINETY

METHODS IN ENZYMOLOGY

The Unfolded Protein Response and Cellular Stress, Part B

METHODS IN ENZYMOLOGY

Editors-in-Chief

JOHN N. ABELSON AND MELVIN I. SIMON

Division of Biology
California Institute of Technology
Pasadena, California

Founding Editors

SIDNEY P. COLOWICK AND NATHAN O. KAPLAN

VOLUME FOUR HUNDRED AND NINETY

Methods in ENZYMOLOGY

The Unfolded Protein Response and Cellular Stress, Part B

EDITED BY

P. MICHAEL CONN
*Divisions of Reproductive Sciences and Neuroscience (ONPRC)
Departments of Pharmacology and Physiology
Cell and Developmental Biology, and
Obstetrics and Gynecology (OHSU)
Beaverton, OR, USA*

AMSTERDAM • BOSTON • HEIDELBERG • LONDON
NEW YORK • OXFORD • PARIS • SAN DIEGO
SAN FRANCISCO • SINGAPORE • SYDNEY • TOKYO
Academic Press is an imprint of Elsevier

Academic Press is an imprint of Elsevier
525 B Street, Suite 1900, San Diego, CA 92101-4495, USA
30 Corporate Drive, Suite 400, Burlington, MA 01803, USA
32 Jamestown Road, London NW1 7BY, UK

First edition 2011

Copyright © 2011, Elsevier Inc. All Rights Reserved.

No part of this publication may be reproduced, stored in a retrieval system or transmitted in any form or by any means electronic, mechanical, photocopying, recording or otherwise without the prior written permission of the publisher

Permissions may be sought directly from Elsevier's Science & Technology Rights Department in Oxford, UK: phone (+44) (0) 1865 843830; fax (+44) (0) 1865 853333; email: permissions@elsevier.com. Alternatively you can submit your request online by visiting the Elsevier web site at http://elsevier.com/locate/permissions, and selecting *Obtaining permission to use Elsevier material*

Notice

No responsibility is assumed by the publisher for any injury and/or damage to persons or property as a matter of products liability, negligence or otherwise, or from any use or operation of any methods, products, instructions or ideas contained in the material herein. Because of rapid advances in the medical sciences, in particular, independent verification of diagnoses and drug dosages should be made

For information on all Academic Press publications
visit our website at elsevierdirect.com

ISBN: 978-0-12-385114-7
ISSN: 0076-6879

Printed and bound in United States of America
11 12 13 10 9 8 7 6 5 4 3 2 1

**Working together to grow
libraries in developing countries**

www.elsevier.com | www.bookaid.org | www.sabre.org

ELSEVIER   BOOK AID International   Sabre Foundation

Contents

Contributors	xi
Preface	xvii
Volumes in Series	xix

1. Methods for Investigating the UPR in Filamentous Fungi — 1
Thomas Guillemette, Arthur F. J. Ram, Neuza D. S. P. Carvalho, Aymeric Joubert, Philippe Simoneau, and David B. Archer

1. Introduction	2
2. Conditions to Study the UPR in Filamentous Fungi	5
3. Analysis of Expression During the ER Stress	11
4. Modifying the UPR Signaling by Targeted Gene Replacement	14
Acknowledgments	25
References	26

2. Assays for Detecting the Unfolded Protein Response — 31
Karen Cawley, Shane Deegan, Afshin Samali, and Sanjeev Gupta

1. Introduction	33
2. Experimental Approaches for the Detection of ER Stress	36
3. Concluding Remarks	47
Acknowledgments	49
References	49

3. Analysis of the Role of Nerve Growth Factor in Promoting Cell Survival During Endoplasmic Reticulum Stress in PC12 Cells — 53
Koji Shimoke, Harue Sasaya, and Toshihiko Ikeuchi

1. Introduction	54
2. Induction of ER Stress in PC12 Cells	55
3. Measurement of Caspase-3 Activity	62
4. Fragmentation of Caspase-12	63
5. Hyperinduction of GRP78 by NGF	65
Acknowledgments	68
References	68

4. **Measuring ER Stress and the Unfolded Protein Response Using Mammalian Tissue Culture System**   71
   Christine M. Oslowski and Fumihiko Urano

   1. Introduction   72
   2. ER Stress and the UPR   73
   3. Cell Culture System and ER Stress Induction   76
   4. Measuring ER Stress   78
   5. Studying the UPR Activation   79
   6. Studying UPR Downstream Markers and Responses   84
   Acknowledgments   88
   References   88

5. **Real-Time Monitoring of ER Stress in Living Cells and Animals Using ESTRAP Assay**   93
   Masanori Kitamura and Nobuhiko Hiramatsu

   1. Introduction   94
   2. SEAP Reporter System   95
   3. Monitoring of ER Stress in Culture Cells by ESTRAP   96
   4. Monitoring of ER Stress *In Vivo* by ESTRAP   101
   5. Conclusion   104
   References   105

6. **HIV Protease Inhibitors Induce Endoplasmic Reticulum Stress and Disrupt Barrier Integrity in Intestinal Epithelial Cells**   107
   Huiping Zhou

   1. The Pathophysiological Relevance of HIV PI-Induced ER Stress and UPR Activation in IECs   108
   2. Analysis of HIV PI-Induced ER Stress and UPR Activation in Cultured IECs   109
   3. Analysis of HIV PI-Induced ER Stress and Disruption of Barrier Integrity in the Intestine   114
   4. Summary   118
   Acknowledgments   118
   References   118

7. **Dexamethasone Induction of a Heat Stress Response**   121
   José A. Guerrero, Vicente Vicente, and Javier Corral

   1. Introduction   122
   2. Evaluation of the Heat Stress Response in Patients Treated with Dexamethasone   124
   3. Heat Stress Response and UPR in Mice and Cellular Models   126
   References   134

8. Detecting and Quantitating Physiological Endoplasmic Reticulum Stress — 137

Ling Qi, Liu Yang, and Hui Chen

1. Introduction — 138
2. Detecting UPR at the Level of UPR Sensors — 139
3. Quantitating ER Stress at the Level of UPR Sensors — 141
4. Detecting Levels of Other Common UPR Targets — 142
5. Important Tips — 143
6. Concluding Remarks — 144
Acknowledgments — 145
References — 145

9. PI 3-Kinase Regulatory Subunits as Regulators of the Unfolded Protein Response — 147

Jonathon N. Winnay and C. Ronald Kahn

1. The Unfolded Protein Response — 148
2. PI 3-Kinase Regulatory Subunits as Modulators of the UPR — 150
3. Assessing UPR Pathway Activation — 151
References — 157

10. The Emerging Role of Histone Deacetylases (HDACs) in UPR Regulation — 159

Soumen Kahali, Bhaswati Sarcar, and Prakash Chinnaiyan

1. HDAC Enzymes and Cancer — 159
2. HDAC Inhibitors — 160
3. HDAC Inhibition and the UPR — 162
4. Future Directions — 172
References — 173

11. Immunohistochemical Detection of Activating Transcription Factor 3, a Hub of the Cellular Adaptive–Response Network — 175

Tsonwin Hai, Swati Jalgaonkar, Christopher C. Wolford, and Xin Yin

1. Introduction — 176
2. An IHC Protocol for ATF3 — 181
Acknowledgments — 192
References — 192

12. Experimental Approaches for Elucidation of Stress-Sensing Mechanisms of the Ire1 Family Proteins — 195

Daisuke Oikawa and Yukio Kimata

1. Introduction — 196

2.	Monitoring *In Vivo* Activity of Mutated Ire1 Family Proteins	199
3.	Interaction Between BiP and the Ire1 Family Proteins	206
4.	Direct Interaction of Unfolded Proteins with Yeast Ire1 but not with Mammalian IRE1α	210
5.	Conclusion and perspective	213
Acknowledgments		214
References		214

13. Measurement and Modification of the Expression Level of the Chaperone Protein and Signaling Regulator GRP78/BiP in Mammalian Cells — 217

Wan-Ting Chen and Amy S. Lee

1.	Introduction	218
2.	Detection of Total GRP78	220
3.	Detection of Cytosolic GRP78 Isoform	224
4.	Detection of Cell Surface GRP78	228
Acknowledgments		231
References		232

14. The Endoplasmic Reticulum-Associated Degradation and Disulfide Reductase ERdj5 — 235

Ryo Ushioda and Kazuhiro Nagata

1.	Introduction	236
2.	Productive Folding of Newly Synthesized Proteins in the ER	238
3.	ERAD: A Strategy for ERQC Mechanism	242
4.	Conclusion	249
References		252

15. Structural Insight into the Protective Role of P58(IPK) during Unfolded Protein Response — 259

Jiahui Tao and Bingdong Sha

1.	Endoplasmic Reticulum Stress and Unfolded Protein Response	260
2.	P58(IPK) Might be a Dual-Function Protein	261
3.	Knocking Out P58(IPK) in Mouse Models	261
4.	P58(IPK) is an ER-Resident Hsp40	262
5.	Crystal Structure of Mouse P58(IPK) TPR Domain	263
6.	P58(IPK) Functions as a Molecular Chaperone to Bind Unfolded Proteins Using Subdomain I	264
7.	Structural Basis for P58(IPK) J Domain–BiP Interaction	267
8.	Working Model of P58(IPK) During UPR	267
9.	Future Research	267
References		268

Contents ix

16. **Principles of IRE1 Modulation Using Chemical Tools**   271

    Kenneth P. K. Lee and Frank Sicheri

    1. Introduction: The Unfolded Protein Response   272
    2. Structural Biology of IRE1–XBP1 Signaling   276
    3. Chemical Approaches to Modulate IRE1 Function   282
    References   292

17. **Methods to Study Stromal-Cell Derived Factor 2 in the Context of ER Stress and the Unfolded Protein Response in *Arabidopsis thaliana***   295

    Andrea Schott and Sabine Strahl

    1. Introduction   296
    2. *Arabidopsis sdf2* T-DNA Insertion Lines   297
    3. Inducing and Monitoring ER Stress in *Arabidopsis sdf2* Mutants   299
    4. Analyzing the ER Stress-Induced Expression Pattern of *SDF2* in *Arabidopsis*   304
    5. The Subcellular Localization of SDF2   309
    6. Purification of Recombinant SDF2 Protein from *E. coli*   313
    Acknowledgments   316
    References   317

18. **Nitrosative Stress-Induced *S*-Glutathionylation of Protein Disulfide Isomerase**   321

    Joachim D. Uys, Ying Xiong, and Danyelle M. Townsend

    1. Introduction   322
    2. Identification and Confirmation of *S*-Glutathionylated Proteins in Cells   324
    3. Identification of Target Cysteine Residues   327
    4. Characterization of Structural and Functional Consequences of *S*-Glutathionylated PDI   329
    5. Summary   331
    Acknowledgments   331
    References   331

19. **Methods for Analyzing eIF2 Kinases and Translational Control in the Unfolded Protein Response**   333

    Brian F. Teske, Thomas D. Baird, and Ronald C. Wek

    1. Introduction to Translational Control in the UPR   334
    2. Measuring Activation of the eIF2 Kinase Pathway During ER Stress   338
    3. Investigating General and Gene-Specific Translation Control   345

Acknowledgments 353
References 353

Author Index *357*
Subject Index *375*

Contributors

David B. Archer
School of Biology, University of Nottingham, University Park, Nottingham, United Kingdom

Thomas D. Baird
Department of Biochemistry and Molecular Biology, Indiana University School of Medicine, Indianapolis, Indiana, USA

Neuza D. S. P. Carvalho
Molecular Microbiology and Biotechnology, Kluyver Centre for Genomics of Industrial Fermentation, Institute of Biology Leiden, Leiden University, Leiden, The Netherlands

Karen Cawley
Apoptosis Research Centre, School of Natural Sciences (Biochemistry), National University of Ireland Galway, Ireland

Hui Chen
Division of Nutritional Sciences, Cornell University, Ithaca, New York, USA

Wan-Ting Chen
Department of Biochemistry and Molecular Biology, USC Norris Comprehensive Cancer Center, University of Southern California Keck School of Medicine, Los Angeles, California, USA

Prakash Chinnaiyan
Department of Experimental Therapeutics and Radiation Oncology, H. Lee Moffitt Cancer Center and Research Institute, Tampa, FL, USA

Javier Corral
Centro Regional de Hemodonación, University of Murcia, Spain

Shane Deegan
Apoptosis Research Centre, School of Natural Sciences (Biochemistry), National University of Ireland Galway, Ireland

José A. Guerrero
Centro Regional de Hemodonación, University of Murcia, Spain

Thomas Guillemette
UMR PAVE No 77, IFR 149 QUASAV, Angers Cedex, France

Sanjeev Gupta
Apoptosis Research Centre, School of Medicine (Pathology), Clinical Science Institute, National University of Ireland Galway, Ireland

Tsonwin Hai
Department of Molecular and Cellular Biochemistry; Center for Molecular Neurobiology; Molecular, Cellular, and Developmental Biology Program; and Ohio State Biochemistry Program, Ohio State University, Columbus, Ohio, USA

Nobuhiko Hiramatsu
Department of Pathology, University of California San Diego, School of Medicine, La Jolla, California, USA

Toshihiko Ikeuchi
Laboratory of Neurobiology, Department of Life Science and Biotechnology, Faculty of Chemistry, Materials and Bioengineering, Kansai University, Suita, Osaka, Japan

Swati Jalgaonkar
Department of Molecular and Cellular Biochemistry; Center for Molecular Neurobiology; Molecular, Cellular, and Developmental Biology Program, Ohio State University, Columbus, Ohio, USA

Aymeric Joubert
UMR PAVE No 77, IFR 149 QUASAV, Angers Cedex, France

Soumen Kahali
Department of Experimental Therapeutics and Radiation Oncology, H. Lee Moffitt Cancer Center and Research Institute, Tampa, FL, USA

C. Ronald Kahn
Section on Integrative Physiology and Metabolism, Research Division, Joslin Diabetes Center, Harvard Medical School, Boston, Massachusetts, USA

Yukio Kimata
Graduate School of Biological Sciences, Nara Institute of Science and Technology, Ikoma, Nara, Japan

Masanori Kitamura
Department of Molecular Signaling, Interdisciplinary Graduate School of Medicine and Engineering, University of Yamanashi, Chuo, Yamanashi, Japan

Amy S. Lee
Department of Biochemistry and Molecular Biology, USC Norris Comprehensive Cancer Center, University of Southern California Keck School of Medicine, Los Angeles, California, USA

Kenneth P. K. Lee
Program in Systems Biology, Samuel Lunenfeld Research Institute, Mount Sinai Hospital, and Department of Molecular and Medical Genetics, University of Toronto, Toronto, Ontario, Canada

Kazuhiro Nagata
Laboratory of Molecular and Cellular Biology, Department of Molecular Biosciences, Faculty of Life Sciences, Kyoto Sangyo University, Kyoto, Japan

Daisuke Oikawa
Iwawaki Initiative Research Unit, Advanced Science Institute, RIKEN, Wako, Saitama, and Graduate School of Biological Sciences, Nara Institute of Science and Technology, Ikoma, Nara, Japan

Christine M. Oslowski
Program in Gene Function and Expression, University of Massachusetts, Medical School, Worcester, Massachusetts, USA

Ling Qi
Division of Nutritional Sciences; and Graduate Program in Biochemistry, Molecular and Cell Biology, Cornell University, Ithaca, New York, USA

Arthur F. J. Ram
Molecular Microbiology and Biotechnology, Kluyver Centre for Genomics of Industrial Fermentation, Institute of Biology Leiden, Leiden University, Leiden, The Netherlands

Afshin Samali
Apoptosis Research Centre, School of Natural Sciences (Biochemistry), National University of Ireland Galway, Ireland

Bhaswati Sarcar
Department of Experimental Therapeutics and Radiation Oncology, H. Lee Moffitt Cancer Center and Research Institute, Tampa, FL, USA

Harue Sasaya
Laboratory of Neurobiology, Department of Life Science and Biotechnology, Faculty of Chemistry, Materials and Bioengineering, Kansai University, Suita, Osaka, Japan

Andrea Schott
Department of Cell Chemistry, Center for Organismal Studies (COS), University of Heidelberg, Heidelberg, Germany

Bingdong Sha
Department of Cell Biology, University of Alabama at Birmingham, Birmingham, Alabama, USA

Koji Shimoke
Laboratory of Neurobiology, Department of Life Science and Biotechnology, Faculty of Chemistry, Materials and Bioengineering, Kansai University, Suita, Osaka, Japan

Frank Sicheri
Program in Systems Biology, Samuel Lunenfeld Research Institute, Mount Sinai Hospital, and Department of Molecular and Medical Genetics, University of Toronto, Toronto, Ontario, Canada

Philippe Simoneau
UMR PAVE No 77, IFR 149 QUASAV, Angers Cedex, France

Sabine Strahl
Department of Cell Chemistry, Center for Organismal Studies (COS), University of Heidelberg, Heidelberg, Germany

Jiahui Tao
Department of Cell Biology, University of Alabama at Birmingham, Birmingham, Alabama, USA

Brian F. Teske
Department of Biochemistry and Molecular Biology, Indiana University School of Medicine, Indianapolis, Indiana, USA

Danyelle M. Townsend
Department of Pharmaceutical and Biomedical Sciences, Medical University of South Carolina, Charleston, South Carolina, USA

Fumihiko Urano
Program in Gene Function and Expression, and Program in Molecular Medicine, University of Massachusetts, Medical School, Worcester, Massachusetts, USA

Ryo Ushioda
Laboratory of Molecular and Cellular Biology, Department of Molecular Biosciences, Faculty of Life Sciences, Kyoto Sangyo University, Kyoto, Japan

Joachim D. Uys
Department of Pharmaceutical and Biomedical Sciences, Medical University of South Carolina, Charleston, South Carolina, USA

Vicente Vicente
Centro Regional de Hemodonación, University of Murcia, Spain

Ronald C. Wek
Department of Biochemistry and Molecular Biology, Indiana University School of Medicine, Indianapolis, Indiana, USA

Jonathon N. Winnay
Section on Integrative Physiology and Metabolism, Research Division, Joslin Diabetes Center, Harvard Medical School, Boston, Massachusetts, USA

Christopher C. Wolford
Department of Molecular and Cellular Biochemistry; and Center for Molecular Neurobiology, Ohio State University, Columbus, Ohio, USA

Ying Xiong
Department of Pharmaceutical and Biomedical Sciences, Medical University of South Carolina, Charleston, South Carolina, USA

Liu Yang
Graduate Program in Biochemistry, Molecular and Cell Biology, Cornell University, Ithaca, New York, USA

Xin Yin
Department of Molecular and Cellular Biochemistry; Center for Molecular Neurobiology; and Ohio State Biochemistry Program, Ohio State University, Columbus, Ohio, USA

Huiping Zhou
Department of Microbiology and Immunology, School of Medicine, Virginia Commonwealth University, Richmond, Virginia, USA

Preface

The observation that the living cell contains a mechanism to sense and correct the accumulation of unfolded (or incorrectly folded) proteins in the endoplasmic reticulum was formidable in organizing thoughts about cellular integration. This mechanism both halts further protein synthesis and promotes the production of chaperone proteins that act to relieve this problem. If this problem cannot be corrected, the mechanism can initiate programmed cell death. Aspects of this unfolded protein response (UPR) are conserved from yeast to man, an observation that suggests a key role in the process of maintaining a living cell.

The UPR presents a way of understanding cellular regulation, a mechanism for disease, and a therapeutic opportunity.

The present volume provides descriptions of the occurrence of the UPR, methods used to assess it, and pharmacological tools and other methodological approaches to analyze its impact on cellular regulation. The authors explain how these methods are able to provide important biological insights.

Authors were selected based on research contributions in the area about which they have written and based on their ability to describe their methodological contribution in a clear and reproducible way. They have been encouraged to make use of graphics, comparisons to other methods, and to provide tricks and approaches not revealed in prior publications that make it possible to adapt methods to other systems.

The editor expresses appreciation to the contributors for providing their contributions in a timely fashion, to the senior editors for guidance, and to the staff at Academic Press for helpful input.

September, 2010

P. Michael Conn
Portland, Oregon, USA

Methods in Enzymology

Volume I. Preparation and Assay of Enzymes
Edited by Sidney P. Colowick and Nathan O. Kaplan

Volume II. Preparation and Assay of Enzymes
Edited by Sidney P. Colowick and Nathan O. Kaplan

Volume III. Preparation and Assay of Substrates
Edited by Sidney P. Colowick and Nathan O. Kaplan

Volume IV. Special Techniques for the Enzymologist
Edited by Sidney P. Colowick and Nathan O. Kaplan

Volume V. Preparation and Assay of Enzymes
Edited by Sidney P. Colowick and Nathan O. Kaplan

Volume VI. Preparation and Assay of Enzymes *(Continued)*
Preparation and Assay of Substrates
Special Techniques
Edited by Sidney P. Colowick and Nathan O. Kaplan

Volume VII. Cumulative Subject Index
Edited by Sidney P. Colowick and Nathan O. Kaplan

Volume VIII. Complex Carbohydrates
Edited by Elizabeth F. Neufeld and Victor Ginsburg

Volume IX. Carbohydrate Metabolism
Edited by Willis A. Wood

Volume X. Oxidation and Phosphorylation
Edited by Ronald W. Estabrook and Maynard E. Pullman

Volume XI. Enzyme Structure
Edited by C. H. W. Hirs

Volume XII. Nucleic Acids (Parts A and B)
Edited by Lawrence Grossman and Kivie Moldave

Volume XIII. Citric Acid Cycle
Edited by J. M. Lowenstein

Volume XIV. Lipids
Edited by J. M. Lowenstein

Volume XV. Steroids and Terpenoids
Edited by Raymond B. Clayton

VOLUME XVI. Fast Reactions
Edited by KENNETH KUSTIN

VOLUME XVII. Metabolism of Amino Acids and Amines (Parts A and B)
Edited by HERBERT TABOR AND CELIA WHITE TABOR

VOLUME XVIII. Vitamins and Coenzymes (Parts A, B, and C)
Edited by DONALD B. MCCORMICK AND LEMUEL D. WRIGHT

VOLUME XIX. Proteolytic Enzymes
Edited by GERTRUDE E. PERLMANN AND LASZLO LORAND

VOLUME XX. Nucleic Acids and Protein Synthesis (Part C)
Edited by KIVIE MOLDAVE AND LAWRENCE GROSSMAN

VOLUME XXI. Nucleic Acids (Part D)
Edited by LAWRENCE GROSSMAN AND KIVIE MOLDAVE

VOLUME XXII. Enzyme Purification and Related Techniques
Edited by WILLIAM B. JAKOBY

VOLUME XXIII. Photosynthesis (Part A)
Edited by ANTHONY SAN PIETRO

VOLUME XXIV. Photosynthesis and Nitrogen Fixation (Part B)
Edited by ANTHONY SAN PIETRO

VOLUME XXV. Enzyme Structure (Part B)
Edited by C. H. W. HIRS AND SERGE N. TIMASHEFF

VOLUME XXVI. Enzyme Structure (Part C)
Edited by C. H. W. HIRS AND SERGE N. TIMASHEFF

VOLUME XXVII. Enzyme Structure (Part D)
Edited by C. H. W. HIRS AND SERGE N. TIMASHEFF

VOLUME XXVIII. Complex Carbohydrates (Part B)
Edited by VICTOR GINSBURG

VOLUME XXIX. Nucleic Acids and Protein Synthesis (Part E)
Edited by LAWRENCE GROSSMAN AND KIVIE MOLDAVE

VOLUME XXX. Nucleic Acids and Protein Synthesis (Part F)
Edited by KIVIE MOLDAVE AND LAWRENCE GROSSMAN

VOLUME XXXI. Biomembranes (Part A)
Edited by SIDNEY FLEISCHER AND LESTER PACKER

VOLUME XXXII. Biomembranes (Part B)
Edited by SIDNEY FLEISCHER AND LESTER PACKER

VOLUME XXXIII. Cumulative Subject Index Volumes I-XXX
Edited by MARTHA G. DENNIS AND EDWARD A. DENNIS

VOLUME XXXIV. Affinity Techniques (Enzyme Purification: Part B)
Edited by WILLIAM B. JAKOBY AND MEIR WILCHEK

VOLUME XXXV. Lipids (Part B)
Edited by JOHN M. LOWENSTEIN

VOLUME XXXVI. Hormone Action (Part A: Steroid Hormones)
Edited by BERT W. O'MALLEY AND JOEL G. HARDMAN

VOLUME XXXVII. Hormone Action (Part B: Peptide Hormones)
Edited by BERT W. O'MALLEY AND JOEL G. HARDMAN

VOLUME XXXVIII. Hormone Action (Part C: Cyclic Nucleotides)
Edited by JOEL G. HARDMAN AND BERT W. O'MALLEY

VOLUME XXXIX. Hormone Action (Part D: Isolated Cells, Tissues, and Organ Systems)
Edited by JOEL G. HARDMAN AND BERT W. O'MALLEY

VOLUME XL. Hormone Action (Part E: Nuclear Structure and Function)
Edited by BERT W. O'MALLEY AND JOEL G. HARDMAN

VOLUME XLI. Carbohydrate Metabolism (Part B)
Edited by W. A. WOOD

VOLUME XLII. Carbohydrate Metabolism (Part C)
Edited by W. A. WOOD

VOLUME XLIII. Antibiotics
Edited by JOHN H. HASH

VOLUME XLIV. Immobilized Enzymes
Edited by KLAUS MOSBACH

VOLUME XLV. Proteolytic Enzymes (Part B)
Edited by LASZLO LORAND

VOLUME XLVI. Affinity Labeling
Edited by WILLIAM B. JAKOBY AND MEIR WILCHEK

VOLUME XLVII. Enzyme Structure (Part E)
Edited by C. H. W. HIRS AND SERGE N. TIMASHEFF

VOLUME XLVIII. Enzyme Structure (Part F)
Edited by C. H. W. HIRS AND SERGE N. TIMASHEFF

VOLUME XLIX. Enzyme Structure (Part G)
Edited by C. H. W. HIRS AND SERGE N. TIMASHEFF

VOLUME L. Complex Carbohydrates (Part C)
Edited by VICTOR GINSBURG

VOLUME LI. Purine and Pyrimidine Nucleotide Metabolism
Edited by PATRICIA A. HOFFEE AND MARY ELLEN JONES

VOLUME LII. Biomembranes (Part C: Biological Oxidations)
Edited by SIDNEY FLEISCHER AND LESTER PACKER

VOLUME LIII. Biomembranes (Part D: Biological Oxidations)
Edited by SIDNEY FLEISCHER AND LESTER PACKER

VOLUME LIV. Biomembranes (Part E: Biological Oxidations)
Edited by SIDNEY FLEISCHER AND LESTER PACKER

VOLUME LV. Biomembranes (Part F: Bioenergetics)
Edited by SIDNEY FLEISCHER AND LESTER PACKER

VOLUME LVI. Biomembranes (Part G: Bioenergetics)
Edited by SIDNEY FLEISCHER AND LESTER PACKER

VOLUME LVII. Bioluminescence and Chemiluminescence
Edited by MARLENE A. DELUCA

VOLUME LVIII. Cell Culture
Edited by WILLIAM B. JAKOBY AND IRA PASTAN

VOLUME LIX. Nucleic Acids and Protein Synthesis (Part G)
Edited by KIVIE MOLDAVE AND LAWRENCE GROSSMAN

VOLUME LX. Nucleic Acids and Protein Synthesis (Part H)
Edited by KIVIE MOLDAVE AND LAWRENCE GROSSMAN

VOLUME 61. Enzyme Structure (Part H)
Edited by C. H. W. HIRS AND SERGE N. TIMASHEFF

VOLUME 62. Vitamins and Coenzymes (Part D)
Edited by DONALD B. MCCORMICK AND LEMUEL D. WRIGHT

VOLUME 63. Enzyme Kinetics and Mechanism (Part A: Initial Rate and Inhibitor Methods)
Edited by DANIEL L. PURICH

VOLUME 64. Enzyme Kinetics and Mechanism
(Part B: Isotopic Probes and Complex Enzyme Systems)
Edited by DANIEL L. PURICH

VOLUME 65. Nucleic Acids (Part I)
Edited by LAWRENCE GROSSMAN AND KIVIE MOLDAVE

VOLUME 66. Vitamins and Coenzymes (Part E)
Edited by DONALD B. MCCORMICK AND LEMUEL D. WRIGHT

VOLUME 67. Vitamins and Coenzymes (Part F)
Edited by DONALD B. MCCORMICK AND LEMUEL D. WRIGHT

VOLUME 68. Recombinant DNA
Edited by RAY WU

VOLUME 69. Photosynthesis and Nitrogen Fixation (Part C)
Edited by ANTHONY SAN PIETRO

VOLUME 70. Immunochemical Techniques (Part A)
Edited by HELEN VAN VUNAKIS AND JOHN J. LANGONE

VOLUME 71. Lipids (Part C)
Edited by JOHN M. LOWENSTEIN

VOLUME 72. Lipids (Part D)
Edited by JOHN M. LOWENSTEIN

VOLUME 73. Immunochemical Techniques (Part B)
Edited by JOHN J. LANGONE AND HELEN VAN VUNAKIS

VOLUME 74. Immunochemical Techniques (Part C)
Edited by JOHN J. LANGONE AND HELEN VAN VUNAKIS

VOLUME 75. Cumulative Subject Index Volumes XXXI, XXXII, XXXIV–LX
Edited by EDWARD A. DENNIS AND MARTHA G. DENNIS

VOLUME 76. Hemoglobins
Edited by ERALDO ANTONINI, LUIGI ROSSI-BERNARDI, AND EMILIA CHIANCONE

VOLUME 77. Detoxication and Drug Metabolism
Edited by WILLIAM B. JAKOBY

VOLUME 78. Interferons (Part A)
Edited by SIDNEY PESTKA

VOLUME 79. Interferons (Part B)
Edited by SIDNEY PESTKA

VOLUME 80. Proteolytic Enzymes (Part C)
Edited by LASZLO LORAND

VOLUME 81. Biomembranes (Part H: Visual Pigments and Purple Membranes, I)
Edited by LESTER PACKER

VOLUME 82. Structural and Contractile Proteins (Part A: Extracellular Matrix)
Edited by LEON W. CUNNINGHAM AND DIXIE W. FREDERIKSEN

VOLUME 83. Complex Carbohydrates (Part D)
Edited by VICTOR GINSBURG

VOLUME 84. Immunochemical Techniques (Part D: Selected Immunoassays)
Edited by JOHN J. LANGONE AND HELEN VAN VUNAKIS

VOLUME 85. Structural and Contractile Proteins (Part B: The Contractile Apparatus and the Cytoskeleton)
Edited by DIXIE W. FREDERIKSEN AND LEON W. CUNNINGHAM

VOLUME 86. Prostaglandins and Arachidonate Metabolites
Edited by WILLIAM E. M. LANDS AND WILLIAM L. SMITH

VOLUME 87. Enzyme Kinetics and Mechanism (Part C: Intermediates, Stereo-chemistry, and Rate Studies)
Edited by DANIEL L. PURICH

VOLUME 88. Biomembranes (Part I: Visual Pigments and Purple Membranes, II)
Edited by LESTER PACKER

VOLUME 89. Carbohydrate Metabolism (Part D)
Edited by WILLIS A. WOOD

VOLUME 90. Carbohydrate Metabolism (Part E)
Edited by WILLIS A. WOOD

VOLUME 91. Enzyme Structure (Part I)
Edited by C. H. W. HIRS AND SERGE N. TIMASHEFF

VOLUME 92. Immunochemical Techniques (Part E: Monoclonal Antibodies and General Immunoassay Methods)
Edited by JOHN J. LANGONE AND HELEN VAN VUNAKIS

VOLUME 93. Immunochemical Techniques (Part F: Conventional Antibodies, Fc Receptors, and Cytotoxicity)
Edited by JOHN J. LANGONE AND HELEN VAN VUNAKIS

VOLUME 94. Polyamines
Edited by HERBERT TABOR AND CELIA WHITE TABOR

VOLUME 95. Cumulative Subject Index Volumes 61–74, 76–80
Edited by EDWARD A. DENNIS AND MARTHA G. DENNIS

VOLUME 96. Biomembranes [Part J: Membrane Biogenesis: Assembly and Targeting (General Methods; Eukaryotes)]
Edited by SIDNEY FLEISCHER AND BECCA FLEISCHER

VOLUME 97. Biomembranes [Part K: Membrane Biogenesis: Assembly and Targeting (Prokaryotes, Mitochondria, and Chloroplasts)]
Edited by SIDNEY FLEISCHER AND BECCA FLEISCHER

VOLUME 98. Biomembranes (Part L: Membrane Biogenesis: Processing and Recycling)
Edited by SIDNEY FLEISCHER AND BECCA FLEISCHER

VOLUME 99. Hormone Action (Part F: Protein Kinases)
Edited by JACKIE D. CORBIN AND JOEL G. HARDMAN

VOLUME 100. Recombinant DNA (Part B)
Edited by RAY WU, LAWRENCE GROSSMAN, AND KIVIE MOLDAVE

VOLUME 101. Recombinant DNA (Part C)
Edited by RAY WU, LAWRENCE GROSSMAN, AND KIVIE MOLDAVE

VOLUME 102. Hormone Action (Part G: Calmodulin and Calcium-Binding Proteins)
Edited by ANTHONY R. MEANS AND BERT W. O'MALLEY

VOLUME 103. Hormone Action (Part H: Neuroendocrine Peptides)
Edited by P. MICHAEL CONN

VOLUME 104. Enzyme Purification and Related Techniques (Part C)
Edited by WILLIAM B. JAKOBY

VOLUME 105. Oxygen Radicals in Biological Systems
Edited by LESTER PACKER

VOLUME 106. Posttranslational Modifications (Part A)
Edited by FINN WOLD AND KIVIE MOLDAVE

VOLUME 107. Posttranslational Modifications (Part B)
Edited by FINN WOLD AND KIVIE MOLDAVE

VOLUME 108. Immunochemical Techniques (Part G: Separation and Characterization of Lymphoid Cells)
Edited by GIOVANNI DI SABATO, JOHN J. LANGONE, AND HELEN VAN VUNAKIS

VOLUME 109. Hormone Action (Part I: Peptide Hormones)
Edited by LUTZ BIRNBAUMER AND BERT W. O'MALLEY

VOLUME 110. Steroids and Isoprenoids (Part A)
Edited by JOHN H. LAW AND HANS C. RILLING

VOLUME 111. Steroids and Isoprenoids (Part B)
Edited by JOHN H. LAW AND HANS C. RILLING

VOLUME 112. Drug and Enzyme Targeting (Part A)
Edited by KENNETH J. WIDDER AND RALPH GREEN

VOLUME 113. Glutamate, Glutamine, Glutathione, and Related Compounds
Edited by ALTON MEISTER

VOLUME 114. Diffraction Methods for Biological Macromolecules (Part A)
Edited by HAROLD W. WYCKOFF, C. H. W. HIRS, AND SERGE N. TIMASHEFF

VOLUME 115. Diffraction Methods for Biological Macromolecules (Part B)
Edited by HAROLD W. WYCKOFF, C. H. W. HIRS, AND SERGE N. TIMASHEFF

VOLUME 116. Immunochemical Techniques
(Part H: Effectors and Mediators of Lymphoid Cell Functions)
Edited by GIOVANNI DI SABATO, JOHN J. LANGONE, AND HELEN VAN VUNAKIS

VOLUME 117. Enzyme Structure (Part J)
Edited by C. H. W. HIRS AND SERGE N. TIMASHEFF

VOLUME 118. Plant Molecular Biology
Edited by ARTHUR WEISSBACH AND HERBERT WEISSBACH

VOLUME 119. Interferons (Part C)
Edited by SIDNEY PESTKA

VOLUME 120. Cumulative Subject Index Volumes 81–94, 96–101

VOLUME 121. Immunochemical Techniques (Part I: Hybridoma Technology and Monoclonal Antibodies)
Edited by JOHN J. LANGONE AND HELEN VAN VUNAKIS

VOLUME 122. Vitamins and Coenzymes (Part G)
Edited by FRANK CHYTIL AND DONALD B. MCCORMICK

VOLUME 123. Vitamins and Coenzymes (Part H)
Edited by FRANK CHYTIL AND DONALD B. MCCORMICK

VOLUME 124. Hormone Action (Part J: Neuroendocrine Peptides)
Edited by P. MICHAEL CONN

VOLUME 125. Biomembranes (Part M: Transport in Bacteria, Mitochondria, and Chloroplasts: General Approaches and Transport Systems)
Edited by SIDNEY FLEISCHER AND BECCA FLEISCHER

VOLUME 126. Biomembranes (Part N: Transport in Bacteria, Mitochondria, and Chloroplasts: Protonmotive Force)
Edited by SIDNEY FLEISCHER AND BECCA FLEISCHER

VOLUME 127. Biomembranes (Part O: Protons and Water: Structure and Translocation)
Edited by LESTER PACKER

VOLUME 128. Plasma Lipoproteins (Part A: Preparation, Structure, and Molecular Biology)
Edited by JERE P. SEGREST AND JOHN J. ALBERS

VOLUME 129. Plasma Lipoproteins (Part B: Characterization, Cell Biology, and Metabolism)
Edited by JOHN J. ALBERS AND JERE P. SEGREST

VOLUME 130. Enzyme Structure (Part K)
Edited by C. H. W. HIRS AND SERGE N. TIMASHEFF

VOLUME 131. Enzyme Structure (Part L)
Edited by C. H. W. HIRS AND SERGE N. TIMASHEFF

VOLUME 132. Immunochemical Techniques (Part J: Phagocytosis and Cell-Mediated Cytotoxicity)
Edited by GIOVANNI DI SABATO AND JOHANNES EVERSE

VOLUME 133. Bioluminescence and Chemiluminescence (Part B)
Edited by MARLENE DELUCA AND WILLIAM D. MCELROY

VOLUME 134. Structural and Contractile Proteins (Part C: The Contractile Apparatus and the Cytoskeleton)
Edited by RICHARD B. VALLEE

VOLUME 135. Immobilized Enzymes and Cells (Part B)
Edited by KLAUS MOSBACH

VOLUME 136. Immobilized Enzymes and Cells (Part C)
Edited by KLAUS MOSBACH

VOLUME 137. Immobilized Enzymes and Cells (Part D)
Edited by KLAUS MOSBACH

VOLUME 138. Complex Carbohydrates (Part E)
Edited by VICTOR GINSBURG

VOLUME 139. Cellular Regulators (Part A: Calcium- and
Calmodulin-Binding Proteins)
Edited by ANTHONY R. MEANS AND P. MICHAEL CONN

VOLUME 140. Cumulative Subject Index Volumes 102–119, 121–134

VOLUME 141. Cellular Regulators (Part B: Calcium and Lipids)
Edited by P. MICHAEL CONN AND ANTHONY R. MEANS

VOLUME 142. Metabolism of Aromatic Amino Acids and Amines
Edited by SEYMOUR KAUFMAN

VOLUME 143. Sulfur and Sulfur Amino Acids
Edited by WILLIAM B. JAKOBY AND OWEN GRIFFITH

VOLUME 144. Structural and Contractile Proteins (Part D: Extracellular Matrix)
Edited by LEON W. CUNNINGHAM

VOLUME 145. Structural and Contractile Proteins (Part E: Extracellular Matrix)
Edited by LEON W. CUNNINGHAM

VOLUME 146. Peptide Growth Factors (Part A)
Edited by DAVID BARNES AND DAVID A. SIRBASKU

VOLUME 147. Peptide Growth Factors (Part B)
Edited by DAVID BARNES AND DAVID A. SIRBASKU

VOLUME 148. Plant Cell Membranes
Edited by LESTER PACKER AND ROLAND DOUCE

VOLUME 149. Drug and Enzyme Targeting (Part B)
Edited by RALPH GREEN AND KENNETH J. WIDDER

VOLUME 150. Immunochemical Techniques (Part K: *In Vitro* Models of B and T Cell Functions and Lymphoid Cell Receptors)
Edited by GIOVANNI DI SABATO

VOLUME 151. Molecular Genetics of Mammalian Cells
Edited by MICHAEL M. GOTTESMAN

VOLUME 152. Guide to Molecular Cloning Techniques
Edited by SHELBY L. BERGER AND ALAN R. KIMMEL

VOLUME 153. Recombinant DNA (Part D)
Edited by RAY WU AND LAWRENCE GROSSMAN

VOLUME 154. Recombinant DNA (Part E)
Edited by RAY WU AND LAWRENCE GROSSMAN

VOLUME 155. Recombinant DNA (Part F)
Edited by RAY WU

VOLUME 156. Biomembranes (Part P: ATP-Driven Pumps and Related Transport: The Na, K-Pump)
Edited by SIDNEY FLEISCHER AND BECCA FLEISCHER

VOLUME 157. Biomembranes (Part Q: ATP-Driven Pumps and Related Transport: Calcium, Proton, and Potassium Pumps)
Edited by SIDNEY FLEISCHER AND BECCA FLEISCHER

VOLUME 158. Metalloproteins (Part A)
Edited by JAMES F. RIORDAN AND BERT L. VALLEE

VOLUME 159. Initiation and Termination of Cyclic Nucleotide Action
Edited by JACKIE D. CORBIN AND ROGER A. JOHNSON

VOLUME 160. Biomass (Part A: Cellulose and Hemicellulose)
Edited by WILLIS A. WOOD AND SCOTT T. KELLOGG

VOLUME 161. Biomass (Part B: Lignin, Pectin, and Chitin)
Edited by WILLIS A. WOOD AND SCOTT T. KELLOGG

VOLUME 162. Immunochemical Techniques (Part L: Chemotaxis and Inflammation)
Edited by GIOVANNI DI SABATO

VOLUME 163. Immunochemical Techniques (Part M: Chemotaxis and Inflammation)
Edited by GIOVANNI DI SABATO

VOLUME 164. Ribosomes
Edited by HARRY F. NOLLER, JR., AND KIVIE MOLDAVE

VOLUME 165. Microbial Toxins: Tools for Enzymology
Edited by SIDNEY HARSHMAN

VOLUME 166. Branched-Chain Amino Acids
Edited by ROBERT HARRIS AND JOHN R. SOKATCH

VOLUME 167. Cyanobacteria
Edited by LESTER PACKER AND ALEXANDER N. GLAZER

VOLUME 168. Hormone Action (Part K: Neuroendocrine Peptides)
Edited by P. MICHAEL CONN

VOLUME 169. Platelets: Receptors, Adhesion, Secretion (Part A)
Edited by JACEK HAWIGER

VOLUME 170. Nucleosomes
Edited by PAUL M. WASSARMAN AND ROGER D. KORNBERG

VOLUME 171. Biomembranes (Part R: Transport Theory: Cells and Model Membranes)
Edited by SIDNEY FLEISCHER AND BECCA FLEISCHER

VOLUME 172. Biomembranes (Part S: Transport: Membrane Isolation and Characterization)
Edited by SIDNEY FLEISCHER AND BECCA FLEISCHER

VOLUME 173. Biomembranes [Part T: Cellular and Subcellular Transport: Eukaryotic (Nonepithelial) Cells]
Edited by SIDNEY FLEISCHER AND BECCA FLEISCHER

VOLUME 174. Biomembranes [Part U: Cellular and Subcellular Transport: Eukaryotic (Nonepithelial) Cells]
Edited by SIDNEY FLEISCHER AND BECCA FLEISCHER

VOLUME 175. Cumulative Subject Index Volumes 135–139, 141–167

VOLUME 176. Nuclear Magnetic Resonance (Part A: Spectral Techniques and Dynamics)
Edited by NORMAN J. OPPENHEIMER AND THOMAS L. JAMES

VOLUME 177. Nuclear Magnetic Resonance (Part B: Structure and Mechanism)
Edited by NORMAN J. OPPENHEIMER AND THOMAS L. JAMES

VOLUME 178. Antibodies, Antigens, and Molecular Mimicry
Edited by JOHN J. LANGONE

VOLUME 179. Complex Carbohydrates (Part F)
Edited by VICTOR GINSBURG

VOLUME 180. RNA Processing (Part A: General Methods)
Edited by JAMES E. DAHLBERG AND JOHN N. ABELSON

VOLUME 181. RNA Processing (Part B: Specific Methods)
Edited by JAMES E. DAHLBERG AND JOHN N. ABELSON

VOLUME 182. Guide to Protein Purification
Edited by MURRAY P. DEUTSCHER

VOLUME 183. Molecular Evolution: Computer Analysis of Protein and Nucleic Acid Sequences
Edited by RUSSELL F. DOOLITTLE

VOLUME 184. Avidin-Biotin Technology
Edited by MEIR WILCHEK AND EDWARD A. BAYER

VOLUME 185. Gene Expression Technology
Edited by DAVID V. GOEDDEL

VOLUME 186. Oxygen Radicals in Biological Systems (Part B: Oxygen Radicals and Antioxidants)
Edited by LESTER PACKER AND ALEXANDER N. GLAZER

VOLUME 187. Arachidonate Related Lipid Mediators
Edited by ROBERT C. MURPHY AND FRANK A. FITZPATRICK

VOLUME 188. Hydrocarbons and Methylotrophy
Edited by MARY E. LIDSTROM

VOLUME 189. Retinoids (Part A: Molecular and Metabolic Aspects)
Edited by LESTER PACKER

VOLUME 190. Retinoids (Part B: Cell Differentiation and Clinical Applications)
Edited by LESTER PACKER

VOLUME 191. Biomembranes (Part V: Cellular and Subcellular Transport: Epithelial Cells)
Edited by SIDNEY FLEISCHER AND BECCA FLEISCHER

VOLUME 192. Biomembranes (Part W: Cellular and Subcellular Transport: Epithelial Cells)
Edited by SIDNEY FLEISCHER AND BECCA FLEISCHER

VOLUME 193. Mass Spectrometry
Edited by JAMES A. MCCLOSKEY

VOLUME 194. Guide to Yeast Genetics and Molecular Biology
Edited by CHRISTINE GUTHRIE AND GERALD R. FINK

VOLUME 195. Adenylyl Cyclase, G Proteins, and Guanylyl Cyclase
Edited by ROGER A. JOHNSON AND JACKIE D. CORBIN

VOLUME 196. Molecular Motors and the Cytoskeleton
Edited by RICHARD B. VALLEE

VOLUME 197. Phospholipases
Edited by EDWARD A. DENNIS

VOLUME 198. Peptide Growth Factors (Part C)
Edited by DAVID BARNES, J. P. MATHER, AND GORDON H. SATO

VOLUME 199. Cumulative Subject Index Volumes 168–174, 176–194

VOLUME 200. Protein Phosphorylation (Part A: Protein Kinases: Assays, Purification, Antibodies, Functional Analysis, Cloning, and Expression)
Edited by TONY HUNTER AND BARTHOLOMEW M. SEFTON

VOLUME 201. Protein Phosphorylation (Part B: Analysis of Protein Phosphorylation, Protein Kinase Inhibitors, and Protein Phosphatases)
Edited by TONY HUNTER AND BARTHOLOMEW M. SEFTON

VOLUME 202. Molecular Design and Modeling: Concepts and Applications (Part A: Proteins, Peptides, and Enzymes)
Edited by JOHN J. LANGONE

VOLUME 203. Molecular Design and Modeling: Concepts and Applications (Part B: Antibodies and Antigens, Nucleic Acids, Polysaccharides, and Drugs)
Edited by JOHN J. LANGONE

VOLUME 204. Bacterial Genetic Systems
Edited by JEFFREY H. MILLER

VOLUME 205. Metallobiochemistry (Part B: Metallothionein and Related Molecules)
Edited by JAMES F. RIORDAN AND BERT L. VALLEE

VOLUME 206. Cytochrome P450
Edited by MICHAEL R. WATERMAN AND ERIC F. JOHNSON

VOLUME 207. Ion Channels
Edited by BERNARDO RUDY AND LINDA E. IVERSON

VOLUME 208. Protein–DNA Interactions
Edited by ROBERT T. SAUER

VOLUME 209. Phospholipid Biosynthesis
Edited by EDWARD A. DENNIS AND DENNIS E. VANCE

VOLUME 210. Numerical Computer Methods
Edited by LUDWIG BRAND AND MICHAEL L. JOHNSON

VOLUME 211. DNA Structures (Part A: Synthesis and Physical Analysis of DNA)
Edited by DAVID M. J. LILLEY AND JAMES E. DAHLBERG

VOLUME 212. DNA Structures (Part B: Chemical and Electrophoretic Analysis of DNA)
Edited by DAVID M. J. LILLEY AND JAMES E. DAHLBERG

VOLUME 213. Carotenoids (Part A: Chemistry, Separation, Quantitation, and Antioxidation)
Edited by LESTER PACKER

VOLUME 214. Carotenoids (Part B: Metabolism, Genetics, and Biosynthesis)
Edited by LESTER PACKER

VOLUME 215. Platelets: Receptors, Adhesion, Secretion (Part B)
Edited by JACEK J. HAWIGER

VOLUME 216. Recombinant DNA (Part G)
Edited by RAY WU

VOLUME 217. Recombinant DNA (Part H)
Edited by RAY WU

VOLUME 218. Recombinant DNA (Part I)
Edited by RAY WU

VOLUME 219. Reconstitution of Intracellular Transport
Edited by JAMES E. ROTHMAN

VOLUME 220. Membrane Fusion Techniques (Part A)
Edited by NEJAT DÜZGÜNEŞ

VOLUME 221. Membrane Fusion Techniques (Part B)
Edited by NEJAT DÜZGÜNEŞ

VOLUME 222. Proteolytic Enzymes in Coagulation, Fibrinolysis, and Complement Activation (Part A: Mammalian Blood Coagulation Factors and Inhibitors)
Edited by LASZLO LORAND AND KENNETH G. MANN

VOLUME 223. Proteolytic Enzymes in Coagulation, Fibrinolysis, and Complement Activation (Part B: Complement Activation, Fibrinolysis, and Nonmammalian Blood Coagulation Factors)
Edited by LASZLO LORAND AND KENNETH G. MANN

VOLUME 224. Molecular Evolution: Producing the Biochemical Data
Edited by ELIZABETH ANNE ZIMMER, THOMAS J. WHITE, REBECCA L. CANN, AND ALLAN C. WILSON

VOLUME 225. Guide to Techniques in Mouse Development
Edited by PAUL M. WASSARMAN AND MELVIN L. DEPAMPHILIS

VOLUME 226. Metallobiochemistry (Part C: Spectroscopic and Physical Methods for Probing Metal Ion Environments in Metalloenzymes and Metalloproteins)
Edited by JAMES F. RIORDAN AND BERT L. VALLEE

VOLUME 227. Metallobiochemistry (Part D: Physical and Spectroscopic Methods for Probing Metal Ion Environments in Metalloproteins)
Edited by JAMES F. RIORDAN AND BERT L. VALLEE

VOLUME 228. Aqueous Two-Phase Systems
Edited by HARRY WALTER AND GÖTE JOHANSSON

VOLUME 229. Cumulative Subject Index Volumes 195–198, 200–227

VOLUME 230. Guide to Techniques in Glycobiology
Edited by WILLIAM J. LENNARZ AND GERALD W. HART

VOLUME 231. Hemoglobins (Part B: Biochemical and Analytical Methods)
Edited by JOHANNES EVERSE, KIM D. VANDEGRIFF, AND ROBERT M. WINSLOW

VOLUME 232. Hemoglobins (Part C: Biophysical Methods)
Edited by JOHANNES EVERSE, KIM D. VANDEGRIFF, AND ROBERT M. WINSLOW

VOLUME 233. Oxygen Radicals in Biological Systems (Part C)
Edited by LESTER PACKER

VOLUME 234. Oxygen Radicals in Biological Systems (Part D)
Edited by LESTER PACKER

VOLUME 235. Bacterial Pathogenesis (Part A: Identification and Regulation of Virulence Factors)
Edited by VIRGINIA L. CLARK AND PATRIK M. BAVOIL

VOLUME 236. Bacterial Pathogenesis (Part B: Integration of Pathogenic Bacteria with Host Cells)
Edited by VIRGINIA L. CLARK AND PATRIK M. BAVOIL

VOLUME 237. Heterotrimeric G Proteins
Edited by RAVI IYENGAR

VOLUME 238. Heterotrimeric G-Protein Effectors
Edited by RAVI IYENGAR

VOLUME 239. Nuclear Magnetic Resonance (Part C)
Edited by THOMAS L. JAMES AND NORMAN J. OPPENHEIMER

VOLUME 240. Numerical Computer Methods (Part B)
Edited by MICHAEL L. JOHNSON AND LUDWIG BRAND

VOLUME 241. Retroviral Proteases
Edited by LAWRENCE C. KUO AND JULES A. SHAFER

VOLUME 242. Neoglycoconjugates (Part A)
Edited by Y. C. LEE AND REIKO T. LEE

VOLUME 243. Inorganic Microbial Sulfur Metabolism
Edited by HARRY D. PECK, JR., AND JEAN LEGALL

VOLUME 244. Proteolytic Enzymes: Serine and Cysteine Peptidases
Edited by ALAN J. BARRETT

VOLUME 245. Extracellular Matrix Components
Edited by E. RUOSLAHTI AND E. ENGVALL

VOLUME 246. Biochemical Spectroscopy
Edited by KENNETH SAUER

VOLUME 247. Neoglycoconjugates (Part B: Biomedical Applications)
Edited by Y. C. LEE AND REIKO T. LEE

VOLUME 248. Proteolytic Enzymes: Aspartic and Metallo Peptidases
Edited by ALAN J. BARRETT

VOLUME 249. Enzyme Kinetics and Mechanism (Part D: Developments in Enzyme Dynamics)
Edited by DANIEL L. PURICH

VOLUME 250. Lipid Modifications of Proteins
Edited by PATRICK J. CASEY AND JANICE E. BUSS

VOLUME 251. Biothiols (Part A: Monothiols and Dithiols, Protein Thiols, and Thiyl Radicals)
Edited by LESTER PACKER

VOLUME 252. Biothiols (Part B: Glutathione and Thioredoxin; Thiols in Signal Transduction and Gene Regulation)
Edited by LESTER PACKER

VOLUME 253. Adhesion of Microbial Pathogens
Edited by RON J. DOYLE AND ITZHAK OFEK

VOLUME 254. Oncogene Techniques
Edited by PETER K. VOGT AND INDER M. VERMA

VOLUME 255. Small GTPases and Their Regulators (Part A: Ras Family)
Edited by W. E. BALCH, CHANNING J. DER, AND ALAN HALL

VOLUME 256. Small GTPases and Their Regulators (Part B: Rho Family)
Edited by W. E. BALCH, CHANNING J. DER, AND ALAN HALL

VOLUME 257. Small GTPases and Their Regulators (Part C: Proteins Involved in Transport)
Edited by W. E. BALCH, CHANNING J. DER, AND ALAN HALL

VOLUME 258. Redox-Active Amino Acids in Biology
Edited by JUDITH P. KLINMAN

VOLUME 259. Energetics of Biological Macromolecules
Edited by MICHAEL L. JOHNSON AND GARY K. ACKERS

VOLUME 260. Mitochondrial Biogenesis and Genetics (Part A)
Edited by GIUSEPPE M. ATTARDI AND ANNE CHOMYN

VOLUME 261. Nuclear Magnetic Resonance and Nucleic Acids
Edited by THOMAS L. JAMES

VOLUME 262. DNA Replication
Edited by JUDITH L. CAMPBELL

VOLUME 263. Plasma Lipoproteins (Part C: Quantitation)
Edited by WILLIAM A. BRADLEY, SANDRA H. GIANTURCO, AND JERE P. SEGREST

VOLUME 264. Mitochondrial Biogenesis and Genetics (Part B)
Edited by GIUSEPPE M. ATTARDI AND ANNE CHOMYN

VOLUME 265. Cumulative Subject Index Volumes 228, 230–262

VOLUME 266. Computer Methods for Macromolecular Sequence Analysis
Edited by RUSSELL F. DOOLITTLE

VOLUME 267. Combinatorial Chemistry
Edited by JOHN N. ABELSON

VOLUME 268. Nitric Oxide (Part A: Sources and Detection of NO; NO Synthase)
Edited by LESTER PACKER

VOLUME 269. Nitric Oxide (Part B: Physiological and Pathological Processes)
Edited by LESTER PACKER

VOLUME 270. High Resolution Separation and Analysis of Biological Macromolecules (Part A: Fundamentals)
Edited by BARRY L. KARGER AND WILLIAM S. HANCOCK

VOLUME 271. High Resolution Separation and Analysis of Biological Macromolecules (Part B: Applications)
Edited by BARRY L. KARGER AND WILLIAM S. HANCOCK

VOLUME 272. Cytochrome P450 (Part B)
Edited by ERIC F. JOHNSON AND MICHAEL R. WATERMAN

VOLUME 273. RNA Polymerase and Associated Factors (Part A)
Edited by SANKAR ADHYA

VOLUME 274. RNA Polymerase and Associated Factors (Part B)
Edited by SANKAR ADHYA

VOLUME 275. Viral Polymerases and Related Proteins
Edited by LAWRENCE C. KUO, DAVID B. OLSEN, AND STEVEN S. CARROLL

VOLUME 276. Macromolecular Crystallography (Part A)
Edited by CHARLES W. CARTER, JR., AND ROBERT M. SWEET

VOLUME 277. Macromolecular Crystallography (Part B)
Edited by CHARLES W. CARTER, JR., AND ROBERT M. SWEET

VOLUME 278. Fluorescence Spectroscopy
Edited by LUDWIG BRAND AND MICHAEL L. JOHNSON

VOLUME 279. Vitamins and Coenzymes (Part I)
Edited by DONALD B. MCCORMICK, JOHN W. SUTTIE, AND CONRAD WAGNER

VOLUME 280. Vitamins and Coenzymes (Part J)
Edited by DONALD B. MCCORMICK, JOHN W. SUTTIE, AND CONRAD WAGNER

VOLUME 281. Vitamins and Coenzymes (Part K)
Edited by DONALD B. MCCORMICK, JOHN W. SUTTIE, AND CONRAD WAGNER

VOLUME 282. Vitamins and Coenzymes (Part L)
Edited by DONALD B. MCCORMICK, JOHN W. SUTTIE, AND CONRAD WAGNER

VOLUME 283. Cell Cycle Control
Edited by WILLIAM G. DUNPHY

VOLUME 284. Lipases (Part A: Biotechnology)
Edited by BYRON RUBIN AND EDWARD A. DENNIS

VOLUME 285. Cumulative Subject Index Volumes 263, 264, 266–284, 286–289

VOLUME 286. Lipases (Part B: Enzyme Characterization and Utilization)
Edited by BYRON RUBIN AND EDWARD A. DENNIS

VOLUME 287. Chemokines
Edited by RICHARD HORUK

VOLUME 288. Chemokine Receptors
Edited by RICHARD HORUK

VOLUME 289. Solid Phase Peptide Synthesis
Edited by GREGG B. FIELDS

VOLUME 290. Molecular Chaperones
Edited by GEORGE H. LORIMER AND THOMAS BALDWIN

VOLUME 291. Caged Compounds
Edited by GERARD MARRIOTT

VOLUME 292. ABC Transporters: Biochemical, Cellular, and Molecular Aspects
Edited by SURESH V. AMBUDKAR AND MICHAEL M. GOTTESMAN

VOLUME 293. Ion Channels (Part B)
Edited by P. MICHAEL CONN

VOLUME 294. Ion Channels (Part C)
Edited by P. MICHAEL CONN

VOLUME 295. Energetics of Biological Macromolecules (Part B)
Edited by GARY K. ACKERS AND MICHAEL L. JOHNSON

VOLUME 296. Neurotransmitter Transporters
Edited by SUSAN G. AMARA

VOLUME 297. Photosynthesis: Molecular Biology of Energy Capture
Edited by LEE MCINTOSH

VOLUME 298. Molecular Motors and the Cytoskeleton (Part B)
Edited by RICHARD B. VALLEE

VOLUME 299. Oxidants and Antioxidants (Part A)
Edited by LESTER PACKER

VOLUME 300. Oxidants and Antioxidants (Part B)
Edited by LESTER PACKER

VOLUME 301. Nitric Oxide: Biological and Antioxidant Activities (Part C)
Edited by LESTER PACKER

VOLUME 302. Green Fluorescent Protein
Edited by P. MICHAEL CONN

VOLUME 303. cDNA Preparation and Display
Edited by SHERMAN M. WEISSMAN

VOLUME 304. Chromatin
Edited by PAUL M. WASSARMAN AND ALAN P. WOLFFE

VOLUME 305. Bioluminescence and Chemiluminescence (Part C)
Edited by THOMAS O. BALDWIN AND MIRIAM M. ZIEGLER

VOLUME 306. Expression of Recombinant Genes in Eukaryotic Systems
Edited by JOSEPH C. GLORIOSO AND MARTIN C. SCHMIDT

VOLUME 307. Confocal Microscopy
Edited by P. MICHAEL CONN

VOLUME 308. Enzyme Kinetics and Mechanism (Part E: Energetics of Enzyme Catalysis)
Edited by DANIEL L. PURICH AND VERN L. SCHRAMM

VOLUME 309. Amyloid, Prions, and Other Protein Aggregates
Edited by RONALD WETZEL

VOLUME 310. Biofilms
Edited by RON J. DOYLE

VOLUME 311. Sphingolipid Metabolism and Cell Signaling (Part A)
Edited by ALFRED H. MERRILL, JR., AND YUSUF A. HANNUN

VOLUME 312. Sphingolipid Metabolism and Cell Signaling (Part B)
Edited by ALFRED H. MERRILL, JR., AND YUSUF A. HANNUN

VOLUME 313. Antisense Technology
(Part A: General Methods, Methods of Delivery, and RNA Studies)
Edited by M. IAN PHILLIPS

VOLUME 314. Antisense Technology (Part B: Applications)
Edited by M. IAN PHILLIPS

VOLUME 315. Vertebrate Phototransduction and the Visual Cycle (Part A)
Edited by KRZYSZTOF PALCZEWSKI

VOLUME 316. Vertebrate Phototransduction and the Visual Cycle (Part B)
Edited by KRZYSZTOF PALCZEWSKI

VOLUME 317. RNA–Ligand Interactions (Part A: Structural Biology Methods)
Edited by DANIEL W. CELANDER AND JOHN N. ABELSON

VOLUME 318. RNA–Ligand Interactions (Part B: Molecular Biology Methods)
Edited by DANIEL W. CELANDER AND JOHN N. ABELSON

VOLUME 319. Singlet Oxygen, UV-A, and Ozone
Edited by LESTER PACKER AND HELMUT SIES

VOLUME 320. Cumulative Subject Index Volumes 290–319

VOLUME 321. Numerical Computer Methods (Part C)
Edited by MICHAEL L. JOHNSON AND LUDWIG BRAND

VOLUME 322. Apoptosis
Edited by JOHN C. REED

VOLUME 323. Energetics of Biological Macromolecules (Part C)
Edited by MICHAEL L. JOHNSON AND GARY K. ACKERS

VOLUME 324. Branched-Chain Amino Acids (Part B)
Edited by ROBERT A. HARRIS AND JOHN R. SOKATCH

VOLUME 325. Regulators and Effectors of Small GTPases
(Part D: Rho Family)
Edited by W. E. BALCH, CHANNING J. DER, AND ALAN HALL

VOLUME 326. Applications of Chimeric Genes and Hybrid Proteins
(Part A: Gene Expression and Protein Purification)
Edited by JEREMY THORNER, SCOTT D. EMR, AND JOHN N. ABELSON

VOLUME 327. Applications of Chimeric Genes and Hybrid Proteins
(Part B: Cell Biology and Physiology)
Edited by JEREMY THORNER, SCOTT D. EMR, AND JOHN N. ABELSON

VOLUME 328. Applications of Chimeric Genes and Hybrid Proteins (Part C: Protein–Protein Interactions and Genomics)
Edited by JEREMY THORNER, SCOTT D. EMR, AND JOHN N. ABELSON

VOLUME 329. Regulators and Effectors of Small GTPases (Part E: GTPases Involved in Vesicular Traffic)
Edited by W. E. BALCH, CHANNING J. DER, AND ALAN HALL

VOLUME 330. Hyperthermophilic Enzymes (Part A)
Edited by MICHAEL W. W. ADAMS AND ROBERT M. KELLY

VOLUME 331. Hyperthermophilic Enzymes (Part B)
Edited by MICHAEL W. W. ADAMS AND ROBERT M. KELLY

VOLUME 332. Regulators and Effectors of Small GTPases (Part F: Ras Family I)
Edited by W. E. BALCH, CHANNING J. DER, AND ALAN HALL

VOLUME 333. Regulators and Effectors of Small GTPases (Part G: Ras Family II)
Edited by W. E. BALCH, CHANNING J. DER, AND ALAN HALL

VOLUME 334. Hyperthermophilic Enzymes (Part C)
Edited by MICHAEL W. W. ADAMS AND ROBERT M. KELLY

VOLUME 335. Flavonoids and Other Polyphenols
Edited by LESTER PACKER

VOLUME 336. Microbial Growth in Biofilms (Part A: Developmental and Molecular Biological Aspects)
Edited by RON J. DOYLE

VOLUME 337. Microbial Growth in Biofilms (Part B: Special Environments and Physicochemical Aspects)
Edited by RON J. DOYLE

VOLUME 338. Nuclear Magnetic Resonance of Biological Macromolecules (Part A)
Edited by THOMAS L. JAMES, VOLKER DÖTSCH, AND ULI SCHMITZ

VOLUME 339. Nuclear Magnetic Resonance of Biological Macromolecules (Part B)
Edited by THOMAS L. JAMES, VOLKER DÖTSCH, AND ULI SCHMITZ

VOLUME 340. Drug–Nucleic Acid Interactions
Edited by JONATHAN B. CHAIRES AND MICHAEL J. WARING

VOLUME 341. Ribonucleases (Part A)
Edited by ALLEN W. NICHOLSON

VOLUME 342. Ribonucleases (Part B)
Edited by ALLEN W. NICHOLSON

VOLUME 343. G Protein Pathways (Part A: Receptors)
Edited by RAVI IYENGAR AND JOHN D. HILDEBRANDT

VOLUME 344. G Protein Pathways (Part B: G Proteins and Their Regulators)
Edited by RAVI IYENGAR AND JOHN D. HILDEBRANDT

VOLUME 345. G Protein Pathways (Part C: Effector Mechanisms)
Edited by RAVI IYENGAR AND JOHN D. HILDEBRANDT

VOLUME 346. Gene Therapy Methods
Edited by M. IAN PHILLIPS

VOLUME 347. Protein Sensors and Reactive Oxygen Species (Part A: Selenoproteins and Thioredoxin)
Edited by HELMUT SIES AND LESTER PACKER

VOLUME 348. Protein Sensors and Reactive Oxygen Species (Part B: Thiol Enzymes and Proteins)
Edited by HELMUT SIES AND LESTER PACKER

VOLUME 349. Superoxide Dismutase
Edited by LESTER PACKER

VOLUME 350. Guide to Yeast Genetics and Molecular and Cell Biology (Part B)
Edited by CHRISTINE GUTHRIE AND GERALD R. FINK

VOLUME 351. Guide to Yeast Genetics and Molecular and Cell Biology (Part C)
Edited by CHRISTINE GUTHRIE AND GERALD R. FINK

VOLUME 352. Redox Cell Biology and Genetics (Part A)
Edited by CHANDAN K. SEN AND LESTER PACKER

VOLUME 353. Redox Cell Biology and Genetics (Part B)
Edited by CHANDAN K. SEN AND LESTER PACKER

VOLUME 354. Enzyme Kinetics and Mechanisms (Part F: Detection and Characterization of Enzyme Reaction Intermediates)
Edited by DANIEL L. PURICH

VOLUME 355. Cumulative Subject Index Volumes 321–354

VOLUME 356. Laser Capture Microscopy and Microdissection
Edited by P. MICHAEL CONN

VOLUME 357. Cytochrome P450, Part C
Edited by ERIC F. JOHNSON AND MICHAEL R. WATERMAN

VOLUME 358. Bacterial Pathogenesis (Part C: Identification, Regulation, and Function of Virulence Factors)
Edited by VIRGINIA L. CLARK AND PATRIK M. BAVOIL

VOLUME 359. Nitric Oxide (Part D)
Edited by ENRIQUE CADENAS AND LESTER PACKER

VOLUME 360. Biophotonics (Part A)
Edited by GERARD MARRIOTT AND IAN PARKER

VOLUME 361. Biophotonics (Part B)
Edited by GERARD MARRIOTT AND IAN PARKER

VOLUME 362. Recognition of Carbohydrates in Biological Systems (Part A)
Edited by YUAN C. LEE AND REIKO T. LEE

VOLUME 363. Recognition of Carbohydrates in Biological Systems (Part B)
Edited by YUAN C. LEE AND REIKO T. LEE

VOLUME 364. Nuclear Receptors
Edited by DAVID W. RUSSELL AND DAVID J. MANGELSDORF

VOLUME 365. Differentiation of Embryonic Stem Cells
Edited by PAUL M. WASSAUMAN AND GORDON M. KELLER

VOLUME 366. Protein Phosphatases
Edited by SUSANNE KLUMPP AND JOSEF KRIEGLSTEIN

VOLUME 367. Liposomes (Part A)
Edited by NEJAT DÜZGÜNEŞ

VOLUME 368. Macromolecular Crystallography (Part C)
Edited by CHARLES W. CARTER, JR., AND ROBERT M. SWEET

VOLUME 369. Combinational Chemistry (Part B)
Edited by GUILLERMO A. MORALES AND BARRY A. BUNIN

VOLUME 370. RNA Polymerases and Associated Factors (Part C)
Edited by SANKAR L. ADHYA AND SUSAN GARGES

VOLUME 371. RNA Polymerases and Associated Factors (Part D)
Edited by SANKAR L. ADHYA AND SUSAN GARGES

VOLUME 372. Liposomes (Part B)
Edited by NEJAT DÜZGÜNEŞ

VOLUME 373. Liposomes (Part C)
Edited by NEJAT DÜZGÜNEŞ

VOLUME 374. Macromolecular Crystallography (Part D)
Edited by CHARLES W. CARTER, JR., AND ROBERT W. SWEET

VOLUME 375. Chromatin and Chromatin Remodeling Enzymes (Part A)
Edited by C. DAVID ALLIS AND CARL WU

VOLUME 376. Chromatin and Chromatin Remodeling Enzymes (Part B)
Edited by C. DAVID ALLIS AND CARL WU

VOLUME 377. Chromatin and Chromatin Remodeling Enzymes (Part C)
Edited by C. DAVID ALLIS AND CARL WU

VOLUME 378. Quinones and Quinone Enzymes (Part A)
Edited by HELMUT SIES AND LESTER PACKER

VOLUME 379. Energetics of Biological Macromolecules (Part D)
Edited by JO M. HOLT, MICHAEL L. JOHNSON, AND GARY K. ACKERS

VOLUME 380. Energetics of Biological Macromolecules (Part E)
Edited by JO M. HOLT, MICHAEL L. JOHNSON, AND GARY K. ACKERS

VOLUME 381. Oxygen Sensing
Edited by CHANDAN K. SEN AND GREGG L. SEMENZA

VOLUME 382. Quinones and Quinone Enzymes (Part B)
Edited by HELMUT SIES AND LESTER PACKER

VOLUME 383. Numerical Computer Methods (Part D)
Edited by LUDWIG BRAND AND MICHAEL L. JOHNSON

VOLUME 384. Numerical Computer Methods (Part E)
Edited by LUDWIG BRAND AND MICHAEL L. JOHNSON

VOLUME 385. Imaging in Biological Research (Part A)
Edited by P. MICHAEL CONN

VOLUME 386. Imaging in Biological Research (Part B)
Edited by P. MICHAEL CONN

VOLUME 387. Liposomes (Part D)
Edited by NEJAT DÜZGÜNEŞ

VOLUME 388. Protein Engineering
Edited by DAN E. ROBERTSON AND JOSEPH P. NOEL

VOLUME 389. Regulators of G-Protein Signaling (Part A)
Edited by DAVID P. SIDEROVSKI

VOLUME 390. Regulators of G-Protein Signaling (Part B)
Edited by DAVID P. SIDEROVSKI

VOLUME 391. Liposomes (Part E)
Edited by NEJAT DÜZGÜNEŞ

VOLUME 392. RNA Interference
Edited by ENGELKE ROSSI

VOLUME 393. Circadian Rhythms
Edited by MICHAEL W. YOUNG

VOLUME 394. Nuclear Magnetic Resonance of Biological Macromolecules (Part C)
Edited by THOMAS L. JAMES

VOLUME 395. Producing the Biochemical Data (Part B)
Edited by ELIZABETH A. ZIMMER AND ERIC H. ROALSON

VOLUME 396. Nitric Oxide (Part E)
Edited by LESTER PACKER AND ENRIQUE CADENAS

VOLUME 397. Environmental Microbiology
Edited by JARED R. LEADBETTER

VOLUME 398. Ubiquitin and Protein Degradation (Part A)
Edited by RAYMOND J. DESHAIES

VOLUME 399. Ubiquitin and Protein Degradation (Part B)
Edited by RAYMOND J. DESHAIES

VOLUME 400. Phase II Conjugation Enzymes and Transport Systems
Edited by HELMUT SIES AND LESTER PACKER

VOLUME 401. Glutathione Transferases and Gamma Glutamyl Transpeptidases
Edited by HELMUT SIES AND LESTER PACKER

VOLUME 402. Biological Mass Spectrometry
Edited by A. L. BURLINGAME

VOLUME 403. GTPases Regulating Membrane Targeting and Fusion
Edited by WILLIAM E. BALCH, CHANNING J. DER, AND ALAN HALL

VOLUME 404. GTPases Regulating Membrane Dynamics
Edited by WILLIAM E. BALCH, CHANNING J. DER, AND ALAN HALL

VOLUME 405. Mass Spectrometry: Modified Proteins and Glycoconjugates
Edited by A. L. BURLINGAME

VOLUME 406. Regulators and Effectors of Small GTPases: Rho Family
Edited by WILLIAM E. BALCH, CHANNING J. DER, AND ALAN HALL

VOLUME 407. Regulators and Effectors of Small GTPases: Ras Family
Edited by WILLIAM E. BALCH, CHANNING J. DER, AND ALAN HALL

VOLUME 408. DNA Repair (Part A)
Edited by JUDITH L. CAMPBELL AND PAUL MODRICH

VOLUME 409. DNA Repair (Part B)
Edited by JUDITH L. CAMPBELL AND PAUL MODRICH

VOLUME 410. DNA Microarrays (Part A: Array Platforms and Web-Bench Protocols)
Edited by ALAN KIMMEL AND BRIAN OLIVER

VOLUME 411. DNA Microarrays (Part B: Databases and Statistics)
Edited by ALAN KIMMEL AND BRIAN OLIVER

VOLUME 412. Amyloid, Prions, and Other Protein Aggregates (Part B)
Edited by INDU KHETERPAL AND RONALD WETZEL

VOLUME 413. Amyloid, Prions, and Other Protein Aggregates (Part C)
Edited by INDU KHETERPAL AND RONALD WETZEL

VOLUME 414. Measuring Biological Responses with Automated Microscopy
Edited by JAMES INGLESE

VOLUME 415. Glycobiology
Edited by MINORU FUKUDA

VOLUME 416. Glycomics
Edited by MINORU FUKUDA

VOLUME 417. Functional Glycomics
Edited by MINORU FUKUDA

VOLUME 418. Embryonic Stem Cells
Edited by IRINA KLIMANSKAYA AND ROBERT LANZA

VOLUME 419. Adult Stem Cells
Edited by IRINA KLIMANSKAYA AND ROBERT LANZA

VOLUME 420. Stem Cell Tools and Other Experimental Protocols
Edited by IRINA KLIMANSKAYA AND ROBERT LANZA

VOLUME 421. Advanced Bacterial Genetics: Use of Transposons and Phage for Genomic Engineering
Edited by KELLY T. HUGHES

VOLUME 422. Two-Component Signaling Systems, Part A
Edited by MELVIN I. SIMON, BRIAN R. CRANE, AND ALEXANDRINE CRANE

VOLUME 423. Two-Component Signaling Systems, Part B
Edited by MELVIN I. SIMON, BRIAN R. CRANE, AND ALEXANDRINE CRANE

VOLUME 424. RNA Editing
Edited by JONATHA M. GOTT

VOLUME 425. RNA Modification
Edited by JONATHA M. GOTT

VOLUME 426. Integrins
Edited by DAVID CHERESH

VOLUME 427. MicroRNA Methods
Edited by JOHN J. ROSSI

VOLUME 428. Osmosensing and Osmosignaling
Edited by HELMUT SIES AND DIETER HAUSSINGER

VOLUME 429. Translation Initiation: Extract Systems and Molecular Genetics
Edited by JON LORSCH

VOLUME 430. Translation Initiation: Reconstituted Systems and Biophysical Methods
Edited by JON LORSCH

VOLUME 431. Translation Initiation: Cell Biology, High-Throughput and Chemical-Based Approaches
Edited by JON LORSCH

VOLUME 432. Lipidomics and Bioactive Lipids: Mass-Spectrometry–Based Lipid Analysis
Edited by H. ALEX BROWN

VOLUME 433. Lipidomics and Bioactive Lipids: Specialized Analytical Methods and Lipids in Disease
Edited by H. ALEX BROWN

VOLUME 434. Lipidomics and Bioactive Lipids: Lipids and Cell Signaling
Edited by H. ALEX BROWN

VOLUME 435. Oxygen Biology and Hypoxia
Edited by HELMUT SIES AND BERNHARD BRÜNE

VOLUME 436. Globins and Other Nitric Oxide-Reactive Protiens (Part A)
Edited by ROBERT K. POOLE

VOLUME 437. Globins and Other Nitric Oxide-Reactive Protiens (Part B)
Edited by ROBERT K. POOLE

VOLUME 438. Small GTPases in Disease (Part A)
Edited by WILLIAM E. BALCH, CHANNING J. DER, AND ALAN HALL

VOLUME 439. Small GTPases in Disease (Part B)
Edited by WILLIAM E. BALCH, CHANNING J. DER, AND ALAN HALL

VOLUME 440. Nitric Oxide, Part F Oxidative and Nitrosative Stress in Redox Regulation of Cell Signaling
Edited by ENRIQUE CADENAS AND LESTER PACKER

VOLUME 441. Nitric Oxide, Part G Oxidative and Nitrosative Stress in Redox Regulation of Cell Signaling
Edited by ENRIQUE CADENAS AND LESTER PACKER

VOLUME 442. Programmed Cell Death, General Principles for Studying Cell Death (Part A)
Edited by ROYA KHOSRAVI-FAR, ZAHRA ZAKERI, RICHARD A. LOCKSHIN, AND MAURO PIACENTINI

VOLUME 443. Angiogenesis: *In Vitro* Systems
Edited by DAVID A. CHERESH

VOLUME 444. Angiogenesis: *In Vivo* Systems (Part A)
Edited by DAVID A. CHERESH

VOLUME 445. Angiogenesis: *In Vivo* Systems (Part B)
Edited by DAVID A. CHERESH

VOLUME 446. Programmed Cell Death, The Biology and Therapeutic Implications of Cell Death (Part B)
Edited by ROYA KHOSRAVI-FAR, ZAHRA ZAKERI, RICHARD A. LOCKSHIN, AND MAURO PIACENTINI

VOLUME 447. RNA Turnover in Bacteria, Archaea and Organelles
Edited by LYNNE E. MAQUAT AND CECILIA M. ARRAIANO

VOLUME 448. RNA Turnover in Eukaryotes: Nucleases, Pathways and Analysis of mRNA Decay
Edited by LYNNE E. MAQUAT AND MEGERDITCH KILEDJIAN

VOLUME 449. RNA Turnover in Eukaryotes: Analysis of Specialized and Quality Control RNA Decay Pathways
Edited by LYNNE E. MAQUAT AND MEGERDITCH KILEDJIAN

VOLUME 450. Fluorescence Spectroscopy
Edited by LUDWIG BRAND AND MICHAEL L. JOHNSON

VOLUME 451. Autophagy: Lower Eukaryotes and Non-Mammalian Systems (Part A)
Edited by DANIEL J. KLIONSKY

VOLUME 452. Autophagy in Mammalian Systems (Part B)
Edited by DANIEL J. KLIONSKY

VOLUME 453. Autophagy in Disease and Clinical Applications (Part C)
Edited by DANIEL J. KLIONSKY

VOLUME 454. Computer Methods (Part A)
Edited by MICHAEL L. JOHNSON AND LUDWIG BRAND

VOLUME 455. Biothermodynamics (Part A)
Edited by MICHAEL L. JOHNSON, JO M. HOLT, AND GARY K. ACKERS (RETIRED)

VOLUME 456. Mitochondrial Function, Part A: Mitochondrial Electron Transport Complexes and Reactive Oxygen Species
Edited by WILLIAM S. ALLISON AND IMMO E. SCHEFFLER

VOLUME 457. Mitochondrial Function, Part B: Mitochondrial Protein Kinases, Protein Phosphatases and Mitochondrial Diseases
Edited by WILLIAM S. ALLISON AND ANNE N. MURPHY

VOLUME 458. Complex Enzymes in Microbial Natural Product Biosynthesis, Part A: Overview Articles and Peptides
Edited by DAVID A. HOPWOOD

VOLUME 459. Complex Enzymes in Microbial Natural Product Biosynthesis, Part B: Polyketides, Aminocoumarins and Carbohydrates
Edited by DAVID A. HOPWOOD

VOLUME 460. Chemokines, Part A
Edited by TRACY M. HANDEL AND DAMON J. HAMEL

VOLUME 461. Chemokines, Part B
Edited by TRACY M. HANDEL AND DAMON J. HAMEL

VOLUME 462. Non-Natural Amino Acids
Edited by TOM W. MUIR AND JOHN N. ABELSON

VOLUME 463. Guide to Protein Purification, 2nd Edition
Edited by RICHARD R. BURGESS AND MURRAY P. DEUTSCHER

VOLUME 464. Liposomes, Part F
Edited by NEJAT DÜZGÜNEŞ

VOLUME 465. Liposomes, Part G
Edited by NEJAT DÜZGÜNEŞ

VOLUME 466. Biothermodynamics, Part B
Edited by MICHAEL L. JOHNSON, GARY K. ACKERS, AND JO M. HOLT

VOLUME 467. Computer Methods Part B
Edited by MICHAEL L. JOHNSON AND LUDWIG BRAND

VOLUME 468. Biophysical, Chemical, and Functional Probes of RNA Structure, Interactions and Folding: Part A
Edited by DANIEL HERSCHLAG

VOLUME 469. Biophysical, Chemical, and Functional Probes of RNA Structure, Interactions and Folding: Part B
Edited by DANIEL HERSCHLAG

VOLUME 470. Guide to Yeast Genetics: Functional Genomics, Proteomics, and Other Systems Analysis, 2nd Edition
Edited by GERALD FINK, JONATHAN WEISSMAN, AND CHRISTINE GUTHRIE

VOLUME 471. Two-Component Signaling Systems, Part C
Edited by MELVIN I. SIMON, BRIAN R. CRANE, AND ALEXANDRINE CRANE

VOLUME 472. Single Molecule Tools, Part A: Fluorescence Based Approaches
Edited by NILS G. WALTER

VOLUME 473. Thiol Redox Transitions in Cell Signaling, Part A Chemistry and Biochemistry of Low Molecular Weight and Protein Thiols
Edited by ENRIQUE CADENAS AND LESTER PACKER

VOLUME 474. Thiol Redox Transitions in Cell Signaling, Part B Cellular Localization and Signaling
Edited by ENRIQUE CADENAS AND LESTER PACKER

VOLUME 475. Single Molecule Tools, Part B: Super-Resolution, Particle Tracking, Multiparameter, and Force Based Methods
Edited by NILS G. WALTER

VOLUME 476. Guide to Techniques in Mouse Development, Part A Mice, Embryos, and Cells, 2nd Edition
Edited by PAUL M. WASSARMAN AND PHILIPPE M. SORIANO

VOLUME 477. Guide to Techniques in Mouse Development, Part B Mouse Molecular Genetics, 2nd Edition
Edited by PAUL M. WASSARMAN AND PHILIPPE M. SORIANO

VOLUME 478. Glycomics
Edited by MINORU FUKUDA

VOLUME 479. Functional Glycomics
Edited by MINORU FUKUDA

VOLUME 480. Glycobiology
Edited by MINORU FUKUDA

VOLUME 481. Cryo-EM, Part A: Sample Preparation and Data Collection
Edited by GRANT J. JENSEN

VOLUME 482. Cryo-EM, Part B: 3-D Reconstruction
Edited by GRANT J. JENSEN

VOLUME 483. Cryo-EM, Part C: Analyses, Interpretation, and Case Studies
Edited by GRANT J. JENSEN

VOLUME 484. Constitutive Activity in Receptors and Other Proteins, Part A
Edited by P. MICHAEL CONN

VOLUME 485. Constitutive Activity in Receptors and Other Proteins, Part B
Edited by P. MICHAEL CONN

VOLUME 486. Research on Nitrification and Related Processes, Part A
Edited by MARTIN G. KLOTZ

VOLUME 487. Computer Methods, Part C
Edited by MICHAEL L. JOHNSON AND LUDWIG BRAND

VOLUME 488. Biothermodynamics, Part C
Edited by MICHAEL L. JOHNSON , JO M. HOLT AND GARY K. ACKERS

VOLUME 489. The Unfolded Protein Response and Cellular Stress, Part A
Edited by P. MICHAEL CONN

VOLUME 490. The Unfolded Protein Response and Cellular Stress, Part B
Edited by P. MICHAEL CONN

CHAPTER ONE

Methods for Investigating the UPR in Filamentous Fungi

Thomas Guillemette,* Arthur F. J. Ram,[†] Neuza D. S. P. Carvalho,[†] Aymeric Joubert,* Philippe Simoneau,* *and* David B. Archer[‡]

Contents

1. Introduction	2
2. Conditions to Study the UPR in Filamentous Fungi	5
2.1. Culture conditions	5
2.2. Induction of ER-associated stress	8
3. Analysis of Expression During the ER Stress	11
3.1. Transcriptomic analysis	11
3.2. Polysome analysis	12
4. Modifying the UPR Signaling by Targeted Gene Replacement	14
4.1. Construction of UPR-deficient strains	14
4.2. Constitutive activation of the UPR	18
4.3. PEG-mediated transformation of protoplasts	21
4.4. Mutant verification	23
4.5. Phenotypic analysis	23
Acknowledgments	25
References	26

Abstract

Filamentous fungi have a high-capacity secretory system and are therefore widely exploited for the industrial production of native and heterologous proteins. However, in most cases, the yields of nonfungal proteins are significantly lower than those obtained for fungal proteins. One well-studied bottleneck appears to be the result of slow or aberrant folding of heterologous proteins in the ER during the early stages of secretion within the endoplasmic reticulum, leading to stress responses in the host, including the unfolded protein response (UPR). Most of the key elements constituting the signal transduction pathway of the UPR in *Saccharomyces cerevisiae* have been identified in filamentous fungi,

* UMR PAVE No 77, IFR 149 QUASAV, Angers Cedex, France
[†] Molecular Microbiology and Biotechnology, Kluyver Centre for Genomics of Industrial Fermentation, Institute of Biology Leiden, Leiden University, Leiden, The Netherlands
[‡] School of Biology, University of Nottingham, University Park, Nottingham, United Kingdom

including the central activation mechanism of the pathway, that is, the stress-induced splicing of an unconventional (nonspliceosomal) intron in orthologs of the *HAC1* mRNA. This splicing event relieves a translational block in the *HAC1* mRNA, allowing for the translation of the bZIP transcription factor Hac1p that regulates the expression of UPR target genes. The UPR is involved in regulating the folding, yield, and delivery of secretory proteins and that has consequences for fungal lifestyles, including virulence and biotechnology. The recent releases of genome sequences of several species of filamentous fungi and the availability of DNA arrays, GeneChips, and deep sequencing methodologies have provided an unprecedented resource for exploring expression profiles in response to secretion stresses. Furthermore, genome-wide investigation of translation profiles through polysome analyses is possible, and here, we outline methods for the use of such techniques with filamentous fungi and, principally, *Aspergillus niger*. We also describe methods for the batch and controlled cultivation of *A. niger* and for the replacement and study of its *hacA* gene, which provides either a UPR-deficient strain or a constitutively activated UPR strain for comparative analysis with its wild type. Although we focus on *A. niger*, the utility of the *hacA*-deletion strategy is also described for use in investigating the virulence of the plant pathogen *Alternaria brassicicola*.

1. INTRODUCTION

Filamentous fungi have conquered an astonishingly wide range of habitats, and individual species may be saprobic, pathogenic, or mutualistic partners (e.g., in lichenous or mycorrhizal associations). Their dispersal may be facilitated by the production and release of numerous spores (sexual or asexual), and colonization of food sources can also occur either more locally or over large areas by growth as a system of branched tubular bodies called hyphae. The expression and secretion of proteins underpin the saprobic lifestyle (e.g., by expressing hydrolase enzymes) and may also be important in virulence (e.g., by cell-surface presentation of adhesins and by secretion of lytic enzymes) and play an essential role in decomposition and recycling of organic matter in nature.

Although there are still deficits in our understanding of the fundamental mechanisms of the secretory pathway in filamentous fungi, the recent availability of the whole genome sequences of fungi (currently, several hundred yeasts and filamentous fungi, either sequenced or in progress), including *Aspergillus* species widely exploited commercially for their secreted enzymes (Machida *et al.*, 2005; Pel *et al.*, 2007), has provided a wealth of information. It is generally accepted that this pathway does not differ fundamentally from those in *S. cerevisiae* and higher eukaryotes, even though there are differences of detail (Geysens *et al.*, 2009). Key differences arise from the hyphal and branching phenotype of filamentous fungi and the polarity of both growth and protein secretion that occurs primarily at the hyphal tips (Conesa et al., 2001; Fischer et al., 2008; Harris, 2008; Shoji et al., 2008). As in other eukaryotes, the secretory route begins with the entry of secretory proteins into the endoplasmic reticulum (ER) either during or after translation. During transit through the ER, proteins undergo assisted folding and additional modifications such as signal peptide processing, glycosylation, disulfide bond formation, phosphorylation, and subunit assembly. Correctly folded proteins are sorted into coating-protein II vesicles (COPII) and then delivered to the Golgi (or fungal equivalent) for other modifications such as further glycosylation and peptide processing. Finally, the mature proteins are packed again into secretory vesicles and are delivered by exocytosis to the extracellular space at the hyphal tip.

This high-capacity secretory system has driven the commercial exploitation of filamentous fungi as cell factories for provision of native or heterologous enzymes that are used in a wide variety of applications. The progress in genetic manipulation and the availability of gene-transfer systems, combined with process improvements, have led to enhanced yields of native proteins and provided possibilities for improving the yields of heterologous proteins of both fungal and nonfungal origins. However, fungi often fail to secrete the heterologous proteins to the same high level as their own proteins, and this is especially so when the gene donor is not a fungus. Although factors negatively influencing the yield seem to be multiple (Gouka et al., 1997), protein maturation in the ER is regarded as the major bottleneck to achieving high-secreted yields of at least some heterologous proteins from filamentous fungi. Many strategies have been applied to address those limitations and include manipulations of the secretory pathway and, in particular, the ER lumenal environment by overexpression of genes encoding foldases and chaperones or enzymes of the glycosylation machinery. Furthermore, increased gene dosage, optimized codons, use of protease-deficient host strains, expression as translational fusions with an efficiently secreted protein, and introduction of efficient secretion signals or prosequences—have all been used to some advantage (Archer and Turner, 2006; Conesa et al., 2001; Lubertozzi and Keasling, 2009).

In yeast, the ER plays a pivotal role for quality control of proteins by ensuring that correctly folded proteins are delivered to subsequent cellular compartments. A variety of adverse physiological and environmental conditions can disturb ER homeostasis and lead to the accumulation of misfolded proteins. This is the case in expression systems, where there is a high flux of proteins being translocated into the ER. The folding, assembly, and secretion machinery may become saturated, leading to improperly folded structures or protein aggregates that are not secreted. To cope with ER stress, the intracellular signaling pathway, termed the unfolded protein response (UPR), is activated and triggers an extensive transcriptional response (Travers et al., 2000) that increases the protein-folding capacity within the ER. The UPR is also intimately linked to proteolytic systems that deliver misfolded proteins to vacuoles (autophagy) or to proteasomes (ER-associated degradation—ERAD) for degradation. In yeast, the basic leucine zipper (bZIP)-type transcription factor Hac1p is the transcriptional regulator of the UPR. Hac1p synthesis is dependent on the splicing of an unconventional (nonspliceosomal) intron in the *HAC1* mRNA initiated by the ER-located transmembrane kinase and endoribonuclease Ire1p (Ruegsegger *et al.*, 2001). This splicing event is activated in response to ER stress and relieves a translational block, allowing for the synthesis of Hac1p that regulates the expression of UPR target genes. Most of the key elements constituting the signal transduction pathway of the yeast UPR have been identified in filamentous fungi, and the central activation mechanism of the pathway, that is the stress-induced splicing of an unconventional intron, is conserved among filamentous fungi, yeast, and even mammalian cells (Kohno, 2010). An additional feedback mechanism that leads to selective transcriptional downregulation of some genes that encode secreted enzymes is termed repression under secretion stress (RESS) and has been reported in *Trichoderma reesei* and *Aspergillus niger* (Al-Sheikh *et al.*, 2004; Pakula *et al.*, 2003). Taken together, these ER-stress responses diminish the pool of newly synthesized proteins and provide homeostatic protection for the host cell. Thus, understanding more in detail about the process of protein maturation and secretion-related stress in filamentous fungi may hold the major key to improving their use as cell factories.

The genome sequences of filamentous fungi (including several *Aspergillus* spp.) and the availability of DNA GeneChips have provided an unprecedented resource for exploring expression profiles in response to particular environmental cues, including various secretion stresses (Breakspear and Momany, 2007). Transcriptomics to investigate the UPR in fungi was first used in *Aspergillus nidulans* and *T. reesei* (Arvas *et al.*, 2006; Sims *et al.*, 2005). In *A. nidulans*, the authors reported the effects of recombinant protein secretion by comparing a bovine chymosin-producing strain with its parental wild-type strain by using expressed sequence tag microarrays, which covered approximately one-third of the predicted open reading frames. In *T. reesei*, cultures expressing the heterologous protein tissue plasminogen activator (t-PA) and cultures treated with the reducing agent dithiothreitol (DTT) were analyzed with cDNA subtraction and cDNA-amplified fragment length polymorphism (AFLP). A genome-wide expression analysis of secretion stress responses in *A. niger* was reported by Guillemette *et al.* (2007). In that study, ER-associated stress was induced either by chemical treatment of the wild-type cells with DTT or tunicamycin, or by expressing t-PA. The predicted proteins encoded by most of the upregulated genes functioned as part of the secretory system including chaperones, foldases, glycosylation enzymes, vesicle transport proteins, and ERAD proteins. The authors also investigated translational regulation under ER stress by polysomal fractionation. Combining proteomic and transcriptomic profiling of the events following protein secretion stress should lead to a better understanding of the molecular basis of protein secretion and provide targets for strain improvement. Indeed, further transcriptomic studies of *A. niger* in relation to hyphal development and branching, and to protein secretion, have been described and illustrate the value of transcriptomic data and, more broadly, integration of data from the transcriptome and proteome (Jacobs *et al.*, 2009; Jorgensen *et al.*, 2009; Meyer *et al.*, 2009).

We describe some of the methodologies that have been used in the genome-wide analysis of protein secretion stress in filamentous fungi, taking *A. niger* as our model organism. We give less emphasis to transcriptomic analyses because methods are well described in several papers already cited but are also evolving rapidly, for example, with the advent of deep sequencing. We therefore devote more coverage to the preparation of samples for analysis, and include cultivation and the construction and use of *hacA*-deletion strains and of constitutive *hacA* mutants. Deletion of *hacA* from *A. niger* has been reported (Carvalho *et al.*, 2010; Mulder and Nikolaev, 2009) and shown to induce a severe phenotype that emphasizes the important role for HacA in the biology of the fungus. Deletion of *HAC1* was also achieved in the dimorphic fungus *Candida albicans* (Wimalasena *et al.*, 2008), and the deletion strain was less able to produce filaments and was down-regulated for expression of some cell-surface adhesins, suggesting that Hac1p would have a role in the invasive virulence of *C. albicans*. The *hacA* gene was deleted from another fungal pathogen, *Aspergillus fumigatus*, and virulence tests showed that the knockout strain was indeed attenuated (Richie *et al.*, 2009). Because of the importance of Hac1/A in virulence, we also provide an example of deleting the gene from a plant pathogen, *Alternaria brassicicola*.

2. Conditions to Study the UPR in Filamentous Fungi

In this section, we discuss the growth conditions and ER-stress-inducing methods that can be used with *A. niger* for producing fungal material destined for transcriptomic or polysome analyses.

2.1. Culture conditions

Filamentous fungi can grow either in pellets or as a dispersed mycelium. For our transcriptomic studies on UPR, a dispersed growth of the mycelium is preferred as this reduces heterogeneity of the mycelia. Growth in pellets is caused by an early aggregation of conidial spores and, after subsequent germination of the spores, micropellets are formed. Upon further growth, these pellets increase in size which results quickly in low oxygen concentration in the center of the pellet, which increases heterogeneity. With the protocol indicated below, reproducible dispersed growth of mycelia can be obtained.

In our UPR studies, we have used two methods of cultivation to perform transcriptomic studies. The choice of cultivation, either a batch culture, or a chemostat culture, depends on the purpose. We have used controlled batch cultivations to examine the response of *A. niger* cells to drugs that interfere with protein folding. This protocol is of course not limited to study the effect of UPR-inducing compounds, but can also be used to study the effect of other compounds that affect fungal growth (Meyer *et al.*, 2007). For these studies, spores were germinated for 5 h which resulted in the formation of a small (20 µm long) germtube, providing a more uniform population of cells than older, morphologically more heterogeneous, cells. At this stage, the antifungal compound is added.

Controlled batch cultivations can also be used to compare gene expression between different mutants. One important consideration here is that for a reliable comparison, the maximum growth rate during batch growth of both strains should be identical. Before performing transcriptomic studies on different mutants, it is important to establish that both mutants have an identical maximum specific growth rate. If not, the transcriptomic study will not only identify differences related to gene expression directly caused by the mutation, but will also identify differences related to a different growth rate. To prevent difference in growth rate to be reflected in the transcriptome, continuous cultivations should be performed in which the growth rate can be controlled (Jorgensen *et al.*, 2009). Here, we provide detailed conditions about controlled batch cultivation of *A. niger* in bioreactors.

2.1.1. Required materials

2.1.1.1. Devices of the bioreactor apparatus
Bioreactor cultivations are performed in Bioflo3000 bioreactors (New Brunswick Scientific). Using the NBS Biocommend software, the pH, temperature-dissolved oxygen tension and agitation can be controlled and monitored. The pH is measured using an autoclavable glass electrode (Mettler Toledo) and the dissolved oxygen tension is measured with an InPro-6000 series O_2 sensor (Metller Toledo). For measuring the content of CO_2 and O_2 in the exhaust gas, a Xentra 4100C Gas Purity analyzer (Servomex BV, the Netherlands) is used. To minimize wall growth in the head space of the reactor, the glass surface of the head space is cooled by tubing through which cold water is flowing.

2.1.1.2. Additional materials
- *Strains of interest*: The *A. niger* strains used in our UPR studies are all derived from the N402 strain. This strain is a UV-mutagenized derivative

of strain N400 that produces short conidiophores and is often used as a starting strain for genetic modifications (Guillemette et al., 2007; van Hartingsveldt et al., 1987). In our bioreactor controlled cultivations that are used for transcriptomic studies, only prototrophic strains are used to prevent effects on gene expression as a result from supplementation of any auxotrophy, for example, by uridine.

- *Growth medium*: A defined minimal medium (MM) is used to perform bioreactor cultivation. The growth-limiting component of the medium is the carbon source (0.75%, w/v). The MM contains the following: 7.5-g glucose, 4.5-g NH_4Cl, 1.5-g KH_2PO_4, 0.5-g KCl, 0.5-g $MgSO_4 \cdot 7H_2O$, and 1-ml trace metal solution per liter. The trace metal solution contains per liter, 10-g EDTA, 4.4-g $ZnSO_4 \cdot 7H_2O$, 1.01-g $MnCl_2 \cdot 4H_2O$, 0.32-g $CoCl_2 \cdot 6H_2O$, 0.315-g $CuSO_4 \cdot 5H_2O$, 0.22-g $(NH_4)_6Mo_7O_{24} \cdot 4H_2O$, 1.47-g $CaCl_2 \cdot 2H_2O$, and 1-g $FeSO_4 \cdot 7H_2O$ (modified from composition given by Vishniac and Santer (1957). The pH is adjusted to 3. The carbon source is heat-sterilized separately from the MM. Germination of conidial spores is improved by adding 0.003% (w/w) yeast extract to the culture medium.

Complete medium plates contain per liter 10-g glucose, 6.0-g $NaNO_3$, 1.5-g KH_2PO_4, 0.5-g KCl, 0.5-g $MgSO_4 \cdot 7H_2O$, 1.0-g casamino acids, 5.0-g yeast extract, 20-g agar, and 1-ml trace metal solution.

2.1.1.3. Other reagents

- Detergent solution (Tween80 0.05%, w/v, NaCl 0.9%, w/v)

2.1.1.4. Disposables

- Petri dishes

2.1.2. Inoculation

Conidia for inoculation of bioreactor cultures are harvested from solidified complete medium plates with sterile detergent solution and sterile cotton sticks to scrape off the spores. Bioreactors containing 5 l of MM are inoculated with 1×10^9 conidia/l. During cultivation, the temperature is maintained at 30 or 37 °C and the pH is kept at 3, by computer-controlled addition of 2-M NaOH or 1-M HCl. During the first 6 h of cultivation, the culture is aerated through the head space of the reactor vessel and the stirrer speed is kept low at 250 rpm to prevent hydrophobic conidia escaping from the water. After 6 h, allowing for germination of the conidia (now hydrophilic), sterile air is sparged into the culture and mixing is intensified (750 rpm). During the growth, the dissolved oxygen tension is kept above 40% of air saturation at any time, ensuring sufficient oxygen for growth.

2.1.2.1. Short batch cultivations To examine transcriptomic responses to growth-disturbing compounds or antifungals, germlings are grown for 5 h at 37 °C before addition of the compound. Small volume sampling (10 ml) can be performed at various times after exposure, and we normally sample at 15-min intervals for microscopic analysis. Large sampling volumes (400 ml) of culture are taken after 1 or 2 h after addition of the antifungals, and mycelial samples are quickly harvested via filtration and frozen using liquid nitrogen. Sampling of a 400-ml culture yields enough biomass for RNA extraction to perform Affymetrix microarray analysis (Meyer *et al.*, 2007). DDT and tunicamycin are the most commonly used compounds to induce an UPR response. In common with other antifungals, the effect of the response is dependent on the concentration of the compound and the concentration of the cells. To study transcriptomics responses, we use sublethal concentrations of the antifungal compound and monitor the response of the cells shortly after adding the compound (within 1 h) to minimize secondary effects. Concentrations of the antifungal drug used should be empirically determined.

2.1.2.2. Prolonged batch cultivation After 6 h of cultivation and the start of sparger aeration and increasing the stirrer speed to 750 rpm, 0.01% polypropyleneglycol (PPG 2000, Fluka Chemika) is added to the medium as an antifoam agent. Acidification of the culture and the amount of NaOH used to maintain the pH at 3 is used as an indirect measure for growth. Dissolved oxygen tension is always above 40% of air saturation at any time ensuring oxygen-sufficient growth. During cultivation, samples are drawn regularly to monitor culture growth. Medium can be analyzed for residual carbon source and extracellular protein concentrations are determined to calculate protein production yields. Dry weight biomass concentration in the culture is determined by weighing lyophilized mycelia from a known weight of culture. Mycelium is separated from the culture broth by filtration through a GF/C glass microfiber filter (Whatman) and dried. The end of the batch phase (depletion of glucose) can by monitored by plotting biomass concentrations as well as monitoring the alkali addition. An example of a typical growth profile of a batch culture of *A. niger* is shown in Fig. 1.1.

2.2. Induction of ER-associated stress

The UPR is classically induced by treating mycelium cultures with two chemical agents which disrupt protein folding in the ER. DTT is a strong reducing agent that disturbs the oxidative environment of the ER and prevents disulfide bond formation. Tunicamycin is a drug produced by *Streptomyces lysosuperificus*, which inhibits N-linked glycosylation by preventing core oligosaccharide addition to nascent polypeptides and thereby blocks protein folding and transit through the ER. To assess the UPR under

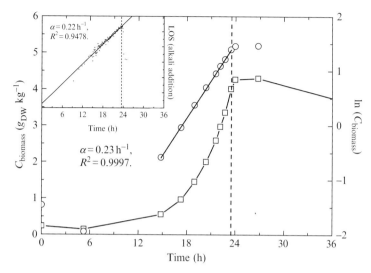

Figure 1.1 Growth profile of *A. niger* wild-type batch culture. Dry weight biomass concentration (g_{DW} kg^{-1}) as a function of time (h) illustrates the growth of the cultures. The maximum specific growth rate is determined from the slope (α) of the ln transformation of biomass ($C_{biomass}$) in the exponential growth phase as a function of time (h), as well from log transformation of alkali addition as a function of time (h) (see insert). Dash-line represents the end of the exponential growth phase (depletion of glucose).

conditions of heterologous protein production, expression profiles from a recombinant protein-producing strain and its parental wild-type strain can be compared. Thus, we have used a strain-producing recombinant t-PA, a serine protease. We recommend the use of at least two distinct UPR inducer treatments since discrepancy between responses is sometimes obtained. We know in particular that DTT has a variety of other effects on the cell and triggers a relatively large overall variation in gene expression levels compared to other treatments. The responses to each stress are then compared and the overlaps common to these conditions should lead to the identification of robust sets of UPR-regulated genes.

Before performing transcriptomic or polysome analysis, we confirm that each of the stress conditions leads to induction of the UPR by examining the transcriptional induction (by northern hybridization or real-time PCR) of genes known to be affected by ER stress, such as the ER chaperone-encoding gene *bipA* and the foldase-encoding gene *pdiA* (data not shown). The splicing of the *hacA* transcript also needs to be examined. This could be done from RNAs derived from treated or untreated cultures by using different methods: (i) by conventional RT-PCR with primers designed across the atypical intron (Guillemette *et al.*, 2007), (ii) by northern hybridization by using a cDNA fragment comprising the region from the start codon up to the unconventional intron as a probe (Mulder *et al.*, 2004), (iii) by real-time RT-PCR by using a specific primer designed to span the atypical 20-bp intron (Fig. 1.2). A quantitative RT-PCR method was

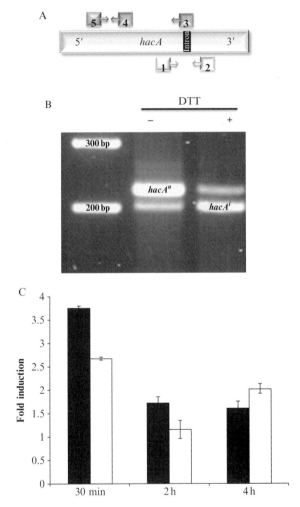

Figure 1.2 Analysis of the splicing of the *hacA* transcript in *A. niger* by PCR technology. (A) Schematic representation of the *hacA* ORF showing the positions of primers used for RT-PCR analysis. (B) Conventional RT-PCR analysis of *hacA* processing with primers 1 and 2 designed across the intron in the presence (+) or absence (−) of 20-mM DTT for 1 h. (B) Analysis of *hacA* expression by means of real-time RT-PCR in a wild-type strain during DTT exposure (20 mM) for 0.5, 2, and 4 h. The white box represents the fold induction observed with primers 4 and 5 designed for amplifying the cDNA derived from *hacA* mRNA (spliced and unspliced). The black boxes represent the fold induction observed with primers 1 and 3 designed to span the atypical 20-bp intron and are therefore specific for the spliced form, $hacA^i$.

initially developed to measure the splicing ratio of the mammalian *XBP1* mRNA (Hirota *et al.*, 2006) and was also recently succesfully applied to filamentous fungi (Joubert *et al.*, unpublished). The results in Fig. 1.2 clearly show an increase in the abundance of a shorter product following ER stress, indicating that treatment (DTT in this case) induces the conversion of the unspliced *hacA* mRNA (*hacA^u*) into the spliced form (*hacA^i*).

ER-stress reagents (DTT or tunicamycin) are added to fungal cultures. DTT stocks are prepared in water and the tunicamycin stock in DMSO. DTT or tunicamycin is added to the liquid medium at a final concentration of 20 mM or 10 µg/ml, respectively. Note that these concentrations are 10-fold higher than those routinely used with yeast. Control cultures have an equivalent volume of sterile water or DMSO added. In *A. niger*, expression analyses show that both upregulation of UPR target genes and appearance of the active *hacA* mRNA emerge within 30 min after exposure of the mycelium to DTT, and are still observed 2 h after adding the DTT (Guillemette *et al.*, 2007; Mulder *et al.*, 2004). Like DTT treatment, tunicamycin treatment and production of the recombinant t-PA lead to the activation of the UPR, although the effect appears later (1 h after induction of the ER stress) than in the DTT-treated mycelia.

3. Analysis of Expression During the ER Stress

3.1. Transcriptomic analysis

The genome sequences of two different strains of *A. niger* have underpinned the construction of two different microarrays in the Affymetrix format (Andersen *et al.*, 2008; Pel *et al.*, 2007), and one of those arrays is a trispecies array affording comparsions between *A. niger*, *A. nidulans*, and *Aspergillus oryzae* (Andersen *et al.*, 2008). The Affymetrix protocols for transcriptomics and analysis are available on-line and have proved to be very effective for transcriptomic studies with *A. niger* (Guillemette *et al.*, 2007; Jorgensen *et al.*, 2009; Vongsangnak *et al.*, 2010). Despite the availability of Gene-Chips for *A. niger*, other array formats have been used for *A. niger* and many other species of filamentous fungi and so, it is not feasible to present a preferred method here. Instead, we advise that the cited papers are referred to for *A. niger*, and methods for other formats and fungal species are explored through the literature. The advent of nonarray methodologies for genome-wide transcriptomics brings new opportunities and these methods are beginning to be applied to filamentous fungi. There are various platforms available for the so-called deep sequencing (Bashir *et al.*, 2010) and, while such methods have been applied in the sequencing of fungal DNA, for example, in the comparison of mutant strains (e.g., Le Crom *et al.*, 2009), it is the use of RNA sequencing that concerns us here.

Transcript sequencing holds some advantages over array approaches by providing, for example, quantitative comparisons of the abundance of different transcripts and in the detection and quantification of transcript variations, for example, in splicing (Wang *et al.*, 2009). We anticipate reports of using such approaches to investigate further the UPR in filamentous fungi.

3.2. Polysome analysis

Expression profiling data are more meaningful when mRNA samples are enriched for transcripts that are being translated (Pradet-Balade *et al.*, 2001). This can be achieved by fractionation of cytoplasmic extracts in sucrose gradients, based on the methods described for polysome analysis by Arava (2003). This method involves size separation of large cellular components and monitoring the A_{254} across the gradient. It enables the separation of free mRNPs (ribonucleoprotein particles) from mRNAs fully loaded with ribosomes (i.e., polysomes). Polysomes represent actively translated transcripts and this fraction is directly correlated with the set of *de novo* synthesized proteins in a particular cellular state, enabling the determination of the translation efficiencies that are characteristic for each transcript in a cell (Smith *et al.*, 1999). In addition, changes in the distribution of a given mRNA indicate how this translational efficiency can vary under different conditions. Because it is generally accepted that translational control predominantly occurs at the initiation step (McCarthy, 1998), the number of mRNA molecules engaged in polysomes should be a robust indicator of the synthesis rate of the corresponding protein. To examine the effect of secretion stress on net translational activity in *A. niger*, we provide a detailed description of the method to analyze the global relative distribution of ribosomes between polysomes, 80S monosomes, and dissociated 40S and 60S subunits.

3.2.1. Required materials
3.2.1.1. Devices and materials of the density gradient fractionator apparatus
RNA fractionation is performed with the Foxy Jr. Fraction Collector (TELEDYNE ISCO). RNA samples are loaded onto prepared gradients and spun in a ultracentrifuge. Once the spin is completed, the system allows you to fractionate and quantitate centrifuged zones with precision. The ISCO Density Gradient System produces a continuous absorbance profile as the gradient is collected in precisely measured fractions. Fractionation is performed by introducing a dense chase solution into the bottom of the centrifuged tube, raising the gradient intact by bulk flow. Chase solution is injected by piercing the bottom of the tube. System includes tube piercing stand, peristaltic pump, UA-6 Detector with 254

and 280 nm filters, density gradient flow cell, and Foxy Jr. Fraction Collector.

3.2.1.2. Additional materials

- Ultracentrifugation is performed in an Optima™ MAX-XP Benchtop Ultracentrifuge or an Optima™ MAX High-Capacity Personal Micro-Ultracentrifuge (Beckman Coulter) with an MLA-80 rotor. Alternatively, you can use a Beckman SW41 Ti rotor.
- Mortar and pestle, needle

3.2.1.3. Other reagents

- Polysome extraction buffer (20-mM Tris–HCl, pH 8.0, 140-mM KCl, 1.5-mM MgCl$_2$, 1% Triton X-100, 0.1-mg/ml cycloheximide, 1.0-mg/ml heparin, 0.5-mM DTT)
- Chase solution (80% sucrose solutions prepared in 50-mM Tris–acetate, pH 7.0, 50-mM NH$_4$Cl, 12-mM MgCl$_2$, 1-mM DTT)
- TE (10-mM Tris–HCl, pH 8.0, 0,1-mM EDTA))

3.2.1.4. Disposables

- 8-ml centrifuge tubes, 2-ml (Eppendorf) tubes

3.2.2. Sample preparation

Ribosomal fractions are prepared according to the method described for polysome analysis (Arava, 2003) modified as follows: At the time of harvest, cycloheximide is added to a final concentration of 0.1 mg/ml to trap elongating ribosomes. The cultures are swirled rapidly and chilled on ice for 10 min. Fungal material is pelleted by centrifugation at 11,000×g for 10 min at 4 °C. The pellet is then resuspended in 5 ml of polysome extraction buffer and sedimented. This washing step is repeated, and cells are frozen in liquid nitrogen and stored at −80 °C. Approximately 0.25 g of cells are ground in liquid nitrogen with a mortar and pestle, and the powder is resuspended in 750 μl of ice-cold polysome extraction buffer. Excess cell debris are removed by sedimentation at 4000×g for 5 min at 4 °C. The supernatant is clarified by further centrifugation (12,000×g, 10 min, 4 °C).

3.2.3. Sucrose gradient preparation

We prepare 15–70% sucrose density gradients for ultracentrifugation without using a gradient maker. Sucrose solutions are prepared in 50-mM Tris–acetate (pH 7.0), 50-mM NH$_4$Cl, 12-mM MgCl$_2$, and 1-mM DTT. 1.3 ml of the lighter sucrose solution (15%) is laid on the bottom of the centrifuge tubes using a long needle. 2.6 ml each of 30% and 50% sucrose, and finally, 1.3 ml of 70% sucrose are then carefully underlaid without disturbing the interfaces. The tubes are closed with parafilm and stored at 4 °C for at least 12 h before use to allow the gradients to thaw and diffuse to create a continuous gradient.

3.2.4. RNA fractionation

Each sample is loaded on an 8-ml 15–70% (w/v) sucrose gradient and sedimented at 150,000×g (55,000 rpm) and 4 °C in a Beckman MLA-80 rotor for 135 min. The gradient is fractionated with the density gradient fractionator while monitoring the absorbance at 254 nm. 0.5 ml fractions are collected from the top of the gradient directly into a 1-ml volume of 6-M guanidine hydrochloride. RNA is precipitated by adding an equal volume of 100% ethanol and resuspended in TE. The RNA is again precipitated by the addition of 50 μl of 3-M sodium acetate (pH 5.2) and 1 ml of 100% ethanol, and resuspended in TE. For RNAs destined for microarray analyses, the polysomal fractions and nonpolysomal fractions are pooled, respectively. Each fraction is treated with RQ1 RNase-Free DNase (Promega) and is again cleaned up by applying the samples to an RNeasy mini column (Qiagen). An aliquot of the RNA solution is extracted from each fraction and subjected to electrophoresis through a formaldehyde gel (Fig. 1.3).

4. Modifying the UPR Signaling by Targeted Gene Replacement

4.1. Construction of UPR-deficient strains

As described in Section 1 in more detail, key proteins in UPR signaling are conserved in eukaryotic cells. They include the ER-localized transmembrane protein Ire1/IreA and transcription factor Hac1/HacA. Genes homologous to *IRE1* and *HAC1* are usually recognized by bidirectional Blast searches. Gene disruptions are often used to study the UPR, and genes encoding Hac1/HacA and Ire1/IreA are common targets for disruption. The most commonly used selection markers for making gene disruptions in *A. niger* include the counter selectable markers *pyrG* and *amdS*, and the hygromycin resistance marker (the *hph* gene from *Escherichia coli*). The use of hygromycin as a selection strategy is widely used for transformation of filamentous fungi as it does not require the isolation of an auxotrophic mutant (e.g., the isolation of a uridine/uracil requiring *pyrG* mutant).

Deletion of the *hacA* gene with the hygromycin resistance marker has been reported in both *A. niger* and *A. fumigatus* (Mulder and Nikolaev, 2009; Richie *et al.*, 2009) and in *A. niger* using the *pyrG* gene (Carvalho *et al.*, 2010). Several approaches are available for constructing gene deletion cassettes which include; (i) traditional PCR amplification and cloning of

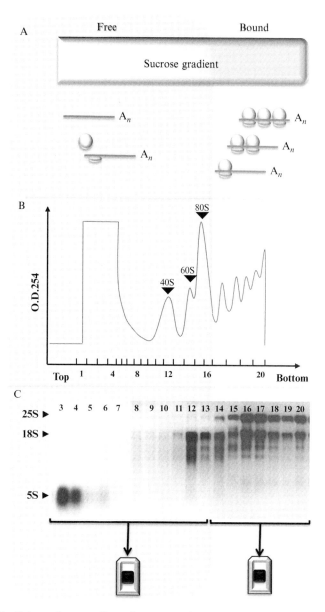

Figure 1.3 Schematic steps for polysome analysis. (A) Cytoplasmic extracts from *A. niger* cells are fractionated in 15–70% sucrose gradients. It enables the separation between free mRNAs and those loaded with ribosomes and engaged in translation. (B) Representative absorbance profile for RNA separated by velocity sedimentation through the sucrose gradient. For each fraction, absorbance at 254 nm is monitored using a UA-6 UV detector. The positions of the 40S, 60S, 80S, and polysomal peaks are indicated. The expected peaks for the tRNAs and other small RNAs in fractions 1–4 are

flanking regions on both sites of the selection markers, (ii) fusion PCR approaches to generate the entire deletion cassette by PCR, (iii) the use of Multisite Gateway recombination (Invitrogen) for the *in vitro* recombination of flanking regions and the selection markers or, (iv) the use of *S. cerevisiae* for *in vivo* recombination. It is beyond the scope of this review to discuss the different approaches to generate disruption cassettes, but in principle, all approaches are suitable. In this chapter, we include a Fusion PCR protocol which is routinely used by the authors to construct gene deletion cassettes and has been used to disrupt the *hacA* gene in *A. brassicicola* (Joubert *et al.* unpublished).

4.1.1. Required materials
4.1.1.1. Materials

- Thermocycler

4.1.1.2. Other reagents

- High-Fidelity DNA Polymerase
- Plasmid pCB1636 (Sweigard *et al.*, 1995)

4.1.1.3. Disposables

- 0.7-ml (Eppendorf) tubes

4.1.2. Construction of a gene replacement cassette for use in *A. brassicicola*

Deletion of *hacA* in *A. brassicicola* was accomplished by replacing the *hacA* ORF with a hygromycin B resistance cassette. Gene replacement occurs via homologous double crossover between a linear construct and the target genomic locus. The linear construct flanks the resistance cassette (that contains the *hph* gene under control of the *trpC* fungal promoter) with two fragments representing 5′ and 3′ regions of the target gene, and is constructed by using a PCR-assisted DNA assembly called double-joint PCR (Fig. 1.4; Yu *et al.*, 2004). For all the PCR steps, we highly

---

obscured by the high absorbance, presumably from proteins and detergents used in the preparation. (C) RNA is then extracted from each fraction and an aliquot is subjected to electrophoresis through a formaldehyde gel. As expected, 25-S and 18-S ribosomal RNAs were the prominent species. Before processing mRNA to cDNA, labeling, hybridization to *A. niger* Affymetrix GeneChips (Affymetrix, Inc., Santa Clara, CA), the polysomal fractions and nonpolysomal fractions can be pooled, respectively. Each pooled fraction is then treated with RQ1 RNase-Free DNase (Promega) and is again cleaned up by applying the samples to an RNeasy mini column (Qiagen).

Methods for Investigating the UPR in Filamentous Fungi 17

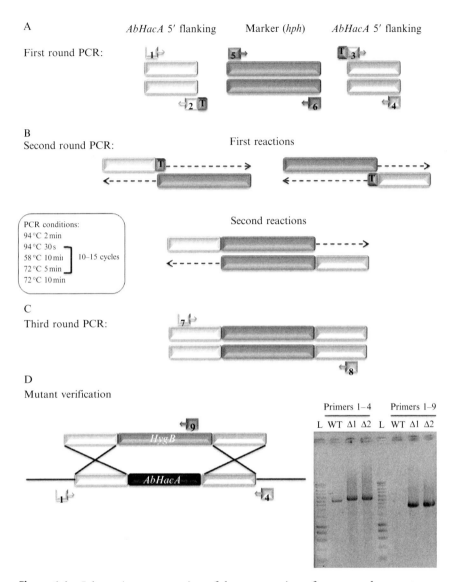

Figure 1.4 Schematic representation of the construction of a gene replacement cassette and verification of the mutants. The arrows numbered from 1 to 9 represent primers for the PCRs and primers 2 and 3 are 45–60 bases-long chimeric primers. (A) First-round PCR: amplification of the components using the specific and chimeric primers. A typical reaction will fuse DNA fragments of a 5′ flanking sequence, a 3′ flanking sequence, and a selectable marker (*hph* gene). Primers 2 and 3 carry 24 bases of homologous sequence overlapping with the ends of the selectable marker of choice. (B) Second-round PCR: the assembly reaction is carried out without using any specific primers, as the overhanging chimeric extensions act as primers. The first two cycles are shown in detail. (C) Third-round PCR: amplification of the final product using nested

recommend the use of a High-Fidelity DNA Polymerase. A typical reaction fuses DNA fragments of a 5′ flanking sequence (amplified with the primers 1 and 2), a 3′ flanking sequence (amplified with the primers 3 and 4), and the selectable marker. Here, the marker corresponds to the hygromycin resistance cassette (1436 bp) that is amplified with the primer M13F and M13R from the plasmid pCB1636 (Sweigard et al., 1995). Primers 2 and 3 carry 24 bases complementary to M13F and M13R sequences, respectively, overlapping with the ends of the resistance cassette. During the first round of PCR, amplifications of 5′ flanking region, marker, and 3′ flanking region are separately carried out. The amplicons are then purified using a commercially available PCR cleanup kit. The three amplicons are mixed in 1:3:1 molar ratio, the total DNA amount of the three components being between 100 and 1000 ng. These three DNA fragments are specifically joined together during the second-round PCR. The assembly reaction is carried out without adding any specific primers, the overhanging chimeric extensions acting as primers. In the third round of PCR, the final PCR construct is amplified with a nested primer pair (primers 5 and 6). We can also use the first-round primer pair (primers 1 and 4) but the yield of amplicons is often lower and we also get PCR artifacts more easily. For round 3 fusion PCR, we generally use 1–2 µl of the purified round 1 products as template. The final amplicon is purified and further concentrated in a volume of 10 µl to a concentration of at least 1 µg/µl.

4.2. Constitutive activation of the UPR

Activation of the HacA transcription factor includes the unconventional splicing of an intron from the $hacA^u$ mRNA, creating a transcriptionally active form of HacA. This mechanism allows a straightforward way to construct strains with a constitutively activated form of HacA and has been used in several *Aspergillus* species to generate strains with a constitutively activated UPR. Valkonen et al. (2003) reported the construction of an *Aspergillus awamori* strain which expressed the $hacA^i$ cDNA lacking the 20 nt unconventional intron and including a 150-bp truncation at the 5′ end of the mRNA. In that study, the active form of *hacA* was expressed under

---

primers 7 and 8. (D) Confirmation of gene replacement: transformants are randomly picked and examined for double crossover-mediated gene replacement pattern by PCR amplification of the *hacA* locus using the primer pairs 1/4 or 1/9. The primer 9 is designed inside the *hph* sequence. As shown, amplicons of wild-type and deletion alleles of *hacA* obtained with primer pairs 1/4 differ in size. No amplicon is obtained from wild-type matrix when using the primer 9. The results are shown for two mutants called Δ1 and Δ2. Molecular sizes (kb) were estimated based on a 1-kb ladder (lane L, Eurogentec, Seraing, Belgium).

control of the highly expressed and inducible glucoamylase promoter. Mulder and Nikolaev (2009) constructed a constitutive HacA strain by expressing *hacA^i*, lacking the 20 nucleotide intron, in a *hacA* deletion background. The site of integration of the introduced *hacA^i* was controlled by using the *pyrG**approach (van Gorcom and van den Hondel, 1988). Finally, we recently used an approach to replace the wild-type *hacA* gene with the active *hacA^i* form that lacks the 20 nucleotide intron at the *hacA* locus (Carvalho and Ram, unpublished). In all the three studies, the expression of the active form of HacA resulted in constitutive activation of the UPR pathway. Activation of the UPR pathway is often monitored by examining the expression level of UPR target genes such as *bipA* and *pdiA*. The latter approach in which the wild-type gene (*hacA^u*) is replaced by a mutated and constitutively active form (*hacA^i*) of the gene at the endogenous locus is an approach of general interest and is therefore described in this chapter.

4.2.1. Required materials
4.2.1.1. Materials

- Thermocycler

4.2.1.2. Additional materials

- *Strain of interest*: Use of *Ku70* mutant dramatically increases the frequency of homologous recombination and reduces the number of transformants to be analyzed in order to find a transformant in which the cassette is integrated at the endogenous locus.

4.2.1.3. Other reagents

- High-fidelity DNA polymerase
- Plasmid PBluescript-SK
- Restriction enzymes

4.2.1.4. Disposables

- 0.7-ml (Eppendorf) tubes

4.2.2. Replacement of *hacA^u* by a constitutive *hacA^i* at the endogeneous locus in *A. niger*

To construct the replacement cassette, primers are designed to amplify various fragments, as depicted in Fig. 1.5. Three fragments, consisting of the *hacA^u* gene, encoding promoter and terminator regions, the *Aspergillus oryzae pyrG* selection marker, and the *hacA* terminator region are amplified

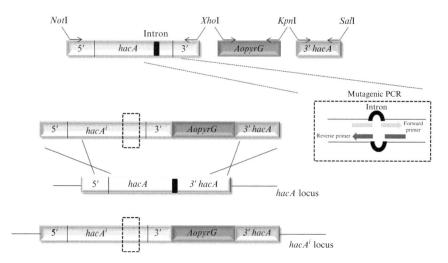

Figure 1.5 Schematic representation of the construction of a cassette for obtaining a constitutive UPR strain. Three fragments containing the *hacA* locus, the *Aspergillus oryzae pyrG* gene, and the *hacA* terminator region are amplified by PCR and ligated into a standard cloning vector (pBluescript-SK) using appropiate restriction enzymes. This plasmid is used as a template for site-directed mutagenesis using two complementary primers lacking the intron sequence. Transformants containing the construct lacking the intron (*hacA^i*) at the *hacA* locus can be selected by Southern blot analysis and the lacking of the intron can then be confirmed by PCR analysis.

by PCR and cloned into pBluescript-SK using appropriate restriction enzymes. The *A. oryzae pyrG* gene and flanking regions (de Ruiter-Jacobs et al., 1989) are used to prevent homologous recombination of the final construct at the *A. niger pyrG* locus. The plasmid containing the *hacA^u* ORF is used as a template to introduce the *hacA^i* gene by site-directed mutagenesis according to the Quick Change II site-directed mutagenesis protocol (Stratagene). Overlapping complementary primers lacking the intron sequence are used to obtain the *hacA^i* gene in which the intron is deleted. To prevent single crossover of the construct, the construct should be linearized before transformation into an *A. niger ku70* mutant (Meyer et al., 2007). Southern blot analysis should be performed to confirm integration of *hacA^i* at the endogenous locus. As depicted in Fig. 1.5, the *A. oryzae pyrG* marker is flanked by repeats of the *hacA* terminator region. The direct repeat flanking the *AopyrG* gene allows efficient looping out of the *pyrG* marker after subjecting the strain to counter-selection using 5-fluoro-orotic acid (5-FOA). A detailed protocol for 5-FOA counter-selection has been published recently(Meyer et al., 2010). The removal of the *pyrG* gene from the *hacA* locus completely restores the wild-type context of the *hacA* locus, and the *pyrG* auxotrophy can be used for transform additional constructs into this strain.

4.3. PEG-mediated transformation of protoplasts

The establishment of an effective transformation method for the filamentous fungi is of crucial importance to examine gene functions. At the moment, protoplast-mediated transformation and *Agrobacterium*-mediated transformation are the most commonly used techniques. As a detailed protocol for *Agrobacterium*-mediated transformation of *Aspergillus* sp. has been published recently (Michielse et al., 2008), we describe here a protocol that we routinely use for the transformation of *A. brassicicola* protoplasts. This protocol can be used for many other filamentous fungi, providing that the enzyme solution is adapted to the targeted fungus. Indeed, the yield of protoplasts in fungi depends mainly of the composition of the lytic enzyme batch used. *Alternaria* spp. are members of the *Dothideomycetes* and they commonly present a highly melanized cell wall that requires specific enzyme mixture.

4.3.1. Required materials
4.3.1.1. Materials

- Thoma counting chamber
- 250-ml flasks

4.3.1.2. Other reagents

- Enzyme-osmoticum is prepared in 0.7-M NaCl and contains kitalase (Wako Chemicals, Richmond, VA, USA) at 10 mg/ml and driselase (Interspex) at 20 mg/ml.
- STC buffer (1.2-M sorbitol, 10-mM Tris–HCl, pH 7.5, and 50-mM CaCl$_2$)
- PEG solution (MW 3350–4000) 60%, 10-mM Tris–HCl, pH 7.5, and 50-mM CaCl$_2$)
- *Regeneration medium*: you need to prepare the flasks A, B, C, and autoclave them separately. Combine them after autoclaving and hold at 55 °C.
  - Flask A: 1-g yeast extract, 1-g casein hydrolysate, water to 50 ml
  - Flask B: 342 g sucrose, water to 500 ml
  - Flask C: 16-g agar, water to 450 ml

4.3.1.3. Disposables

- Petri dishes
- 50-ml tubes and 10-ml tubes

4.3.2. Protoplast-mediated transformation of A. brassicicola

Transformation is carried out with linear PCR products based on the transformation protocol for *A. alternata* (Akamatsu *et al.*, 1997), with modifications. Approximately 5×10^6 fungal conidia are harvested from a potato dextrose (PD) agar culture plate and introduced into 50 ml of PD broth media. They are cultured for 16 h at 25 °C with shaking at 100 rpm. The fungal material is harvested by centrifugation at $3500-4000 \times g$ for 10 min, washed twice with 0.7-M NaCl, followed by centrifugation again under the same conditions as before. During centrifugation, start making the enzyme-osmoticum. The enzyme-osmoticum is added to the mycelia in Falcon tubes and the cell walls of the germlings are digested for 3–4 h at 32 °C with gentle shaking every 30 min during incubation. The protoplasts are separated from undigested mycelia and cell-wall debris by filtering the suspension through 2–4 layers of cheesecloth. The protoplasts are then collected by centrifugation at $2500 \times g$ for 10 min. The enzyme mixture can be reused by collecting the supernatant in other tubes and storing at -80 °C. The pellet is washed twice with 10 ml of 0.7 M NaCl and then with 10 ml of STC buffer. The protoplasts are gently resuspended in 500-µl–1-ml STC to reach a concentration of 10^6–10^7 protoplasts per ml. The counting is performed by using a Thoma chamber.

At least 10 µg of PCR products in 10 µl of ddH$_2$O is added to the protoplast suspension in 12-ml Greiner tubes and gently mixed. The transformation mix is incubated on ice for 20 min. PEG solution is then added in three aliquots of 200, 200, and 800 µl each. The tubes are warmed by hand before each addition and drops must be added slowly (drop by drop). The tubes are also mixed after each addition by rolling. Incubations for 5 min on ice are required between each addition. Finally, the transformation mix is diluted with 1-ml STC. Then, 200 µl of the transformation mixture is added to 25 ml of molten regeneration medium in a 50-ml tube and subsequently poured into a 100×15-mm Petri dish. Allow the medium to solidify and then incubate the plates at 32 °C. After 24 h, the plates are overlaid with 10 ml of 1% agar containing hygromycin B (Sigma-Aldrich, St. Louis, USA). The final concentration of hygromycin B should be 15 µg/ml. Please note that sensitivity to hygromycin B varies according to strain, so strain sensitivity must be assessed in advance. Individual hygB-resistant transformants are transferred to a fresh hygB-containing plate between 5 and 15 days after each transformation. Each transformant is purified further by picking isolated conidia under the microscope and transferring them to a fresh hygB-containing plate. Three successive rounds of single-spore isolation are performed.

4.4. Mutant verification

Two approaches, diagnostic PCR and Southern analysis of genomic DNA, are used to analyze mutants at the gene level and to verify that linear DNA products have inserted correctly. For both approaches, we routinely use high-throughput methods for isolation of genomic DNA from different fungal species. Detailed protocols have been reported by Meyer et al. (2010). We carry out a PCR screen by using primers homologous to the selection marker and genomic sequence outside the flanking regions, to confirm that integration of the replacement constructs occurred by homologous recombination at the targeted loci (Fig. 1.4D). The complete deletion of the coding region in mutants is further confirmed using two internal primers. Moreover, genomic Southern blot genotyping is performed to check single-copy deletion of the target gene in mutants that will be used for further phenotypic analysis (data not shown).

4.5. Phenotypic analysis

Detailed analysis of the phenotypes of both the wild-type and mutants affected in the process under investigation is essential to determine gene function or investigate the importance of a pathway in a cell. We summarize in Table 1.1 the main phenotypic criteria that have been investigated in $\Delta hacA$ UPR-deficient strains obtained in *A. niger* (Mulder and Nikolaev, 2009), *A. fumigatus* (Richie et al., 2009), and in *A. brassicicola* (Joubert et al., unpublished).

Among all the criteria that could be considered for establishing the phenotypic pattern of null mutants in filamentous fungi, growth characteristics and susceptibility to different drugs are key elements that we routinely monitor on agar plates as well as in liquid medium using laser nephelometry (NEPHELOstar Galaxy, BMG Labtech, Offenburg, Germany). By contrast with photometry (i.e., the measurement of light transmitted through a particle suspension), nephelometry is a direct method of measuring light scattered by particles in suspension. As the scattered light intensity is directly proportional to the suspended particle concentration, nephelometry is a powerful method for recording microbial growth and especially for studying filamentous fungi, which cannot be efficiently investigated through spectrophotometric assays. The advantages of nephelometry compared to analysis of colony expansion rates on solid media or spectrophotometric assays are discussed in a recent paper (Joubert et al., 2010), in which we described a filamentous fungi-tailored procedure based on microscale liquid cultivation and automated nephelometric recording of growth.

Table 1.1 Main phenotypic criteria investigated in fungal Δ*hacA* strains

Phenotypic analysis	Fungal organism	Methods
Growth rate	*A. niger* *A. fumigatus* *A. brassicicola*	– Measure of colony diameter on different agar media (minimal or complete media, IMA, PDA, skim milk)
Sporulation	*A. niger* *A. brassicicola*	– Microscopic observation of conidia (using stereomicroscope or scanning electron microscopy and quantification of sporulation using a Thoma counting chamber
Hyphal morphology	*A. niger* *A. brassicicola*	– Microscopic observations of mycelia in liquid culture or grown between glass slides
Susceptibility to ER stress	*A. niger* *A. fumigatus* *A. brassicicola*	– Incubation in the presence of DTT, tunicamycin, or brefeldin A – Measure of colony diameter (from solid media) or growth monitoring using nephelometry (from liquid media)
Susceptibility to thermal stress	*A. fumigatus*	– Measure of colony diameter from solid media incubated at 37 and 45 °C – Measure of the percentage of surviving CFUs
Susceptibility to antifungal drugs	*A. fumigatus*	– Use of Etest strips impregnated with caspofungin, fluconazole, amphotericin B, and itraconazole
Expression of UPR target genes	*A. niger* *A. fumigatus* *A. brassicicola*	– Northern hybridization, real-time RT-PCR
Susceptibility to host defense metabolites	*A. brassicicola*	– Growth monitoring using nephelometry in the presence of cruciferous phytoalexins
Cell wall structure and composition	*A. fumigatus* *A. brassicicola*	– Treatment with calcofluor white and Congo red – Biochemical analysis of the cell wall – Observations with transmission electron microscopy
Protein secretion	*A. fumigatus* *A. brassicicola*	– SDS-PAGE analysis of culture supernatants – Quantification of secreted proteolytic activity with the Azocoll assay

Table 1.1 (continued)

Phenotypic analysis	Fungal organism	Methods
Virulence	A. fumigatus A. brassicicola	– Quantification of esterase activity with the artificial substrate p-nitrophenyl butyrate (PNB) – Use of an outbred and inbred mouse model of invasive aspergillosis – Inoculation of spore suspension on intact and prewounded leaves of Arabidopsis thaliana
Visualization of infection structures	A. brassicicola	– Staining of fungal material with Solophenyl Flavine 7GFE 500 followed by microscopic observations – Observations with scanning electron microscopy

We recommend this technique for the evaluation of antifungal activity and for large-scale phenotypic profiling.

Figure 1.6A shows a representative growth curve for *A. brassicicola* wild-type strain and $\Delta hacA$ mutant plotted from the nephelometric assays. Nephelometric monitoring confirms that, at least for early steps of the growth kinetics, the $\Delta hacA$ mutant is growth-impaired in PD broth. In another example (Fig. 1.6B), disruption of ER homeostasis is triggered by exposure to brefeldin A, which is an ER stress-inducing agent. As expected, we observe that the *hacA* null strain is growth-impaired in the presence of concentrations of brefeldin A that could be tolerated by the wild-type strain, indicating that *hacA* inactivation increased the sensitivity of *A. brassicicola* cells to these treatments.

ACKNOWLEDGMENTS

We thank Thomas. R. Jørgensen for his guidance in bioreactor cultivations of *A. niger* and preparing Fig. 1.1. This work was partly supported by the French Région Pays de la Loire (QUALISEM research program), the UK Biotechnology and Biological Sciences Research Council (BioResearch Industry Club), and the research program of the Kluyver Centre for Genomics of Industrial Fermentation which is part of the Netherlands Genomics Initiative/Netherlands Organization for Scientific Research.

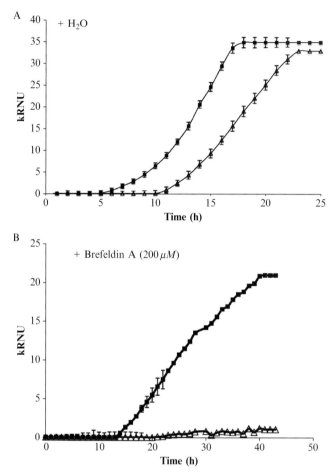

Figure 1.6 Nephelometric monitoring of growth of wild-type and $\Delta hacA$ strain. Conidial suspensions (10^5 spores/ml, final concentration) of wild-type (black squares) and $\Delta hacA$ (open triangles) are used to inoculate microplate wells containing a standard PDB medium that is supplemented (B) or not (A) with 200-μM brefeldin A. Microplates are placed in a laser-based microplate nephelometer (NEPHELOstar Galaxy, BMG Labtech) and growth is monitored automatically over a 30-h period. Each genotype is analyzed in triplicate and the experiments are repeated three times per growth condition.

REFERENCES

Akamatsu, H., Itoh, Y., Kodama, M., Otani, H., and Kohmoto, K. (1997). AAL-toxin-deficient mutants of alternaria alternata tomato pathotype by restriction enzyme-mediated integration. *Phytopathology* **87,** 967–972.

Al-Sheikh, H., Watson, A. J., Lacey, G. A., Punt, P. J., MacKenzie, D. A., Jeenes, D. J., Pakula, T., Penttila, M., Alcocer, M. J., and Archer, D. B. (2004). Endoplasmic reticulum stress leads to the selective transcriptional downregulation of the glucoamylase gene in *Aspergillus niger. Mol. Microbiol.* **53**, 1731–1742.

Andersen, M. R., Vongsangnak, W., Panagiotou, G., Salazar, M. P., Lehmann, L., and Nielsen, J. (2008). A trispecies *Aspergillus* microarray: Comparative transcriptomics of three *Aspergillus* species. *Proc. Natl. Acad. Sci. USA* **105**, 4387–4392.

Arava, Y. (2003). Isolation of polysomal RNA for microarray analysis. *Methods Mol. Biol.* **224**, 79–87.

Archer, D. B., and Turner, G. (2006). Genomics of protein secretion and hyphal growth in *Aspergillus. In* "The Mycota XIII, Fungal Genomics," (A. Brown, ed.), pp. 75–96. Springer, New york.

Arvas, M., Pakula, T., Lanthaler, K., Saloheimo, M., Valkonen, M., Suortti, T., Robson, G., and Penttila, M. (2006). Common features and interesting differences in transcriptional responses to secretion stress in the fungi *Trichoderma reesei* and *Saccharomyces cerevisiae*. *BMC Genomics* **7**, 32.

Bashir, A., Bansal, V., and Bafna, V. (2010). Designing deep sequencing experiments: Detecting structural variation and estimating transcript abundance. *BMC Genomics* **11**, 385.

Breakspear, A., and Momany, M. (2007). The first fifty microarray studies in filamentous fungi. *Microbiology* **153**, 7–15.

Carvalho, N. D., Arentshorst, M., Jin Kwon, M., Meyer, V., and Ram, A. F. (2010). Expanding the ku70 toolbox for filamentous fungi: Establishment of complementation vectors and recipient strains for advanced gene analyses. *Appl. Microbiol. Biotechnol.* **87**, 1463–1473.

Conesa, A., Punt, P. J., van Luijk, N., and van den Hondel, C. A. (2001). The secretion pathway in filamentous fungi: A biotechnological view. *Fungal Genet. Biol.* **33**, 155–171.

de Ruiter-Jacobs, Y. M., Broekhuijsen, M., Unkles, S. E., Campbell, E. I., Kinghorn, J. R., Contreras, R., Pouwels, P. H., and van den Hondel, C. A. (1989). A gene transfer system based on the homologous pyrG gene and efficient expression of bacterial genes in *Aspergillus oryzae. Curr. Genet.* **16**, 159–163.

Fischer, R., Zekert, N., and Takeshita, N. (2008). Polarized growth in fungi—Interplay between the cytoskeleton, positional markers and membrane domains. *Mol. Microbiol.* **68**, 813–826.

Geysens, S., Whyteside, G., and Archer, D. B. (2009). Genomics of protein folding in the endoplasmic reticulum, secretion stress and glycosylation in the aspergilli. *Fungal Genet. Biol.* **46**, S121–S140.

Gouka, R. J., Punt, P. J., and van den Hondel, C. A. (1997). Efficient production of secreted proteins by Aspergillus: Progress, limitations and prospects. *Appl. Microbiol. Biotechnol.* **47**, 1–11.

Guillemette, T., van Peij, N. N., Goosen, T., Lanthaler, K., Robson, G. D., van den Hondel, C. A., Stam, H., and Archer, D. B. (2007). Genomic analysis of the secretion stress response in the enzyme-producing cell factory *Aspergillus niger. BMC Genomics* **8**, 158.

Harris, S. D. (2008). Branching of fungal hyphae: Regulation, mechanisms and comparison with other branching systems. *Mycologia* **100**, 823–832.

Hirota, M., Kitagaki, M., Itagaki, H., and Aiba, S. (2006). Quantitative measurement of spliced XBP1 mRNA as an indicator of endoplasmic reticulum stress. *J. Toxicol. Sci.* **31**, 149–156.

Jacobs, D. I., Olsthoorn, M. M., Maillet, I., Akeroyd, M., Breestraat, S., Donkers, S., van der Hoeven, R. A., van den Hondel, C. A., Kooistra, R., Lapointe, T., Menke, H., Meulenberg, R., *et al.* (2009). Effective lead selection for improved protein production in *Aspergillus niger* based on integrated genomics. *Fungal Genet. Biol.* **46**(Suppl. 1), S141–S152.

Jorgensen, T. R., Goosen, T., Hondel, C. A., Ram, A. F., and Iversen, J. J. (2009). Transcriptomic comparison of *Aspergillus niger* growing on two different sugars reveals coordinated regulation of the secretory pathway. *BMC Genomics* **10**, 44.

Joubert, A., Calmes, B., Berruyer, R., Pihet, M., Bouchara, J. P., Simoneau, P., and Guillemette, T. (2010). Laser nephelometry applied in an automated microplate system to study filamentous fungus growth. *Biotechniques* **48**, 399–404.

Kohno, K. (2010). Stress-sensing mechanisms in the unfolded protein response: Similarities and differences between yeast and mammals. *J. Biochem.* **147**, 27–33.

Le Crom, S., Schackwitz, W., Pennacchio, L., Magnuson, J. K., Culley, D. E., Collett, J. R., Martin, J., Druzhinina, I. S., Mathis, H., Monot, F., Seiboth, B., Cherry, B., *et al.* (2009). Tracking the roots of cellulase hyperproduction by the fungus *Trichoderma reesei* using massively parallel DNA sequencing. *Proc. Natl. Acad. Sci. USA* **106**, 16151–16156.

Lubertozzi, D., and Keasling, J. D. (2009). Developing Aspergillus as a host for heterologous expression. *Biotechnol. Adv.* **27**, 53–75.

Machida, M., Asai, K., Sano, M., Tanaka, T., Kumagai, T., Terai, G., Kusumoto, K., Arima, T., Akita, O., Kashiwagi, Y., Abe, K., Gomi, K., *et al.* (2005). Genome sequencing and analysis of *Aspergillus oryzae*. *Nature* **438**, 1157–1161.

McCarthy, J. E. (1998). Posttranscriptional control of gene expression in yeast. *Microbiol. Mol. Biol. Rev.* **62**, 1492–1553.

Meyer, V., Damveld, R. A., Arentshorst, M., Stahl, U., van den Hondel, C. A., and Ram, A. F. (2007). Survival in the presence of antifungals: Genome-wide expression profiling of *Aspergillus niger* in response to sublethal concentrations of caspofungin and fenpropimorph. *J. Biol. Chem.* **282**, 32935–32948.

Meyer, V., Arentshorst, M., Flitter, S. J., Nitsche, B. M., Kwon, M. J., Reynaga-Pena, C. G., Bartnicki-Garcia, S., van den Hondel, C. A., and Ram, A. F. (2009). Reconstruction of signaling networks regulating fungal morphogenesis by transcriptomics. *Eukaryot. Cell* **8**, 1677–1691.

Meyer, V., Ram, A., and Punt, P. J. (2010). Genetics, genetic manipulation, and approaches to strain improvement of filamentous fungi. *In* "Manual of Industrial Microbiology and Biotechnology," (A. L. Demain and J. L. Davies, eds.), 3rd edn., pp. 318–329. Wiley, New York.

Michielse, C. B., Hooykaas, P. J., van den Hondel, C. A., and Ram, A. F. (2008). Agrobacterium-mediated transformation of the filamentous fungus *Aspergillus awamori*. *Nat. Protoc.* **3**, 1671–1678.

Mulder, H. J., and Nikolaev, I. (2009). HacA-dependent transcriptional switch releases hacA mRNA from a translational block upon endoplasmic reticulum stress. *Eukaryot. Cell* **8**, 665–675.

Mulder, H. J., Saloheimo, M., Penttila, M., and Madrid, S. M. (2004). The transcription factor HACA mediates the unfolded protein response in *Aspergillus niger*, and up-regulates its own transcription. *Mol. Genet. Genomics* **271**, 130–140.

Pakula, T. M., Laxell, M., Huuskonen, A., Uusitalo, J., Saloheimo, M., and Penttila, M. (2003). The effects of drugs inhibiting protein secretion in the filamentous fungus *Trichoderma reesei*. Evidence for down-regulation of genes that encode secreted proteins in the stressed cells. *J. Biol. Chem.* **278**, 45011–45020.

Pel, H. J., de Winde, J. H., Archer, D. B., Dyer, P. S., Hofmann, G., Schaap, P. J., Turner, G., de Vries, R. P., Albang, R., Albermann, K., Andersen, M. R., Bendtsen, J. D., *et al.* (2007). Genome sequencing and analysis of the versatile cell factory *Aspergillus niger* CBS 513.88. *Nat. Biotechnol.* **25**, 221–231.

Pradet-Balade, B., Boulme, F., Beug, H., Mullner, E. W., and Garcia-Sanz, J. A. (2001). Translation control: Bridging the gap between genomics and proteomics? *Trends Biochem. Sci.* **26**, 225–229.

Richie, D. L., Hartl, L., Aimanianda, V., Winters, M. S., Fuller, K. K., Miley, M. D., White, S., McCarthy, J. W., Latge, J. P., Feldmesser, M., Rhodes, J. C., and Askew, D. S. (2009). A role for the unfolded protein response (UPR) in virulence and antifungal susceptibility in *Aspergillus fumigatus*. *PLoS Pathog.* **5,** e1000258.

Ruegsegger, U., Leber, J. H., and Walter, P. (2001). Block of HAC1 mRNA translation by long-range base pairing is released by cytoplasmic splicing upon induction of the unfolded protein response. *Cell* **107,** 103–114.

Shoji, J. Y., Arioka, M., and Kitamoto, K. (2008). Dissecting cellular components of the secretory pathway in filamentous fungi: Insights into their application for protein production. *Biotechnol. Lett.* **30,** 7–14.

Sims, A. H., Gent, M. E., Lanthaler, K., Dunn-Coleman, N. S., Oliver, S. G., and Robson, G. D. (2005). Transcriptome analysis of recombinant protein secretion by *Aspergillus nidulans* and the unfolded-protein response in vivo. *Appl. Environ. Microbiol.* **71,** 2737–2747.

Smith, C. W., Klaasmeyer, J. G., Edeal, J. B., Woods, T. L., and Jones, S. J. (1999). Effects of serum deprivation, insulin and dexamethasone on polysome percentages in C2C12 myoblasts and differentiating myoblasts. *Tissue Cell* **31,** 451–458.

Sweigard, J. A., Carroll, A. M., Kang, S., Farrall, L., Chumley, F. G., and Valent, B. (1995). Identification, cloning, and characterization of PWL2, a gene for host species specificity in the rice blast fungus. *Plant Cell* **7,** 1221–1233.

Travers, K. J., Patil, C. K., Wodicka, L., Lockhart, D. J., Weissman, J. S., and Walter, P.\ (2000). Functional and genomic analyses reveal an essential coordination between the unfolded protein response and ER-associated degradation. *Cell* **101,** 249–258.

Valkonen, M., Ward, M., Wang, H., Penttila, M., and Saloheimo, M. (2003). Improvement of foreign-protein production in *Aspergillus niger* var. awamori by constitutive induction of the unfolded-protein response. *Appl. Environ. Microbiol.* **69,** 6979–6986.

van Gorcom, R. F., and van den Hondel, C. A. (1988). Expression analysis vectors for *Aspergillus niger*. *Nucleic Acids Res.* **16,** 9052.

van Hartingsveldt, W., Mattern, I. E., van Zeijl, C. M., Pouwels, P. H., and van den Hondel, C. A. (1987). Development of a homologous transformation system for *Aspergillus niger* based on the pyrG gene. *Mol. Gen. Genet.* **206,** 71–75.

Vishniac, W., and Santer, M. (1957). The thiobacilli. *Bacteriol. Rev.* **21,** 195–213.

Vongsangnak, W., Nookaew, I., Salazar, M., and Nielsen, J. (2010). Analysis of genome-wide coexpression and coevolution of *Aspergillus oryzae* and *Aspergillus niger*. *OMICS* **14,** 165–175.

Wang, Z., Gerstein, M., and Snyder, M. (2009). RNA-Seq: A revolutionary tool for transcriptomics. *Nat. Rev. Genet.* **10,** 57–63.

Wimalasena, T. T., Enjalbert, B., Guillemette, T., Plumridge, A., Budge, S., Yin, Z., Brown, A. J., and Archer, D. B. (2008). Impact of the unfolded protein response upon genome-wide expression patterns, and the role of Hac1 in the polarized growth, of *Candida albicans*. *Fungal Genet. Biol.* **45,** 1235–1247.

Yu, J. H., Hamari, Z., Han, K. H., Seo, J. A., Reyes-Dominguez, Y., and Scazzocchio, C. (2004). Double-joint PCR: A PCR-based molecular tool for gene manipulations in filamentous fungi. *Fungal Genet. Biol.* **41,** 973–981.

CHAPTER TWO

Assays for Detecting the Unfolded Protein Response

Karen Cawley,*,[1] Shane Deegan,*,[1] Afshin Samali,* *and* Sanjeev Gupta[†]

Contents

1. Introduction	33
2. Experimental Approaches for the Detection of ER Stress	36
2.1. Detecting sXBP-1 mRNA	36
2.2. mRNA and protein levels of UPR target genes	39
2.3. Reporter assays for activity of XBP-1 and ATF-6	41
2.4. Detection of IRE1 activation and ATF-6 translocation from the ER to the nucleus with fluorescent microscopy	44
3. Concluding Remarks	47
Acknowledgments	49
References	49

Abstract

The endoplasmic reticulum (ER) is the site for folding of membrane and secreted proteins in the cell. Physiological or pathological processes that disturb protein folding in the ER cause ER stress and activate a set of signaling pathways termed the unfolded protein response (UPR). The UPR leads to transcriptional activation of genes encoding ER-resident chaperones, oxidoreductases, and ER-associated degradation (ERAD) components. Thus, UPR promotes cellular repair and adaptation by enhancing protein-folding capacity, reducing the secretory protein load, and promoting degradation of misfolded proteins. In mammalian cells, the UPR also triggers apoptosis, perhaps when adaptive responses fail. Research into ER stress and the UPR continues to grow at a rapid rate as many new investigators are entering the field. Here, we describe the experimental methods that we have used to study UPR in tissue culture cells. These methods can be used by researchers to plan and interpret experiments aimed at evaluating whether the UPR and related processes are

* Apoptosis Research Centre, School of Natural Sciences (Biochemistry), National University of Ireland Galway, Ireland
[†] Apoptosis Research Centre, School of Medicine (Pathology), Clinical Science Institute, National University of Ireland Galway, Ireland
[1] Contributed equally

activated or not. It is important to note that these are general guidelines for monitoring the UPR and not all assays will be appropriate for every model system.

Abbreviations

ARE	Antioxidant response element
ATF3	activating transcription factor 3
ATF4	activating transcription factor 4
ATF6	activating transcription factor 6
BFA	Brefeldin A
BiP	binding immunoglobulin protein
b-ZIP	basic leucine-zipper domain
CHOP	CAAT/enhancer binding protein
CMV	cytomegalovirus
CRE	cAMP-response element
CRELD2	cysteine-rich with EGF-like domains 2
CSSR	core stress sensing region
EDEM1	ER degradation enhancer, mannosidase alpha-like 1
eIF2α	eukaryotic initiation factor 2α
ERAD	ER-associated degradation machinery
ERAI	ER stress–activator indicator
ERSE	ER stress–response element
GCN2	general control nonderepressible-2
GFP	green fluorescent protein
GLS1/GLS2	Golgi localization signal 1/2
GRP78	glucose-regulated protein 78
GRP94	glucose-regulated protein 94
HERP	homocysteine-induced ER protein
HO-1	heme oxygenase 1
HRDI	HMG-coA reductase degradation 1
HRI	heme-regulated Inhibitor
HSP	heat shock protein
IRE1	Inositol-requiring enzyme 1
KEAP1	Kelch-like ECH-associated protein 1
MHC	major histocompatibility complex
NRF2	NF-E2-related factor 2
PCR	polymerase chain reaction
PDI	protein disulfide isomerase
PERK	PKR-like ER kinase
PKR	protein kinase double-stranded RNA-dependent

RFP	red fluorescent protein
RT	reverse transcription
S1P	site-1 protease
S2P	site-2 protease
TG	Thapsigargin
TLR	toll-like receptors
TM	Tunicamycin
UPR	unfolded protein response
UPRE	unfolded protein response element
XBP1	X box-binding protein-1

1. Introduction

The endoplasmic reticulum (ER) is a complex organelle consisting of an extensive membrane network and is equipped with machinery for the folding and maturation of secreted and transmembrane proteins. Compared to the cytosol, the ER lumen has greater calcium concentration and dedicated enzymes for protein glycosylation. The ER maintains an oxidizing redox potential to promote disulfide bond formation in maturing secretory proteins. These conditions are necessary for the optimum functioning of ER-resident chaperones and enzymes, such as BiP/glucose-regulated protein 78 (GRP78), protein disulfide isomerase (PDI), calnexin, calreticulin, and peptidylpropyl isomerase required for the folding and processing of nascent proteins in the ER lumen (Ma and Hendershot, 2004; Tu and Weissman, 2004). A physiological or pathological stress inflicted on a cell disrupts the ER homeostatic environment, in turn disabling its protein-folding capacity. The unfolded protein response (UPR) is activated in response to this ER stress (Ron and Walter, 2007). The UPR mediates its response via three ER transmembrane receptors: PKR-like ER kinase (PERK), activating transcription factor 6 (ATF6), and Inositol-requiring enzyme 1 (IRE1). The activation of these ER transmembrane receptors leads to the induction of a battery of transcription factors that target an ER stress–response element (ERSE) in the promoter region of genes encoding for ER chaperone proteins, folding enzymes, and components of the ER-associated degradation (ERAD) machinery. The activation of these genes enhances the cell's folding capacity and the degradation of misfolded and aggregated proteins, allowing the cell to relieve the stress and restore ER homeostasis (Szegezdi *et al.*, 2006; Yamamoto *et al.*, 2004).

In nonstressed cells, the ER transmembrane receptors remain in an inactive state due to their association with the ER chaperone protein,

GRP78. Upon ER stress, the accumulation of unfolded proteins in the ER lumen results in the dissociation of GRP78 and the activation of all the three receptors (Szegezdi et al., 2006); however, recent research has shown increasing complexity to the activation of these receptors and requires further events after GRP78 dissociation. It has been shown in yeast that a mutation of GRP78 binding site in the luminal domain of IRE1 does not result in constitutive activation of the receptors, suggesting that other factors are required for its activation (Oikawa et al., 2007). Dimerization of IRE1 core stress sensing region (CSSR) in the ER lumen creates a major histocompatibility complex (MHC)-like groove. It is believed that unfolded proteins bind to this MHC-like grove, promoting the formation of higher order oligomers required for UPR activation. The luminal domain of PERK and IRE1 shows similar features and thus are believed to be activated in a similar manner (Zhou et al., 2006). In nonstressed cells, ATF6 is present on the ER membrane in an oxidized form as a monomer, dimer, and oligomer due to intra- and intermolecular disulfide bridges (Nadanaka et al., 2007). During ER stress, GRP78 dissociates, exposing a region on ATF6 containing two Golgi localization signals, GLS1 and GLS2, which signals ATF6 for translocation to the Golgi body. During ER stress, the oxidized ATF6 is reduced and only the reduced monomeric form of ATF6 is recognized by S1 and S2 proteases for proteolytic processing into its active form; oxidized ATF6 is not processed and retrotranslocated back to the ER membrane (Shen et al., 2002). This differential activation of the three ER stress sensor molecules may explain why different ER stimuli can induce a differential array of UPR target genes and, in some cases, can specifically activate just one arm of the UPR.

PERK is a serine/threonine kinase and is a member of the eIF2 kinase family (Wek et al., 2006). Upon release of GRP78, PERK homodimerizes via its luminal domain and subsequently *trans*-autophosphorylates its respective cytoplasmic domain at residue threonine 980, resulting in its activation (Ron and Walter, 2007; Schroder and Kaufman, 2005). Activated PERK phosphorylates eukaryotic translation initiation factor 2α (eIF2α), resulting in translation attenuation of transcripts with a 5′ cap, thus eIF2α phosphorylation reduces the protein load on the ER membrane and attenuates cell growth and proliferation (Harding et al., 2000; Jiang and Wek, 2005). Phosphorylation of eIF2α leads to increased translation of ATF4 mRNA which encodes for a cAMP response element binding transcription factor which transcriptionally activates a number of UPR target genes. The ATF4 gene carries a regulatory sequence in its 5′ untranslated region which allows it to be translated in a cap-independent manner (Harding et al., 2000, 2003; Szegezdi et al., 2006). PERK has also been shown to directly phosphorylate NF-E2-related factor 2 (NRF2), resulting in its activation via its dissociation from Kelch-like ECH-associated protein 1 (KEAP1). KEAP1 sequesters NRF2 in the cytosol, rendering NRF2 inactive during nonstressed conditions. NRF2 is a b-ZIP cap "n" collar TF which binds to an antioxidant response element (ARE) in the promoter region of genes encoding for detoxifying enzymes such as heme oxygenase (Cullinan and Diehl, 2006; Zhang, 2006). IRE1 is activated by autophosphorylation on serine 724 residue (Ron and Walter, 2007; Schroder and Kaufman, 2005). IRE1 has two domains, a serine/threonine kinase domain and an endoribonuclease (Kimata et al., 2007). IRE1 endoribonuclease activity induces cleavage of X box-binding protein 1 (XBP1) mRNA. The XBP1 mRNA is then ligated by an uncharacterized RNA ligase and translated to produce spliced XBP1 (sXBP1) (Calfon et al., 2002). sXBP1 is a highly active transcription factor and plays a key role in reinstating homeostasis during ER stress (Lee et al., 2003). The 90-kDa full-length basic leucine-zipper transcription factor ATF6 is released from GRP78 upon ER stress and translocated to the Golgi body where it is cleaved by site 1 and site 2 proteases to produce a 50-kDa active transcription factor. ATF6 transcriptionally upregulates XBP1 expression which then undergoes IRE1 processing to produce an active transcription factor (Haze et al., 1999). sXBP1 acting in concert with ATF6 activates genes encoding for ER chaperone proteins, folding enzymes and components of the ERAD machinery (Yamamoto et al., 2007). Moreover, the UPR has been shown to tightly regulate autophagy during ER stress, and PERK seems to be a key player in its regulation (Rouschop et al., 2010).

The posttranslational changes that the three ER stress sensors undergo upon activation of the UPR can serve as markers for their activation: phosphorylation of IRE1 on serine 724, phosphorylation of PERK on threonine 980, and the proteolytic cleavage of ATF6 (Haze et al., 1999; Ron and Walter, 2007; Schroder and Kaufman, 2005). However, as phospho-IRE1, phospho-PERK, and the proteolytically processed ATF6 proteins are expressed at very low levels in the cells, there has been a lack of effective commercial antibodies to detect them. A commonly used indicator of ER stress is the upregulation and nuclear translocation of the transcription factor C/EBP homologous protein (CHOP), as it is a downstream target of all the three arms of the UPR (Wang et al., 1996). It was recently reported, however, that three out of seven commercially available antibodies gave false positive by Western blotting and immunohistochemistry and thus is advised to validate the specificity of the antibody before carrying out experiments (Haataja et al., 2008).

This draws our attention to look at downstream targets of the UPR to monitor its induction, such as CHOP, ATF4, XBP1, GRP78, and HERP. When monitoring the activation of the canonical UPR, it is essential to evaluate the induction of a UPR target gene downstream of all the three arms of the UPR. It has been reported that components of the UPR can be activated/induced independently of the classical UPR. For example, the innate immune system initiates its response via a group of receptors known as the Toll-like receptors (TLR). It was recently reported that TLR activates IRE1 and subsequent XBP1 splicing independent of UPR activation

(Martinon et al., 2010). Phosphorylation of eIF2α and enhanced translation of the transcription factors ATF4, ATF3, and CHOP can occur via eIF2 kinases other then PERK, such as GCN2 (general control nonderepressible-2) activated in response to amino acid deprivation, UV irradiation, and proteasome inhibition, PKR (protein kinase double-stranded RNA-dependent) activated in response to viral infection, and HRI (heme-regulated inhibitor) activated in response to low heme levels as well as heat and oxidative stresses (Wek et al., 2006). It was reported that following brain ischemia, PERK was activated independent of unfolded nascent proteins (Sanderson et al., 2010). The pharmacological compound OSU-03012 is a derivative of the COX2 inhibitor Celecoxib; however, it lacks the COX2 inhibitory activity. The exact mechanism of this compound is still being elucidated; however, it has been shown that OSU-03012 induces cell death in a PERK-dependent manner and is greatly attenuated in PERK$^{-/-}$ cells (Park et al., 2008).

The ER's sensitive environment is prone to many physiological and pathological stresses such as hypoxia, glucose deprivation, and viral infection. ER stress and the UPR play integral roles in many human diseases such as cancer, autoimmune diseases, diabetes, atherosclerosis, ischemia, stroke, and most neurodegenerative diseases; however, the UPR's role in these diseases are unclear and have not been completely elucidated (Kim et al., 2008a; Lin et al., 2008). Research into the UPR and its role in disease is under constant dissection; thus, it is important to set out a standard set of criteria to follow when monitoring the UPR (Samali et al., 2010). It is important to understand that there is no one assay that is best suited to monitor the UPR in all model systems and that some assays may work better than others depending on the circumstances. Therefore, it is essential to always use multiple assays when monitoring the UPR; as discussed, different stresses may induce a different UPR cascade and some stresses may induce a nonclassical UPR response independent of ER stress.

2. Experimental Approaches for the Detection of ER Stress

2.1. Detecting sXBP-1 mRNA

The dissociation of GRP78 from IRE1 results in its dimerization and subsequent activation (Szegezdi et al., 2006). Full-length XBP1 requires the endoribonuclease domain of active IRE1 for processing into active sXBP1; thus, the splicing of XBP1 is a key marker for IRE activation. In mammals, a 26 nucleotide intron is spliced out by activated IRE1, leading to a shift in the codon reading frame (Fig. 2.1A). A similar mechanism is

Assays for Detecting the Unfolded Protein Response

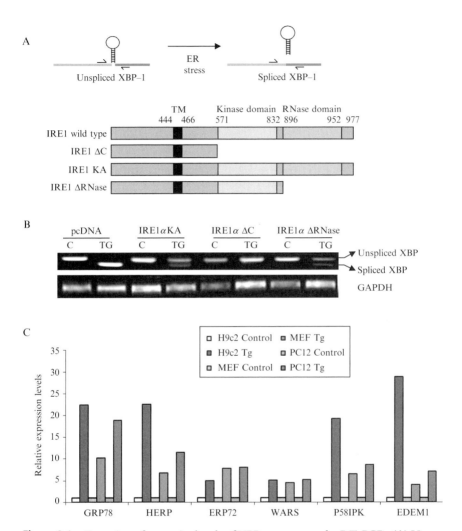

Figure 2.1 Detection of transcript levels of UPR target genes by RT-PCR. (A) Upper panel, cartoon of XBP1 splicing during ER stress. Lower panel, schematic representation of various mutant constructs of IRE1. (B) Modulation of XBP1 splicing by mutant IRE1. Total RNA was isolated from HEK 293 cells were transfected with IRE1 mutants, either untreated or treated with thapsigargin (0.5 μM) 6 h, and RT-PCR analysis of total RNA was performed to simultaneously detect both spliced and unspliced XBP1 mRNA and GADPH. (C) Induction of UPR target genes upon exposure to thapsigargin. Total RNA was isolated from indicated cells after treatment thapsigargin (Tg) and the expression levels of the indicated genes were determined by real-time RT-PCR, normalizing against GAPDH expression. Adapted from Samali et al., 2010. (See Color Insert.)

observed in yeast with the splicing of HAC1 by ire1p. The XBP1 protein encoded by the spliced mRNA is more stable and is a potent transcription factor of the basic leucine-zipper (b-ZIP) family and one of the key regulators of ER folding capacity. sXBP1 can activate the expression of many UPR- and ERAD-associated proteins such as p58^{ipk}, ERDJ4, EDEM1, and HSP40 (Kawahara et al., 1998; Lee et al., 2003). This particular splicing event is considered to be unconventional when compared to classical splicing via the spliceosome. The spliceosome recognizes a sequence at an exon–intron boundary that generally follows chambon's rule, that is, splices a site beginning with AU that ends in AG, GU–AG, or AU–AC (Tarn and Steitz, 1997; Uemura et al., 2009). On the contrary, unconventional splicing is completely independent from spliceosome processing and depends solely on IRE1 and an RNA ligase which recognizes a stem loop structure that replaces the typical chambon's sequence. The splicing of XBP1 mRNA can be detected by semiquantitative RT-PCR using primers specific for XBP1 which will detect both unspliced and spliced isoforms (Fig. 2.1A and B). The ER stress-mediated splicing of XBP1 requires activation of IRE1, and if the function of IRE1 is compromised, ER stress-mediated splicing of XBP1 is attenuated (Fig. 2.1B). We have detected IRE1-dependent splicing of XBP1mRNA under conditions of ER stress by using various mutants of IRE1 (Fig. 2.1A and B; Gupta et al., 2010).

2.1.1. Protocol for conventional RT-PCR for sliced XBP1

1. Cells are maintained in a 25-cm^2 Flask (Sarstedt) in the appropriate growth medium with required supplements. They are trypsinized using 1× Trypsin/EDTA (Sigma), when approximately 80% confluent.
2. For experiments, seed cells to the appropriate density in a 6-well plate.
3. The next day, transfect with each IRE1 mutant.
4. Commence treatments 24 h posttransfection, with a known ER stress inducer such as tunicamycin, the inhibitor of N-linked glycosylation; thapsigargin, the SERCA pump inhibitor or brefeldin A, which blocks transport of protein from the ER to the Golgi apparatus. All the three compounds are available from Sigma Aldrich Ltd. A dose and time response for the chosen compound should be done prior to commencement of experiments, as this will vary from cell type to cell type.
5. Isolate RNA using Qiagens RNeasy Kit or Sigma's TRI reagent by following the manufacturer's instructions. Then, 2 μg RNA is reverse-transcribed into cDNA using invitrogen's Oligo dT and superscript III reverse transcriptase. Conventional PCR can then be used for amplification of spliced and unspliced XBP1 using the following primers and

Promega's GoTaq mastermix. A PCR for GAPDH or another appropriate internal control should be carried out on the same cDNA.

Rat XBP1

Forward primer: TTACGAGAGAAAACTCATGGGC
Reverse primer: GGGTCCAACTTGTCCAGAATGC
Size of PCR products: Unspliced XBP1 = 289 bp, Spliced XBP1 = 263 bp

Human XBP1

Forward primer: TTACGAGAGAAAACTCATGGCC
Reverse primer: GGGTCCAAGTTGTCCAGAATGC
Size of PCR products: Unspliced XBP1 = 289 bp, Spliced XBP1 = 263 bp

Mouse XBP1

Forward primer: GAACCAGGAGTTAAGAACACG
Reverse primer: AGGCAACAGTGTCAGAGTCC
Size of PCR products: Unspliced XBP1 = 205 bp, Spliced XBP1 = 179 bp

2.2. mRNA and protein levels of UPR target genes

2.2.1. qRT-PCR for UPR target genes

The three arms of the UPR pathway all lead to the transcriptional activation of genes that encode for chaperone proteins, folding enzymes, and proteins associated with ERAD to help restore the ER to its normal functioning state (Kim et al., 2008a; Szegezdi et al., 2006). The transcription of these genes is regulated via the *cis*-acting ERSE, ERSE II, and UPRE (ER stress–response element I and II, unfolded protein response element), which were identified in the promoters of several UPR-associated genes. The mammalian ERSE has the consensus sequence CCAAT-N9-CCACG and is known to regulate GRP genes such as GRP78 (BiP), ERP72 (Endoplasmic reticulum resident protein 72), GRP94, Calreticulin, and CRELD2 (cysteine-rich with EGF-like domains 2). It has been shown that ATF6 and XBP1 are translocated to the nucleus upon ER stress, and only after the binding of transcription factor NF-Y (Nuclear transcription factor-Y) to the sequence CCAAT, ATF6 and XBP1 can bind to the CCACG region of the ERSE and initiate transcription (Roy and Lee, 1999; Yamamoto et al., 2004). Both ERSE I and ERSE II were identified in the promoter of HERP, and it was shown that the first part of the consensus sequence of ERSE II has reverse complementarity to that of ERSE I, ATTGG-N-CCACG instead of CCAAT-N9-CCACG, and promoter activity was shown to be greatest when both types of ERSE are present. Activation of ERSE II by ATF6 is

NF-Y dependent, while XBP1 activates it independently (Kokame et al., 2001; Yamamoto et al., 2004). HERP is a protein found in the ER membrane and exposed to the cytoplasm where it interacts with ubiquilins and ERAD machinery in a prosurvival role to perhaps, direct unfolded proteins to the proteasome for degradation.(Kim et al., 2008b) The UPRE has a consensus sequence of TGACGTGG/A and thus far only has two known activators, XBP1 and LUMAN, a b-ZIP transcription factor that structurally resembles ATF6 but is not activated by ER stress inducers like tunicamycin and thapsigargin, even though it is located in the ER membrane and activates EDEM, a known component of ERAD. As with ERSE II, XBP1 induces the transcription of UPRE-containing genes in the absence of NF-Y (DenBoer et al., 2005; Yamamoto et al., 2004). By determining the levels of UPR-associated transcripts in a particular model system, it can elucidate the mechanism behind an observed response. There are several strategies used to detect differential mRNA levels, such as Northern blotting which allows the levels, size, and the presence of spliced forms of an mRNA to be determined; however, this technique is more labor-intensive and less sensitive. Conventional RT-PCR can be used, but only one product can be detected at any one time, and quantification of mRNA levels is a problem. We prefer real-time RT-PCR with Taqman chemistry (also known as "fluorogenic 5′ nuclease chemistry") because of its sensitivity, specificity, speed, and ease of handling. In our laboratory, the induction of mRNA of UPR target genes has been detected in a variety of mammalian cell lines using real-time RT-PCR (Fig. 2.1C).

2.2.2. Protocol for qRT-PCR for UPR genes

1. Cells are maintained as previously described.
2. Firstly, cells are seeded at an appropriate density in a 25–75 cm^2 flask.
3. Treat cells with an ER stress-inducing agent and harvest cells using TRI reagent or Qiagens RNeasy Kit.
4. As described previously, 2 μg of RNA is reverse-transcribed into cDNA, and a PCR for GAPDH is carried out using Promega GoTaq 2× master mix to check the efficiency of the RT reaction. GAPDH levels should be comparable in each sample.
5. For qRT-PCR, cDNA may need to be diluted or a higher concentration generated if many genes are to be analyzed at once; it is then combined with 2× Taqman fast universal PCR master mix and 20× Taqman gene expression assays and accurately dispensed to a fast optical MicroAmp 96-well plate in triplicate (all supplied by Applied Biosystems). The PCR is run for 40 cycles on an Applied Biosystems fast 7500 machine using the default program for fast reagents.
6. The relative expression levels (relative to GAPDH or an appropriate internal control gene, the control should not vary by more than one Ct value for each

sample analyzed) are calculated using the $2^{\Delta\Delta Ct}$ method when comparing a treated sample to an untreated sample. See Table 2.1 for a list of Taqman assays that have worked consistently well in our experiments.

2.2.3. Western blotting for UPR target genes

We recommend the determination of the protein levels of established UPR target genes whose expression has been reported to increase during conditions of ER stress. Here, we describe a set of proteins that are useful to monitor when determining the activation of the UPR. When carrying out Western blotting, we suggest the performance of standard procedures for determining the protein levels of bona fide UPR target genes within protein samples (Fig. 2.2). The *trans*-autophosphorylation of PERK on threonine 980 is a key marker for the activation of the UPR. Another common protein upregulated in response to the UPR is CHOP. As mentioned earlier, three out of seven commercially available antibodies gave false positive by Western blotting and immunohistochemistry; however, two CHOP antibodies (Table 2.2) we have validated work very well. Commercially available antibodies for phospho-IRE1 and cleaved ATF6 have possibly not given reproducible results; because these proteins are expressed at such low levels in the cell, it is difficult to detect with immunoblotting. Table 2.2 provides a list of antibodies that have worked best in our experience to detect several UPR marker proteins in Western blotting and immunohistochemistry. It is recommended that blot be done for more than one protein to detect UPR activation.

2.3. Reporter assays for activity of XBP-1 and ATF-6

The most salient feature of the UPR is an increase in the transactivation function of a number of transcription factors such as ATF6, ATF4, XBP1, CHOP, and NRF2. As mentioned above, the induction of UPR genes is mediated by *cis*-acting ERSE, ERSE II, or the UPRE present in the promoter of UPR target genes. There are several reporter systems that can be used to detect ATF6 and XBP1 activation. In the p5xATF6-GL3 reporter, the luciferase gene is under the control of the c-fos minimal promoter and five tandem copies of the ATF6 consensus binding site identified by *in vitro* gel mobility shift assays with recombinant ATF6 (Wang *et al.*, 2000). In p4xXBPGL3 reporter, the luciferase gene is under the control of four tandem copies of the XBP1 consensus binding site 5′-CGCG(TGGATGACGTGTACA)$_4$-3′ (Lee *et al.*, 2003). In addition, there are several other ERSE reporters that have promoter regions of GRP78, GRP94, Calreticulin, and XBP1 (Roy and Lee, 1999) and an ERSE II reporter that has the HERP promoter upstream of the luciferase reporter gene (Kokame *et al.*, 2001). These reporters should be used in combination

Table 2.1 List of Taqman assays that reproducibly detect markers of UPR

Target gene	Accession number	Assay number	Accession number	Assay number
GRP78	NM_013083.1	Rn01435771_g1	NM_022310.2	Mm01333324_g1
HERP	NM_053523.1	Rn01536690_m1	NM_022331.1	Mm01249592_m1
ERP72	NM_053849.1	Rn01451754_m1	NM_009787.2	Mm00437958_m1
WARS	NM_001013170.2	Rn01429998_g1	NM_011710.2	Mm00457097_m1
P58IPK	NM_022232	Rn00573712_m1	NM_008929	Mm00515299_m1
EDEM1	XM_238366.4	Rn01765441_m1	NM_138677.2	Mm00551797_m1

Assays for Detecting the Unfolded Protein Response 43

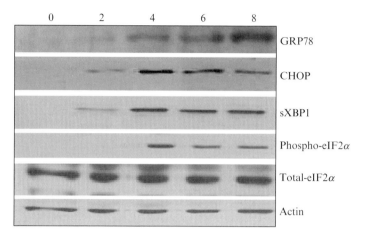

Figure 2.2 Detection of the protein levels of UPR target genes PC12 cells were treated with 0.25 µM of Tg for 0, 2, 4, 6, and 8 h. Whole cell lyzates were analyzed by Western blot for GRP78, CHOP, spliced XBP1, phospho-eIF-2α, total-eIF-2α. β-Actin was used to determine equal loading of samples. Adapted from Samali *et al.*, 2010.

Table 2.2 List of antibodies that reproducibly detect markers of UPR

Target name	Supplier	Applications
phospho-PERK	#3191; Cell Signaling Technology,	WB (1:2000), IHC (1:100)
phospho-PERK	#3179; Cell Signaling Technology,	WB
CHOP	MA1-250, Affinity bioreagents	WB
CHOP	sc-793; Santa Cruz Biotechnology	WB (1:1000), IHC (1:400–1:800)
spliced XBP-1	sc-7160; Santa Cruz Biotechnology	WB (1:2000), IHC (1:100)
ATF4	ARP37017_P050; Aviva Systems Biology	WB (1:5000)
Grp78	SPA-926; Stressgen; AB32618; Abcam	WB (1:1000), IHC (1:200)
phospho-eIF2 alpha	#9721; Cell Signaling Technology	WB (1:2500), IHC (1:100)
total-PERK antibody	sc-9477; Santa Cruz Biotechnology	WB (1:1000), IP
IRE1-alpha	#3294; Cell Signaling Technology	WB (1:1000)

with the corresponding mutant promoter where the functional *cis*-elements have been mutated.

2.3.1. Protocol for Reporter assays for ATF6 and XBP1

1. Cells are seeded at 70–80% confluency in a 24-well plate or for smaller cells in a 6-well plate.
2. Twenty-four hours after seeding, the reporter construct is transfected into the cells using the appropriate transfection reagent and optimal DNA–lipid ratio, which should be optimized for your model system in advance. Along with the luciferase reporter, a plasmid construct with the gene encoding Renilla luciferase or β-galactosidase should also be transfected to serve as an internal control for transfection efficiency.
3. Treatment with ER stress-inducing agents such as brefeldin A, tunicamycin, or thapsigargin is carried out 24 h after transfection and left for the appropriate time point. Again, the dose and duration of treatment should be determined for different model systems before commencing experiments, but generally, treatments range between 6 and 48 h.
4. Cells are lysed and the bioluminescence of luciferase is measured. We generally use Promega's Dual glow luciferase assay kit. By normalizing to the control plasmid, the fold induction of the reporter's activity can be determined.

2.4. Detection of IRE1 activation and ATF-6 translocation from the ER to the nucleus with fluorescent microscopy

2.4.1. IRE1 activation using XBP1–*venus* reporter

The IRE1-dependent splicing of XBP1 has been taken advantage of to generate a fluorescent reporter construct to monitor IRE1 activation and subsequent XBP1 activation by fusing the full-length XBP1 sequence to the green fluorescent protein, venus (Brunsing *et al.*, 2008; Iwawaki *et al.*, 2004). The design of the XBP1–*venus* reporter is shown in Fig. 2.3A. In this construct, the gene encoding *venus* is cloned downstream the 26-nt ER stress-specific intron of human *XBP1* (Iwawaki *et al.*, 2004). Under normal conditions, the mRNA of the fusion gene is not spliced, and its translation terminates at the stop codon near the joint between the *XBP1* and *Venus* gene. However, during ER stress, the 26-nt intron is spliced out, leading to a frame shift of the chimeric XBP1–*venus* mRNA, similar to that of the endogenous XBP1 mRNA. Translation of the spliced mRNA produces an XBP1–*venus* fusion protein and cells experiencing ER stress can be detected by monitoring the fluorescence activity of *Venus*. As *Venus* expression can only occur from the spliced form of the XBP1–GFP mRNA, its presence

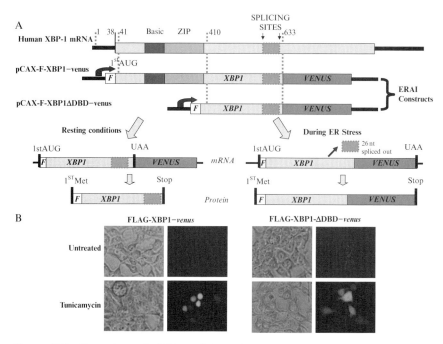

Figure 2.3 Detection of IRE1 activity using XBP1–*venus* reporter constructs. (A) Schematic presentation of ERAI plasmid obtained by fusing XBP1 and *venus*, a variant of the green fluorescent protein (adapted from Iwawaki et al., 2004). (B) Twenty-four hours after transfection with F-XBP1–*venus* and F-XBP1ΔDBD–*venus*, 293T cells were left untreated or treated with (1 μg/ml) tunicamycin for 24 h and then analyzed by fluorescence microscopy. Adapted from Samali et al., 2010.

signals the activation of IRE1 (Fig. 2.3B). Upon transfection of the XBP1–GFP reporter into cells, tunicamycin treatment results in detectable fluorescence in the nucleus, whereas negligible fluorescence is detected in any compartment under normal conditions (Brunsing et al., 2008; Iwawaki et al., 2004). Moreover, *Venus* expression during tunicamycin treatment has been shown in splicing assays to correlate with the extent of splicing of the UPR intron from XBP1/GFP mRNA (Brunsing et al., 2008; Iwawaki et al., 2004). One important point to note is that overexpression of F-XBP1–*venus* construct interferes with the induction of UPR target genes in a dominant-negative manner (Iwawaki et al., 2004). In F-XBP1ΔDBD–*venus* construct, DNA-binding domain (DBD) of XBP1 is deleted. F-XBP1ΔDBD–*venus* construct is recommended for use, as overexpression of F-XBP1ΔDBD–*venus* does not affect the induction of UPR target genes and can be used to detect the activation of IRE1 similar to F-XBP1–*venus* construct.

2.4.2. Protocol for IRE1 activation by fluorescent microscopy

1. 293T cells were seeded at 75,000 cells/well in a 24-well plate.
2. Twenty-four hours after seeding, the cells were transiently transfected with F- XBP1–*venus* or F- XBP1ΔDBD–*venus* using the transfection reagent JET PEI (Polyplus Transfection #101-01N) at a ratio of 1 μg DNA/2 μl JET PEI per well/24-well plate.
3. The DNA and JET PEI is diluted in 50 μl of NaCl independently and then pooled and incubated for 20 min before being added to the cells. Transfections are done in an antibiotic-free medium.
4. The transfection media is removed after 6–8 h and replaced with culture media containing antibiotic.
5. The cells are allowed to recover for 24 h posttransfection before ER stress-inducing compounds are added.
6. The cells were treated with 1 μg/ml of tunicamycin for 24 h.
7. The media was removed 24 h posttreatment, and the cells were stained with the nuclear dye DAPI and visualized using fluorescent microscopy.

Figure 2.3B shows accumulation of XBP1–*venus* in the nucleus and the accumulation of XBP1ΔDBD–*venus* in the cytosol upon tunicamycin treatment. It is important to note that the overexpression of XBP1–*venus* disrupts the induction of its UPR target genes in a dominant-negative manner, whereas XBP1ΔDBD–*venus* accumulates in the cytosol and does not interfere with endogenous XBP1 regulation of its target genes (Iwawaki *et al.*, 2004).

2.4.3. ATF6 activation using ATF6–GFP fusion constructs

A key regulatory step in ATF6 activation is its transport from the ER to the Golgi body, where it is processed by S1P and S2P proteases (Chen *et al.*, 2002; Nadanaka *et al.*, 2004). The cytoplasmic fragment of ATF6, thereby liberated from the membrane, translocates into the nucleus and activates transcription of its target genes (Chen *et al.*, 2002; Nadanaka *et al.*, 2004). A GFP–ATF6 fusion protein, which relocates from the ER to the nucleus via the Golgi apparatus in response to ER stress, can be used to monitor the activation of ATF6 by fluorescent microscopy (Chen *et al.*, 2002; Nadanaka *et al.*, 2004). One limitation of this approach, however, is that overexpression can sometimes alter the subcellular localization and kinetics of protein trafficking. This problem has been addressed to some extent by expressing GFP–ATF6 from a shortened CMV promoter which has a deletion of 430 base pairs from the 5′ side. The short promoter possesses considerably lower activity than does the full promoter, and GFP–ATF6 expressed using the short CMV promoter is localized exclusively to the ER and translocates to the nucleus similarly to that of endogenous ATF6 (Nadanaka *et al.*, 2004).

2.4.4. Protocol for ATF6 activation by fluorescent microscopy

1. 293T cells were seeded at 75,000 cells/well in a 24-well plate.
2. Twenty-four hours after seeding, the cells were transiently transfected with pCMVshort-EGFP–ATF6 (WT), pCMVshort-EGFP–ATF6 (S1P$^-$), and pCMVshort-EGFP–ATF6 (S2P$^-$) plasmids using the transfection reagent JET PEI (Polyplus Transfection #101-01N) at a ratio of 1 μg DNA/2 μl JET PEI per well/24-well plate.
3. The DNA and JET PEI is diluted in 50 μl of NaCl independently and then pooled and incubated for 20 min before being added to the cells. Transfections are done in an antibiotic-free medium.
4. The transfection media is removed after 6–8 h and replaced with culture media containing antibiotic.
5. The cells are allowed to recover for 24 h posttransfection before ER stress-inducing compounds are added.
6. The cells were treated with 1 μg/ml of tunicamycin for 24 h.
7. The media was removed 24 h posttreatment, and the cells were stained with the nuclear dye DAPI and visualized using fluorescent microscopy.

As shown in Fig. 2.4A, the wild-type GFP–ATF6 was translocated to the nucleus via the Golgi apparatus. Both EGFP–ATF6 (S1P$^-$) and EGFP–ATF6 (S2P$^-$) were localized in 293T cells similarly to the wild-type GFP–ATF6 (Fig. 2.4B-a–c). In contrast to wild-type GFP–ATF6 (Fig. 2.4B-a, d, and g), GFP–ATF6_(S1P$^-$) (Fig. 2.4B-b, e, and h) and EGFP–ATF6 (S2P$^-$) (Fig. 2.4B-c, f, and i) remained associated with the Golgi apparatus even 4 h after tunicamycin treatment. These results demonstrate that cleavage by S1P and S2P is critical for the processing of GFP–ATF6 and that only the processed product, GFP–ATF6, can enter the nucleus.

3. Concluding Remarks

The UPR is activated in response to numerous physiological and pathological stresses and has also shown to be essential for the differentiation of B-cells into antibody secreting-plasma cells, to enhance the secretory capacity of the cells. Activation of the UPR can be classed into two groups: (1) the canonical UPR, which demonstrates the classical accumulation of unfolded proteins in the ER lumen due to a disturbance in the ER's redox state or calcium flux, resulting in GRP78 dissociation and activation of the three UPR sensors, and (2) noncanonical UPR, which shows partial activation of the UPR such as phosphorylation of eIF2α and enhanced translation of the transcription factors ATF4, ATF3, and CHOP via eIF2 kinases other than PERK (GCN2, PKR, HRI); activation of IRE1 and subsequent XBP1 splicing downstream of the innate immune response via the TLR

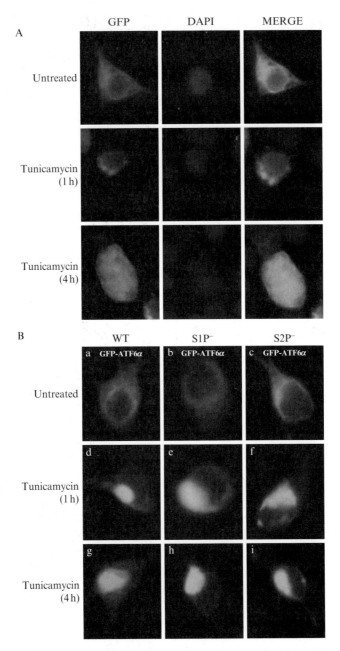

Figure 2.4 ER stress-induced processing and nuclear translocation of GFP–ATF6. (A) Twenty-four hours after transfection with pCMVshort-EGFP–ATF6 (WT), 293T cells were left untreated or treated with 1-μg/ml tunicamycin for the indicated periods. Cells were fixed in 4% paraformaldehyde, stained with DAPI, and then analyzed by fluorescence microscopy. (B) Twenty-four hours after transfection with pCMVshort-EGFP–ATF6 (WT), pCMVshort-EGFP–ATF6(S1P⁻), or pCMVshort-EGFP–ATF6 (S2P⁻), 293T cells were left untreated or treated with 1-μg/ml tunicamycin for the indicated periods and then analyzed by fluorescence microscopy. Adapted from Samali et al., 2010. (See Color Insert.)

receptors; activation of PERK following brain ischemia independent of unfolded nascent proteins, and activation of the PERK receptor with the pharmacological compound OSU-03024. ER stress and the UPR should not be used interchangeably; although ER stress will most likely always induce the UPR, activation of the UPR does not always mean that the ER is under stress. Therefore, it is useful to follow a set of established guidelines when monitoring the UPR and is essential that multiple assays are used to ensure correct interpretation of results.

ACKNOWLEDGMENTS

This publication has emanated from research conducted with the financial support of Science Foundation, Ireland under grant number 09/RFP/BIC2371 and Health Research Board (HRB), Ireland under grant number HRA_HSR/2010/24.

REFERENCES

Brunsing, R., et al. (2008). B- and T-cell development both involve activity of the unfolded protein response pathway. *J. Biol. Chem.* **283,** 17954–17961.

Calfon, M., et al. (2002). IRE1 couples endoplasmic reticulum load to secretory capacity by processing the XBP-1 mRNA. *Nature* **415,** 92–96.

Chen, X., et al. (2002). The luminal domain of ATF6 senses endoplasmic reticulum (ER) stress and causes translocation of ATF6 from the ER to the Golgi. *J. Biol. Chem.* **277,** 13045–13052.

Cullinan, S. B., and Diehl, J. A. (2006). Coordination of ER and oxidative stress signaling: The PERK/Nrf2 signaling pathway. *Int. J. Biochem. Cell Biol.* **38,** 317–332.

DenBoer, L. M., et al. (2005). Luman is capable of binding and activating transcription from the unfolded protein response element. *Biochem. Biophys. Res. Commun.* **331,** 113–119.

Gupta, S., et al. (2010). HSP72 protects cells from ER stress-induced apoptosis via enhancement of IRE1alpha-XBP1 signaling through a physical interaction. *PLoS Biol.* **8**e1000410.

Haataja, L., et al. (2008). Many commercially available antibodies for detection of CHOP expression as a marker of endoplasmic reticulum stress fail specificity evaluation. *Cell Biochem. Biophys.* **51,** 105–107.

Harding, H. P., et al. (2000). Perk is essential for translational regulation and cell survival during the unfolded protein response. *Mol. Cell* **5,** 897–904.

Harding, H. P., et al. (2003). An integrated stress response regulates amino acid metabolism and resistance to oxidative stress. *Mol. Cell* **11,** 619–633.

Haze, K., et al. (1999). Mammalian transcription factor ATF6 is synthesized as a transmembrane protein and activated by proteolysis in response to endoplasmic reticulum stress. *Mol. Biol. Cell* **10,** 3787–3799.

Iwawaki, T., et al. (2004). A transgenic mouse model for monitoring endoplasmic reticulum stress. *Nat. Med.* **10,** 98–102.

Jiang, H. Y., and Wek, R. C. (2005). Phosphorylation of the alpha-subunit of the eukaryotic initiation factor-2 (eIF2alpha) reduces protein synthesis and enhances apoptosis in response to proteasome inhibition. *J. Biol. Chem.* **280,** 14189–14202.

Kawahara, T., et al. (1998). Unconventional splicing of HAC1/ERN4 mRNA required for the unfolded protein response. *J. Biol. Chem.* **273**, 1802–1807.

Kim, I., et al. (2008a). Cell death and endoplasmic reticulum stress: Disease relevance and therapeutic opportunities. *Nat. Rev. Drug Discov.* **7**, 1013–1030.

Kim, T.-Y., et al. (2008b). Herp enhances ER-associated protein degradation by recruiting ubiquilins. *Biochem. Biophys. Res. Commun.* **369**, 741–746.

Kimata, Y., et al. (2007). Two regulatory steps of ER-stress sensor Ire1 involving its cluster formation and interaction with unfolded proteins. *J. Cell Biol.* **179**, 75–86.

Kokame, K., et al. (2001). Identification of ERSE-II, a new cis-acting element responsible for the ATF6-dependent mammalian unfolded protein response. *J. Biol. Chem.* **276**, 9199–9205.

Lee, A.-H., Iwakoshi, N. N., and Glimcher, L. H. (2003). XBP-1 Regulates a Subset of Endoplasmic Reticulum Resident Chaperone Genes in the Unfolded Protein Response. *Mol. Cell. Biol.* **23**, 7448–7459.

Lin, J. H., et al. (2008). Endoplasmic reticulum stress in disease pathogenesis. *Annu. Rev. Pathol.* **3**, 399–425.

Ma, Y., and Hendershot, L. M. (2004). ER chaperone functions during normal and stress conditions. *J. Chem. Neuroanat.* **28**, 51–65.

Martinon, F., et al. (2010). TLR activation of the transcription factor XBP1 regulates innate immune responses in macrophages. *Nat. Immunol.* **11**, 411–418.

Nadanaka, S., et al. (2004). Activation of mammalian unfolded protein response is compatible with the quality control system operating in the endoplasmic reticulum. *Mol. Biol. Cell* **15**, 2537–2548.

Nadanaka, S., et al. (2007). Role of disulfide bridges formed in the luminal domain of ATF6 in sensing endoplasmic reticulum stress. *Mol. Cell. Biol.* **27**, 1027–1043.

Oikawa, D., et al. (2007). Self-association and BiP dissociation are not sufficient for activation of the ER stress sensor Ire1. *J. Cell Sci.* **120**, 1681–1688.

Park, M. A., et al. (2008). OSU-03012 stimulates PKR-like endoplasmic reticulum-dependent increases in 70-kDa heat shock protein expression, attenuating its lethal actions in transformed cells. *Mol. Pharmacol.* **73**, 1168–1184.

Ron, D., and Walter, P. (2007). Signal integration in the endoplasmic reticulum unfolded protein response. *Nat. Rev. Mol. Cell Biol.* **8**, 519–529.

Rouschop, K. M., et al. (2010). The unfolded protein response protects human tumor cells during hypoxia through regulation of the autophagy genes MAP1LC3B and ATG5. *J. Clin. Invest.* **120**, 127–141.

Roy, B., and Lee, A. (1999). The mammalian endoplasmic reticulum stress response element consists of an evolutionarily conserved tripartite structure and interacts with a novel stress-inducible complex. *Nucleic Acids Res.* **27**, 1437–1443.

Samali, A., et al. (2010). Methods for monitoring endoplasmic reticulum stress and the unfolded protein response. *Int. J. Cell Biol.* 830307.

Sanderson, T. H., et al. (2010). PERK activation following brain ischemia is independent of unfolded nascent proteins. *Neuroscience* **169**, 1307–1314.

Schroder, M., and Kaufman, R. J. (2005). The mammalian unfolded protein response. *Annu. Rev. Biochem.* **74**, 739–789.

Shen, J., et al. (2002). ER stress regulation of ATF6 localization by dissociation of BiP/GRP78 binding and unmasking of Golgi localization signals. *Dev. Cell* **3**, 99–111.

Szegezdi, E., et al. (2006). Mediators of endoplasmic reticulum stress-induced apoptosis. *EMBO Rep.* **7**, 880–885.

Tarn, W.-Y., and Steitz, J. A. (1997). Pre-mRNA splicing: The discovery of a new spliceosome doubles the challenge. *Trends Biochem. Sci.* **22**, 132–137.

Tu, B. P., and Weissman, J. S. (2004). Oxidative protein folding in eukaryotes: Mechanisms and consequences. *J. Cell Biol.* **164**, 341–346.

Uemura, A., et al. (2009). Unconventional splicing of XBP1 mRNA occurs in the cytoplasm during the mammalian unfolded protein response. *J. Cell Sci.* **122,** 2877–2886.

Wang, X. Z., et al. (1996). Signals from the stressed endoplasmic reticulum induce C/EBP-homologous protein (CHOP/GADD153). *Mol. Cell. Biol.* **16,** 4273–4280.

Wang, Y., Shen, J., Arenzana, N., Tirasophon, W., Kaufman, R. J., and Prywes, R. (2000). Activation of ATF6 and an ATF6 DNA binding site by the endoplasmic reticulum stress response. *J. Biol. Chem.* **275,** 27013–27020.

Wek, R. C., et al. (2006). Coping with stress: eIF2 kinases and translational control. *Biochem. Soc. Trans.* **34,** 7–11.

Yamamoto, K., et al. (2004). Differential contributions of ATF6 and XBP1 to the activation of endoplasmic reticulum stress-responsive cis-acting elements ERSE, UPRE and ERSE-II. *J. Biochem.* **136,** 343–350.

Yamamoto, K., et al. (2007). Transcriptional induction of mammalian ER quality control proteins is mediated by single or combined action of ATF6alpha and XBP1. *Dev. Cell* **13,** 365–376.

Zhang, D. D. (2006). Mechanistic studies of the Nrf2-Keap1 signaling pathway. *Drug Metab. Rev.* **38,** 769–789.

Zhou, J., et al. (2006). The crystal structure of human IRE1 luminal domain reveals a conserved dimerization interface required for activation of the unfolded protein response. *Proc. Natl. Acad. Sci. USA* **103,** 14343–14348.

CHAPTER THREE

Analysis of the Role of Nerve Growth Factor in Promoting Cell Survival During Endoplasmic Reticulum Stress in PC12 Cells

Koji Shimoke,[1] Harue Sasaya, *and* Toshihiko Ikeuchi

Contents

1. Introduction	54
2. Induction of ER Stress in PC12 Cells	55
2.1. Culture of PC12 cells and treatment with a chemical inducer of ER stress	56
2.2. Use of XBP1 mRNA and protein to upregulate GRP78 expression	57
2.3. Upregulation of BH3-only protein expression	60
3. Measurement of Caspase-3 Activity	62
3.1. Fluorogenic substrates and their specificity	62
4. Fragmentation of Caspase-12	63
4.1. Detection of fragmented caspase-12 through the formation of a modified avidin–biotin antibody complex	64
5. Hyperinduction of GRP78 by NGF	65
5.1. Detection of GRP78 expression by Northern blotting	66
5.2. Detection of GRP78 expression by Western blotting	67
Acknowledgments	68
References	68

Abstract

Nerve growth factor (NGF) was first described by Rita Levi-Montalcini in the early 1960s from her studies of peripheral neurons. It has since been reported that NGF has the potential to elongate neurites or to prevent apoptosis via specific intracellular mechanisms. It has further been reported that as a component of these mechanisms, NGF binds to a specific receptor, TrkA, and thereby contributes to peripheral nerve cell functions or neuronal functions. It is noteworthy in this regard

Laboratory of Neurobiology, Department of Life Science and Biotechnology, Faculty of Chemistry, Materials and Bioengineering, Kansai University, Suita, Osaka, Japan
[1] Corresponding author: E-mail: shimoke@kansai-u.ac.jp, Tel:+81-6-6368-0853, Fax:+81-6-6330-3770

that pheochromocytoma 12 (PC12) cells express TrkA and respond to neurite outgrowth or anti-apoptotic signals by binding to NGF. Hence, PC12 cells have been used as an *in vitro* model system for the study of neuronal functions. It has been reported that endoplasmic reticulum (ER) stress is involved in neurodegenerative disorders, including Alzheimer's, Parkinson's, and Huntington's disease. The common link with regard to ER stress is that the neuronal cells die in these pathologies via specific intracellular mechanisms. This type of cell death, if it is apoptotic in nature, is termed ER stress-mediated apoptosis. In the process of ER stress-mediated apoptosis, the cleavage of pro-caspase-12 residing on the ER and the expression of glucose-regulated protein 78 (GRP78) can be observed. The expression of GRP78 protein is a characteristic of an unfolded protein response (UPR) via specific signal transduction pathways mediated by the unfolded protein response element (UPRE) in the upstream region of the *grp78* gene so on. In ER stress-mediated apoptosis, a caspase cascade is also observed. To further clarify the mechanisms underlying ER stress-mediated apoptosis, a better understanding of the UPR is therefore important. In our current study, we describe a method for detecting gene induction via the UPR, focusing on GRP78 and caspase activities as the measurement end-points. The information generated by our method will accelerate our understanding of the pathophysiological processes leading to ER stress-mediated apoptosis.

1. Introduction

Nerve growth factor (NGF) belongs to the neurotrophin (NT) family which also contains brain-derived neurotrophic factor (BDNF), NT-4/5, and NT-3. The receptors for these neurotrophic factors are also members of a protein family: TrkA for NGF, TrkB for BDNF and NT-4/5, and TrkC for NT-3 (Barbacid, 1995; Suter *et al.*, 1992). Upon binding of the NTs to their specific Trk receptor, the cytosolic domain of this receptor is autophosphorylated at specific tyrosine residues, leading to signal transduction within the cell. Phenotypic analyses have revealed that neurite outgrowth, cell survival, and neurotransmitter release are critically dependent on NT-induced signal transduction in neuronal cells. For example, it has been reported that the mitogen-activated kinase (MAPK; Borasio *et al.*, 1993; Price *et al.*, 2003), phosphatidylinositol 3-kinase (PI3-K; Kim *et al.*, 2005; Nonomura *et al.*, 1996; Pyle *et al.*, 2006; Shimoke *et al.*, 1997, 2004a,b, 2005), and phospholipase C-gamma (PLC-γ) signaling pathways (Matsumoto *et al.*, 2006) generally contribute to neurite outgrowth, cell survival, and neurotransmitter release, respectively. The importance of PI3-K and its downstream serine/threonine kinase Akt in the cell has been confirmed by treatments with specific inhibitors and the ectopic expression of dominant-negative proteins (Shimoke and Chiba, 2001). The existence of signaling pathways for cell survival was revealed through searches for the substrates of Akt. In addition, although the precise mechanisms are not fully clear, interesting evidence has emerged that the signaling pathways for cell survival attenuate those for cell death, specifically apoptosis. For instance, Bax, a member of the B cell CLL/lymphoma-2 (Bcl-2) family of proteins which induces apoptosis, is phosphorylated by Akt (Gardai et al., 2004). This suggests that cell fates are determined via a balance between anti-apoptotic and pro-apoptotic signal pathways. Hence, cell viability can only be maintained or increased if pro-apoptotic signals are suppressed (Billen et al., 2008; Orrenius, 2004; Szegezdi et al., 2009).

In addition to the ordinary apoptotic response networks, it has been reported that endoplasmic reticulum (ER) stress resulting from the accumulation of unfolded proteins in the ER lumen also triggers apoptotic cell death (Imaizumi et al., 2001; Marciniak and Ron, 2006; Rasheva and Domingos, 2009; Szegezdi et al., 2009). This is referred to as ER stress-mediated apoptosis. Many previous reports have indicated that this type of apoptotic response is involved in the processes leading to various neurodegenerative diseases (Imaizumi et al., 2001; Marciniak and Ron, 2006; Nakayama et al., 2008). It has not been fully clarified, however, whether specific pro-apoptotic signals involving specific caspase cascades or the expression of aberrant proteins via the unfolded protein response (UPR) are necessary for the onset of ER stress-mediated apoptosis (Nakagawa and Yuan, 2000; Nakagawa et al., 2000; Orrenius, 2004; Rasheva and Domingos, 2009). Intriguingly, the anti-apoptotic signals generated via UPR are as prominent as the pro-apoptotic mechanisms that arise from this process. For instance, the chaperone protein glucose-regulated protein 78 (GRP78) is induced through the unfolded protein response element (UPRE) in the upstream region of its gene and corrects the structure of unfolded proteins upon the expression of XBP1, whose mRNA is spliced by Ire1, in order to promote cell survival (Kishi et al., 2010; Rasheva and Domingos, 2009; Yoshida et al., 2001). Thus, GRP78 residing in the ER is also used as an ER stress marker (Shimoke et al., 2003). We have reported previously that ER stressors in the presence of NGF induce GRP78 (Kishi et al., 2010).

Given that ER stress-mediated apoptosis involves many factors, it will be necessary to optimize the way of detection of these molecules, because the result of the experiment is not matched depending on the miscellaneous conditions or reagents. Strategies to better understand ER stress-mediated apoptosis will greatly assist the development of future therapies for neurodegenerative diseases.

2. INDUCTION OF ER STRESS IN PC12 CELLS

The types of reagents that can be used to induce ER stress in PC12 cells and the evaluation of the molecular dynamics of GRP78 expression via XBP1 expression are presented in this section. The detection of GRP78 has previously been problematic, but the development of an effective commercially available antibody (Santa Cruz Biotechnology, Santa Cruz, CA) has resolved this matter. We describe the use of this reagent in detail in Section 5. In addition, it has been established that specific BH3 only protein (p53-upregulated modulator of apoptosis—PUMA or Bcl-2-interacting mediator—Bim) is upregulated by ER stress. The use of siRNAs is a very powerful way to analyze specific molecular mechanisms, and we utilize siRNAs against PUMA and Bim in our method.

2.1. Culture of PC12 cells and treatment with a chemical inducer of ER stress

Pheochromocytoma 12 (PC12) cells were first established by L.A. Greene in 1976 (Greene and Tischler, 1976) and many subclones with slight phenotypic variations have since been established (Hatanaka, 1981; Ikenaka *et al.*, 1990; Sano *et al.*, 1988). We used the PC12h cell line in our current method, which is one of the established PC12 subclones and is resistant to serum-free (SF) conditions for at least 24 h. We refer to these cells hereafter as PC12 as they are similar to the originally described parental cells in all respects apart from their tolerance to SF conditions.

2.1.1. On the day of plating

The PC12 cells are maintained in Dulbecco's modified eagle's medium (DMEM) supplemented with 5% (v/v) fetal bovine serum (FBS), 5% (v/v) heat-inactivated horse serum (HS), and 0.1% (v/v) penicillin–streptomycin solution (Gibco BRL, Grand Island, NY). For the general analytic usage of PC12 cells, such as the measurement of cell viability or immunocytochemical analysis, the cells should be seeded onto collagen coated 96-well plates (BD Biosciences, San Jose, CA) or 8-well chamber slides (Thermo Fisher Scientific Inc., Rochester, NY) at 1.0–1.5×10^5 cells/cm^2. The collagen was purified from rat tail and diluted in 0.15 M acetic acid. The coating is performed under a 30% (v/v) aqueous ammonia (1–2 ml per glassware) solution in adequate spill-proof glassware for 30 min.

2.1.2. On the day of treatment with ER stress inducers

At 1 day after plating, the culture medium is changed to serum-free DMEM (SF-DMEM) containing 0.1% (v/v) penicillin–streptomycin solution, and either 1 μg/ml tunicamycin (Tm; Sigma, St. Louis, MO), 0.3 μM thapsigargin (Tg; Wako, Osaka, Japan), or 24.5 mM 2-deoxy-D-glucose (2DG; Wako, Osaka, Japan). Under these conditions, the cell viability will be 30–50% of the nontreatment control cells after 24 h. A 100 ng/ml concentration of NGF, which has been exenterated and purified from a mouse submaxillary grand, is added to the medium just described above to yield mixed SF-DMEM with ER stress inducer (Tm, Tg, or 2DG). The medium is then replaced with 20% (v/v) alamarBlue® (TREK Diagnostics Systems, Cleveland, OH)-containing SF-DMEM. The cells are incubated for several hours if the cell viability needs to be measured.

2.1.3. Additionally required materials and reagents for cell culture

For cell growth, 260-ml Nunc easy flasks (Thermo Fisher Scientific Inc., Rochester, NY) were used. For passaging and cell manipulation, 10 ml disposable measuring pipettes (BD Biosciences, San Jose, CA) attached to a silicon dropper are prepared. When the growth medium is changed, about 10–20 passes should be used to properly disperse the cells in the flask.

2.2. Use of XBP1 mRNA and protein to upregulate GRP78 expression

Following treatment with ER stress inducers, the chaperone protein GRP78 is upregulated in the ER in order to facilitate the correct refolding of different proteins (Rasheva and Domingos, 2009). Thus, GRP78 is a useful marker of ER stress. The localized overexpression of GRP78 in the ER can be attributed to the UPR if ER stress occurs due to an accumulation of unfolded proteins. The signal response to the accumulation of unfolded proteins in the ER is mediated by three proteins, PERK, Ire1, and ATF6, all of which have different molecular mechanisms but operate as a sensor network (see Fig. 3.1).

Ire1 is found in the cytosol and contains an RNase domain. The pre-mRNA of the transcription factor, XBP1, is spliced into mature transcripts, which are translated immediately, in the cytosol if ER stress is sensed and phosphorylated by a dimmer form (see Fig. 3.1). This splicing process of the pre-mRNA can be monitored experimentally.

2.2.1. Isolation of total RNA for the detection of pre- and mature XBP1 mRNA

The PC12 cells are plated onto collagen-coated 6 cm dishes (Sumitomo Bakelite Co. Ltd., Tokyo, Japan) at 1.0–1.5×10^5 cells/cm^2 prior to treatment with specific reagents. The following day, the reagents are added to the cells for 8–24 h after which the total RNA is isolated into a 1.5-ml tube (LMS Co. Ltd., Tokyo, Japan) using 1 ml of isogen (Nippongene, Tokyo, Japan). Subsequently, a 200-µl aliquot of 100% chloroform is added to the isolated solution which is then vigorously vortexed. After centrifugation (10,000 rpm, $9100 \times g$) for 10 min at room temperature, the upper aqueous solution is moved to a 1.5-ml fresh tube which contains 500 µl of ice-cold 100% isopropanol. The solution is vigorously shaken and centrifuged at 10,000 rpm ($9100 \times g$) for 15 min at room temperature. A dry pellet (total RNA) is obtained after washing in ice-cold 70% (v/v) ethanol.

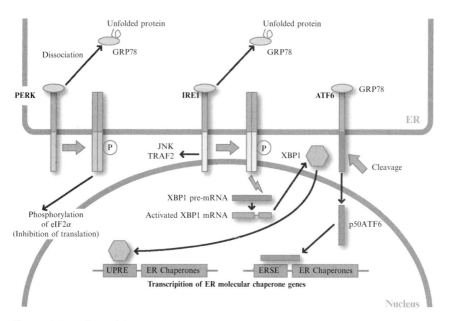

Figure 3.1 *Effects of three sensor proteins on ER stress.* There are at least three sensor proteins on the ER membrane that detect unfolded proteins. GRP78 binds to these sensors and then dissociates when ER stress occurs. At the same time, PERK is dimerized or oligomerized and phosphorylated in the cytoplasm to phosphorylate eIF2α, which inhibits translation. Ire1 is also phosphorylated in the cytoplasm and XBP1 mRNA resides in the cytoplasm is thereby activated and translated by the RNase domain of Ire1. XBP1 protein then binds to the UPRE to induce GRP78. Ire1 also recruits TRAF2, JNK, and caspase-12, etc. to form a proapoptotic protein complex. ATF6 (p90ATF6) is cleaved by SRBP1 and SRBP2 to produce p50ATF6 which induces GRP78 through its binding to a specific *cis*-element in this gene.

2.2.2. Transfection of PC12 cells with a human ATF6 expression plasmid

The expression plasmid pCGN-human-ATF6 harbors a hemagglutinin (HA)-tagged full-length human ATF6 cDNA driven by the cytomegalovirus promoter 10 and was kindly provided by Dr Ron Prywes (Department of Biological Science, Columbia University, NY). We used this construct to analyze the contribution of p50ATF6 to the attenuation of ER stress. PC12 cells are grown to 50–60% confluence in 6 cm dishes and are then transfected with the pCGN-human-ATF6 plasmid by Lipofectamine 2000 (Invitrogen, Carlsbad, CA). The OPTI-MEM (Sigma, St. Louis, MO) mixture used for the transfection was optimized as follows: the quantity of plasmid versus Lipofectamine 2000 per 100 µl OPTI-MEM is 1 µg versus 3 µl. Note that the plasmid and Lipofectamine 2000 mixture in

OPTI-MEM should be made as two separate solutions of 50 μl OPTI-MEM, each in a 1.5-ml sterilized tube. The tubes are mixed and vortexed, incubated for 5–20 min at room temperature and then added to 3 ml of fresh growth medium in a 6 cm dish. The transfected cells are cultured for a further 48 h before exposure to different reagents.

2.2.3. Isolation of p50ATF6 from the nucleus

After treatment with different reagents, the PC12 cells are cultured for 4–8 h and collected. Nuclear fractions are then extracted from whole cell lysates using the NE-PER nuclear and cytoplasmic extraction reagent (Pierce Biotechnology, Rockford, IL). The manufacturer's instructions are available at http://www.piercenet.com/files/0872as8.pdf.

2.2.4. Detection of p50ATF6 by Western blotting

The detection of nuclear p50ATF6 was performed by Western blotting using a HA antibody (Sigma, St. Louis, MO). Cleared cellular lysate is dissolved in a sample buffer after quantification of the protein levels (Protein assay kit, Bio-Rad Laboratories Inc., Hercules, CA). Proteins are resolved by sodium dodecyl sulfate-polyacrylamide gel electrophoresis (SDS-PAGE) and blotted onto a polyvinylidene fluoride (PVDF) membrane (Millipore, Billerica, MA) using a semidry electrophoretic transfer system (Atto, Tokyo, Japan). The membrane is then incubated with a primary ATF6 antibody (Santa Cruz Biotechnology, Santa Cruz, CA; 1:1000 dilution) after blocking with skim milk (Nakarai, Kyoto, Japan) at 4 °C overnight. The PVDF membrane is then incubated with a horse radish peroxidase (HRP)-conjugated secondary antibody (Santa Cruz Biotechnology, Santa Cruz, CA; 1:1000) for 1 h. The bands are detected using enhanced chemiluminescence (SuperSignal West Femto; Pierce, Rockford, IL) and visualized with a light-capture system (Atto, Tokyo, Japan).

2.2.5. Required equipments

- Centrifuge: TOMY, Tokyo, Japan, MX-300
- Vortex: Voltex Mixer, Scientific Industries Inc., Bohemia, NY, G-560
- CO_2 incubator: SANYO, Osaka, Japan, MCO-175
- Clean bench: SANYO, Osaka, Japan, MCV-131BNF
- Power supplier: Atto, Tokyo, Japan, AE-8450
- Semidry electrophoretic transfer system: Atto, Tokyo, Japan, AE-6677
- Light-capture system (for chemiluminescence detection): Atto, Tokyo, Japan, AE-6962N

2.3. Upregulation of BH3-only protein expression

The Bcl-2 family of proteins can be divided into two groups, the pro- and anti-apoptotic proteins (Orrenius, 2004; Rasheva and Domingos, 2009). The pro-apoptotic subfamily belongs to the Bcl-2 homology domain 3 (BH3)-only proteins, which have a potential to negate the effects of the anti-apoptotic Bcl-2 family proteins. These factors can also activate Bcl-2-associated X protein (Bax)/Bcl-2-antagonist/killer (Bak), which initiates the caspase cascade through its binding to the BH domain. It has also been reported that BH3-only proteins induce apoptosis under ER stress and through other pathways (Billen et al., 2008; Orrenius, 2004; Rasheva and Domingos, 2009). A number of recent reports have suggested that ER stress-mediated apoptosis is dependent on the effects of specific BH3-only protein(s). In a study from our laboratory also, we have shown that PUMA and Bim of cell death are involved in ER stress-mediated apoptosis in PC12 cells. A method for analyzing the proapoptotic functions of these two proteins is described below.

2.3.1. On the day of plating

PC12 cells are maintained in DMEM supplemented with 5% (v/v) FBS, 5% (v/v) heat-inactivated HS, and 0.1% (v/v) penicillin–streptomycin solution (Gibco BRL, Grand Island, NY), as described (see also Section 2.1.1). For the detection of PUMA or Bim by Western blotting, cells are seeded onto collagen-coated 6 cm plastic dishes (Sumitomo Bakelite Co. Ltd., Tokyo, Japan) at 0.3–0.5×10^5 cells/cm^2. For fluorescence microscopy, the cells should be seeded onto collagen-coated 8-well chamber slides (Thermo Fisher Scientific Inc., Rochester, NY) at 0.3–0.5×10^5 cells/cm^2.

2.3.2. On the day of treatment with ER stress inducers

The day after plating, the medium is replaced with serum-free DMEM (SF-DMEM) supplemented with 0.1% (v/v) penicillin–streptomycin solution, and 1 μg/ml tunicamycin (Tm; Sigma, St. Louis, MO), 0.3 μM thapsigargin (Tg; Wako, Osaka, Japan), or 24.5 mM 2-deoxy-D-glucose (2DG; Wako, Osaka, Japan), as described in Section 2.1.2.

2.3.3. siRNA

PC12 cells are plated at 0.3–0.5×10^5 cells/cm^2 prior to treatment with reagents onto a 6 cm diameter for Western blotting or an 8-well chamber slide for the terminal deoxynucleotide transferase-mediated deoxyuridine triphosphate nick end-labeling (TUNEL) staining or propidium iodide (PI) staining. The following day, the medium is replaced with fresh serum-containing DMEM and the cells are transfected with commercially synthesized siRNAs (Applied Biosystems by life technologies (ABI), Carlsbad, CA) using Lipofectamine™ RNAiMAX (Invitrogen, Carlsbad, CA).

Briefly, 3 μg pEGFP-N1 or 3 μl pDsREd-N1 (Invitrogen, Carlsbad, CA) are diluted with 100 n*M* siRNA (including negative control siRNA) per sample in 50 μl OPTI-MEM at room temperature for 10 min with gentle vortexing. At the same time, 4 μl Lipofectamine™ RNAiMAX is diluted in 50 μl OPTI-MEM per sample. These two solutions are then mixed and incubated for 10–20 min after vortexing. The resulting 100-μl aliquots of siRNA transfection solution are then added to 6 cm dishes or 8-well chamber slides containing fresh growth medium. After 2 days, the medium should be replaced with SF-DMEM containing fresh siRNA transfection solution in the presence of the desired test reagents (ER stress inducer plus inhibitors and/or NGF, etc.) within 24 h.

Lysates are then made and Western blotting with an anti-PUMA antibody is performed, as described in Section 2.2.4. to confirm the effects of siRNA (Cell Signaling, Beverly, MA) or anti-Bim antibody (Becton Dickinson, Franklin Lakes, NJ). The sequences of the synthesized siRNA (Silencer® Select Predesigned siRNA; ABI, Carlsbad, CA) are as follows:

- Control siRNA No. 1: not made public, but validated as a negative control
- PUMA: 5′-GGAGGGUCAUGUAUAAUCUTT-3′ and 5′-AGAUUA UACAUGACCCUCCAG-3′
- Bim: 5′-CGCAAAUGGUUAUCUUACATT-3′ and 5′-UGUAAGA UAACCAUUUGCGGG-3′

2.3.4. TUNEL and PI staining

To address whether the PC12 cells undergo apoptotic cell death, the nuclei in the cells should be stained using the TUNEL method (DNA fragmentation detection kit, Promega, G3250, Madison, WI) or with PI dye (Sigma, St. Louis, MO). This should be performed in an 8-well chamber plate, not a culture dish or flask, to minimize waste and facilitate fluorescence microscopy (Nikon, Tokyo, Japan). The cells are fixed in 4% (w/v) paraformaldehyde (Wako, Osaka, Japan) in phosphate-buffered saline (PBS) after treatment with the staining reagents and then permeabilized using 0.05% Triton X-100 (Wako, Osaka, Japan) at 0 °C for 5 min. The cells should then be reacted with the terminal deoxynucleotidyl transferase using fluorescent nucleotides at 37 °C for 1 h after treatment with proteinase K. This reaction is terminated by adding 2× saline sodium citrate (SSC). For PI staining, it is not necessary to permeabilize the cells, which are simply incubated with 10 μg/ml PI in PBS at room temperature for 30 min in an 8-well chamber slide after fixation in paraformaldehyde. Apoptotic nuclei can be observed and photographed under a fluorescence microscope at an excitation of 470 nm and emission of 535 nm. If the pDsRed-N1 plasmid is cotransfected with siRNA, TUNEL staining, which can detect apoptotic nuclei via a green signal, has to be performed in order to detect the merged positive image as yellow. Accordingly, PI, which can detect the apoptotic nucleus via a red signal, has to be used if the pEGFP-N1 plasmid is cotransfected to distinguish the staining.

3. Measurement of Caspase-3 Activity

During ER stress-mediated apoptosis, specific caspases and cysteine protease are activated (Rasheva and Domingos, 2009; Shimoke et al., 2004a, b). The critical role of caspase-12 as an ER stress-specific caspase has also been reported (Nakagawa and Yuan, 2000; Nakagawa et al., 2000). This caspase has been shown to be activated by calpain during the process of ER stress-mediated apoptosis, although this evidence remains controversial. Caspase-3, which is involved in most apoptotic pathways, is also activated during ER stress-mediated apoptosis, suggesting that disruptions to mitochondrial function by specific Bcl-2 family proteins and caspase cascades are important in this response (see Fig. 3.2). We outline a protocol for the measurement of caspase-3 activity in the following sections.

3.1. Fluorogenic substrates and their specificity

Caspase-3 has many substrates but a fluorogenic peptide substrate has been generally successful in measuring caspase-3-like activity experimentally (Shimoke and Chiba, 2001; Shimoke et al., 2004a,b).

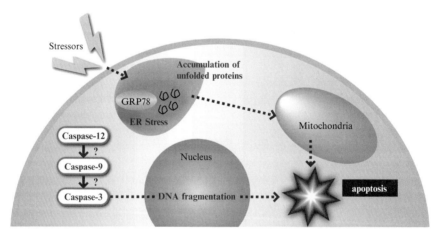

Figure 3.2 *Schema of ER stress-mediated apoptosis.* The accumulation of unfolded proteins in the ER occurs when the cells are exposed to chemical stressors such as tunicamycin, thapsigargin, or 2-deoxy-D-glucose. ER stress-mediated apoptosis occurs through mitochondria-mediated and/or caspase-mediated pathways when ER stress is severe.

3.1.1. On the day of plating

PC12 cells are maintained as described in Section 2.1.1. For the measurement of caspase-3 activity, the cells are seeded onto collagen coated 6-well plates (BD Biosciences, San Jose, CA) at a density of 1.0–1.5 × 10^5 cells/cm^2.

3.1.2. On the day of treatment with ER stress inducers

The day after plating, the medium needs to be replaced, as described in Section 2.1.2.

3.1.3. Preparation of lysates and measurement of caspase-3 activity

The PC12 cells are collected and lysed in a buffer containing 0.1% 3-((3-cholamidopropyl)dimethylammonio)-1-propanesulfonate (CHAPS), 10 mM Hepes–KOH (pH 7.4), 2 mM ethylenediamine tetraacetic acid (EDTA), and 1 mM phenylmethylsulfonyl fluoride (PMSF). After centrifugation (10,000 rpm, 9100×g) for 10 min at 4 °C, a 50-μl aliquot of each supernatant is mixed with 50 μl of 2× ICE buffer containing 20 mM Hepes–KOH (pH 7.4), 20% glycerol (v/v), 2 mM PMSF, 4 mM dithiothreitol (DTT), and 50 μM acetyl-Asp-Glu-Val-Asp-aminomethylcoumarin (Ac-DEVD-MCA; Peptide Institute, Osaka, Japan). This will enable the emission of a fluorescent signal by cleaving peptide-MCA to peptide plus MCA, and incubations should proceed in a 96-well plate (BD Biosciences, San Jose, CA) at 37 °C for 1 h. After the addition of 100 μl of distilled water, fluorescence is detected using a spectrofluorometer (Dainippon Sumitomo Pharma Co. Ltd., Osaka, Japan). The excitation and emission wavelengths are 355 and 460 nm, respectively. (*Note*: it is important to confirm whether the caspase-3 activity of all tested samples increases linearly for at least 2 h).

3.1.4. Required equipments

- Centrifuge: TOMY, Tokyo, Japan, MX-300
- CO_2 incubator: SANYO, Osaka, Japan, MCO-175
- Clean bench: SANYO, Osaka, Japan, MCV-131BNF
- Spectrofluorometer: Dainippon Sumitomo Pharma Co. Ltd., Osaka, Japan, Acent

4. FRAGMENTATION OF CASPASE-12

Caspase-12 was identified as the initiator caspase of ER stress-mediated apoptosis in neurons in a knockout mouse study (Nakagawa *et al.*, 2000). However, the activation mechanism remains unclear unlike other initiator caspases, although the fragment is produced in a likewise manner to other caspase pathways (Nakagawa et al., 2000; Shimoke et al., 2004a,b). Recently, the candidate molecule to activate caspase-12 has been reported (Song et al., 2008). We optimized the detection method for caspase-12 by using a modified Western blotting method for PC12 cells as described below.

4.1. Detection of fragmented caspase-12 through the formation of a modified avidin–biotin antibody complex

Pro-caspase-12 is present on the ER and becomes fragmented when ER stress is initiated. This also causes the activation of caspase-3, a known initiator caspase. Hence, it is believed that caspase-12 is an initiator caspase for ER stress-mediated apoptosis (Nakagawa and Yuan, 2000). On the other hand, it has also been reported that caspase-12 is not necessary for ER stress-mediated apoptosis. One of the likely reasons for this contradiction is that the regulatory mechanisms for caspase-12 activity have yet to be elucidated. We describe below a method to detect fragmented and active caspase-12 by a modified Western blotting protocol (Shimoke et al., 2004a,b, 2005).

4.1.1. On the day of plating
The PC12 cells are maintained as described in Section 2.1.1. To assay for the fragmentation of caspase-12, the cells should be seeded onto collagen-coated 6-cm dishes at a density of $1.0–1.5 \times 10^5$ cells/cm^2.

4.1.2. On the day of treatment with ER stress inducers
The day after plating, the medium is replaced as described in Section 2.1.2.

4.1.3. Modified Western blotting method for the detection of fragmented caspase-12
The cells are lysed in a buffer containing 50 mM Tris–HCl (pH 7.8), 150 mM NaCl, 1% sodium SDS, 1 mM EDTA, 1 mM PMSF, and 2 μg/ml aprotinin. Total lysates (20–30 μg per lane) are resolved by SDS-PAGE and blotted onto PVDF membrane (Millipore, Billerica, MA) using a semidry blotter (Atto, Osaka, Japan). The primary antibody is an anti-caspase-12 (Santa Cruz Biotechnology, Santa Cruz, CA; 1:1000 dilution in Tris-buffered saline–TBS) which is applied in solution to the membrane after blocking with Block-Ace (Dainippon Sumitomo Pharma Co. Ltd., Osaka, Japan) overnight. The membrane is then incubated with a biotin-conjugated secondary antibody at room temperature for 1 h. To enhance the signals from the fragmented bands, an avidin–biotin complex (ABC) kit (VECTASTAIN ABC kit; Vector Laboratories, Burlingame, CA) should

be used. The kit contains a biotin-conjugated secondary antibody and avidin-conjugated HRP which form an avidin–biotin conjugated antibody complex. This complex, (one drop of biotin and 10 ml avidin-conjugated HRP in TBS) must be allowed to form for at least for 1 h before its use with the membrane-attached secondary antibody. The complex is incubated with the membrane at room temperature for 1 h. The bands on the membrane are detected by enhanced chemiluminescence (SuperSignal West Femto, Pierce, Rockford, IL) and visualized using a light-capture system (Atto, Tokyo, Japan). This immunoblotting method was also used for cultured cerebral cortical neurons.

4.1.4. Required reagents and equipments

- Antibodies: caspase-12 (Santa Cruz Biotechnology, Santa Cruz, CA; sc-5627)
  - GAPDH (Millipore, Billerica, MA; MAB375)
  - antimouse antibody conjugated with biotin (included in the ABC kit)
- ABC VECTASTAIN mouse kit: Vector Laboratories, Burlingame, CA, PK-4002
- Centrifuge: TOMY, Tokyo, Japan, MX-300
- Vortex: Voltex Mixer, Scientific Industries Inc., Bohemia, NY, G-560
- CO_2 incubator: SANYO, Osaka, Japan, MCO-175
- Clean bench: SANYO, Osaka, Japan, MCV-131BNF
- Power supplier: Atto, Tokyo, Japan, AE-8450
- Semidry electrophoretic transfer system: Atto, Tokyo, Japan, AE-6677
- Light-capture system (for chemiluminescence detection): Atto, Tokyo, Japan, AE-6962N

5. HYPERINDUCTION OF GRP78 BY NGF

As mentioned in Section 2, GRP78 residing in the ER is upregulated via its UPRE to promote the correct folding of proteins in the genome (see Section 2.2). The key molecule in this process is XBP1 which binds to UPRE (Yoshida et al., 2001). Interestingly, NGF also upregulates GRP78 in all circumstances, except under conditions of ER stress where it cannot do so alone (Kishi et al., 2010). A new interpretation of this molecular regulatory mechanism is thus needed, given that GRP78 is an indicator of ER stress which leads to ER stress-mediated apoptosis that is in turn attenuated by NGF (Fig. 3.3).

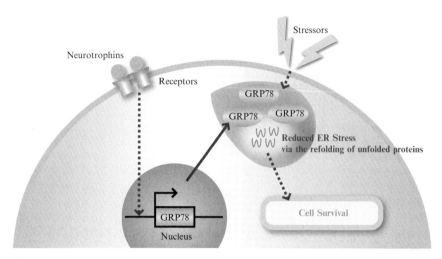

Figure 3.3 Proposed mechanism for the promotion of cell survival by neurotrophins via UPRE. PC12 cells contain an NGF receptor, TrkA, which upon binding of this neurotrophin, transduces signals to the UPRE upstream of the GRP78 gene. The *de novo* synthesized GRP78 localizes in the ER and refolds the incorrect structures of proteins, thus alleviating ER stress and promoting cell survival. This mechanism may be applied to neurons also which harbor members of the Trk receptor family of proteins.

5.1. Detection of GRP78 expression by Northern blotting

Due to the lack of an effective antibody at that time, we had to assay for GRP78 expression using Northern blotting (Shimoke *et al.*, 2003).

5.1.1. Preparation of radioisotope-labeled GRP78 probe

We generated a ^{32}P-GRP78 probe using a random primer DNA labeling kit (TaKaRa, Shiga, Japan) with a slightly modified procedure. A GRP78 DNA fragment of 377 bp was obtained by RT-PCR (using the primers 5′-ACAAAACTGATTCCGAGGAACA-3′ and 5′-ATGCGCTCTTTG AGCTTTTTGT-3′) from tunicamycin-treated PC12 cells (see also Section 2.2.1) and subcloned into the pGEM-T Easy vector. The fragment was then produced by *Eco*RI digestion and purified from an agarose gel with a Geneclean II kit (Q-Biogene, Solon, OH). The isolated and purified DNA fragment is then suspended in 14 µl of aqueous solution in a 1.5-ml tube and denatured by heating at 95 °C for 3 min and cooling on ice for 5 min. To the solution is then added 2.5 µl 10× buffer, 2.5 µl dNTP mixture, 1.85 MBq [α-^{32}P] dCTP, and subsequently, 1 µl Exo-free Klenow fragment. The reaction then proceeds at 37 °C for 10 min and the enzyme is inactivated at 65 °C for 5 min. Just prior to use, the labeled probe must be heated at 95 °C for 3 min and cooled on ice. A control ^{32}P-28S RNA probe is generated as described above using a 28S RNA fragment.

5.1.2. Northern blotting procedure

PC12 cells are seeded onto collagen-coated 6-cm dishes and the medium is replaced with serum-free DMEM supplemented with the reagents described in Sections 4.1.1 and 4.1.2. Total RNA is then isolated with isogen (Nippongene, Tokyo, Japan), and 20 µg of this preparation is electrophoresed on a 1% (w/v) agarose/2 M formaldehyde gel and transferred onto a nitrocellulose membrane (NEN Life Science Products, Inc., Boston, MA) overnight. Hybridization should be performed overnight at 65 °C by slow rotation in 250-ml glass tubes containing a 50-ml solution containing 10% (v/v) Denhardt's solution (Wako, Japan), 0.5% (w/v) SDS, 6× SSC, 100 µg/ml of salmon sperm DNA, and the radiolabeled ^{32}P-GRP78 probe (0.4 MBq). The membrane is then washed twice with 25 ml of 2× SSC containing 0.5% SDS at 65 °C for 15 min in the incubator and specific bands are visualized using BAS2000 (Fuji, Tokyo, Japan). To estimate the RNA content, the membrane is rehybridized with a radiolabeled ^{32}P-28S rRNA probe in an identical manner to that described above.

5.1.3. Required reagents and equipments

- Cell culture: reagents (Section 2)
- Kit: Random primer DNA labeling kit, TaKaRa, Shiga, Japan, 6045
  - GENECLEAN II kit, Q-Biogene, Solon, OH, 1001-400
- Hybridization: DNA OVEN, KURBO Industries Ltd., Osaka, Japan, MI-100
  - 250-ml glass tube, KURBO Industries Ltd., Osaka, Japan, MI-BS

5.2. Detection of GRP78 expression by Western blotting

There are issues with the effectiveness of many of the available GRP78 antibodies at present, but we describe our Western blotting procedure for GRP78 below using a commercially available antibody that has been found to be useful.

5.2.1. Immunoblotting procedure for detection of GRP78 protein expression

PC12 cell culture and treatments with specific reagents are described in Sections 2, 3, and 4. PC12 cells are seeded onto 6-cm dishes and are lysed with buffer containing 50 mM Tris–HCl (pH7.8), 150 mM NaCl, 1% SDS, 1 mM EDTA, 1 mM PMSF, and 2 µg/mL of aprotinin. The supernatant is used for the analysis after the elimination of debris by centrifugation (10,000 rpm, 9100×g) for 15 min at room temperature. The lysates (20 µg of protein per lane) are then subjected to SDS-PAGE in Laemmli buffer after a protein assay to normalize the loading in each lane, and the resolved proteins are blotted onto a PVDF membrane using a semidry blotter (Atto, Tokyo, Japan). The GRP78 primary antibody was commercially sourced from Santa Cruz Biotechnology (sc-13968; Santa Cruz, CA) and incubated at a dilution of 1:1000, after blocking treatment with skim milk (Nakarai, Kyoto, Japan), at 4 °C overnight. The membrane was then incubated with a HRP-conjugated secondary antibody (Santa Cruz, Santa Cruz, CA) for 1 h at room temperature. The bands can then be detected by enhanced chemiluminescence (SuperSignal West Femto, Pierce, Rockford, IL) visualized with a light-capture system (Atto, Tokyo, Japan). Note that the ABC method to enhance the signal as described in Section 4.1.3 does not apply for this detection method.

5.2.2. Required reagents and equipments

- Antibodies: GRP78 (Santa Cruz, CA, sc-13968)
    - GAPDH (Millipore, Billerica, MA, MAB375)
    - Anti-rabbit antibody conjugated with HRP, Santa Cruz, CA, sc-2004
    - Anti-mouse antibody conjugated with HRP, Santa Cruz, CA, sc-2005
- Centrifuge: TOMY, Tokyo, Japan, MX-300
- Vortex: Voltex Mixer, Scientific Industries Inc., Bohemia, NY, G-560
- CO_2 incubator: SANYO, Osaka, Japan, MCO-175
- Clean bench: SANYO, Osaka, Japan, MCV-131BNF
- Power supplier: Atto, Tokyo, Japan, AE-8450
- Semidry electrophoretic transfer system: Atto, Tokyo, Japan, AE-6677
- Light-capture system (for chemiluminescence detection): Atto, Tokyo, Japan, AE-6962N

ACKNOWLEDGMENTS

We thank Dr. Ron Prywes (Columbia University, NY) for providing the human ATF6 coding plasmid. We also thank Dr Soichiro Kishi, Dr. Masashi Yamada and Dr. Kiyotoshi Sekiguchi (Osaka University, Japan) for their assistance with the experiments. This study was supported, in part, by grants-in-aid for scientific research (KAKENHI 21570152) and a "Strategic Project to Support the Formation of Research Bases at Private Universities (SENRYAKU)" (2008–2012) from MEXT (Ministry of Education, Culture, Sports, Science and Technology of Japan). This study was performed, in part, under the Cooperative Research Program of Institute for Protein Research, Osaka University in Japan.

REFERENCES

Barbacid, M. (1995). Neurotrophic factors and their receptors. *Curr. Opin. Cell Biol.* **7**, 148–155.
Billen, L. P., Shamas-Din, A., and Andrews, D. W. (2008). Bid: A Bax-like BH3 protein. *Oncogene* **27**(Suppl. 1), S93–S104.

Borasio, G. D., Markus, A., Wittinghofer, A., Barde, Y. A., and Heumann, R. (1993). Involvement of ras p21 in neurotrophin-induced response of sensory, but not sympathetic neurons. *J. Cell Biol.* **121,** 665–672.

Gardai, S. J., Hildeman, D. A., Frankel, S. K., Whitlock, B. B., Frasch, S. C., Borregaard, N., Marrack, P., Bratton, D. L., and Henson, P. M. (2004). Phosphorylation of Bax Ser184 by Akt regulates its activity and apoptosis in neutrophils. *J. Biol. Chem.* **279,** 21085–21095. .

Greene, L. A., and Tischler, A. S. (1976). Establishment of a noradrenergic clonal line of rat adrenal pheochromocytoma cells which respond to nerve growth factor. *Proc. Natl. Acad. Sci. USA* **73,** 2424–2428.

Hatanaka, H. (1981). Nerve growth factor-mediated stimulation of tyrosine hydroxylase activity in a clonal rat pheochromocytoma cell line. *Brain Res.* **222,** 225–233.

Ikenaka, K., Nakahira, K., Takayama, C., Wada, K., Hatanaka, H., and Mikoshiba, K. (1990). Nerve growth factor rapidly induces expression of the 68-kDa neurofilament gene by posttranscriptional modification in PC12h-R cells. *J. Biol. Chem.* **265,** 19782–19785.

Imaizumi, K., Miyoshi, K., Katayama, T., Yoneda, T., Taniguchi, M., Kudo, T., and Tohyama, M. (2001). The unfolded protein response and Alzheimer's disease. *Biochim. Biophys. Acta* **1536,** 85–96.

Kim, S. J., Winter, K., Nian, C., Tsuneoka, M., Koda, Y., and McIntosh, C. H. (2005). Glucose-dependent insulinotropic polypeptide (GIP) stimulation of pancreatic beta-cell survival is dependent upon phosphatidylinositol 3-kinase (PI3K)/protein kinase B (PKB) signaling, inactivation of the forkhead transcription factor Foxo1, and down-regulation of bax expression. *J. Biol. Chem.* **280,** 22297–22307.

Kishi, S., Shimoke, K., Nakatani, Y., Shimada, T., Okumura, N., Nagai, K., Shin-Ya, K., and Ikeuchi, T. (2010). Nerve growth factor attenuates 2-deoxy-d-glucose-triggered endoplasmic reticulum stress-mediated apoptosis via enhanced expression of GRP78. *Neurosci. Res.* **66,** 14–21.

Marciniak, S. J., and Ron, D. (2006). Endoplasmic reticulum stress signaling in disease. *Physiol. Rev.* **86,** 1133–1149.

Matsumoto, T., Numakawa, T., Yokomaku, D., Adachi, N., Yamagishi, S., Numakawa, Y., Kunugi, H., and Taguchi, T. (2006). Brain-derived neurotrophic factor-induced potentiation of glutamate and GABA release: Different dependency on signaling pathways and neuronal activity. *Mol. Cell. Neurosci.* **31,** 70–84.

Nakagawa, T., and Yuan, J. (2000). Cross-talk between two cysteine protease families. Activation of caspase-12 by calpain in apoptosis. *J. Cell Biol.* **150,** 887–894.

Nakagawa, T., Zhu, H., Morishima, N., Li, E., Xu, J., Yankner, B. A., and Yuan, J. (2000). Caspase-12 mediates endoplasmic-reticulum-specific apoptosis and cytotoxicity by amyloid-beta. *Nature* **403,** 98–103.

Nakayama, H., Hamada, M., Fujikake, N., Nagai, Y., Zhao, J., Hatano, O., Shimoke, K., Isosaki, M., Yoshizumi, M., and Ikeuchi, T. (2008). ER stress is the initial response to polyglutamine toxicity in PC12 cells. *Biochem. Biophys. Res. Commun.* **377,** 550–555.

Nonomura, T., Kubo, T., Oka, T., Shimoke, K., Yamada, M., Enokido, Y., and Hatanaka, H. (1996). Signaling pathways and survival effects of BDNF and NT-3 on cultured cerebellar granule cells. *Brain Res. Dev. Brain Res.* **97,** 42–50.

Orrenius, S. (2004). Mitochondrial regulation of apoptotic cell death. *Toxicol. Lett.* **149,** 19–23.

Price, R. D., Yamaji, T., and Matsuoka, N. (2003). FK506 potentiates NGF-induced neurite outgrowth via the Ras/Raf/MAP kinase pathway. *Br. J. Pharmacol.* **140,** 825–829.

Pyle, A. D., Lock, L. F., and Donovan, P. J. (2006). Neurotrophins mediate human embryonic stem cell survival. *Nat. Biotechnol.* **24,** 344–350.

Rasheva, V. I., and Domingos, P. M. (2009). Cellular responses to endoplasmic reticulum stress and apoptosis. *Apoptosis* **14,** 996–1007.

Sano, M., Kato, K., Totsuka, T., and Katoh-Semba, R. (1988). A convenient bioassay for NGF using a new subline of PC12 pheochromocytoma cells (PC12D). *Brain Res.* **459,** 404–406.

Shimoke, K., Kubo, T., Numakawa, T., Abiru, Y., Enokido, Y., Takei, N., Ikeuchi, T., and Hatanaka, H. (1997). Involvement of phosphatidylinositol-3 kinase in prevention of low K(+)-induced apoptosis of cerebellar granule neurons. *Brain Res. Dev. Brain Res.* **101,** 197–206.

Shimoke, K., and Chiba, H. (2001). Nerve growth factor prevents 1-methyl-4-phenyl-1, 2, 3, 6-tetrahydropyridine-induced cell death via the Akt pathway by suppressing caspase-3-like activity using PC12 cells: Relevance to therapeutical application for Parkinson's disease. *J. Neurosci. Res.* **63,** 402–409.

Shimoke, K., Kudo, M., and Ikeuchi, T. (2003). MPTP-induced reactive oxygen species promote cell death through a gradual activation of caspase-3 without expression of GRP78/Bip as a preventive measure against ER stress in PC12 cells. *Life Sci.* **73,** 581–593.

Shimoke, K., Amano, H., Kishi, S., Uchida, H., Kudo, M., and Ikeuchi, T. (2004a). Nerve growth factor attenuates endoplasmic reticulum stress-mediated apoptosis via suppression of caspase-12 activity. *J. Biochem.* **135,** 439–446.

Shimoke, K., Utsumi, T., Kishi, S., Nishimura, M., Sasaya, H., Kudo, M., and Ikeuchi, T. (2004b). Prevention of endoplasmic reticulum stress-induced cell death by brain-derived neurotrophic factor in cultured cerebral cortical neurons. *Brain Res.* **1028,** 105–111.

Shimoke, K., Kishi, S., Utsumi, T., Shimamura, Y., Sasaya, H., Oikawa, T., Uesato, S., and Ikeuchi, T. (2005). NGF-induced phosphatidylinositol 3-kinase signaling pathway prevents thapsigargin-triggered ER stress-mediated apoptosis in PC12 cells. *Neurosci. Lett.* **389,** 124–128.

Song, S., Lee, H., Kam, T. I., Tai, M. L., Lee, J. Y., Noh, J. Y., Shim, S. M., Seo, S. J., Kong, Y. Y., Nakagawa, T., Chung, C. W., Choi, D. Y., *et al.* (2008). E2-25K/Hip-2 regulates caspase-12 in ER stress-mediated Abeta neurotoxicity. *J. Cell Biol.* **182,** 675–684.

Suter, U., Angst, C., Tien, C. L., Drinkwater, C. C., Lindsay, R. M., and Shooter, E. M. (1992). NGF/BDNF chimeric proteins: Analysis of neurotrophin specificity by homolog-scanning mutagenesis. *J. Neurosci.* **12,** 306–318.

Szegezdi, E., Macdonald, D. C., Ni Chonghaile, T., Gupta, S., and Samali, A. (2009). Bcl-2 family on guard at the ER. *Am. J. Physiol. Cell Physiol.* **296,** C941–C953.

Yoshida, H., Matsui, T., Yamamoto, A., Okada, T., and Mori, K. (2001). XBP1 mRNA is induced by ATF6 and spliced by IRE1 in response to ER stress to produce a highly active transcription factor. *Cell* **107,** 881–891.

CHAPTER FOUR

MEASURING ER STRESS AND THE UNFOLDED PROTEIN RESPONSE USING MAMMALIAN TISSUE CULTURE SYSTEM

Christine M. Oslowski* *and* Fumihiko Urano*,[†]

Contents

1. Introduction — 72
2. ER Stress and the UPR — 72
    2.1. ER and ER stress — 73
    2.2. The unfolded protein response — 73
    2.3. IRE1, PERK, and ATF6 signaling pathways — 74
3. Cell Culture System and ER Stress Induction — 76
    3.1. Mammalian cells as a model system for studying ER stress and the UPR — 76
    3.2. Examples of cell lines and primary cells that are used as a model — 76
    3.3. ER stress inducers — 77
4. Measuring ER Stress — 78
    4.1. ER dilation (EM) — 79
    4.2. Real-time redox measurements during ER stress — 79
5. Studying the UPR Activation — 79
    5.1. Methods for measuring IRE1α activation — 79
    5.2. Methods for measuring PERK activation — 83
    5.3. Methods for measuring ATF6α activation — 83
6. Studying UPR Downstream Markers and Responses — 84
    6.1. Immunostaining and immunofluorescence for downstream markers of the UPR — 84
    6.2. Measuring transcriptional activation of the UPR — 85
    6.3. Measuring translational attenuation of the UPR — 86
    6.4. Measuring ERAD and protein stability — 86

* Program in Gene Function and Expression, University of Massachusetts, Medical School, Worcester, Massachusetts, USA
[†] Program in Molecular Medicine, University of Massachusetts, Medical School, Worcester, Massachusetts, USA

6.5. Measuring mRNA degradation   87
6.6. Measuring ER stress-mediated apoptosis   87
Acknowledgments   88
References   88

Abstract

The endoplasmic reticulum (ER) functions to properly fold and process secreted and transmembrane proteins. Environmental and genetic factors that disrupt ER function cause an accumulation of misfolded and unfolded proteins in the ER lumen, a condition termed ER stress. ER stress activates a signaling network called the Unfolded Protein Response (UPR) to alleviate this stress and restore ER homeostasis, promoting cell survival and adaptation. However, under unresolvable ER stress conditions, the UPR promotes apoptosis. Here, we discuss the current methods to measure ER stress levels, UPR activation, and subsequent pathways in mammalian cells. These methods will assist us in understanding the UPR and its contribution to ER stress-related disorders such as diabetes and neurodegeneration.

1. INTRODUCTION

The endoplasmic reticulum (ER) is a multifunctional organelle essential for the synthesis, folding, and processing of secretory and transmembrane proteins. For proteins to fold properly, a balance between the ER protein load and the folding capacity to process this load must be established. However, physiological and pathological stimuli can disrupt this ER homeostasis, resulting in an accumulation of misfolded and unfolded proteins, a condition known as ER stress. ER stress activates a complex signaling network, referred to as the unfolded protein response (UPR), to reduce ER stress and restore homeostasis. However, if the UPR fails to reestablish the ER to normality, ER stress causes cell dysfunction and death (Kim *et al.*, 2008). Recent evidence further indicates that ER stress-mediated cell dysfunction and death is involved in the pathogenesis of human chronic disorders, including diabetes and neurodegeneration (Ron and Walter, 2007). This chapter discusses the methods for measuring and quantifying ER stress levels, UPR activation, and the subsequent downstream outcomes. We will mainly focus on the tissue culture system. Studying ER stress and the UPR will help us understand the pathophysiology and develop novel therapeutic modalities for ER stress-related disorders.

2. ER Stress and the UPR

2.1. ER and ER stress

The ER has an important role in the folding and maturation of newly synthesized secretory and transmembrane proteins. To ensure proper protein folding, the ER lumen maintains a unique environment to establish a balance between the ER protein load and the capacity to handle this load. This ER homeostasis can be perturbed by physiological and pathological insults such as high protein demand, viral infections, environmental toxins, inflammatory cytokines, and mutant protein expression, resulting in an accumulation of misfolded and unfolded proteins in the ER lumen, a condition termed as ER stress.

2.2. The unfolded protein response

The adaptive response to ER stress is the UPR (Fig. 4.1). The UPR is initiated by three ER transmembrane proteins: Inositol Requiring 1 (IRE1), PKR-like ER kinase (PERK), and Activating Transcription Factor 6 (ATF6). During unstressed conditions, the ER chaperone, immunoglobin binding protein (BiP) binds to the luminal domains of these master regulators, keeping them inactive. Upon ER stress, BiP dissociates from these sensors, resulting in their activation.

The activated UPR regulates downstream effectors with the following three distinct functions: adaptive response, feedback control, and cell fate regulation (Fig. 4.1; Oslowski and Urano, 2010). The UPR adaptive response includes upregulation of molecular chaperones and protein-processing enzymes to increase folding and handling efficiency, translational attenuation to reduce ER workload and prevent further accumulation of unfolded proteins, and an increase in ER-associated protein degradation (ERAD) and autophagy components to promote clearance of unwanted proteins. Feedback control involves the negative regulation of UPR activation as ER homeostasis is being reestablished to prevent harmful hyperactivation. Cell fate regulation by the UPR plays an important role in the pathogenesis of ER stress-related disorder. Our current model is that the UPR directly regulates both apoptotic and antiapoptotic outputs, acting as a binary switch between the life and death of ER-stressed cells (Oslowski and Urano, 2010). When the cell encounters ER stress that the UPR can mitigate, the cell will survive and is primed for future ER stress insults. However, during unresolvable ER stress conditions, the UPR fails to reduce ER stress and restore homeostasis promoting cell death.

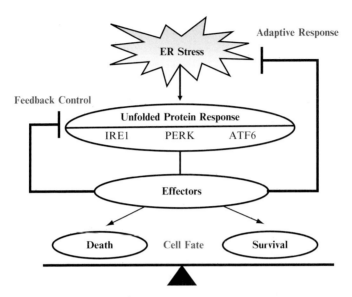

Figure 4.1 *Response categories of the unfolded protein response (UPR)*. Three ER transmembrane proteins, IRE1, PERK, and ATF6, sense ER stress in the ER lumen and become activated, regulating a cascade of signaling pathways collectively termed the UPR. The UPR has three functions: adaptive response, feedback control, and cell fate. Under the adaptive response, the UPR aims to reduce ER stress and restore ER homeostasis. If the UPR is successful, the UPR signaling pathways are turned off by feedback mechanisms. The UPR also regulates both survival and death factors that govern whether the cell will live or not depending on the severity of the ER stress condition.

2.3. IRE1, PERK, and ATF6 signaling pathways

As mentioned previously, the UPR is regulated by the three master regulators, IRE1, PERK, and ATF6 (Fig. 4.2).

IRE1, a type I ER transmembrane kinase, senses ER stress by its N-terminal luminal domain (Urano *et al.*, 2000a). Upon sensing the presence of unfolded or misfolded proteins, IRE1 dimerizes and autophosphorylates to become active. There are two isoforms of IRE1: IRE1α and IRE1β. IRE1α is expressed in all cell types and has been extensively studied. Activated IRE1α splices X-box binding protein 1 (XBP-1) mRNA (Calfon *et al.*, 2002; Shen *et al.*, 2001; Yoshida *et al.*, 2001). Spliced XBP-1 mRNA encodes a basic leucine zipper (b-ZIP) transcription factor that upregulates UPR target genes, including genes that function in ERAD such as ER-degradation-enhancing-α-mannidose-like protein (EDEM; Yoshida *et al.*, 2003), as well as genes that function in folding proteins such as protein disulfide isomerase (PDI; Lee *et al.*, 2003a). High levels of chronic ER stress can lead to the recruitment of TNF-receptor-associated factor 2 (TRAF2) by IRE1 and the activation of apoptosis-signaling-kinase 1 (ASK1).

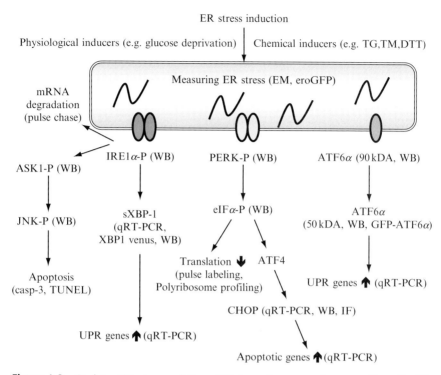

Figure 4.2 *Studying ER stress and the unfolded protein response in mammalian cells.* ER stress can be induced by several chemical and physiological inducers. Actual ER stress has been difficult to measure directly. Currently, observing ER distension by electron microscopy (EM) and measuring oxidative protein folding (eroGFP) are available. The activation of the UPR master regulators has also been challenging but could be attempted by detecting IRE1α and PERK phosphorylation, and ATF6α cleavage by specific antibodies using Western blot (WB). ATF6α translocation can be monitored by fluorescence microscopy of GFP–ATF6α. The downstream outputs of the UPR master regulars are readily measurable. IRE1α splices XBP-1 mRNA which can be detected by quantitative real-time PCR (qRT-PCR), XBP-1 venus, and WB. IRE1α also mediates degradation of ER-localized mRNAs, which can be measured by pulse-chase assays. PERK phosphorylates eIF2α, which can be detected by specific antibodies. Phosphorylated eIF2α triggers global mRNA translation attenuation, which can be measured by standard methods such as pulse labeling and polyribosome profiling. Activated ATF6α is a transcription factor and regulation of its downstream target genes can be measured by qRT-PCR. The UPR in general regulates several transcription factors and in turn, their transcriptional targets can be measured by qRT-PCR and luciferase assays. The UPR also induces expression of ERAD components as well as survival and death components. ERAD and apoptosis can be measured by standard methods.

Activated ASK1 activates c-Jun N-terminal protein kinase (JNK), which in turn plays a role in apoptosis by regulating the BCL2 family of proteins (Nishitoh *et al.*, 1998, 2002; Urano *et al.*, 2000b).

PERK is also a type I ER transmembrane kinase. Similar to IRE1α, when activated by ER stress, PERK oligomerizes, autophosphorylates, and then directly phosphorylates Ser51 on the α subunit of eukaryotic initiation factor 2 (eIF2α; Harding et al., 1999). Phosphorylated eIF2α prevents the formation of ribosomal initiation complexes, leading to global mRNA translational attenuation. This reduction in ER workload protects cells from ER stress-mediated apoptosis (Harding et al., 2000a). Meanwhile, some mRNAs require eIF2α phosphorylation for translation such as the mRNA encoding ATF4. ATF4 is a b-ZIP transcription factor that regulates several UPR target genes, including those involved in ER stress-mediated apoptosis such as C/EBP homologous protein (CHOP; Harding et al., 2000b).

A third regulator of ER stress signaling is the type II ER transmembrane transcription factor, ATF6 (Yoshida et al., 1998). ATF6 has two isoforms, ATF6α and ATF6β. ATF6α has been extensively studied in the context of ER stress. Upon ER stress conditions, ATF6α transits to the Golgi where it is cleaved by site 1 (S1) and site 2 (S2) proteases, generating an activated b-ZIP factor (Ye et al., 2000). This processed form of ATF6α translocates to the nucleus to activate UPR genes involved in protein folding, processing, and degradation (Haze et al., 1999; Yoshida et al., 2000).

3. Cell Culture System and ER Stress Induction

3.1. Mammalian cells as a model system for studying ER stress and the UPR

Protein folding in the ER is essential to the survival of individual cells, explaining the evolution of the UPR in unicellular organisms such as yeast. But as secretion is the basis of multicellularity, ER protein folding homeostasis powerfully impacts the physiology of mammals. Dysregulation of ER homeostasis can cause chronic diseases in humans. Therefore, it is important to study ER stress and the UPR using mammalian cells to understand the UPR and ER stress-related diseases.

3.2. Examples of cell lines and primary cells that are used as a model

There are several mammalian cells lines that demonstrate a response to commonly applied ER stress inducers and the subsequent UPR activation. Mouse embryonic fibroblasts (MEFs) are powerful tools for studying the UPR because MEFs from Ire1α, Ire1β, Perk, ATf6α, ATF6β, Xbp-1, Atf4, and Gadd34 knockout mice are available (Bertolotti et al., 2001; Harding et al., 2001, 2003; Novoa et al., 2003; Reimold et al., 2000; Reimold et al., 2001; Urano et al., 2000b; Wu et al., 2007; Yamamoto et al., 2007). The UPR is particularly important for maintaining ER homeostasis in professional secretory cells such as pancreatic β cells and plasma cells. The mouse β cell line, MIN6, and the rat β cell lines, INS-1 and INS-1 832/13, are often used to study ER stress and the UPR in the context of the β cell (Asfari et al., 1992; Hohmeier et al., 2000; Miyazaki et al., 1990). Human, mouse, and rat primary islets are also great tools available. Multiple myeloma is a cancer derived from plasma cells. J558 is a multiple myloma cell line and has been shown to be useful to develop a therapy targeting the XBP-1 pathways (Lee et al., 2003b). It is important to note that each cell type responds to ER stress and activates the UPR in a unique manner.

3.3. ER stress inducers

3.3.1. Pharmaceutical ER stress inducers

There are several available chemicals to induce ER stress and activate the UPR in a tissue culture system, including tunicamycin, thapsigargin, Brefedin A, dithiothreitol (DTT), and MG132. The concentration and duration of treatment in which ER stress is induced by these compounds should be determined for each particular system. Typically, only a few hours are required to induce ER stress, and long exposures often induce ER stress-mediated cell death. Tunicamycin is an inhibitor of the UDP-N-acetylglucosamine-dolichol phosphate N-acetylglucosamine-1-phosphate transferase (GPT), therefore blocking the initial step of glycoprotein biosynthesis in the ER. Thus, treatment of tunicamycin causes accumulation of unfolded glycoproteins in the ER, leading to ER stress. In many cell types, ER stress can be induced by treating cells with 2.5–5 μg/ml of tunicamycin for 5 h. Thapsigargin is a specific inhibitor of the sarcoplasmic/endoplasmic reticulum Ca2+-ATPase (SERCA). Treatment with thapsigargin results in a decrease in ER calcium levels. When calcium levels are lowered in the ER, the calcium-dependent ER chaperones, such as calnexin, lose their chaperone activity, leading to the accumulation of unfolded proteins. ER stress can be induced by treating cells with 0.1–1 μM of thapsigargin for 5 h. Brefeldin A inhibits transport of proteins from the ER to the Golgi and induces retrograde protein transport from the Golgi apparatus to the ER. This leads to the accumulation of unfolded proteins in the ER. DTT is a strong reducing agent and blocks disulfide-bond formation, quickly leading to ER stress within minutes. As DTT also blocks disulfide-bond formation of newly synthesized proteins in the cytosol, it is not a specific ER stress inducer. MG132, is a specific and cell-permeable proteasome inhibitor. Consequently, MG132 blocks ERAD and causes misfolded proteins to accumulate in the ER. Thus, MG132 induces ER stress indirectly.

3.3.2. Physiological ER stress inducers

Physiological perturbants are commonly used to induce mild ER stress in a tissue culture system. As mentioned with pharmaceutical inducers, the amount and time of exposure to these inducers should be determined in a given system. Glucose deprivation blocks N-linked glycozylation and reduces cellular ATP levels, leading to ER stress in many cell types. ER stress can be induced by treating cells with glucose-free media for 24–48 h. Glucose deprivation is not a strong inducer of cell death.

ER stress and the UPR are unique from cell to cell. Therefore, there are specific ER stress inducers for a given cell type, each activating the UPR in a distinct manner. ER stress and the UPR have been extensively studied in pancreatic β cells. Some specific β cell ER stress inducers include acute and chronic high glucose, cytokines, free fatty acids, and overexpression of mutant insulin-2. Both acute (1–3 h) and chronic (24 h) high glucose (16.7 mM) induce IRE1α phosphorylation in pancreatic β cells. However, acute high glucose does not induce PERK activation. Inflammatory cytokines such as IL-1β and IFN-γ can cause ER stress in β cell lines and primary islets. In rat insulinoma, INS-1E cells and rat primary islets, ER stress can be triggered by IL-1β (50 units/ml) alone or IL-1β (50 units/ml) + IFN-γ (0.036 μg/ml) in 24 h (Cardozo et al., 2005) activating the IRE1 and PERK arms of the UPR. Saturated free fatty acid, palmitate (0.25–5 mM) can cause ER stress and activate the UPR within at least 4 h in INS-1E rat insulinoma cells and MIN6 mouse insulinoma cells as well as in primary rat and human islets (Kharroubi et al., 2004). Finally, β cells specialize in the synthesis, processing, and secretion of insulin. Thus, overexpression of misfolded mutant insulin-2 (C96Y) in β cell lines and pancreatic islets from Akita mice expressing this mutant insulin demonstrate ER stress and activation of the UPR (Oyadomari et al., 2002).

It has been shown that free cholesterol causes ER stress in macrophages and preferentially activates the CHOP branch of the UPR. Acetyl-LDL (100 μg/ml) plus the acyl-CoA:cholesterol acyltransferase inhibitor 58035 induces ER stress in macrophages in 5–10 h (Feng et al., 2003). This method is used to study the role of ER stress-mediated macrophage death during the progression of atherosclerosis (Feng et al., 2003).

In vascular endothelial cells, homocysteine can cause ER stress. In human umbilical vein endothelial cells (HUVEC), GRP78 (BiP) mRNA expression is induced by 1–5 mM of homocysteine in 4–8 h (Outinen et al., 1999).

4. Measuring ER Stress

Commonly, we add an ER stress inducer and measure the activation of the UPR and the consequent downstream responses. However, these results do not directly reflect the accumulation of misfolded and unfolded protein within the ER lumen. It has been challenging to directly measure ER stress levels in cells. Here, we discuss at least two methods that directly measure ER stress.

4.1. ER dilation (EM)

Upon ER stress, the ER lumen is remarkably enlarged in cells and tissues, which can be detected by electron microscopy (Akiyama *et al.*, 2009; Riggs *et al.*, 2005; Wang *et al.*, 1999). This method has been used often to detect ER stress in pancreatic β cells.

4.2. Real-time redox measurements during ER stress

The ER maintains an oxidizing environment to promote disulfide-bond formation in newly synthesized proteins (Travers *et al.*, 2000; Tu and Weissman, 2004). An increase in ER protein load could overwhelm oxidative folding enzymes, preventing proper disulfide formation and therefore inducing ER stress. Feroz Papa's group recently developed a method to monitor the redox state of GFP to reflect ER stress (Merksamer *et al.*, 2008). This reporter named "eroGFP (ER-targeted redox-sensitive GFP)" has been designed to change fluorescence at two maximas, 400 and 490 nm, upon disulfide formation between an engineered cysteine pair. As eroGFP becomes reduced, excitation at 490 nm increases and decreases at 400 nm. The ratio of the fluorescence measured at 490 nm versus 400 nm reports ER redox status in cells. Currently, this method is only available in yeast cells. We are collaborating with Papa's group to adapt this system in mammalian cells (Ishigaki, Marksamer, Lu, Papa, and Urano, unpublished).

5. Studying the UPR Activation

The ability to measure IRE1α and PERK phosphorylation, and ATF6α cleavage would be ideal to determine UPR activation levels. However, endogenous expression levels of these molecules are low and hard to detect with available commercial antibodies. Thus, alternatively, we suggest measuring expression and activation levels of downstream components regulated by these master regulators to determine UPR activation.

5.1. Methods for measuring IRE1α activation

IRE1α is an ER transmembrane serine/threonine kinase undergoing autophosphorylation upon ER stress. Thus, the best way to measure activation levels of IRE1α is to measure its phosphorylation levels. Our group developed anti-phospho-IRE1α-specific antibody from bulk antiserum by affinity purification, followed by adsorption against the nonphospho analog column peptide (Open biosystems, Huntsville, AL). The peptide sequence for generating the antibody was CVGRH (pS) FSRRSG. This phosphopeptide was synthesized, multi-link-conjugated to KLH, and used to immunize 2SPF rabbits. Rabbit anti-total-IRE1α antibody (B9134) was generated using a peptide, EGWIAPEMLSEDCK. Samples are prepared by lysing cells with ice-cold M-PER buffer (PIERCE, Rockford, IL) containing protease inhibitors, incubated on ice for 15 min, and centrifuged at $13,000 \times g$ for 15 min at 4 °C. The supernatant was collected and the total protein concentrations were measured. Ten micrograms of proteins are prepared with a sample buffer and heated for 5 min at 95 °C. Denatured proteins were separated using 4–20% linear gradient SDS-PAGE (Bio Rad, Hercules, CA) and transferred onto a PVDF membrane. Nonspecific binding sites are blocked with 5% milk in 1× TBS + 0.1% Tween-20 (TBST) for 1 h at room temperature, followed by overnight incubation at 4 °C with diluted anti-IRE1α antibody (1:1000) in 5% milk–TBST. The next day, blots are incubated with diluted HRP-conjugated antirabbit antibody (1:3000, Cell Signaling) in 5% BSA–TBST for 1 h. Membrane-bound antibodies are detected by ECL Western Blotting Substrate (Pierce). All of our immunoblots are performed in this manner. Table 4.1 lists working antibodies that we commonly use to detect UPR proteins.

Activated IRE1α functions as an endoribonuclease splicing a 26-base pair intron from XBP-1 mRNA. Spliced XBP-1 mRNA is translated into a stable and active UPR transcription factor. Measuring XBP-1 splicing represents a reliable indirect method of determining IRE1α activation. We have designed primers that can specifically detect spliced and unspliced XBP-1 transcripts by quantitative real-time polymerase chain reaction (PCR; Table 4.2; Allen et al., 2004). Using these primers, we could successfully quantify the expression levels of spliced and unspliced XBP-1 in different cell lines (Fonseca et al., 2010; Ishigaki et al., 2010). Total RNA was isolated from cells using RNeasy Mini Kit (Qiagen) and reverse-transcribed to cDNA using 1 μg of total RNA with ImProm-II reverse transcription system (Promega). Primers and diluted cDNA samples were prepared with Power SYBR Green PCR Master Mix (Applied Biosystems, Foster City, CA) for qRT-PCR. For the thermal cycle reaction, the iQ5 system (Bio Rad, Hercules, CA) was used at 95 °C for 10 min, then 40 cycles at 95 °C for 10 s, and at 55 °C for 30 s. The relative amount for each transcript was calculated by a standard curve of cycle thresholds for serial dilutions of cDNA samples and normalized to the amount of β-actin. The PCR was performed in triplicate for each sample, after which all the experiments were repeated twice. Spliced and unspliced XBP-1 mRNA can also be measured by semiquantitative RT-PCR, and XBP-1 protein can be detected by immunoblot using anti-XBP-1-specific antibody (Santa Cruz).

Table 4.1 List of antibodies for detecting ER stress markers

Antibody	Source	Weight (kDa)	Supplier	Blocking buffer, dilution
IRE1α	Rabbit	130	Cell Signaling, #3294	BSA, WB 1:1000
Phospho-IRE1α	Rabbit	110	Novus, NB100-2323	Milk, WB 1:1000
Spliced XBP-1	Rabbit	54	Santa Cruz, sc-7160	Milk, WB 1:1000
Total PERK	Rabbit	150	Rocklan, 100-401-962	Milk, WB 1:1000
Phospho-PERK	Rabbit	170	Cell Signaling, #3179	BSA, WB 1:1000
eIF2α	Rabbit	36	Santa Cruz, sc-11386	Milk, WB 1:1000
Phospho-eIF2α	Rabbit	38	Cell Signaling, #3597	BSA, WB 1:1000
ATF6α	Rabbit	90	Santa Cruz, sc-22799	Milk, WB 1:100
Cleaved ATF6α	Mouse	50	Imgenex, IMG-273	Milk, WB 1:1000
CHOP/GADD153	Mouse	31	Pierce, MA1-250	Milk, WB 1:2000
BiP/GRP78	Rabbit	78	Stressgen, SPA-826	Milk, WB 1:1000, IF: 1:100
ATF4	Rabbit	39	ProteinTech, 10835-1-AP	Milk, WB 1:1000
PDI	Mouse	58	Stressgen, SPA-891	Milk, WB 1:1000, IF: 1:100

Table 4.2 List of primers for detecting ER stress markers by real-time PCR

Gene	Human	Mouse	Rat
sXBP-1	CTGAGTCCGAATCAGGTGCAG; ATCCATGGGAGATGTTCTGG	CTGAGTCCGAATCAGGTGCAG; GTCCATGGGAAGATGTTCTGG	CTGAGTCCGAATCAGGTGCAG; ATCCATGGGAAGATGTTCTGG
usXBP-1	CAGCACTCAGACTACGTGCA; ATCCATGGGAGATGTTCTGG	CAGCACTCAGACTATGTGCA; GTCCATGGGAAGATGTTCTGG	CAGCACTCAGACTACGTGCG; ATCCATGGGAAGATGTTCTGG
Total XBP-1	TGGCCGGGTCTGCTGAGTCCG; ATCCATGGGAGATGTTCTGG	TGGCCGGGTCTGCTGAGTCCG; GTCCATGGGAAGATGTTCTGG	TGGCCGGGTCTGCTGAGTCCG; ATCCATGGGAAGATGTTCTGG
ATF4	GTTCTCCAGCGACAAGGCTA; ATCCTGCTTGCTGTTGTTGG	GGGTTCTGTCTTCCACTCCA; AAGCAGCAGAGTCAGGCTTTC	AATGATGACCTGGAAACCA; TCTTGGACTAGAGGGGCAAA
CHOP	AGAACCAGGAAACGGAAACAGA; TCTCCCTTCATGCGCTGCTTT	CCACCACACCTGAAAGCAGAA; AGGTGAAAGGCAGGACTCA	AGAGTGGTCAGTGCGCAGC; CTCATTCTCCTGCTCCTTCTCC
BiP	TGTTCAACCAATTATCAGCAAACTC; TTCTGCTGTATCCTCTTCACCAGT	TTCAGCCAATTATCAGCAAACTCT; TTTTCTGATGTATCCTCTTCACCAGT	TGGGTACATTTGATCTGACTGGA; CTCAAAGGTGACTTCAATCTGGG
GRP94	GAAACGGATGCCTGGTGG; GCCCCTTCTTCCTGGGTC	AAGAATGAAGAAAAACAGGACAAAA; CAAATGGAGAAAGATTCCGCC	
EDEM	CAAGTGTGGGTACGCCACG; AAAGAAGCTCTCCATCCGGTC	CTACCTGCGAAGAGGCCG; GTTCATGAGCTGCCCACTGA	

As stated above, a 26-nucleotide intron of XBP-1 mRNA is spliced out under ER stress conditions, leading to a frame shift. Taking advantage of the ER stress-dependent splicing of XBP-1, a fluorescent signal-based ER stress reporter has been developed (Iwawaki *et al.*, 2004). A gene encoding venus, a variant of green fluorescent protein, is fused to human XBP-1 downstream of the 26-nt intron. Under normal conditions, the mRNA of this fusion gene is not spliced, and therefore, its translation is terminated at a stop codon between XBP-1 and venus genes. However, under ER stress conditions, the 26-nt intron is spliced out, leading to a frameshift allowing the production of XBP-1-venus fusion protein which can be detected by monitoring the fluorescence activity of venus. In transgenic mice expressing XBP-1-venus fusion protein, strong fluorescence in the kidney and pancreas was detected when tunicamycin was injected intraperitoneally. Thus, transgenic mice expressing XBP-1-venus could be used to monitor ER stress levels *in vivo*.

IRE1α knockout MEFs are available in the research community (Calfon *et al.*, 2002; Urano *et al.*, 2000b; Zhang *et al.*, 2005). These cells can be used as a negative control for assessing the activation levels of IRE1α and its downstream components.

5.2. Methods for measuring PERK activation

Similar to IRE1α, PERK also undergoes transautophosphorylation upon ER stress. The phosphorylation levels of PERK can be detected by a phospho-specific PERK antibody (Cell Signaling Technologies, Danvers, MA). Upon activation, PERK phosphorylates eIF2α to reduce global mRNA translation. Measuring eIF2α phosphorylation levels by immunoblot using anti-phospho-eIF2α-specific antibody (Cell Signaling, Danvers, MA) indirectly reflects PERK activation. However, it must be noted that other eIF2α kinases exist and therefore, proper controls should be included to confirm PERK dependent-eIF2α phosphorylation.

Salubrinal is a selective inhibitor of cellular complexes that dephosphorylate eIF2α (Boyce *et al.*, 2005). Thus, addition of salubrinal to cells enhances eIF2α phosphorylation. This compound can be used to study the eIF2α-dependent arm of PERK signaling. We treat cells with 75 μM salubrinal for 16 h to observe enhanced eIF2α phosphorylation.

Other available tools include PERK knockout mice and PERK knockout MEFs (Harding *et al.*, 2001; Zhang *et al.*, 2002, 2006).

5.3. Methods for measuring ATF6α activation

As mentioned previously, in response to ER stress, ATF6α (90 kDa) transits to the Golgi apparatus and cleaved by SP1 and S2P, producing a 50-kDa form which translocates to the nucleus to activate transcription of UPR genes (Yoshida et al., 2000). By transfecting cells with a GFP–ATF6α fusion protein, ATF6 translocation events upon ER stress can be monitored by fluorescence microscopy. (Nadanaka et al., 2004) Detection of the cleaved 50-kDa form of ATF6α by immunoblot using anti-ATF6α-specific antibody (Imgenex) can be used as an indicator of ATF6α activation. The immunoblot protocol includes an unmasking step after blocking in order to reveal the antigen for antibody binding. Incubate the membrane in a sealed ziplock bag containing 50 ml of unmasking buffer (2%, w/v, SDS, 62.5 mM Tris–HCl, or standard 1× PBS, 100 mM β-mercaptoethanol) in a 70 °C water bath for 30 min. Discard the unmasking buffer and wash the blot twice with PBS or Tris–HCl. Reblock the membrane and continue the protocol as usual. Another alternative to study ATF6α activation is measuring the mRNA expression levels of genes that are regulated transcriptionally by ATF6α. The details of measuring transcription will be discussed later.

6. Studying UPR Downstream Markers and Responses

6.1. Immunostaining and immunofluorescence for downstream markers of the UPR

Immunostaining can also be used to measure UPR activation. The advantage of immunostaining is that we can study tissues from patients or mouse models with ER stress-related diseases. Many of the antibodies discussed previously could be used for immunocytochemistry. In addition, CHOP, BiP, and PDI antibodies can be used as indicators of cells undergoing ER stress conditions. CHOP is regulated under the PERK-eIF2α-ATF4 pathway and has been shown to have a role in ER stress-mediated apoptosis. However, it must be cautioned that many commercially available antibodies for detection of CHOP expression fail specificity evaluation (Haataja et al., 2008). BiP is a central regulator of the UPR stress sensors as well as an ER chaperone to assist protein folding. BiP is highly expressed in the ER and can be used as an ER marker. PDI is involved in oxidative protein folding in the ER lumen and its expression is induced by ER stress.

To perform immunocytochemistry, we grow our cells typically onto four well Lab-Tek chambers. After ER stress induction, cells are fixed with 4% paraformaldehyde in PBS for 30 min at room temperature, followed by permeabilization with 4% paraformaldehyde and 0.1% Triton for 2 min at room temperature. Chambers are removed and slides are rinsed in PBST (PBS + 0.1% Tween). Nonspecific binding sites are blocked by incubating samples with Image-iT Signal Enhancer (Invitrogen) for 30 min at room temperature. Samples are incubated with a diluted primary antibody in an antibody diluent (Dako) overnight at 4 °C in a humidified chamber. The next day, slides are rinsed with PBST 3 times and then incubated with a diluted secondary antibody in an antibody diluent (1:200–1:1000) for 1 h at room temperature. Slides are rinsed 4 times with PBST and 1 more time with PBS. Finally, the slides are mounted with ProLong Gold antifade mounting medium containing DAPI to stain DNA (Molecular Probes, Invitrogen). The slides could be viewed immediately by fluorescence microscopy or stored in the dark at 4 °C for a month and at −80 °C for several months.

6.2. Measuring transcriptional activation of the UPR

Upon ER stress conditions, activated master regulators of the UPR communicate to the nucleus to regulate the transcription of genes involved in protein folding and processing to increase the ER protein folding capacity, ERAD, and autophagy components to reduce the ER workload, and cell survival and death factors to determine the fate of the cell depending on the ER stress condition.

IRE1α, PERK, and ATF6α are all involved in regulating transcription during ER stress. IRE1α directly regulates the splicing of XBP-1 mRNA to produce a transcriptionally active b-ZIP transcription factor. XBP-1 regulates chaperones, folding catalysts, and ERAD components such as BiP, EDEM, and HRD1. Phosphorylated eIF2α by PERK generally reduces mRNA translation; however, preferentially favors translation of some mRNAs such as the transcription factor ATF4. ATF4 regulates genes involved in antioxidative stress, amino acid biosynthesis, protein folding and degradation, and apoptosis such as CHOP. CHOP of GADD153 is a b-ZIP transcription factor regulating apoptosis-related genes such as death receptor 5 (DR5), tribble 3 (TRB3), and members of the BCL2 family of proteins. One of the most reliable methods to measure transcriptional regulation of the UPR in ER-stressed cells is by quantitative real-time PCR. Table 4.2 lists commonly measured UPR genes and their primers.

Many of the genes regulated by the UPR contain unique *cis*-acting response elements within their promoters. These include ERSE (ER stress response element, 5′-CCAAT-N9-CCACG-3′), ERSE-II (ER stress response element II, 5′-ATTGG-N1-CCACG-3′), and the UPRE (Unfolded Protein Response element, 5′-TGACGTGG/A-3′).

Instead of measuring mRNA expression levels of UPR genes, there are several reporter systems that reflect endogenous UPR activation levels. A luciferase plasmid driven by the human GRP78 promoter is commonly used. The promoter contains three copies of ERSE upstream of the TATA element. Another luciferase reporter often studied contains one or five ATF6α binding sites (Wang *et al.*, 2000). This reporter can be activated by ER stress inducers as well as ATF6α and XBP-1 overexpression. Cells are transfected with luciferase reporters, overexpression vectors, and beta galactosidase internal control using optimized transfection methods. We commonly transfect COS7 or 293 T cells by Lipofectamine 2000 (Invitrogen) for luciferase assays. After 24–48 h posttransfection, cells are treated with ER stress inducers and/or harvested using the Luciferase Assay System kit (Promega). We have determined that low-dose ER stress (e.g., 50 nM Tg and 0.5 μg/ml tunicamycin, 18 h) can activate GRP78 and ATF6α luciferase reporters. Firefly luciferase and beta galactosidase activities (β-gal reporter gene assay, chemiluminescent, Roche) are measured by a standard plate reading luminometer.

6.3. Measuring translational attenuation of the UPR

Translational attenuation is an early UPR response in order to reduce the ER protein workload. In response to ER stress, cells attenuate translation through PERK-mediated eIF2α phosphorylation. As mentioned earlier, eIF2α phosphorylation can be detected by Western blot. Translational attenuation can be measured by metabolic pulse labeling of newly synthesized proteins and polyribosome profiling using standard protocols (Harding et al., 1999). As the UPR restores ER homeostasis, GADD34 interacts with protein phosphatase 1c to dephosphorylate eIF2α, restoring protein synthesis (Novoa et al., 2001). GADD34 expression is induced by ER stress and regulated under the PERK-ATF4 pathway.

6.4. Measuring ERAD and protein stability

The UPR removes harmful proteins by regulating expression of ERAD genes. During ERAD, misfolded and unfolded proteins are recognized by ER chaperones, retrotranslocated out of the ER into the cytosol, and finally ubiquitinated and degraded by the proteasome. ATF6, XBP-1, and ATF4 are all involved in regulating the transcription of ERAD components such as EDEM and HRD1. To study the ERAD pathway, three proteins susceptible to misfolding in the ER, namely, TCRα, mutant alpha-1-antitrypsin NHK3, and the DeltaF508-variant cystic fibrosis transmembrane conductance regulator protein, are often used (Hosokawa et al., 2001, 2003; Yu and Kopito, 1999; Yu et al., 1997). These substrates can be ectopically expressed in cells and their stability can be monitored by cycloheximide chase or metabolic pulse-chase labeling assays. Degradation rates of these proteins reflect the activation levels of ERAD in cells. In addition, ubiquitination of an interested ERAD substrate can be studied by treating cells with the proteasome inhibitor, MG132, immunoprecipitating the protein of interest, followed by immunoblot with antiubiquitin antibody (Cell Signaling).

6.5. Measuring mRNA degradation

It has been shown that ER stress accelerates degradation of mRNAs in cells (Hollien and Weissman, 2006). This is largely dependent on the RNase activity of IRE1α (Hollien and Weissman, 2006; Hollien et al., 2009). Under unresolvable ER stress conditions, the RNase domain of IRE1α plays a role in degrading mRNAs encoding secretory proteins in addition to splicing XBP-1 mRNA. In β cell lines, insulin mRNA has been shown to be a substrate of IRE1 and is quickly degraded under ER stress conditions (Han et al., 2009; Lipson et al., 2008; Pirot et al., 2007). This phenomenon can be used in detecting and quantifying unresolvable ER stress in pancreatic β cells. Cellular mRNA transcription is attenuated by treating β cells with 100 μg/mL actinomycin D for 1 h. Total RNA is isolated at different time points, reverse-transcribed to cDNA, and insulin gene transcripts are measured by real-time PCR as described before. Time point zero for each condition is standardized to 1 and the subsequent rate of degradation of mRNA is measured. Degradation rate of insulin mRNA could be used as a biomarker for β cells experiencing unresolvable ER stress.

6.6. Measuring ER stress-mediated apoptosis

When the UPR fails to restore ER homeostasis and attenuate ER stress, the UPR activation induces apoptosis. ER stress-mediated apoptosis is involved in many human chronic diseases. Thus, measuring ER stress-mediated apoptosis will aid us in understanding the pathogenesis of ER stress-related disorders.

There are several components of the UPR that could contribute to ER stress-mediated apoptosis, including the IRE1α–ASK1–JNK signaling pathways, CHOP regulation of BCL2 protein family members and apoptotic genes, ER localized Bax and Bak, and glycogen synthase kinase 3β (GSK3β). Activated IRE1α binds to the adaptor protein TRAF2 and subsequently activate ASK1, which activates the JNK pathway. The JNK pathway has been shown to play an important role in ER stress-mediated cell death by regulating the BCL2 family of proteins (Nishitoh et al., 2002, 2008; Urano et al., 2000b). IRE1α, ASK1, and JNK are serine/threonine protein kinases and therefore, their activation levels can be measured by phospho-specific antibodies. CHOP is a proapoptotic transcription factor of the UPR (Zinszner et al., 1998). As its baseline expression is low, its upregulation and activation can be measured by immunoblot or real-time PCR as mentioned previously. Cells undergoing ER stress-mediated cell death can also be determined by immunostaining for CHOP. Proapoptotic BCL2 family members, BAX and BAK, are associated with the ER membrane. Upon ER stress, BAK and BAK undergo conformational changes, forming pores in the membrane, causing $Ca2+$ to leak into the cytosol, which in turn stimulate the activation of apoptotic pathways. BAX and BAK double knockout cells are resistant to ER stress-mediated cell death (Zong et al., 2001, 2003). BAX and BAK expression levels can be detected by immunoblot. Finally, GSK3β also plays a role in ER stress-mediated apoptosis. GSK3β is a substrate of the survival kinase, Akt (Cross et al., 1995), and it has been demonstrated that attenuation of Akt phosphorylation during ER stress mediates dephosphorylation of GSK3β, leading to ER stress-mediated apoptosis (Srinivasan et al., 2005). GSK3β phosphorylation levels can be measured by anti-phospho-specific GSK3β antibody.

ER stress-mediated apoptosis can be measured by standard methods. Measuring caspase-3 cleavage (Cell Signaling) by immunoblot, staining the cells with PE Annexin-V (BD Biosciences) followed by FACS analysis, TUNEL staining using the DeadEnd™ Colorimetric TUNEL System (Promega), and cell viability assays using CellTiter-Glo (Promega) are commonly used in our lab.

ACKNOWLEDGMENTS

This work was supported in part by grants from NIH-NIDDK (R01DK067493), the Diabetes and Endocrinology Research Center at the University of Massachusetts Medical School (5 P30 DK32520), and the Juvenile Diabetes Research Foundation International to F. Urano.

REFERENCES

Akiyama, M., Hatanaka, M., Ohta, Y., Ueda, K., Yanai, A., Uehara, Y., Tanabe, K., Tsuru, M., Miyazaki, M., Saeki, S., Saito, T., Shinoda, K., et al. (2009). Increased insulin demand promotes while pioglitazone prevents pancreatic beta cell apoptosis in Wfs1 knockout mice. *Diabetologia* **52**, 653–663.

Allen, J. R., Nguyen, L. X., Sargent, K. E. G., Lipson, K. L., Hackett, A., and Urano, F. (2004). High ER stress in beta-cells stimulates intracellular degradation of misfolded insulin. *Biochem. Biophys. Res. Commun.* **324**, 166–170.

Asfari, M., Janjic, D., Meda, P., Li, G., Halban, P. A., and Wollheim, C. B. (1992). Establishment of 2-mercaptoethanol-dependent differentiated insulin-secreting cell lines. *Endocrinology* **130**, 167–178.

Bertolotti, A., Wang, X., Novoa, I., Jungreis, R., Schlessinger, K., Cho, J. H., West, A. B., and Ron, D. (2001). Increased sensitivity to dextran sodium sulfate colitis in IRE1beta-deficient mice. *J. Clin. Investig.* **107**, 585–593.

Boyce, M., Bryant, K. F., Jousse, C., Long, K., Harding, H. P., Scheuner, D., Kaufman, R. J., Ma, D., Coen, D. M., Ron, D., and Yuan, J. (2005). A selective inhibitor of eIF2alpha dephosphorylation protects cells from ER stress. *Science (New York)* **307**, 935–939.

Calfon, M., Zeng, H., Urano, F., Till, J. H., Hubbard, S. R., Harding, H. P., Clark, S. G., and Ron, D. (2002). IRE1 couples endoplasmic reticulum load to secretory capacity by processing the XBP-1 mRNA. *Nature* **415**, 92–96.

Cardozo, A. K., Ortis, F., Storling, J., Feng, Y. M., Rasschaert, J., Tonnesen, M., Van Eylen, F., Mandrup-Poulsen, T., Herchuelz, A., and Eizirik, D. L. (2005). Cytokines downregulate the sarcoendoplasmic reticulum pump Ca2+ ATPase 2b and deplete endoplasmic reticulum Ca2+, leading to induction of endoplasmic reticulum stress in pancreatic beta-cells. *Diabetes* **54**, 452–461.

Cross, D. A., Alessi, D. R., Cohen, P., Andjelkovich, M., and Hemmings, B. A. (1995). Inhibition of glycogen synthase kinase-3 by insulin mediated by protein kinase B. *Nature* **378**, 785–789.

Feng, B., Yao, P. M., Li, Y., Devlin, C. M., Zhang, D., Harding, H. P., Sweeney, M., Rong, J. X., Kuriakose, G., Fisher, E. A., Marks, A. R., Ron, D., *et al.* (2003). The endoplasmic reticulum is the site of cholesterol-induced cytotoxicity in macrophages. *Nat. Cell Biol.* **5**, 781–792.

Fonseca, S. G., Ishigaki, S., Oslowski, C. M., Lu, S., Lipson, K. L., Ghosh, R., Hayashi, E., Ishihara, H., Oka, Y., Permutt, M. A., and Urano, F. (2010). Wolfram syndrome 1 gene negatively regulates ER stress signaling in rodent and human cells. *J. Clin. Invest.* **17**(2), 107–112.

Haataja, L., Gurlo, T., Huang, C. J., and Butler, P. C. (2008). Many commercially available antibodies for detection of CHOP expression as a marker of endoplasmic reticulum stress fail specificity evaluation. *Cell Biochem. Biophys.* **51**, 105–107.

Han, D., Lerner, A. G., Vande Walle, L., Upton, J. P., Xu, W., Hagen, A., Backes, B. J., Oakes, S. A., and Papa, F. R. (2009). IRE1alpha kinase activation modes control alternate endoribonuclease outputs to determine divergent cell fates. *Cell* **138**, 562–575.

Harding, H. P., Zhang, Y., and Ron, D. (1999). Protein translation and folding are coupled by an endoplasmic-reticulum-resident kinase. *Nature* **397**, 271–274.

Harding, H. P., Zhang, Y., Bertolotti, A., Zeng, H., and Ron, D. (2000a). Perk is essential for translational regulation and cell survival during the unfolded protein response. *Mol. Cell* **5**, 897–904.

Harding, H. P., Novoa, I., Zhang, Y., Zeng, H., Wek, R., Schapira, M., and Ron, D. (2000b). Regulated translation initiation controls stress-induced gene expression in mammalian cells. *Mol. Cell* **6**, 1099–1108.

Harding, H. P., Zeng, H., Zhang, Y., Jungries, R., Chung, P., Plesken, H., Sabatini, D. D., and Ron, D. (2001). Diabetes mellitus and exocrine pancreatic dysfunction in perk-/- mice reveals a role for translational control in secretory cell survival. *Mol. Cell* **7**, 1153–1163.

Harding, H. P., Zhang, Y., Zeng, H., Novoa, I., Lu, P. D., Calfon, M., Sadri, N., Yun, C., Popko, B., Paules, R., Stojdl, D. F., Bell, J. C., *et al.* (2003). An integrated stress response regulates amino acid metabolism and resistance to oxidative stress. *Mol. Cell* **11**, 619–633.

Haze, K., Yoshida, H., Yanagi, H., Yura, T., and Mori, K. (1999). Mammalian transcription factor ATF6 is synthesized as a transmembrane protein and activated by proteolysis in response to endoplasmic reticulum stress. *Mol. Biol. Cell* **10**, 3787–3799.

Hohmeier, H. E., Mulder, H., Chen, G., Henkel-Rieger, R., Prentki, M., and Newgard, C. B. (2000). Isolation of INS-1-derived cell lines with robust ATP-sensitive K+ channel-dependent and -independent glucose-stimulated insulin secretion. *Diabetes* **49**, 424–430.

Hollien, J., and Weissman, J. S. (2006). Decay of endoplasmic reticulum-localized mRNAs during the unfolded protein response. *Science (New York)* **313**, 104–107.

Hollien, J., Lin, J. H., Li, H., Stevens, N., Walter, P., and Weissman, J. S. (2009). Regulated Ire1-dependent decay of messenger RNAs in mammalian cells. *J. Cell Biol.* **186**, 323–331.

Hosokawa, N., Wada, I., Hasegawa, K., Yorihuzi, T., Tremblay, L. O., Herscovics, A., and Nagata, K. (2001). A novel ER alpha-mannosidase-like protein accelerates ER-associated degradation. *EMBO Rep.* **2**, 415–422.

Hosokawa, N., Tremblay, L. O., You, Z., Herscovics, A., Wada, I., and Nagata, K. (2003). Enhancement of endoplasmic reticulum (ER) degradation of misfolded Null Hong Kong alpha1-antitrypsin by human ER mannosidase I. *J. Biol. Chem.* **278,** 26287–26294.

Ishigaki, S., Fonseca, S. G., Oslowski, C. M., Jurczyk, A., Shearstone, J. R., Zhu, L. J., Permutt, M. A., Greiner, D. L., Bortell, R., and Urano, F. (2010). AATF mediates an antiapoptotic effect of the unfolded protein response through transcriptional regulation of AKT1. *Cell Death Differ.* **17,** 774–786.

Iwawaki, T., Akai, R., Kohno, K., and Miura, M. (2004). A transgenic mouse model for monitoring endoplasmic reticulum stress. *Nat. Med.* **10,** 98–102.

Kharroubi, I., Ladriere, L., Cardozo, A. K., Dogusan, Z., Cnop, M., and Eizirik, D. L. (2004). Free fatty acids and cytokines induce pancreatic beta-cell apoptosis by different mechanisms: role of nuclear factor-kappaB and endoplasmic reticulum stress. *Endocrinology* **145,** 5087–5096.

Kim, I., Xu, W., and Reed, J. C. (2008). Cell death and endoplasmic reticulum stress: disease relevance and therapeutic opportunities. *Nat. Rev. Drug Discov.* **7,** 1013–1030.

Lee, A. H., Iwakoshi, N. N., and Glimcher, L. H. (2003a). XBP-1 regulates a subset of endoplasmic reticulum resident chaperone genes in the unfolded protein response. *Mol. Cell. Biol.* **23,** 7448–7459.

Lee, A. H., Iwakoshi, N. N., Anderson, K. C., and Glimcher, L. H. (2003b). Proteasome inhibitors disrupt the unfolded protein response in myeloma cells. *Proc. Natl. Acad. Sci. USA* **100,** 9946–9951.

Lipson, K. L., Ghosh, R., and Urano, F. (2008). The Role of IRE1alpha in the Degradation of Insulin mRNA in Pancreatic beta-Cells. *PLoS ONE* **3,** e1648.

Merksamer, P. I., Trusina, A., and Papa, F. R. (2008). Real-time redox measurements during endoplasmic reticulum stress reveal interlinked protein folding functions. *Cell* **135,** 933–947.

Miyazaki, J., Araki, K., Yamato, E., Ikegami, H., Asano, T., Shibasaki, Y., Oka, Y., and Yamamura, K. (1990). Establishment of a pancreatic beta cell line that retains glucose-inducible insulin secretion: special reference to expression of glucose transporter isoforms. *Endocrinology* **127,** 126–132.

Nadanaka, S., Yoshida, H., Kano, F., Murata, M., and Mori, K. (2004). Activation of mammalian unfolded protein response is compatible with the quality control system operating in the endoplasmic reticulum. *Mol. Biol. Cell* **15,** 2537–2548.

Nishitoh, H., Saitoh, M., Mochida, Y., Takeda, K., Nakano, H., Rothe, M., Miyazono, K., and Ichijo, H. (1998). ASK1 is essential for JNK/SAPK activation by TRAF2. *Mol. Cell* **2,** 389–395.

Nishitoh, H., Matsuzawa, A., Tobiume, K., Saegusa, K., Takeda, K., Inoue, K., Hori, S., Kakizuka, A., and Ichijo, H. (2002). ASK1 is essential for endoplasmic reticulum stress-induced neuronal cell death triggered by expanded polyglutamine repeats. *Genes Dev.* **16,** 1345–1355.

Nishitoh, H., Kadowaki, H., Nagai, A., Maruyama, T., Yokota, T., Fukutomi, H., Noguchi, T., Matsuzawa, A., Takeda, K., and Ichijo, H. (2008). ALS-linked mutant SOD1 induces ER stress- and ASK1-dependent motor neuron death by targeting Derlin-1. *Genes Dev.* **22,** 1451–1464.

Novoa, I., Zeng, H., Harding, H. P., and Ron, D. (2001). Feedback inhibition of the unfolded protein response by GADD34-mediated dephosphorylation of eIF2alpha. *J. Cell Biol.* **153,** 1011–1022.

Novoa, I., Zhang, Y., Zeng, H., Jungreis, R., Harding, H. P., and Ron, D. (2003). Stress-induced gene expression requires programmed recovery from translational repression. *EMBO J.* **22,** 1180–1187.

Oslowski, C. M., and Urano, F. (2010). The binary switch between life and death of endoplasmic reticulum-stressed beta cells. *Curr. Opin. Endocrinol. Diabet. Obes.*

Outinen, P. A., Sood, S. K., Pfeifer, S. I., Pamidi, S., Podor, T. J., Li, J., Weitz, J. I., and Austin, R. C. (1999). Homocysteine-induced endoplasmic reticulum stress and growth arrest leads to specific changes in gene expression in human vascular endothelial cells. *Blood* **94,** 959–967.

Oyadomari, S., Koizumi, A., Takeda, K., Gotoh, T., Akira, S., Araki, E., and Mori, M. (2002). Targeted disruption of the Chop gene delays endoplasmic reticulum stress-mediated diabetes. *J. Clin. Investig.* **109,** 525–532.

Pirot, P., Naamane, N., Libert, F., Magnusson, N. E., Orntoft, T. F., Cardozo, A. K., and Eizirik, D. L. (2007). Global profiling of genes modified by endoplasmic reticulum stress in pancreatic beta cells reveals the early degradation of insulin mRNAs. *Diabetologia* **50,** 1006–1014.

Reimold, A. M., Etkin, A., Clauss, I., Perkins, A., Friend, D. S., Zhang, J., Horton, H. F., Scott, A., Orkin, S. H., Byrne, M. C., Grusby, M. J., and Glimcher, L. H. (2000). An essential role in liver development for transcription factor XBP-1. *Genes Dev.* **14,** 152–157.

Reimold, A. M., Iwakoshi, N. N., Manis, J., Vallabhajosyula, P., Szomolanyi-Tsuda, E., Gravallese, E. M., Friend, D., Grusby, M. J., Alt, F., and Glimcher, L. H. (2001). Plasma cell differentiation requires the transcription factor XBP-1. *Nature* **412,** 300–307.

Riggs, A. C., Bernal-Mizrachi, E., Ohsugi, M., Wasson, J., Fatrai, S., Welling, C., Murray, J., Schmidt, R. E., Herrera, P. L., and Permutt, M. A. (2005). Mice conditionally lacking the Wolfram gene in pancreatic islet beta cells exhibit diabetes as a result of enhanced endoplasmic reticulum stress and apoptosis. *Diabetologia* **48,** 2313–2321.

Ron, D., and Walter, P. (2007). Signal integration in the endoplasmic reticulum unfolded protein response. *Nat. Rev.* **8,** 519–529.

Shen, X., Ellis, R. E., Lee, K., Liu, C. Y., Yang, K., Solomon, A., Yoshida, H., Morimoto, R., Kurnit, D. M., Mori, K., and Kaufman, R. J. (2001). Complementary signaling pathways regulate the unfolded protein response and are required for C. *Cell* **107,** 893–903.

Srinivasan, S., Ohsugi, M., Liu, Z., Fatrai, S., Bernal-Mizrachi, E., and Permutt, M. A. (2005). Endoplasmic reticulum stress-induced apoptosis is partly mediated by reduced insulin signaling through phosphatidylinositol 3-kinase/Akt and increased glycogen synthase kinase-3beta in mouse insulinoma cells. *Diabetes* **54,** 968–975.

Travers, K. J., Patil, C. K., Wodicka, L., Lockhart, D. J., Weissman, J. S., and Walter, P. (2000). Functional and genomic analyses reveal an essential coordination between the unfolded protein response and ER-associated degradation. *Cell* **101,** 249–258.

Tu, B. P., and Weissman, J. S. (2004). Oxidative protein folding in eukaryotes: mechanisms and consequences. *J. Cell Biol.* **164,** 341–346.

Urano, F., Bertolotti, A., and Ron, D. (2000a). IRE1 and efferent signaling from the endoplasmic reticulum. *J. Cell Sci.* **113,** 3697–3702.

Urano, F., Wang, X., Bertolotti, A., Zhang, Y., Chung, P., Harding, H. P., and Ron, D. (2000b). Coupling of stress in the ER to activation of JNK protein kinases by transmembrane protein kinase IRE1. *Science (New York, N.Y.)* **287,** 664–666.

Wang, J., Takeuchi, T., Tanaka, S., Kubo, S. K., Kayo, T., Lu, D., Takata, K., Koizumi, A., and Izumi, T. (1999). A mutation in the insulin 2 gene induces diabetes with severe pancreatic beta-cell dysfunction in the Mody mouse. *J. Clin. Investig.* **103,** 27–37.

Wang, Y., Shen, J., Arenzana, N., Tirasophon, W., Kaufman, R. J., and Prywes, R. (2000). Activation of ATF6 and an ATF6 DNA binding site by the endoplasmic reticulum stress response. *J. Biol. Chem.* **275,** 27013–27020.

Wu, J., Rutkowski, D. T., Dubois, M., Swathirajan, J., Saunders, T., Wang, J., Song, B., Yau, G. D., and Kaufman, R. J. (2007). ATF6alpha optimizes long-term endoplasmic reticulum function to protect cells from chronic stress. *Dev. Cell* **13,** 351–364.

Yamamoto, K., Sato, T., Matsui, T., Sato, M., Okada, T., Yoshida, H., Harada, A., and Mori, K. (2007). Transcriptional induction of mammalian ER quality control proteins is mediated by single or combined action of ATF6alpha and XBP1. *Dev. Cell* **13**, 365–376.

Ye, J., Rawson, R. B., Komuro, R., Chen, X., Dave, U. P., Prywes, R., Brown, M. S., and Goldstein, J. L. (2000). ER stress induces cleavage of membrane-bound ATF6 by the same proteases that process SREBPs. *Mol. Cell* **6**, 1355–1364.

Yoshida, H., Haze, K., Yanagi, H., Yura, T., and Mori, K. (1998). Identification of the cis-acting endoplasmic reticulum stress response element responsible for transcriptional induction of mammalian glucose-regulated proteins. Involvement of basic leucine zipper transcription factors. *J. Biol. Chem.* **273**, 33741–33749.

Yoshida, H., Okada, T., Haze, K., Yanagi, H., Yura, T., Negishi, M., and Mori, K. (2000). ATF6 activated by proteolysis binds in the presence of NF-Y (CBF) directly to the cis-acting element responsible for the mammalian unfolded protein response. *Mol. Cell. Biol.* **20**, 6755–6767.

Yoshida, H., Matsui, T., Yamamoto, A., Okada, T., and Mori, K. (2001). XBP1 mRNA is induced by ATF6 and spliced by IRE1 in response to ER stress to produce a highly active transcription factor. *Cell* **107**, 881–891.

Yoshida, H., Matsui, T., Hosokawa, N., Kaufman, R. J., Nagata, K., and Mori, K. (2003). A time-dependent phase shift in the mammalian unfolded protein response. *Dev. Cell* **4**, 265–271.

Yu, H., and Kopito, R. R. (1999). The role of multiubiquitination in dislocation and degradation of the alpha subunit of the T cell antigen receptor. *J. Biol. Chem.* **274**, 36852–36858.

Yu, H., Kaung, G., Kobayashi, S., and Kopito, R. R. (1997). Cytosolic degradation of T-cell receptor alpha chains by the proteasome. *J. Biol. Chem.* **272**, 20800–20804.

Zhang, P., McGrath, B., Li, S., Frank, A., Zambito, F., Reinert, J., Gannon, M., Ma, K., McNaughton, K., and Cavener, D. R. (2002). The PERK eukaryotic initiation factor 2 alpha kinase is required for the development of the skeletal system, postnatal growth, and the function and viability of the pancreas. *Mol. Cell. Biol.* **22**, 3864–3874.

Zhang, K., Wong, H. N., Song, B., Miller, C. N., Scheuner, D., and Kaufman, R. J. (2005). The unfolded protein response sensor IRE1alpha is required at 2 distinct steps in B cell lymphopoiesis. *J. Clin. Invest.* **115**, 268–281.

Zhang, W., Feng, D., Li, Y., Iida, K., McGrath, B., and Cavener, D. R. (2006). PERK EIF2AK3 control of pancreatic beta cell differentiation and proliferation is required for postnatal glucose homeostasis. *Cell Metab.* **4**, 491–497.

Zinszner, H., Kuroda, M., Wang, X., Batchvarova, N., Lightfoot, R. T., Remotti, H., Stevens, J. L., and Ron, D. (1998). CHOP is implicated in programmed cell death in response to impaired function of the endoplasmic reticulum. *Genes Dev.* **12**, 982–995.

Zong, W. X., Lindsten, T., Ross, A. J., MacGregor, G. R., and Thompson, C. B. (2001). BH3-only proteins that bind pro-survival Bcl-2 family members fail to induce apoptosis in the absence of Bax and Bak. *Genes Dev.* **15**, 1481–1486.

Zong, W. X., Li, C., Hatzivassiliou, G., Lindsten, T., Yu, Q. C., Yuan, J., and Thompson, C. B. (2003). Bax and Bak can localize to the endoplasmic reticulum to initiate apoptosis. *J. Cell Biol.* **162**, 59–69.

CHAPTER FIVE

REAL-TIME MONITORING OF ER STRESS IN LIVING CELLS AND ANIMALS USING ESTRAP ASSAY

Masanori Kitamura[*,1] and Nobuhiko Hiramatsu[†]

Contents

1. Introduction	94
2. SEAP Reporter System	95
3. Monitoring of ER Stress in Culture Cells by ESTRAP	96
3.1. Establishment of reporter cells	96
3.2. Treatment of reporter cells with ER stress inducers	98
3.3. Chemiluminescent assay and formazan assay	99
3.4. Data analysis	101
4. Monitoring of ER Stress *In Vivo* by ESTRAP	101
4.1. ESTRAP mice	101
4.2. Reporter cell-implanted mice	103
5. Conclusion	104
References	105

Abstract

Endoplasmic reticulum (ER) stress is involved in a wide range of pathologies. Detection and monitoring of the unfolded protein response are required to disclose the link between ER stress and diseases. Assessment of ER stress is also essential for evaluation of therapeutic drugs *in vitro* and *in vivo*; that is, their therapeutic utility as well as adverse effects. For detection and monitoring of ER stress in living cells and animals, ER stress-responsive alkaline phosphatase (ESTRAP), also called secreted alkaline phosphatase (SEAP), serves as a useful indicator. In cells genetically engineered to express SEAP, secretion of SEAP is quickly downregulated in response to ER stress. This phenomenon is observed in a wide range of cell types triggered by various ER stress inducers. The magnitude of the decrease in extracellular SEAP is proportional to the intensity of ER stress, which is inversely correlated with the induction of

[*] Department of Molecular Signaling, Interdisciplinary Graduate School of Medicine and Engineering, University of Yamanashi, Chuo, Yamanashi, Japan
[†] Department of Pathology, University of California San Diego, School of Medicine, La Jolla, California, USA
[1] Corresponding author

endogenous ER stress markers. In contrast to SEAP, the activity of intracellular luciferase is not affected by ER stress. ER stress causes a decrease in SEAP activity not via transcriptional suppression but via abnormal posttranslational modification, accelerated degradation, and reduced secretion of SEAP protein. In mice constitutively producing SEAP, *in vivo* induction of ER stress similarly causes rapid reduction in serum SEAP activity. Using SEAP as an indicator, real-time monitoring of ER stress in living cells and animals is feasible. The ESTRAP method provides a powerful tool to investigate the pathogenesis of ER stress-associated diseases, to assess toxicity and the adverse effects of drugs, and to develop therapeutic agents for the treatment of ER stress-related disorders.

1. INTRODUCTION

Endoplasmic reticulum (ER) stress is involved in a wide range of pathologies, including ischemic/hypoxic injury, viral/bacterial infection, malignant diseases, neurodegenerative disorders, diabetes mellitus, atherosclerosis, inflammation, heavy metal intoxication, and toxicity of various drugs (Kitamura, 2008). Detection, quantification, and monitoring of ER stress are essential to disclose the link between ER stress and various pathophysiological events, to investigate molecular mechanisms underlying diseases, to assess toxicity and the adverse effects of medical drugs, and to screen effective agents for the treatment of ER stress-related disorders.

Several methods have been used for the assessment of ER stress, as recently reviewed by Samali *et al.* (2010). Endogenous biomarkers, for example, 78 kDa glucose-regulated protein (GRP78) and CCAAT/enhancer-binding protein-homologous protein (CHOP), are most commonly used as indicators for ER stress. Phosphorylation of RNA-dependent protein kinase-like ER kinase (PERK) and eukaryotic translation initiation factor 2α (eIF2α), cleavage of activating transcription factor 6 (ATF6), and splicing of X-box-binding protein 1 (*XBP1*) mRNA are also used as endogenous markers for ER stress. Alternatively, reporter assays using the ER stress response element (ERSE) or the unfolded protein response (UPR) element (UPRE) fused to a *lacZ* gene or a *luciferase* gene have been used for monitoring ER stress (Samali *et al.*, 2010). However, these systems require extraction of RNA or protein and do not allow for continuous or successive monitoring of ER stress in living cells and animals. The use of green fluorescence protein (GFP) or luciferase may be useful for *in vitro* imaging of ER stress (Iwawaki *et al.*, 2004; Hosoda *et al.*, 2010), but it is still not competent for quantitative, continuous assessment of ER stress in the internal organs of living animals.

Secretory proteins enter the subcellular pathway through the ER. In the ER, the proteins are folded into native conformation and undergo a multitude of posttranslational modifications. Only correctly folded proteins are exported to the Golgi apparatus. Based on this current knowledge, perturbation of ER function (i.e., ER stress) should cause disturbance of protein secretion, which can be monitored using secreted marker proteins. One candidate for this purpose is secreted alkaline phosphatase (SEAP). As we reported previously, SEAP serves as a sensitive, quantitative biomarker for ER stress (Hiramatsu *et al.*, 2005, 2006b). In cells engineered to express SEAP, secretion of SEAP is rapidly suppressed in response to ER stress. This phenomenon is observed in a wide range of cell types triggered by various ER stress inducers. The magnitude of the decrease in extracellular SEAP is proportional to the intensity of ER stress and inversely correlated with the induction of endogenous ER stress markers such as GRP78 and CHOP. In contrast to SEAP, activity of non-secreted luciferase and fluorescence intensity of GFP are not suppressed by ER stress (Hiramatsu *et al.*, 2006a,b; Yamazaki *et al.*, 2009). ER stress down-regulates SEAP activity independently of transcriptional regulation. It is via abnormal posttranslational modification, accelerated degradation, and reduced secretion of SEAP protein (Hiramatsu *et al.*, 2006b). This SEAP-based monitoring method has been designated as the ER stress-responsive alkaline phosphatase (ESTRAP) assay (Hiramatsu *et al.*, 2006b).

The ESTRAP system is also applied for *in vivo* monitoring of ER stress. We generated transgenic sensor mice, ESTRAP mice, systemically producing SEAP. In the ESTRAP mice, induction of ER stress causes rapid reductions in serum SEAP activity (Hiramatsu *et al.*, 2005, 2006b, 2007). Using the *in vitro* and *in vivo* ESTRAP systems, it is feasible to monitor ER stress in living cells and animals by simple sampling of culture media and small amounts of blood.

Using secreted luciferase (*Metridia* luciferase and *Gaussia* luciferase) as indicators, similar systems have been developed for monitoring of ER stress in culture cells (Badr *et al.*, 2007; Hiramatsu *et al.*, 2005). It is based on the fact that, like SEAP, extracellular activity of secreted luciferase is suppressed under ER stress conditions. However, in contrast to SEAP, activity of secreted luciferase may be significantly interfered in the presence of serum, especially by serum albumin (Hiramatsu *et al.*, 2005), suggesting a crucial disadvantage of using secreted luciferase as *in vivo* indicators for ER stress.

2. SEAP Reporter System

Normally, alkaline phosphatase is not secreted, but the recombinant SEAP derived from human placental alkaline phosphatase is efficiently secreted from eukaryotic cells. In SEAP-transfected cells, the activity of extracellular SEAP is directly proportional to changes in intracellular SEAP mRNA and protein (Berger *et al.*, 1988; Cullen and Malim, 1992). SEAP has several important advantages over other reporter proteins. Because preparation of cell lysates is not required, the activity of SEAP can be monitored continuously in identical living cells or animals by sampling of culture media or blood. Only 5 μl of samples are sufficient for the assessment. The assay is fast and easy, and the activity of SEAP can be evaluated sensitively and quantitatively using chemiluminescent assays. No special instruments or special pieces of equipment are needed. Only a conventional luminometer is required. Another important advantage is that background signals due to endogenous alkaline phosphatases are nearly absent. It is because unlike most endogenous alkaline phosphatases, SEAP is heat stable and resistant to inhibition by L-homoarginine (Cullen and Malim, 1992). The activity of endogenous alkaline phosphatase present in samples can be eliminated by preheating the samples at 65 °C and assaying in the presence of L-homoarginine. SEAP is an enzyme with high stability under cell culture conditions ($t_{1/2} = 500$ h), and special attention or carefulness is unnecessary for collection, storage, and handling of samples. It is worthwhile to note that, to our experience, high levels of SEAP expression do not induce expression of *GRP78*, suggesting that overexpression of SEAP *per se* does not trigger ER stress.

3. Monitoring of ER Stress in Culture Cells by ESTRAP

The *in vitro* ESTRAP assay is based on transfection of target cells with a plasmid that introduces the *SEAP* gene under the control of a constitutively active promoter. Viral promoters [e.g., simian virus 40 (SV40) promoter/enhancer, cytomegalovirus promoter/enhancer] or housekeeping gene promoters (e.g., β-actin promoter, elongation factor 1α promoter) serve for this purpose. pSEAP2-Control plasmid is commercially available from Clontech. This expression vector introduces the *SEAP* gene under the control of the SV40 promoter and enhancer. pGL3-Control (Promega) that introduces firefly luciferase under the control of the SV40 promoter/enhancer can be used as an internal control.

The ESTRAP assay is applicable for any cells of interest. For the purposes of continuous monitoring of ER stress for sustained periods, we recommend that stable transfectants be established. However, if it is difficult (e.g., neuron), transiently transfected cells can also be used for detection and short-term monitoring of ER stress.

3.1. Establishment of reporter cells

In this section, we describe a protocol for establishment of reporter cells using LLC-PK1 cells, the porcine renal tubular epithelial cells. However, the following protocol can be applied for various cell types with minor modifications.

3.1.1. Dual reporting cells

LLC-PK1 cells (2×10^6 cells; American Type Culture Collection) are washed twice with PBS, suspended in 500 µl PBS, and transferred into a GenePulser Xcell compatible cuvette (0.4 cm gap; BM Equipment, Japan). Add 10 µg pSEAP2-Control (Clontech), 10 µg pGL3-Control (Promega), and 3 µg pcDNA3.1 (Invitrogen) encoding neomycin phosphotransferase. Mix the cell suspension and leave on ice for 10 min. The cells are then subjected to electroporation (GenePulser Xcell, Bio-Rad; 950 µF, 135 V) and left at room temperature for 10 min. The cells (1/3–1/10) are then seeded onto 100-mm plates that contain 10 ml of 10% fetal bovine serum (FBS). After incubation overnight, the cells are fed with fresh growth medium containing 500 µg/ml G418. The medium containing G418 is replaced every 3 days. After stable transfectants form colonies, individual clones are trypsinized using cloning rings and transferred to two 96-well plates. At this step, individual clones should be serially numbered. After the clones become confluent in 96-well plates, cells in one plate are fed with fresh medium (1% FBS, 100 µl) and incubated for 8–24 h. The culture media and cells are subjected to chemiluminescent assays to evaluate SEAP activity and luciferase activity, respectively (see Section 3.3.1 and 3.3.2). The clones constitutively producing both SEAP and luciferase (LL/SV-SEAP-Luc cells) are selected as dual reporting cells, and those cells are propagated individually from another 96-well plate. In the established reporter cells, luciferase is used as an internal control for normalization of data. It is based on the fact that the activity of nonsecreted luciferase is not affected by ER stress. Of note, stable transfectants should be maintained in the presence of G418, but studies are performed in its absence.

3.1.2. Singular reporting cells

If establishment of double transfectants is unsuccessful, SEAP reporter cells and luciferase reporter cells may be established separately. LLC-PK1 cells are stably transfected with 10 µg pSEAP2-Control and 2 µg pcDNA3.1, or 10 µg pGL3-Control and 2 µg pcDNA3.1, as described in Section 3.1.1. Isolated colonies are transferred to 96-well plates (one plate for LL/SV-SEAP, two plates for LL/SV-Luc). After individual clones become confluent, the cells are fed with fresh medium and incubated. The culture media (from LL/SV-SEAP) and cells (LL/SV-Luc) are subjected to chemiluminescent assays to evaluate SEAP and luciferase activity. Individual positive clones are propagated and used for experiments.

3.1.3. Confirmation of lack of ER stress in reporter cells

High expression levels of SEAP and/or luciferase could induce ER stress. Lack of ER stress in the established reporter cells should be confirmed by Northern blot analysis. Nontransfected LLC-PK1 cells and stable transfectants with different expression levels of SEAP and/or luciferase are seeded in 12-well plates at 2×10^5 cells/well. After incubation for 24–48 h, the total RNA is extracted and subjected to Northern blot analysis of *GRP78* and *CHOP*. We recommend the use of reporter cells that substantially express *SEAP* and/or *luciferase* without induction of *GRP78* and *CHOP*, when compared with that in nontransfected cells.

3.1.4. Use of transient transfection

In some cells of interest, it may be difficult to establish stable transfectants. In such cases, transiently transfected cells may be used for the ESTRAP assay. Either electroporation or lipofection can be used for this purpose. After transfection with reporter plasmids, the cells are incubated for 24 h and used for experiments. However, in transiently transfected cells, expression levels of reporter proteins alter during the course of cultures, which is not suitable for monitoring of ER stress for sustained periods.

3.2. Treatment of reporter cells with ER stress inducers

3.2.1. Quantitative assessment of ER stress

LL/SV-SEAP-Luc cells or 1:1 mixture of LL/SV-SEAP cells and LL/SV-Luc cells are suspended in Dulbecco's modified eagle's medium/Ham's F-12 (DMEM/F-12) containing 5% FBS (growth medium) and seeded in a 96-well plate at a density of 2×10^4 cells/well ($n = 4$). After 24 h, the cultures are rinsed once with fresh culture medium containing 1% FBS (basal medium; 200 µl) to wash out SEAP and treated for 6 h with basal medium (50–100 µl/well) containing tunicamycin (protein glycozylation inhibitor; 0.01–10 µg/ml). In our experience, incubation for less than 1 h is sufficient to detect ER stress, but longer incubation improves the detection sensitivity. After the incubation, culture media ($\sim$50 µl) are harvested from individual wells, transferred into 0.5-ml microtubes, and centrifuged. The supernatants are then subjected to chemiluminescent assay to evaluate SEAP activity (see Section 3.3.1). At this step, the supernatants can be stored at -20 °C. After the sampling of culture media, cells are washed with PBS, added with lysis buffer, and subjected to luciferase assay (see Section 3.3.2). At this step (after addition of lysis buffer), the 96-well plate can be stored at -80 °C until analysis. The luciferase activity is used for normalization of SEAP activity. Alternatively, if only LL/SV-SEAP cells are available, the cells are subjected to formazan assay (see Section 3.3.3) to evaluate the number of viable cells. It may also be used for normalization of SEAP activity. Figure 5.1 shows dose-dependent reduction in the level of extracellular SEAP activity by the treatment with tunicamycin.

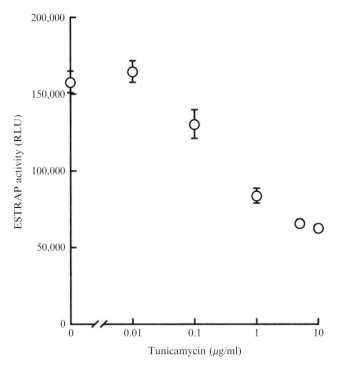

Figure 5.1 *Suppression of SEAP activity by ER stress in ESTRAP cells.* LL/SV-SEAP cells were treated with serial concentrations of tunicamycin (0.01–10 μg/ml) for 6 h, and activity of SEAP in culture medium was evaluated by chemiluminescent assay.

3.2.2. Continuous monitoring of ER stress

LL/SV-SEAP-Luc cells or 1:1 mixture of LL/SV-SEAP cells and LL/SV-Luc cells are seeded in a 96-well plate, as described in Section 3.2.1. After 24 h, individual wells are washed with basal medium to remove extracellular SEAP. The cells are then supplied with fresh basal medium (50 μl) and incubated for 2 h. After the incubation, all culture media are sampled from individual wells and replaced with fresh basal medium (50 μl/well) containing ER stress inducers such as tunicamycin and thapsigargin. Every 2 h, this procedure is repeated, and culture supernatants are stored at $-20\,°C$. At the end of the experiment, cells are subjected to luciferase assay or formazan assay.

3.3. Chemiluminescent assay and formazan assay

3.3.1. SEAP assay

Great EscAPe SEAP Detection Kit (BD Biosciences) is used to evaluate SEAP activity. In brief, 5 μl of cell-free culture medium (or other samples; e.g., serum) is mixed with 15 μl of 1× dilution buffer in a 0.5-ml microtube and incubated at 65 °C for 30 min to inactivate endogenous alkaline phosphatase. After the incubation, the sample (20 μl) is mixed with 20 μl of Assay Buffer containing L-homoarginine in a Gene Light compatible tube (Microtech Nition, Japan), left at room temperature for 5 min, and added with 20 μl of Chemiluminescent Enhancer containing 1.25 mM CSPD substrate. After incubation in the dark for 30 min, the sample is subjected to analysis using a luminometer (Gene Light 55; Microtech Nition). It is important to note that some test reagents might interfere with enzymatic reaction by SEAP. To exclude this possibility, recombinant SEAP (culture supernatant of LL/SV-SEAP cells) is mixed with ER stress inducers or other test reagents used for cell experiments, incubated at 37 °C for the same time period, and subjected to chemiluminescent assay. This assay should also be performed in quadruplicate.

3.3.2. Luciferase assay

Luciferase Assay System (Promega) is used to evaluate luciferase activity. In brief, cells in a 96-well plate are rinsed once with PBS. After removing PBS completely, 1× Reporter Lysis Buffer is dispensed into individual wells (20 μl/well). After freezing and thawing twice and agitation for 5 min, 5 μl of cell lysate is mixed with 25 μl luciferase assay reagent (luciferase assay buffer containing luciferase assay substrate) in a Gene Light compatible tube and subjected to measurement of luciferase activity using Gene Light 55.

3.3.3. Formazan assay

The number of viable cells is assessed using formazan assay. We use Cell Counting Kit-8 (Dojindo, Japan), a water-soluble tetrazolium salt (WST)-based assay, for this purpose. In brief, at the end of the experiments, cells in a 96-well plate are fed with medium containing WST-8 (10% Cell Counting Kit-8 solution). After incubation for 1 h, absorbance is read at 450 nm by a spectrophotometer. The background absorbance is also measured at 650 nm. The absorbance at 450 nm is subtracted by that at 650 nm and used as an indicator for the viable cell number. It is worthwhile to note that formazan assays may be affected by culture conditions or test reagents used. It is known that reductive agents facilitate production of formazans, leading to significant increases in the background. Agents that trigger production of reactive oxygen species may also cause reduction of some tetrazolium salts. On the other hand, acidic culture conditions significantly reduce values in formazan assays. We recommend replacement of culture media to fresh basal medium prior to incubation with tetrazolium salts (Johno et al., 2010).

3.4. Data analysis

The activity of extracellular SEAP produced by reporter cells is down-regulated by ER stress. However, SEAP activity may be reduced by nonspecific factors other than ER stress. For example, long-term exposure to ER stress blocks cell proliferation and may cause apoptosis, leading to reduction in the number of viable cells and consequent decline in the level of extracellular SEAP. ER stress inducers or other test reagents may also affect the SEAP assay *per se*. To exclude these nonspecific influences, we recommend the following procedure for data analysis.

(1) When dual reporting cells (or mixture of singular reporting cells) are used, the SEAP activity in individual wells should be normalized by the activity of luciferase.
(2) If luciferase is not available as an internal control, the SEAP activity should be normalized by the level of viable cells estimated by formazan assay.
(3) It is also important to confirm that the test reagents added to cultures do not affect SEAP assay *per se*, as described in Section 3.3.1.

4. MONITORING OF ER STRESS *IN VIVO* BY ESTRAP

Monitoring of ER stress is required for investigation of various pathophysiological events *in vivo*. However, conventional approaches to assess ER stress require sampling of target tissues and extraction of protein or RNA to evaluate the level of endogenous biomarkers. In contrast, the *in vivo* ESTRAP system allows for noninvasive, real-time monitoring of ER stress in living animals. Because the half-life of serum SEAP is approximately 2–3 h (Hiramatsu *et al.*, 2005, 2007), altered SEAP secretion results in rapid changes in the level of serum SEAP activity. ESTRAP mice are useful for (1) evaluation of ER stress under pathophysiological situations, (2) *in vivo* assessment for adverse effects of drugs that may cause ER stress, and (3) screening of novel therapeutic agents useful for prevention or treatment of ER stress-related disorders.

4.1. ESTRAP mice

4.1.1. Characterization

ESTRAP mice were generated by microinjection of a dioxin-responsive element-controlled *SEAP* gene into fertilized oocytes of C57BL/6 mice (Kasai *et al.*, 2006). A transgenic line showing a constitutive, high level of serum SEAP activity was occasionally isolated from the pool of the offspring and named ESTRAP mice. The ESTRAP mice constitutively express *SEAP* in all tested organs, including the brain, heart, lungs, liver, kidney, and spleen (Hiramatsu et al., 2007). Activity of SEAP is also elevated in all blood-free organs when compared with wild-type mice. In wild-type mice, activity of serum SEAP is $7.5 \pm 1.0 \times 10^2$ relative light unit (RLU). In contrast, ESTRAP mice exhibit high levels of serum SEAP activity, approximately $1.4 \pm 1.9 \times 10^6$ RLU (Hiramatsu et al., 2007). Of note, ESTRAP mice do not exhibit any abnormalities in appearances, development, fertility, and life span. The ESTRAP mice (registered as "*SEAP transgenic mouse*"; RBRC01731) are available from RIKEN Bioresource Center—Experimental Animal Division (Tsukuba, Ibaraki, Japan; URL: http://www.brc.riken.jp/lab/animal/en/).

4.1.2. *In vivo* detection of ER stress

Thapsigargin is one of the most popular inducers of ER stress. Thapsigargin depletes Ca^{2+} store in the ER and thereby causes ER stress. *In vitro*, thapsigargin induces expression of *GRP78* within several hours and rapidly suppresses ESTRAP activity in stably transfected cells (Hiramatsu et al., 2006b). Utility of the ESTRAP mice for *in vivo* monitoring of ER stress can be tested as follows: ESTRAP mice (body weight 20 g) are administered with thapsigargin (1 mg/kg body weight) intraperitoneally, and blood is sampled before and 4–24 h after the injection of thapsigargin. For blood sampling, mice are fastened using Mouse Fastening Adjuster (Sanplatec, Japan), and the tail vein is cut by a razor blade. Blood (20 µl) is collected using a micropipette, transferred into a 0.5-ml microtube, and left at room temperature for 0.5–1 h. After centrifugation, serum (5 µl) is collected and subjected to chemiluminescent assay (see Section 3.3.1). ESTRAP mice administered with thapsigargin exhibit transient, significant decreases in the level of serum SEAP at 8 h, and it is partially recovered after 24 h. Consistent with this result, expression of *GRP78* is induced at 8 h and returns to the basal level at 24 h (Hiramatsu et al., 2007).

4.1.3. Application for endotoxemia

When mice are administered with lipopolysaccharide (LPS) intraperitoneally, expression of *GRP78* is induced in various organs, including the lung, liver, kidney, and spleen (Hiramatsu et al., 2006b). Systemic ER stress under endotoxemia can be monitored as follows: ESTRAP mice are injected with LPS (200 µg/mouse i.p.) and serum is collected periodically up to 48 h. Chemiluminescent assay shows that activity of serum SEAP is rapidly reduced within 2 h, further declines until 8 h, and gradually recovers thereafter. Forty-eight hours after the administration with LPS, the level of serum SEAP returns to the initial level, suggesting that endotoxemia induces transient, reversible ER stress (Fig. 5.2).

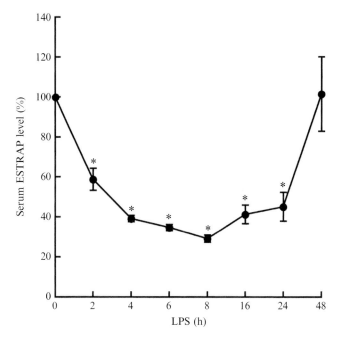

Figure 5.2 *Monitoring of ER stress during endotoxemia.* ESTRAP mice were administered with LPS intraperitoneally, and serum was sampled and subjected to chemiluminescent assay to evaluate SEAP activity.

4.1.4. Application for heavy metal intoxication

Cadmium has the potential to induce ER stress *in vitro* and *in vivo* (Hiramatsu *et al.*, 2007; Yokouchi *et al.*, 2007). Under the cadmium intoxication, ER stress can be monitored as follows: ESTRAP mice are administered with cadmium chloride intraperitoneally (12 mg/kg body weight) and activity of serum SEAP is evaluated every 2 h. After the administration with cadmium, activity of serum SEAP is reduced within 2 h. The decreased SEAP is gradually recovered thereafter and returned to the normal level after 12–24 h (Fig. 5.3). Of note, under this experimental setting, expression of *GRP78* is markedly induced in the liver and kidney at 6 h and recovers to the basal level at 24 h (Hiramatsu *et al.*, 2007).

4.2. Reporter cell-implanted mice

In vivo monitoring of ER stress may be feasible without using transgenic mice, as follows: Using the similar method described in Section 3.1, a mouse hepatoma cell line Hepa-1c1c7 (American Type Culture Collection) derived from the C57BL/6 mouse strain is stably transfected with pSEAP2-

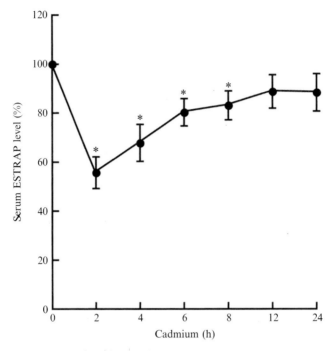

Figure 5.3 *Monitoring of ER stress during acute cadmium intoxication.* ESTRAP mice were injected with cadmium chloride intraperitoneally, and serum was sampled periodically and subjected to chemiluminescent assay.

Control, and Hepa/SV-SEAP cells are established. The established cells are injected into the peritoneal cavity of C57BL/6 mice (2.5×10^5 cells/mouse), and after 24 h, thapsigargin is administered intraperitoneally. Serum is sampled at 2 and 24 h and activity of serum SEAP is evaluated. Like in ESTRAP mice, systemic induction of ER stress depresses the serum level of SEAP within 2 h, and it is partially recovered at 24 h (Hiramatsu *et al.*, 2006b). This result provides evidence for the utility of an *ex vivo* gene transfer approach to monitor ER stress *in vivo*. Of note, however, pSEAP2-Control encodes human SEAP. The implanted reporter cells might be attacked by the host immune system targeting human SEAP protein.

5. Conclusion

Monitoring of ER stress is required for investigation of a broad range of pathophysiologic events *in vitro* and *in vivo*. However, the majority of current approaches require extraction of protein or RNA from cells and tissues. It is a limitation especially for *in vivo* studies where successive monitoring of ER stress is required during the course of diseases. The ESTRAP systems described here have advantages to overcome this problem. *In vitro* ESTRAP system allows for sensitive, quantitative, and continuous monitoring of ER stress in living cells. *In vivo* ESTRAP system is also useful for monitoring of ER stress in living animals under pathologic situations. These methods will be useful for (1) detection, quantification, and monitoring of ER stress under pathophysiological situations, (2) assessment of therapeutic utility or adverse effects of drugs, and (3) development of effective agents for prevention and treatment of ER stress-related disorders.

REFERENCES

Badr, C. E., Hewett, J. W., Breakefield, X. O., and Tannous, B. A. (2007). A highly sensitive assay for monitoring the secretory pathway and ER stress. *PLoS ONE* **2**, e571.

Berger, J., Hauber, J., Hauber, R., Geiger, R., and Cullen, B. R. (1988). Secreted placental alkaline phosphatase: A powerful new quantitative indicator of gene expression in eukaryotic cells. *Gene* **66**, 1–10.

Cullen, B. R., and Malim, M. H. (1992). Secreted placental alkaline phosphatase as a eukaryotic reporter gene. *Methods Enzymol.* **216**, 362–368.

Hiramatsu, N., Kasai, A., Meng, Y., Hayakawa, K., Yao, J., and Kitamura, M. (2005). Alkaline phosphatase vs. luciferase as secreted reporter molecules in vivo. *Anal. Biochem.* **339**, 249–256.

Hiramatsu, N., Kasai, A., Hayakawa, K., Nagai, K., Kubota, T., Yao, J., and Kitamura, M. (2006a). Secreted protein-based reporter systems for monitoring inflammatory events: Critical interference by endoplasmic reticulum stress. *J. Immunol. Methods* **315**, 202–207.

Hiramatsu, N., Kasai, A., Hayakawa, K., Yao, J., and Kitamura, M. (2006b). Real-time detection and continuous monitoring of ER stress in vitro and in vivo by ES-TRAP: Evidence for systemic, transient ER stress during endotoxemia. *Nucleic Acids Res.* **34**, e93.

Hiramatsu, N., Kasai, A., Du, S., Takeda, M., Hayakawa, K., Okamura, M., Yao, J., and Kitamura, M. (2007). Rapid, transient induction of ER stress in the liver and kidney after acute exposure to heavy metal: Evidence from transgenic sensor mice. *FEBS Lett.* **581**, 2055–2059.

Hosoda, A., Tokuda, M., Akai, R., Kohno, K., and Iwawaki, T. (2010). Positive contribution of ERdj5/JPDI to endoplasmic reticulum protein quality control in the salivary gland. *Biochem. J.* **425**, 117–125.

Iwawaki, T., Akai, R., Kohno, K., and Miura, M. (2004). A transgenic mouse model for monitoring endoplasmic reticulum stress. *Nat. Med.* **10**, 98–102.

Johno, H., Takahashi, S., and Kitamura, M. (2010). Influences of acidic conditions on formazan assay: A cautionary note. *Appl. Biochem. Biotech.* **162**, 1529–1535.

Kasai, A., Hiramatsu, N., Hayakawa, K., Yao, J., Maeda, S., and Kitamura, M. (2006). High levels of dioxin-like potential in cigarette smoke evidenced by *in vitro* and *in vivo* biosensing. *Cancer Res.* **66**, 7143–7150.

Kitamura, M. (2008). Endoplasmic reticulum stress and unfolded protein response in renal pathophysiology: Janus faces. *Am. J. Physiol. Ren.* **295**, F323–F342.

Samali, A., Fitzgerald, U., Deegan, S., and Gupta, S. (2010). Methods for monitoring endoplasmic reticulum stress and the unfolded protein response. *Int. J. Cell Biol.* 10.1155/2010/830307.

Yamazaki, H., Hiramatsu, N., Hayakawa, K., Tagaw, Y., Okamura, M., Ogata, R., Huang, T., Nakajima, S., Yao, J., Paton, A. W., Paton, J. C., and Kitamura, M. (2009). Activation of the Akt—NF-κB pathway by subtilase cytotoxin through the ATF6 branch of the unfolded protein response. *J. Immunol.* **183,** 1480–1487.

Yokouchi, M., Hiramatsu, N., Hayakawa, K., Kasai, A., Takano, Y., Yao, J., and Kitamura, M. (2007). Atypical, bidirectional regulation of cadmium-induced apoptosis via distinct signaling of unfolded protein response. *Cell Death Differ.* **14,** 1467–1474.

CHAPTER SIX

HIV Protease Inhibitors Induce Endoplasmic Reticulum Stress and Disrupt Barrier Integrity in Intestinal Epithelial Cells

Huiping Zhou

Contents

1. The Pathophysiological Relevance of HIV PI-Induced ER Stress and UPR Activation in IECs — 108
2. Analysis of HIV PI-Induced ER Stress and UPR Activation in Cultured IECs — 109
   2.1. Detection of HIV PI-induced ER stress using SEAP in IECs — 109
   2.2. Detection of HIV PI-induced ER stress by real-time RT-PCR and Western blot analysis — 110
   2.3. Measurement of paracellular permeability in IECs — 113
3. Analysis of HIV PI-Induced ER Stress and Disruption of Barrier Integrity in the Intestine — 114
   3.1. Measurement of intestinal permeability *in vivo* — 115
   3.2. Detection of HIV PI-induced ER stress in intestine by Western blot analysis — 115
   3.3. Histological examination of HIV PI-induced intestinal tissue injury — 116
4. Summary — 118
Acknowledgments — 118
Reference — 118

Abstract

The integrity of the intestinal epithelial barrier plays a crucial role in maintaining symbiotic homeostasis between microbes in the gut lumen and eukaryotic cells. Disruption of intestinal epithelial barrier function occurs commonly under various pathological conditions, including trauma, inflammatory bowel disease, and drug-induced gastrointestinal toxicity, exhibiting increased intestinal

Department of Microbiology and Immunology, School of Medicine, Virginia Commonwealth University, Richmond, Virginia, USA epithelial paracellular permeability or "leakiness" of the intestinal mucosa. Endoplasmic reticulum (ER) stress has recently been linked to various pathological conditions, including intestinal inflammation. Our previous studies have shown that HIV protease inhibitors (PIs) induce ER stress and activate the unfolded protein response (UPR) in different types of cells, and HIV PI-induced UPR activation contributes to the disruption of barrier function in intestinal epithelial cells and the increase of intestinal permeability. This chapter will discuss the commonly used methods for analysis of ER stress activation and epithelial barrier function. Both *in vitro* cell culture models and *in vivo* animal models are useful tools to examine general drug-induced ER stress and intestinal barrier dysfunction.

1. THE PATHOPHYSIOLOGICAL RELEVANCE OF HIV PI-INDUCED ER STRESS AND UPR ACTIVATION IN IECs

Currently, HIV PIs are among the preferred components used in combination with other drugs as part of highly active antiretroviral therapy (HAART), which has changed HIV infection from an invariably lethal to a manageable chronic condition (Clotet and Negredo, 2003; Dressman *et al.*, 2003; Flexner, 1998; Hui, 2003). However, systemic inflammation persists despite HAART, and the etiology remains incompletely defined and probably multifactorial. Viremia and elevated plasma lipopolysaccharide (LPS) endotoxin, a consequence of microbial translocation from disrupted gut barrier function, are among the possible causes. Cardiovascular disease (CVD) and, perhaps, malignancy and liver disease are among the results (Ancuta *et al.*, 2008; Brenchley and Douek, 2008a,b; Brenchley *et al.*, 2006a,b; Marchetti *et al.*, 2008; Nowroozalizadeh *et al.*, 2010). HIV PIs have been associated with atherosclerotic CVD, which is becoming the leading cause of mortality in HIV-1-infected persons in developed countries in the HAART era (Bode *et al.*, 1999, 2000, 2005; Clotet and Negredo, 2003; Dressman *et al.*, 2003; Flexner, 1998; Koster *et al.*, 2003). The *in vitro* and *in vivo* animal data from our lab and others, have linked HIV PIs with an increase in inflammatory cytokine production from several cell types and disruption of intestinal barrier integrity (Chen *et al.*, 2009; Wu *et al.*, 2010; Zhou *et al.*, 2005, 2006). However, the underlying mechanisms remain to be fully identified and therapeutic strategies are currently unavailable. Our recent study indicates that activation of the unfolded protein response (UPR) represents an important cellular mechanism underlying HIV PI-induced disruption of intestinal barrier integrity.

2. Analysis of HIV PI-Induced ER Stress and UPR Activation in Cultured IECs

IEC-6 cells were purchased from American type culture collection (ATCC, Cat No.: CRL-1592) at passage 13 and used at passages 15–20. This cell line is derived from normal rat intestine and was developed and characterized by Quaroni et al. (1979). They are nontumorigenic and retain the undifferentiated character of epithelial stem cells. The stable IEC-Cdx2L1 cells were developed and characterized by Suh and Traber (1996) and were a kind gift from Dr Peter G. Traber (Baylor College of Medicine, Houston, TX). Before experiments, cells are grown in DMEM containing 4 mM IPTG for 16 days to induce cell differentiation. The differentiated cells have multiple morphological characteristics of villus-type enterocytes with few goblet cells, and sucrase-isomaltase is highly expressed.

HIV PIs, amprenavir (AMPV), lopinavir (LOPV), and ritonavir (RITV), are dissolved in dimethyl sulfoxide (DMSO) and directly added to culture medium (final concentrations 5–25 μM) and incubated for 0.5–24 h.

2.1. Detection of HIV PI-induced ER stress using SEAP in IECs

Activity of secreted alkaline phosphatase (SEAP) is closely associated with endoplasmic reticulum (ER) function and rapidly downregulated by ER stress independent of transcriptional regulation. SEAP has been successfully used to detect the ER stress in both *in vitro* cell culture systems and *in vivo* animal models (Hiramatsu et al., 2006). As IECs are relatively difficult to transfect, we constructed a stable cell line for our studies.

2.1.1. Construction of IEC-6 stable cell line expressing SEAP

i. IEC-6 cells are plated at a density of 1.0×10^6 on a 60-mm plate for overnight.
ii. Cells are transfected with pSEAP2-control and pEGFP-N1 (from Clontech) using Lipofectamine^{TM} 2000 reagent (Invitrogen) for 48 h.
iii. Stable clones containing both pSEAP2 and pEGFP-N1 are selected by adding G418 into culture medium to a final concentration of 500 μg/ml, followed by colonial cloning and flow cytometry cell sorting.
iv. Expression of SEAP and GFP are confirmed by fluorescent microscopy, Western blot analysis, and SEAP enzyme activity assay.
v. The GFP-positive clone with the highest SEAP activity is amplified and maintained in cell culture medium containing G418 (250 μg/ml).

2.1.2. Detection of HIV PI-induced ER stress in IEC-6-SEAP cells

i. IEC-6-SEAP cells are plated at a density of 2×10^4 in a 96-well plate for overnight.
ii. Cells are treated with individual HIV PI or thapsigargin (TG, 100 nM), a known ER stress inducer, for 24 h.
iii. Remove 50 μl of conditioned cell culture medium and transfer to a microcentrifuge tube and centrifuge at $12,000 \times g$ for 30 s to pellet any detached cells.
iv. Take 25 μl of cell culture medium to measure the SEAP activity using Great EscAPe SEAP chemiluminescent Detection Kit according to the manufacturer's instructions (Cat No.: 631701, Clontech).
v. The viability of cells is detected with CellTiter-Glo® Luminescent Cell Viability Assay (Cat No.: G7570, Promega).
vi. The SEAP activity is normalized with the total amount of viable cells (Fig. 6.1).

2.2. Detection of HIV PI-induced ER stress by real-time RT-PCR and Western blot analysis

The ER stress response allows cells to deal with endogenous stress induced by unfolded or misfolded proteins through activation of the UPR. Several signaling molecules of the UPR have been identified, including inositol-

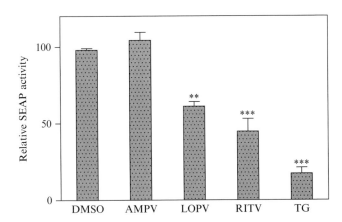

Figure 6.1 Effect of HIV PIs on ER stress induction. IEC-6 cells stably transfected with pSEAP plasmid were treated with HIV PIs (AMPV, LOPV, and RITV; 15 μM) or thapsigargin (TG, 100 nM) for 24 h. Activity of SEAP was measured using the great EscAPe SEAP Detection Kit (Clontech). The viability of cells was detected with CellTiter-Glo Luminescent Cell Viability Assay (Promega). The results were expressed as percent of control. Values were mean ± S.D. of three independent experiments. Statistical significance relative to vehicle control: **$p < 0.01$; ***$p < 0.001$.

requiring transmembrane kinase 1 (IRE1), pancreatic ER kinase (PERK), activating transcription factor 6, and a chaperone protein, glucose-regulated protein 78 (GRP78). These signaling molecules further activate downstream transcription factors such as activating factor 4 (ATF-4), x-box-binding protein 1 (XBP-1), and C/EBP homologous protein (CHOP). CHOP is one of the proteins highly inducible during ER stress and plays a critical role in ER stress-induced apoptosis (Malhotra and Kaufman, 2007; Xu et al., 2005).

2.2.1. Detection of HIV PI-induced UPR activation by real-time RT-PCR

i. IEC-6 cells or IEC-Cdx2L1 cells are plated at a density of 1.2×10^6 on 60-mm plates for overnight.
ii. Treat the cells with individual HIV PIs or TG for 24 h.
iii. Total cellular RNA is isolated using Ambion RNAqueous kit
iv. Total RNA (5 μg) is used for first-strand cDNA synthesis using High-Capacity cDNA Archive Kit (ABI). The mRNA levels of CHOP, ATF-4, sXBP-1 (spliced XBP-1), uXBP-1 (unspliced XBP-1), GRP78, and β-actin are quantified using gene-specific primers (Table 6.1): iQTM SYBR Green Supermix (Bio-Rad Laboratories) is used as a fluorescent dye to detect the presence of double-stranded DNA using iQ5 Multicolor Real-time PCR Detection System. The relative mRNA levels for each gene are normalized to internal control β-actin mRNA using $^{\Delta\Delta}Ct$ method (Fig. 6.2).

Table 6.1 Real-time PCR primers

Gene name	Gene Bank No.	Forward primer (5'–3')	Reverse primer (5'–3')
ATF-4-rat	NM_024403	GGTTCTCCAGCGACAAGG	GGTTTCCAGGTCATCCATTC
CHOP-rat	NM_024134	GGAGCAGGAGAATGAGAG	GACAGACAGGAGGTGATG
sXBP-1-rat	NM_001004210	CTGAGTCCGCAGCAGG	CCAACTTGTCCAGAATGC
uXBP-1-rat	NM_001004210	TCCGCAGCACTCAGACTAC	AGTTCCTCCAGATTAGCAGAC
GRP78-rat	NM_013083	TCCTGCGTCGGTGTATTC	CGTGAGTTGGTCTTGGC
β-Actin-rat	NM_031144	TATCGGCAATGAGCGGTTCC	AGCACTGTGTTGGCATAGAGG

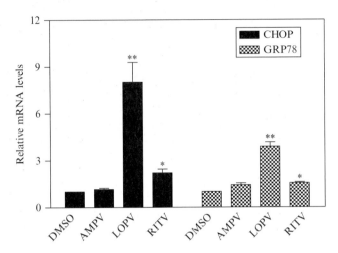

Figure 6.2 Effect of HIV PIs on CHOP and GRP78 mRNA expression. IEC-CdxL1 cells were treated with HIV PIs (AMPV, LOPV, and RITV; 25 μM) for 24 h. The mRNA levels were determined using real-time RT-PCR and normalized to internal control β-actin mRNA using $^{\Delta\Delta}Ct$ method. Statistical significance relative to vehicle control: $*p < 0.05$; $**p < 0.01$.

2.2.2. Detection of HIV PI-induced UPR activation by Western blot analysis

i. IEC-6 cells or IEC-Cdx2L1 cells are plated at a density of 1.2×10^6 on 60-mm plates for overnight.
ii. Treat the cells with individual HIV PIs or TG.
iii. Collect the cells and separate nuclear proteins from cytosolic proteins, as described previously (Zhou et al., 2005).
iv. The nuclear extract (15 μg protein) is resolved on 10% Bis-Tris NuPAGE Novex gels and transferred to nitrocellulose membranes.
v. Block the membranes overnight at 4 °C with 5% nonfat milk in TBS buffer.
vi. Incubate the membranes with antibodies to CHOP (Santa Cruz, sc-575; 1:400), XBP-1 (Santa Cruz, sc-7160; 1:300), ATF-4 (Santa Cruz, sc-200; 1:500), or lamin B (Santa Cruz, sc-6216; 1:500) at RT for 2 h.
vii. Detect the immunoreactive bands using horseradish peroxidase-conjugated goat antirabbit antibody (1:2000) or donkey antigoat antibody (1:5000) and the Western Lightning Chemiluminescence reagent plus. The density of immunoblot is analyzed using ImageJ computer software (NIH; Fig. 6.3).

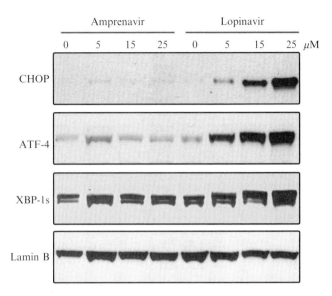

Figure 6.3 Effect of HIV PIs on UPR activation. IEC-6 cells were treated with different concentrations of HIV PIs (AMPV and LOPV, 0–25 μM) for 6 h. The nuclear proteins were isolated. Representative immunoblots against CHOP, ATF-4, XBP-1s, and lamin B are shown. Lamin B was used as loading control.

2.3. Measurement of paracellular permeability in IECs

  i. Plate 4×10^5 IEC-6 cells or IEC-Cdx2L1 cells in 12-well transwell insert (Corning Cat No.: 3460) in 0.5 ml of media (apical chamber).
 ii. Place 1.5 ml of media in basal chamber.
iii. Incubate the cells for 48 h.
 iv. Treat the cells with individual HIV PIs for 24 h.
  v. Prepare 0.5 M mannitol in PBS and sterilize using 0.2-μM filter.
 vi. Prepare fresh medium containing 5 mM mannitol.
vii. Dilute [^{14}C]-mannitol tracer (MW: 184, Amersham Pharmacia Biotech Cat No.: CFA-238; 321 μCi/mg, 250 μl, 3.386 μM) with fresh 5 mM mannitol medium to 1:1000.
viii. Replace the medium in apical wells with 0.5 ml of diluted [^{14}C]-mannitol tracer medium and replace the medium in basal wells with fresh medium containing 5 mM mannitol (NO TRACER).
 ix. Incubate for a predetermined time.
  x. Collect 0.5-ml media from BASAL chamber of each sample and place into a 4.5-ml scintillation fluid to count radioactivity using beta counter.
 xi. Positive control: Ca^{2+}-free medium treatment for 2 h.
xii. Blank count: 0.5-ml tracer-free medium + 4.5-ml scintillation fluid.
xiii. Total count of tracer: 0.5-ml tracer medium + 4.5-ml scintillation fluid.

xiv. Results : (Sample cpm − Blank cpm)/(Total cpm − Blank cpm) × 100% (Fig. 6.4).

3. ANALYSIS OF HIV PI-INDUCED ER STRESS AND DISRUPTION OF BARRIER INTEGRITY IN THE INTESTINE

C57BL/6 wild-type and CHOP$^{-/-}$ mice with C57BL/6 background (male, 6–8 weeks old, Jackson Laboratories, Bar Harbor, ME) are housed and fed standard mouse chow and tap water *ad libitum* throughout the study following protocols approved by the IACUC at Virginia Commonwealth University. Mice are randomly assigned to the following four groups:

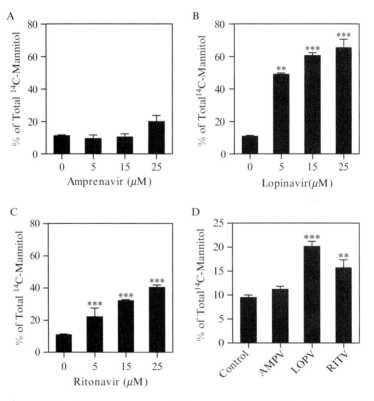

Figure 6.4 Effect of HIV PIs on paracellular permeability in IECs. (A–C) IEC-6 cells were treated with individual HIV PIs (0–25 μM) for 24 h. (D) IEC-Cdx2L1 cells were cultured on the filter in DMEM containing 4 mM isopropyl-β-D-thiogalactopyranoside (IPTG) for 16 days to induce differentiation and treated with individual HIV PIs (25 μM) for 24 h. The paracellular permeability was measured, as described in Section 2.3. Values are mean ± S.D. of three independent experiments and analyzed using one-way ANOVA, ★★$p < 0.01$, ★★★$p < 0.001$.

Control group, AMPV, RITV, and LOPV. Mice are gavaged daily with individual HIV PI (50 mg/kg) or control solution for 2 or 4 weeks. Animals are monitored daily for appearance of diarrhea, body weight loss, and other distress (Wu et al., 2010).

3.1. Measurement of intestinal permeability *in vivo*

i. Gavage mice with FITC-dextran (4 kDa; Sigma-Aldrich, Cat No.: FD4) at a dose of 600 mg/kg body weight 4 h before harvest.
ii. Collect the blood from hepatic portal vein.
iii. Determine the serum concentration of the FITC-dextran using a fluorescence plate reader with an excitation wavelength at 490 nm and an emission wavelength of 530 nm (Fig. 6.5)

3.2. Detection of HIV PI-induced ER stress in intestine by Western blot analysis

i. At the end of treatment, sacrifice the mice by cervical dislocation.
ii. Open the abdominal cavity and remove the intestine and colon to a chilled beaker containing cold PBS.
iii. Rinse out the intestinal contents using cold PBS.
iv. Take a small section of tissue ($\sim$50 mg) and finely mince it using scissors and put into the glass vessel of the $\sim$10-ml Potter-Elvehjem

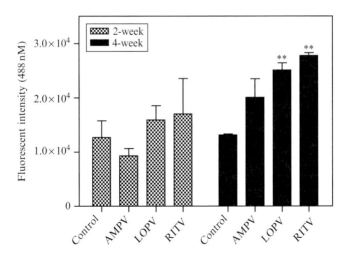

Figure 6.5 Effect of HIV PIs on intestinal permeability *in vivo*. Mice were treated with individual HIV PIs (50 mg/kg) for 2–4 weeks. The intestinal permeability was measured using FITC-dextran, as described in Section 3.1. Statistical significance relative to vehicle control: **$p < 0.01$.

homogenizer containing 2 ml of nuclear extraction buffer A (10 mM HEPES, pH 7.4, 10 mM KCl, 0.1 mM EDTA, 0.1 mM EGTA, 2 mM NaF, 0.5% NP-40) with protease inhibitors (PIs; 2 mM Na$_3$VO$_4$, 25 μg/ml leupeptin, 25 μg/ml aprotinin, 10 μg/ml pepstatin A, and 0.1 mM PMSF). Attach the precooled pestle to the electric motor and homogenize the tissue using 8–10 up-and-down stokes of pestle, roating at ~500 rpm.

v. Transfer the homogenate into a 10-ml centrifuge tube and centrifuge at 16,000×g, 4 °C for 10 min.
vi. Collect the supernatant as cytosolic fraction and resuspend the nuclear pellet in 500 μl of nuclear extraction buffer B (20 mM HEPES, pH 7.4, 400 mM NaCl, 1 mM EDTA, 1 mM EGTA, 2 mM NaF) with PIs (2 mM Na$_3$VO4, 25 μg/ml leupeptin, 25 μg/ml aprotinin, 10 μg/ml pepstatin A, and 0.1 mM PMSF) by pipetting up and down.
vii. Vigorously shake the tubes for 30 min at 4 °C on a shaking platform.
viii. Centrifuge at 8000×g for 5 min at 4 °C.
ix. Aliquot the supernatants and store at −70 °C.
x. Use 50 mg of protein to run Western blot analysis, as described in Section 2.2.2 (Fig. 6.6).

3.3. Histological examination of HIV PI-induced intestinal tissue injury

i. Fix the intestine tissues in 10% neutral-buffered formalin overnight.
ii. Wash the tissue with PBS and transfer to 70% ethanol.
iii. Formalin-fixed tissues are then embedded in paraffin, sectioned at 5 μm, and stained with hematoxylin and eosin (HE) using standard procedures.
iv. Analyze the images using a Motic BA200 microscope. Inflammation and tissue damage are assessed microscopically and histologically based on the epithelial tissue damage, as described previously (Fig. 6.7).

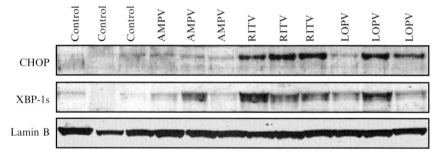

Figure 6.6 HIV PIs activate the UPR in intestine. Mice were treated with individual HIV PIs (50 mg/kg) for 2–4 weeks. Representative immunoblots of nuclear extracts isolated from intestinal tissues against CHOP, XBP-1, and lamin B are shown.

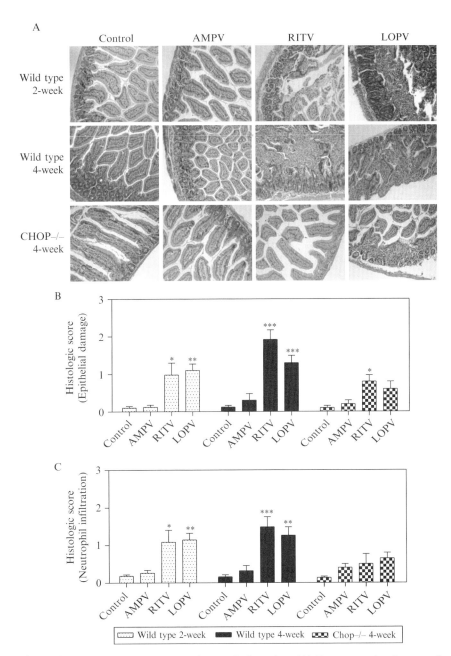

Figure 6.7 HIV PIs induce tissue damage in intestine. (A) Representative images of HE staining for each treatment group are shown. (B and C) Histological score of HIV PI-induced epithelial tissue damage and neutrophil infiltration. Statistical significance relative to vehicle control, $*p < 0.05$; $**p < 0.01$; $***p < 0.01$.

4. Summary

ER stress has been implicated in the pathogenesis of various human diseases, including gastrointestinal diseases such as inflammatory bowel disease (Kaser and Blumberg, 2009, 2010; Lin et al., 2008). IEC dysfunction due to unresolved ER stress as a consequence of HIV PI treatment will disrupt the intestinal immune system and further induce microbial translocation and systemic inflammatory response. Understanding of HIV PI-induced ER stress and UPR activation in IEC will help in the development of novel preventative and therapeutic strategies to ameliorate treatment-associated side effects and complications in HIV patients. The *in vitro* cell culture models and *in vivo* animal models discussed in this chapter will be useful tools to examine general drug-induced gastrointestinal toxicity and barrier dysfunction.

ACKNOWLEDGMENTS

I thank the following companies for providing us with the following compounds used in this research: AIDS Research and Reference Reagent Program, NIH; GlaxoSmithKline (Amprenavir); Abbott Laboratories (ritonavir and lopinavir). This work is supported by grants from the National Institutes of Health (R21AI068432, R01AT004148, R01AI057189, P01DK38030, P30CA16059, and R01DK064240), GlaxoSmithKline research fund, A. D. Williams fund, and Jeffress Memorial Trust.

REFERENCE

Ancuta, P., et al. (2008). Microbial translocation is associated with increased monocyte activation and dementia in AIDS patients. *PLoS ONE* **3**, e2516.

Bode, H., et al. (1999). The HIV protease inhibitors saquinavir, ritonavir, and nelfinavir but not indinavir impair the epithelial barrier in the human intestinal cell line HT-29/B6. *AIDS* **13**, 2595–2597.

Bode, H., et al. (2000). Effects of HIV protease inhibitors on barrier function in the human intestinal cell line HT-29/B6. *Ann. NY Acad. Sci.* **915**, 117–122.

Bode, H., et al. (2005). The HIV protease inhibitors saquinavir, ritonavir, and nelfinavir induce apoptosis and decrease barrier function in human intestinal epithelial cells. *Antivir. Ther.* **10**, 645–655.

Brenchley, J. M., and Douek, D. C. (2008a). HIV infection and the gastrointestinal immune system. *Mucosal Immunol.* **1**, 23–30.

Brenchley, J. M., and Douek, D. C. (2008b). The mucosal barrier and immune activation in HIV pathogenesis. *Curr. Opin. HIV AIDS* **3**, 356–361.

Brenchley, J. M., et al. (2006a). HIV disease: Fallout from a mucosal catastrophe? *Nat. Immunol.* **7**, 235–239.

Brenchley, J. M., et al. (2006b). Microbial translocation is a cause of systemic immune activation in chronic HIV infection. *Nat. Med.* **12**, 1365–1371.

Chen, L., et al. (2009). HIV protease inhibitor lopinavir-induced TNF-alpha and IL-6 expression is coupled to the unfolded protein response and ERK signaling pathways in macrophages. *Biochem. Pharmacol.* **78,** 70–77.

Clotet, B., and Negredo, E. (2003). HIV protease inhibitors and dyslipidemia. *AIDS Rev.* **5,** 19–24.

Dressman, J., et al. (2003). HIV protease inhibitors promote atherosclerotic lesion formation independent of dyslipidemia by increasing CD36-dependent cholesteryl ester accumulation in macrophages. *J. Clin. Invest.* **111,** 389–397.

Flexner, C. (1998). HIV-protease inhibitors. *N Engl J. Med.* **338,** 1281–1292.

Hiramatsu, N., et al. (2006). Real-time detection and continuous monitoring of ER stress in vitro and in vivo by ES-TRAP: Evidence for systemic, transient ER stress during endotoxemia. *Nucleic Acids Res.* **34,** e93.

Hui, D. Y. (2003). Effects of HIV protease inhibitor therapy on lipid metabolism. *Prog. Lipid Res.* **42,** 81–92.

Kaser, A., and Blumberg, R. S. (2009). Endoplasmic reticulum stress in the intestinal epithelium and inflammatory bowel disease. *Semin. Immunol.* **21,** 156–163.

Kaser, A., and Blumberg, R. S. (2010). Endoplasmic reticulum stress and intestinal inflammation. *Mucosal Immunol.* **3,** 11–16.

Koster, J. C., et al. (2003). HIV protease inhibitors acutely impair glucose-stimulated insulin release. *Diabetes* **52,** 1695–1700.

Lin, J. H., et al. (2008). Endoplasmic reticulum stress in disease pathogenesis. *Annu. Rev. Pathol.* **3,** 399–425.

Malhotra, J. D., and Kaufman, R. J. (2007). The endoplasmic reticulum and the unfolded protein response. *Semin. Cell Dev. Biol.* **18,** 716–731.

Marchetti, G., et al. (2008). Microbial translocation is associated with sustained failure in CD4+ T-cell reconstitution in HIV-infected patients on long-term highly active antiretroviral therapy. *AIDS* **22,** 2035–2038.

Nowroozalizadeh, S., et al. (2010). Microbial translocation correlates with the severity of both HIV-1 and HIV-2 infections. *J. Infect. Dis.* **201,** 1150–1154.

Quaroni, A., et al. (1979). Epithelioid cell cultures from rat small intestine. Characterization by morphologic and immunologic criteria. *J. Cell Biol.* **80,** 248–265.

Suh, E., and Traber, P. G. (1996). An intestine-specific homeobox gene regulates proliferation and differentiation. *Mol. Cell. Biol.* **16,** 619–625.

Wu, X., et al. (2010). HIV protease inhibitors induce endoplasmic reticulum stress and disrupt barrier integrity in intestinal epithelial cells. *Gastroenterology* **138,** 197–209.

Xu, C., et al. (2005). Endoplasmic reticulum stress: Cell life and death decisions. *J. Clin. Invest.* **115,** 2656–2664.

Zhou, H., et al. (2005). HIV protease inhibitors activate the unfolded protein response in macrophages: Implication for atherosclerosis and cardiovascular disease. *Mol. Pharmacol.* **68,** 690–700.

Zhou, H., et al. (2006). HIV protease inhibitors activate the unfolded protein response and disrupt lipid metabolism in primary hepatocytes. *Am. J. Physiol. Gastrointest. Liver Physiol.* **291,** G1071–G1080.

CHAPTER SEVEN

Dexamethasone Induction of a Heat Stress Response

José A. Guerrero, Vicente Vicente, *and* Javier Corral

Contents

1. Introduction	122
2. Evaluation of the Heat Stress Response in Patients Treated with Dexamethasone	124
2.1. Direct quantification of plasma HSPs	124
2.2. Evaluation of the heat stress response in peripheral blood mononuclear cells: Measurement of cellular HSP70 levels in freshly isolated monocytes (Madden *et al.*, 2010)	125
3. Heat Stress Response and UPR in Mice and Cellular Models	126
3.1. Dexamethasone in LPS-induced septic shock (Chatterjee *et al.*, 2007)	126
3.2. Dexamethasone in L-ASP-induced conformational disease (Hernandez-Espinosa *et al.*, 2006, 2009)	126
3.3. Protection in cardiomyocytes (Sun *et al.*, 2000)	132
3.4. Protection in epithelial cells (Urayama *et al.*, 1998)	133
References	134

Abstract

Dexamethasone is a potent, synthetic member of the glucocorticoid class of steroid drugs with pleiotropic effects on multiple signaling pathways, and has been widely used in many disorders during the last 50 years. Recent studies sustain a role of this drug in the heat stress response, increasing the levels of heat-shock proteins, particularly under certain stress conditions. More conflictive is the role of dexamethasone on the levels of endoplasmic reticulum chaperons. However, these effects may certainly contribute to explain the therapeutic benefits of dexamethasone in cardiac transplant, sepsis, cancer, and other pathologic disorders associated with stress affecting the folding of proteins. In this chapter, we review the methods that can be used to evaluate the effect of dexamethasone in the heat stress response both in patients and animal and cellular models.

Centro Regional de Hemodonación, University of Murcia, Spain

1. Introduction

Dexamethasone is a potent, synthetic member of the glucocorticoid class of steroid drugs. Dexamethasone has pleiotropic effects on multiple signaling pathways that result in a broad range of activities, being the anti-inflammatory and immunosuppressant the best known (Rhen and Cidlowski, 2005). These broad and strong effects explain that this drug is frequently prescribed for the treatment of various inflammatory diseases, including asthma, chronic obstructive pulmonary disease, and acute respiratory distress syndrome, among others. In addition to these chronic inflammatory diseases, dexamethasone has also been used for the treatment of severe sepsis and septic shock in patients in the intensive care unit. Moreover, dexamethasone is widely used to treat other diseases, such as many tumors (usually in combination with other drugs), migraine, depression, coronary syndromes, to prevent graft rejection, and it has been used as an antiemetic (Rhen and Cidlowski, 2005). However, we cannot forget that this pleiotropic activity can also have adverse effects: growth retardation in children, immunosuppression, hypertension, inhibition of wound repair, osteoporosis, and metabolic disturbances. All these harmful properties contraindicate prolonged glucocorticoid therapy (Rhen and Cidlowski, 2005).

Despite extensive clinical use for more than half a century and exhaustive research, the molecular mechanisms of dexamethasone action remain elusive, although several mechanisms have been postulated. Two major receptors are known to mediate the action of dexamethasone: the glucocorticoid receptor (GR) and the pregnane X receptor (PXR). The activation of GR requires nanomolar, whereas the activation of PXR requires micromolar concentrations of dexamethasone. The GR is a member of the steroid-hormone-receptor family of proteins. It binds with high affinity to cortisol. Within the cell, cortisol regulates the expression of numerous genes through three mechanisms: direct and indirect genomic effects and nongenomic mechanisms (Rhen and Cidlowski, 2005). Thus, the cortisol–GR complex moves to the nucleus, where it binds as a homodimer to DNA sequences called glucocorticoid-responsive elements. The resulting complex recruits either coactivator or corepressor proteins that modify the structure of chromatin, thereby facilitating or inhibiting assembly of the basal transcription machinery and the initiation of transcription by RNA polymerase II. Second, regulation of other glucocorticoid-responsive genes involves interactions between the cortisol–GR complex and other transcription factors such as nuclear factor kB (NFkB). The third mechanism is glucocorticoid signaling through membrane-associated receptors and second messengers (so-called nongenomic pathways).

Additionally, there is increasing evidence that some biological functions of dexamethasone might be mediated through the effects of this molecule in the heat stress response. Actually, the GR-bound cortisol promotes the dissociation of molecular chaperones, including heat-shock proteins (HSPs), from the receptor. Thus, hormone binding induces a temperature-dependent association of GR from HSP90 and their conversion to the DNA-binding form. Reassociation of transformed GR with HSP90 is accompanied by functional reconstitution of the untransformed state of the receptor. The demonstration that HSP90 is necessary for the high affinity steroid binding activity of GR and that HSP90–GR complexes exist in intact cells reinforce interest for the HSP90–GR interaction (Tbarka et al., 1993). Moreover, there are few reports supporting that dexamethasone activates heat-shock factor (HSF-1), the transcription factor that regulates the expression of HSPs in response to stresses such as ischemia, hypoxia, heat, stretch, or injury, and upregulates HSP72 (Knowlton and Sun, 2001). This activation of HSF-1 is not preceded by cellular injury (Sun et al., 2000) and might have potential benefits in different disorders, mainly in the heart, where HSP72, the inducible form of HSP70, has been the most intensely studied. HSP72 is induced by brief ischemia, and overexpression of HSP72 will protect cells and tissues against various forms of stress (Knowlton and Sun, 2001). Moreover, overexpression of other HSPs, including HSP60 together with HSP10 and HSP27, is also protective against cardiac injury (Sun et al., 2000).

We have also demonstrated that the coadministration of dexamethasone with L-asparaginase (L-ASP) in mouse and cellular models increased expression of the three main cytosolic HSPs, HSP27, HSP70, and HSP90, and also endoplasmic reticulum (ER) chaperons, the glucose regulated proteins (GRPs) involved in the unfolded protein response (UPR): GRP78/BIP and GRP94 (Hernandez-Espinosa et al., 2009). The increase in cytosolic HSP rises the GR activity and protects cells from toxic aggregates of misfolded proteins by contributing to the refolding of abnormally folded proteins or targeting them to the ubiquitin–proteasomal system for their destruction, while the boosted expression and clusterization of GRP78 and GRP94 actively contribute to the refolding and degradation of the abnormally folded protein, also improving the secretion of properly folded protein. These effects certainly contribute to explain the reduced amount of intracellular aggregation of antithrombin, a conformationally sensitive anticoagulant, caused by L-ASP. Interestingly, these effects were not shared by prednisone, another corticoid that is also coadministered with L-ASP to patients with acute leukemia. The clinical consequence of these effects is a lower risk of thrombosis in patients treated with dexamethasone (Nowak-Gottl et al., 2003).

Finally, a recent study observed that dexamethasone induces a unique form of cell stress in lymphocytes by quickly upregulating the ER stress gene herp and also inducing the expression of gadd153/chop, an ER stress and apoptosis-related protein. Moreover, dexamethasone treatment rapidly induces a stress response gene, dig2 but did not elevate the expression of GRP78 and GRP94 in lymphocytes (Wang *et al.*, 2003).

All these results suggest that dexamethasone may have different effects on the heat stress response by different pathways of activation, that these effects are cell type-dependent and they may not be completely shared by all corticoids. Certainly, further studies are needed to elucidate the interaction between glucocorticoids and the stress response. To comply with the needs of investigators in the field, we hereby provide detailed methods for successful evaluation of the heat stress response in the presence of dexamethasone (alone or in combination with other drugs), as well as the effect of other corticoids. Thereafter, we describe the procedures of evaluating different elements of the heat stress response and UPR in plasma samples of patients treated with these drugs, as well as in mice and cellular models.

2. Evaluation of the Heat Stress Response in Patients Treated with Dexamethasone

Many patients with different disorders are treated with dexamethasone or other corticoids alone or in combination with other drugs. The analysis of the potential role that these drugs may have on the heat stress response may be easily evaluated at least by two different mechanisms.

2.1. Direct quantification of plasma HSPs

Although HSP are synthesized intracellularly, they can also be mobilized to the plasma membrane or even be released under stress conditions. In these localizations, HSP may have relevant immune and inflammatory roles that have increasing interest (Chen and Cao, 2010).

Quantification of HSP27, HSP60, and HSP70 in plasma of patients can be achieved using different high-sensitivity enzyme-linked immunosorbent assays, some of them commercially available. The total levels of human HSP27, independent of its phosphorylation state, can be quantified using the HSP27 human ELISA KIT from BioSource International, Camarillo, CA (Cat. No. KHO0331) with a sensitivity < 0.3 ng/ml (Brerro-Saby *et al.*, 2010).

Quantification of plasma/serum HSP70 and HSP60 can be performed using the HSP70 high-sensitivity EIA kit and the HSP60 (human) EIA kit from Enzo Life Sciences (Cat. No. ADI-EKS-715 and ADI-EKS-600, respectively), with a sensitivity of 0.09 ng/ml (Fortes and Whitham, 2009). The other HSP70 commercial assay can be purchased from R&D

Systems (Cat. No. KCB1663; Sandstrom *et al.*, 2008). Alternatively, HSP60 and HSP70 levels may also be assessed by coating a 96-well microtiter plate with antihuman HSP60 or HSP70 monoclonal antibodies from Enzo Life Sciences (Cat. No. ADI-SPA-806 and ADI-SPA-822, respectively; Pockley *et al.*, 2000). Firstly, coat a 96-well microtiter plate (Nunc Immunoplate Maxisorp; Life Technologies) with the required antibody (2 μg/ml) resuspended in carbonate buffer (pH 9.5) overnight at 4 °C. Wash plates with 1% Tween 20-PBS (PBS/T) and block wells by incubation with 1% bovine serum albumin (BSA) in PBS/T. Proceed to add serum samples followed by the addition of rabbit polyclonal anti-HSP60 or anti-HSP70 antibody (1:1000; SPA-804 and SPA-812; StressGen). Bound polyclonal antibody is detected with alkaline phosphatase-conjugated murine monoclonal antibody to rabbit immunoglobulins (Sigma Chemical), followed by *p*-nitrophenyl phosphate substrate (Sigma Chemical). Read absorbance at 405 nm. Standard dose–response curves can be also generated in parallel with recombinant human HSP60 (0–2500 ng/ml) or HSP70 (0–20,000 ng/ml), both from Enzo Life Sciences. The interassay variability of the HSP60 and HSP70 immunoassays is 10%. In addition, a commercially available ELISA from Assay Designs (Cat. No. EKS-700A) can be used to determine serum HSP72. The assay sensitivity is reported to be 0.2 ng/ml, with and intraassay coefficient of variance of 4.7% (Yamada *et al.*, 2007).

2.2. Evaluation of the heat stress response in peripheral blood mononuclear cells: Measurement of cellular HSP70 levels in freshly isolated monocytes (Madden *et al.*, 2010)

Draw blood from subjects into CPDA tubes and remove red blood cells by 6% dextran sedimentation. Next, add the resulting leukocyte-rich plasma over a hypertonic gradient Nycoprep 1.068 (Robin Scientific, Solihull, UK) and centrifuge at $650 \times g$ for 15 min at room temperature. Discard the upper fraction, decant the monocyte-containing fraction into a fresh tube, and wash in a 0.3% BSA–PBS solution. Contaminating platelets can be removed by resuspending the cell pellet in 6 ml of 0.3% BSA–PBS and centrifuging at $300 \times g$ for 7 min. Finally, resuspend the cell pellets in culture medium (Roswell Park Memorial Institute medium-1640 containing 0.75 mM L-glutamine, 0.1 mg/ml streptomycin, 0.1 mg/ml penicillin, and 5% autologous serum).

Pellet 10^7 monocytes, rinse them with PBS, and solubilize at 1×10^7 cells/ml in lysis buffer. Briefly vortex the lyzate, incubate on ice for 15 min, and store at lesser than or equal to -80 °C. Before use, thaw lyzates, centrifuge at $2000 \times g$ for 5 min to remove cell debris, and transfer the HSP70-containing supernatant into a clean test tube. Proceed to measure the total levels of HSP70 using the HSP70 EIA kit (DYC1663; R&D Systems, Oxford) following the manufacturer's protocol (Madden et al., 2010).

3. Heat Stress Response and UPR in Mice and Cellular Models

The heat stress response caused by dexamethasone has been evaluated in liver, using different models where hepatocytes can be challenged by exposure to lipopolysaccharide (LPS) or L-ASP.

The effect of dexamethasone on the heat stress response has also been thoroughly evaluated in cardiomyocytes.

Finally, few studies have also showed that dexamethasone treatment also significantly induces HSP protein expression in intestinal cells.

In all these models, the heat stress response induced by dexamethasone may offer some protections with potential therapeutic utility.

3.1. Dexamethasone in LPS-induced septic shock (Chatterjee et al., 2007)

1. Induce acute endotoxemia in Swiss male mice by intraperitoneal (i.p.) injection of 18 mg/kg LPS ($t = 0$ h).
2. Inject sterile saline into control group.
3. Inject LPS-treated mice with 2 mg/kg dexamethasone i.p. at 2 h or vehicle (saline). Sacrifice mice at 6–24 h.

3.2. Dexamethasone in L-ASP-induced conformational disease (Hernandez-Espinosa et al., 2006, 2009)

- *In vivo* studies in mice.
  1. Inject mice i.p. with 10,000 IU/m^2 L-ASP or vehicle (PBS) for 2 consecutive days.
  2. In parallel, inject 10 mg/m^2 dexamethasone i.p. or vehicle. Sacrifice mice on day 2 after injections.

- *In vitro* studies using the human hepatoma cell line HepG2.
  Culture HepG2 in Dulbecco's modified medium (DMEM; Gibco-BRL, Paisley, UK) supplemented with 10% fetal calf serum (Gibco-BRL). At semiconfluency, wash cells with PBS and add fetal calf serum-deprived DMEM (Gibco-BRL) with 70 IU/ml L-ASP (resuspended in sterile PBS) or vehicle and 70 µg/ml dexamethasone or vehicle.

Collect cell culture media samples at 24 h after treatment and freeze at −70 °C until the moment of use.

3.2.1. Histological, immunofluorescence, and electronic microscopy analyses of conformationally sensitive proteins

Extract organs of interest (liver) from injected animals, finely dissect them and fixate in 4% PBS-formaldehyde for histopathology purposes. Embed organs in paraffin and obtain sections of 4 μm to be stained with hematoxylin (Panreac, Catelar del Vallés, Spain), counterstained with eosin (Panreac), dehydrated, and mounted in Dpex mounting medium (Panreac) for light microscopy observation. To detect intracellular β-sheet-mediated deposits of proteins, stain with Congo Red (Merck, Darmstadt, Germany). To visualize glycoproteic deposits, stain with periodic acid Schiff (Panreac). Also, cell viability can be assessed by Hoechst staining (Sigma-Aldrich) and numbers of nuclei may be counted in eight random 40× fields under blue fluorescence (excitation, 365 nm; emission, 465 nm). For HepG2 hepatoma cells, follow procedure as explained above, fixating the cells in 4% PBS-formaldehyde. No dehydration is needed with HepG2 cells. Staining with Congo Red, periodic acid Schiff, and Hoechst staining are performed in a similar manner. Mount preparations in Vectashield mounting medium (Vector Laboratories, Burlingame, CA).

The effect of dexamethasone on hepatic serpins with conformational sensitivity, such as antithrombin and α1-antitrypsin, is recommended. For their evaluation by immunohistochemistry purposes, a peroxidase–antiperoxidase technique can be performed. Block sections in blocking solution (0.1% Tween 20–Tris-buffered saline (TBS), 10% fetal calf serum, 5% BSA) for 30 min and incubate with rabbit antihuman antithrombin polyclonal antibody (1:1000, Dako Diagnostics) or rabbit antihuman α1-antitrypsin polyclonal antibody (1:250, Dako Diagnostics) for 90 min. Then, wash 3 times in 0.1% Tween–TBS, incubate with biotin-labeled antirabbit secondary antibody (1:250; Vector Laboratories), and signal can be further amplified with an ABC kit (Vector Laboratories). Wash sections twice in 0.1% Tween–TBS, rinse, develop with 3,3′-diaminobenzidine (0.5 mg/ml; Sigma-Aldrich) in 50 mmol/l TB (pH 7.6), counterstain with hematoxylin, dehydrate, and mount.

For their evaluation by immunofluorescence, fixate treated HepG2 cells in 4% PBS-formaldehyde. Rinse cells with PBS and block with blocking solution (0.3% Tween–PBS, 10% fetal calf serum, 5% BSA). Incubate cells with rabbit antihuman antithrombin or anti-α1-antitrypsin polyclonal antibodies (1:1000 and 1:250, respectively; Dako Diagnostics) for 90 min and wash 3 times in 0.1% Tween 20-PBS. Then, incubate cells with a fluorescein-labeled antirabbit IgG secondary antibody (1:250, Vector Laboratories) for 1 h, wash 3 times in 0.1% Tween 20-PBS, mount in Vectashield

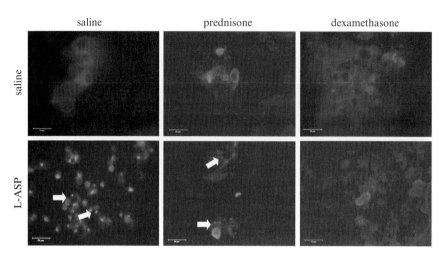

Figure 7.1 Effect of dexamethasone and prednisone on the intracellular accumulation of antithrombin (arrows) caused by L-asparaginase (L-ASP) in HepG2 cells and evaluated by immunofluorescence. (See Color Insert.)

mounting medium (Vector Laboratories), and observe using green fluorescence (excitation, 488 nm; emission, 530 nm). The protective effect of dexamethasone on the intracellular accumulation of antithrombin caused by L-ASP evaluated by this method can be observed in Fig. 7.1.

Finally, these molecules, together with the ultrastructure of the liver can be analyzed by electron microscopy. For this purpose, fix samples of livers from L-ASP-treated or control mice with 4% glutaraldehyde and dehydrate in sequential gradients of ethanol. Infiltrate samples with increasing concentrations of LR White resin (Merck). Polymerization of the resin can be performed under oxygen-free conditions for 24 h at 50 °C. Obtain ultrathin sections by using a Leica ultramicrotome (Leica Microsystems, Heidelberg, Germany) at 50 nm and place them onto copper grids (for ultrastructural analysis) or nickel grids (for immunogold labeling). Fixate copper grid sections with osmium tetroxide and then stain with uranyl acetate and lead citrate. Analyze the ultrastructural morphology of mouse liver using an EM-10 Zeiss transmission electron microscope (Zeiss, Oberkochen, Germany). In parallel, a colloidal gold immunostaining technique for antithrombin and α1-antitrypsin for nickel sections can be prepared. Rinse and block sections in blocking solution (0.1% Tween–TBS, 10% fetal calf serum, 5% BSA) for 45 min. Incubate samples with rabbit antihuman antithrombin and α1-antitrypsin polyclonal antibodies (1:250, Dako Diagnostics) for 90 min. Then, wash sections 3 times in 0.1% Tween–TBS and incubate with 10 nm of protein A-conjugated colloidal gold in Tween–TBS

(1:40; Sigma–Aldrich). Fixate nickel grid sections and proceed to analysis as indicated above.

3.2.2. Analysis of heat stress response and UPR

At sacrifice, extract liver, mince it, and homogenize in 10× volume of ice-cold lysis buffer [10 mmol/l Tris–HCl, 0.5 mmol/l dithiothreitol, 0.035% SDS, 1% Triton X-100, 50 μmol/l protease inhibitor cocktail; Sigma-Aldrich, Poole, UK]. Centrifuge the homogenates at 13,000 rpm for 10 min, store the supernatant, and resuspend the pellets into 5× volume of lysis buffer. Similarly, wash HepG2 cells extensively with sterile PBS and then lyse with 500 μl of lysis buffer. Store hepatocyte and HepG2 samples at −80 °C until the moment of use for Western blot analysis. On the day of the experiment, thaw samples and perform a Bradford assay (Bio-Rad, Denver, CO) to determine protein concentration. Boil supernatants (40 μg) and pellets (40 μg) of liver and cell extracts (40 μg) for 5 min in Laemmli buffer with 1 mmol/l dithiothreitol. Load samples into 10% SDS-PAGE. Use antihuman β-tubulin (Sigma-Aldrich, Dorset, UK) as an internal control.

For immunohistochemistry analysis, we can proceed with the liver, as indicated before.

Finally, quantitative real-time-PCR may be used to quantify the levels of intracellular expression of proteins involved in these responses. For this purpose, embed 50 μg of liver in RNAlater (Qiagen, Valencia, CA). Isolate total RNA from the tissue by using Rneasy Total RNA Isolation kit (Qiagen), with on-column DNase I (Qiagen) digestion, following the manufacturer's instructions. Perform cDNA synthesis with Superscript III (Invitrogen) following the manufacturer's instructions and proceed to quantitative real-time polymerase chain reaction assay with specific primers for HSP70 fw 5′-CTGAACCCGCAGAACAAC-3′ back 5′-CTTCATCTT CGTCAGCACCA-3′; HSP27 fw 5′-CCACTGCCGAGTACGAATTT-3′ back 5′-GGAACTTGCCTTCACTGAGC-3′; HSP90 fw 5′-CTCCTT GGTCTCACC TGTGATA-3′ back 5′-TGTTGCGGTACTACACATC TGC-3′; and gapdh fw 5′-GCCAAGTATGATGACATCAAG-3′ back 5′-AAGGTGGAAGAATGGGAG-3′. Set a threshold in the linear part of the amplification curve (fluorescence $= f$[cycle number]), and for every gene, calculate the number of cycles needed to reach the linear part. Melting curves and agarose gel electrophoresis will establish the purity of the amplified band. Normalize each gene to the housekeeping gene gapdh before calculations of fold change.

To evaluate the heat stress response, we can detect HSPs by using antihuman HSP27, HSP70, and HSP90. Figure 7.2 shows representative immunohistochemistry and Western blot analysis of HSP90 levels in mice treated with L-ASP alone or in combination with dexamethasone or prednisone.

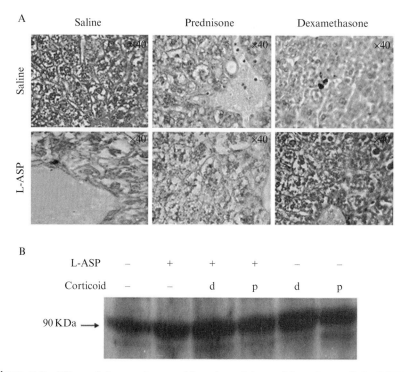

Figure 7.2 Effect of dexamethasone (d) and prednisone (p) on intracellular HSP90 caused by L-asparaginase (L-ASP) in mice and evaluated by immunohistochemistry (A) or Western blot (B).

To evaluate the protein unfolded response, we can detect ER chaperones such as GRP78 or GRP94 by using antihuman GRP78 and antihuman GRP94 by immunohistochemistry or Western blot. Figure 7.3 shows representative immunohistochemistry and Western blot analysis of GRP78 levels in mice treated with L-ASP alone or in combination with dexamethasone or prednisone.

For immunohistochemistry purposes, perform a peroxidase–antiperoxidase technique. Sections were rinsed and endogenous peroxidase inhibited. Rinse sections of liver and incubate in blocking solution (0.1% Tween 20–TBS, 10% fetal calf serum, 5% BSA) for 30 min. Incubate with rabbit antihuman antithrombin polyclonal antibody (1:1000, Dako Diagnostics) or rabbit antihuman α1-antitrypsin polyclonal antibody (1:250, Dako Diagnostics) for 90 min. Then, wash 3 times in 0.1% Tween–TBS, incubate with biotin-labeled antirabbit secondary antibody (1:250; Vector Laboratories), and signal can be further amplified with an ABC kit (Vector Laboratories). Wash sections twice in 0.1% Tween–TBS, rinse, develop with 3,3′-diaminobenzidine (0.5 mg/ml; Sigma-Aldrich) in 50 mmol/l TB

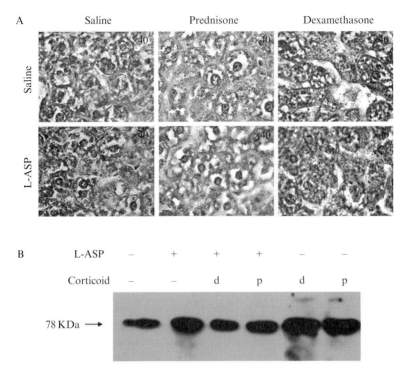

Figure 7.3 Effect of dexamethasone (d) and prednisone (p) on intracellular GRP78 caused by L-asparaginase (L-ASP) in mice and evaluated by immunohistochemistry (A) or Western blot (B).

(pH 7.6), counterstain with hematoxylin, dehydrate, and mount. To analyze the expression of all studied HSPs, assign subjective values to each liver section after screening 6 hepatic fields (40× magnification) and obtain average mean values.

For Western blot analysis, blot proteins to PVDF membranes and immunostain with the same polyclonal antibodies: HSP27, HSP70 (1:500 in both cases), HSP90 (1:1.000), GRP78 (1:250), or GRP94 (1:100; Santa Cruz Biotechnology Inc., CA, U.S.A.). Incubate with mouse antirabbit IgG-horseradish peroxidase conjugate (Amersham Biosciences Ltd., UK) or mouse antigoat IgG-horseadish peroxidase conjugate (Vector Laboratories). Develop bands by means of enhanced chemiluminescence (Amersham Biosciences Ltd., UK). Perform a densitometric analysis using the Quantity-One software (Bio-Rad).

All primary antibodies are able to recognize murine proteins, given the high degree of homology between humans and mice.

3.3. Protection in cardiomyocytes (Sun et al., 2000)

HSPs increase in response to myocardial ischemia and other stresses and have a protective effect against cardiac injury. Dexamethasone activates HSF-1 and induces a 60% increase in HSP72 in adult cardiac myocytes.

3.3.1. Isolation of adult rat cardiac myocytes

Isolate adult rat cardiac myocytes from 3- to 4-month-old male Sprague-Dawley rats weighing 250–300 g. Remove the heart from anesthetized rats and immerse it in ice-cold heparin-Joklik A buffer [modified MEM/Joklik (Gibco, Grand Island, NY), 60 mM taurine, 20 mM creatine, 5 mM HEPES, 0.1% BSA, 1 IU heparin/ml, pH 7.4]. Perfuse on a Langendorff apparatus with heparin-Joklik A to remove any blood for 3 min, followed by a second perfusion step with Joklik A with 0.6 mg/ml collagenase (collagenase type 2, Worthington Biochemical, Lakewood, NJ). Mince the heart and digest twice for 5 min in a shaking water bath at 37 °C. Filter cell suspension and wash twice with Joklik A buffer. Transfer myocytes to the top of a 6% BSA gradient to increase the percent yield of rod-shaped cells. Add 100 mM CaCl$_2$ to a final 1000 μM calcium concentration in 5 steps with 5-min incubations. Resuspend cardiac myocytes in M199 culture media (Gibco) supplemented with 100 U penicillin, 100 μg streptomycin, 20 μl human serum albumin, 5 μg insulin, and 5 μg transferrin per milliliter and transfer to a cell culture flask for 2 h in an incubator at 37 °C to remove fibroblasts. Plate cells on 0.2% laminin (Gibco)-coated dishes in M199 at 37 °C in a humidified incubator with 5% CO_2–95% air. This procedure yields on average 70% rod-shaped cardiac myocytes corresponding >97% to cardiac myocytes.

3.3.2. Dexamethasone treatment

When the cells become adherent to the dishes (after 2–4 h of culture), remove medium and add fresh M199 medium containing either 10 or 100 μM dexamethasone (Sigma, St. Louis, MO), or an equal volume of diluent. Collect samples at different times of treatment for Western blot analysis for HSP levels.

3.3.3. Animal models

Perform and i.p. injection of dexamethasone (1 mg/kg) or an equivalent volume of ethanol to nonfasted rats. At sacrifice, wash hearts from control and dexamethasone-treated rats by flushing buffer through the aorta. Remove atria and other tissues.

3.3.4. Western blot analysis

Wash cells twice with PBS and solubilize them by scraping. Cells or hearts can be lysed as indicated before or using alternative methods. Western blot analysis of HSPs can be performed with the polyclonal antibodies antihuman HSP proteins indicated before. Other antibodies used for these studies included rabbit polyclonal antibody to HSP72 protein (1:5000 dilution), mouse monoclonal antibody to HSP60 protein (1:70,000 dilution), and rabbit polyclonal antibody to HSP27 protein (1:5000 dilution) from Stress-Gen (Victoria, BC, Canada). As a control, a mouse monoclonal antibody to α-actin from Sigma (1:1000 dilution) can be used.

3.3.5. Heat-shock factors analysis

HSP synthesis is controlled by a specific family of transcription factors, HSFs. The primary HSF involved in the regulation of expression of HSPs is HSF-1. Both heat and hypoxia activate HSF-1, which is present in the cytoplasm in an inactive form as a monomer. When stress trimerization occurs as well as phosphorylation, HSF-1 migrates to the nucleus where it binds to the heat-shock element (HSE), which is present in the promoter of the stress response gene, initiating HSP transcription and synthesis.

The effect of dexamethasone on HSF-1 may be evaluated by Western blot or immunohistochemistry analysis using an antihuman HSF-1 monoclonal antibody (1:250; Santa Cruz, U.S.A.) or by Gel shift analysis. For the last purpose, end-labeled [γ-32P]ATP 5'-CTAGAAGCTTCTA-GAAGCTTCT-AG-3' may be used as consensus HSE, and supershift studies can be carried out using a mouse monoclonal anti-HSF-1 (Affinity Bioreagents). Incubate the sample–HSE mixes with antibody at 1:5 and 1:10 dilutions for 30 min. For cold compete experiments, incubate samples with a 50-fold molar excess of cold HSE for 15 min before adding labeled HSE. Collect images using a PhosphoImager (Molecular Dynamics, Sunnyvale, CA).

3.4. Protection in epithelial cells (Urayama *et al.*, 1998)

Dexamethasone stimulates a time- and dose-dependent response in HSP72 protein expression that parallels its effects on cell viability caused by oxidant-induced stress in IEC-18 rat small intestinal cells, as well as in rat intestinal mucosal cells *in vivo*. Culture IEC-18 cells in high-glucose DMEM with 5%, v/v, fetal bovine serum, 0.1 U/ml insulin, 50 mg/ml streptomycin, and 50 U/ml penicillin. Treat cells at or near confluence between passages 16 and 32, with various pharmacologically relevant concentrations of dexamethasone: 10^{-10}–10^{-4} M in complete media. The animal model involves the treatment of Sprague-Dawley rats (250–275 g) for 4 days with i.p. 0.2 mg/kg/d of dexamethasone. Evaluate the expression of HSP72 by Western blot analysis using a specific mouse monoclonal anti-HSP72 antibody, C92 (StressGen, Victoria, BC, Canada).

REFERENCES

Brerro-Saby, C., Delliaux, S., Steinberg, J. G., Boussuges, A., Gole, Y., and Jammes, Y. (2010). Combination of two oxidant stressors suppresses the oxidative stress and enhances the heat shock protein 27 response in healthy humans. *Metabolism* **59,** 879–886.

Chatterjee, S., Premachandran, S., Shukla, J., and Poduval, T. B. (2007). Synergistic therapeutic potential of dexamethasone and L-arginine in lipopolysaccharide-induced septic shock. *J. Surg. Res.* **140,** 99–108.

Chen, T., and Cao, X. (2010). Stress for maintaining memory: HSP70 as a mobile messenger for innate and adaptive immunity. *Eur. J. Immunol.* **40,** 1541–1544.

Fortes, M. B., and Whitham, M. (2009). No endogenous circadian rhythm in resting plasma Hsp72 concentration in humans. *Cell Stress Chaperones* **14,** 273–280.

Hernandez-Espinosa, D., Minano, A., Martinez, C., Perez-Ceballos, E., Heras, I., Fuster, J. L., Vicente, V., and Corral, J. (2006). L-asparaginase-induced antithrombin type I deficiency: Implications for conformational diseases. *Am. J. Pathol.* **169,** 142–153.

Hernandez-Espinosa, D., Minano, A., Ordonez, A., Mota, R., Martinez-Martinez, I., Vicente, V., and Corral, J. (2009). Dexamethasone induces a heat-stress response that ameliorates the conformational consequences on antithrombin of L-asparaginase treatment. *J. Thromb. Haemost.* **7,** 1128–1133.

Knowlton, A. A., and Sun, L. (2001). Heat-shock factor-1, steroid hormones, and regulation of heat-shock protein expression in the heart. *Am. J. Physiol. Heart Circ. Physiol.* **280,** H455–H464.

Madden, J., Coward, J. C., Shearman, C. P., Grimble, R. F., and Calder, P. C. (2010). Hsp70 expression in monocytes from patients with peripheral arterial disease and healthy controls: Monocyte Hsp70 in PAD. *Cell Biol. Toxicol.* **26,** 215–223.

Nowak-Gottl, U., Ahlke, E., Fleischhack, G., Schwabe, D., Schobess, R., Schumann, C., and Junker, R. (2003). Thromboembolic events in children with acute lymphoblastic leukemia (BFM protocols): Prednisone versus dexamethasone administration. *Blood* **101,** 2529–2533.

Pockley, A. G., Wu, R., Lemne, C., Kiessling, R., de Faire, U., and Frostegard, J. (2000). Circulating heat shock protein 60 is associated with early cardiovascular disease. *Hypertension* **36,** 303–307.

Rhen, T., and Cidlowski, J. A. (2005). Antiinflammatory action of glucocorticoids—New mechanisms for old drugs. *N Engl J. Med.* **353,** 1711–1723.

Sandstrom, M. E., Siegler, J. C., Lovell, R. J., Madden, L. A., and McNaughton, L. (2008). The effect of 15 consecutive days of heat-exercise acclimation on heat shock protein 70. *Cell Stress Chaperones* **13,** 169–175.

Sun, L., Chang, J., Kirchhoff, S. R., and Knowlton, A. A. (2000). Activation of HSF and selective increase in heat-shock proteins by acute dexamethasone treatment. *Am. J. Physiol. Heart Circ. Physiol.* **278,** H1091–H1097.

Tbarka, N., Richard-Mereau, C., Formstecher, P., and Dautrevaux, M. (1993). Biochemical and immunological evidence that an acidic domain of hsp 90 is involved in the stabilization of untransformed glucocorticoid receptor complexes. *FEBS Lett.* **322,** 125–128.

Urayama, S., Musch, M. W., Retsky, J., Madonna, M. B., Straus, D., and Chang, E. B. (1998). Dexamethasone protection of rat intestinal epithelial cells against oxidant injury is mediated by induction of heat shock protein 72. *J. Clin. Invest.* **102,** 1860–1865.

Wang, Z., Malone, M. H., Thomenius, M. J., Zhong, F., Xu, F., and Distelhorst, C. W. (2003). Dexamethasone-induced gene 2 (dig2) is a novel pro-survival stress gene induced rapidly by diverse apoptotic signals. *J. Biol. Chem.* **278,** 27053–27058.

Yamada, P. M., Amorim, F. T., Moseley, P., Robergs, R., and Schneider, S. M. (2007). Effect of heat acclimation on heat shock protein 72 and interleukin-10 in humans. *J. Appl. Physiol.* **103,** 1196–1204.

CHAPTER EIGHT

Detecting and Quantitating Physiological Endoplasmic Reticulum Stress

Ling Qi,[*,†] Liu Yang,[†] and Hui Chen[*]

Contents

1. Introduction	138
2. Detecting UPR at the Level of UPR Sensors	139
2.1. Preparation of the Phos-tag gels	139
2.2. Preparation of whole cell lysates	139
2.3. Gel running and transfer	140
2.4. Western blot	140
3. Quantitating ER Stress at the Level of UPR Sensors	141
4. Detecting Levels of Other Common UPR Targets	142
5. Important Tips	143
6. Concluding Remarks	144
Acknowledgments	145
References	145

Abstract

Unfolded protein response (UPR) is a key cellular defense mechanism associated with many human "conformational" diseases, including heart diseases, neurodegeneration, and metabolic syndrome. One of the major obstacles that have hindered our further understanding of physiological UPR and its future therapeutic potential is our inability to detect and quantitate ER stress and UPR activation under physiological and pathological conditions, where ER stress is perceivably very mild. Here, we describe a Phos-tag-based Western blot approach that allows for direct visualization and quantitative assessment of mild ER stress and UPR signaling, directly at the levels of UPR sensors, in various *in vivo* conditions. This method will likely pave the foundation for future studies on physiological UPR, aid in the diagnosis of ER-associated diseases, and facilitate therapeutic strategies targeting UPR *in vivo*.

[*] Division of Nutritional Sciences, Cornell University, Ithaca, New York, USA
[†] Graduate Program in Biochemistry, Molecular and Cell Biology, Cornell University, Ithaca, New York, USA

Methods in Enzymology, Volume 490 © 2011 Elsevier Inc.
ISSN 0076-6879, DOI: 10.1016/B978-0-12-385114-7.00008-8 All rights reserved.

1. Introduction

ER homeostasis is tightly monitored by ER-to-nucleus signaling cascades termed unfolded protein response (UPR) (Ron and Walter, 2007). Recent studies have linked ER stress and UPR activation to many human diseases, including heart complications, neurodegenerative disorders, and metabolic syndrome (Kim et al., 2008; Ron and Walter, 2007; He et al., 2010). Indeed, chemical chaperones and antioxidants aiming to reduce ER stress and UPR activation have been shown to be effective in mouse models of obesity and type-1 diabetes (Back et al., 2009; Basseri et al., 2009; Malhotra et al., 2008). Despite recent advances, our understanding of UPR activation under physiological conditions is still at its infancy, largely due to the lack of sensitive experimental systems that can detect mild UPR sensor activation.

The underlying mechanisms of UPR signaling and activation induced by chemical drugs such as thapsigargin (Tg), tunicamycin, and dithiothreitol (DTT) are becoming increasingly well characterized (Ron and Walter, 2007). Upon ER stress, two key ER-resident transmembrane sensors, inositol-requiring enzyme 1 (IRE1α) and PKR-like ER-kinase (PERK), undergo dimerization or oligomerization and *trans*-autophosphorylation via their C-terminal kinase domains, leading to their activation (Kim et al., 2008; Ron and Walter, 2007). Phosphorylation of IRE1α and PERK has been challenging, if not impossible, to detect under physiological conditions. The mobility-shift of IRE1α shown in many studies is very subtle on regular SDS-PAGE gels and difficult to assess. In addition, commercially available phospho-specific antibodies against P-Ser724A IRE1α and P-Thr980 PERK do not reflect the overall phosphorylation status of the proteins. Importantly, it remains unclear whether Ser724 of IRE1α or Thr980 of PERK is indeed phosphorylated under various physiological and disease conditions.

Alternatively, many studies have used downstream events such as splicing of X-box binding protein 1 (XBP1) mRNA, phosphorylation of eukaryotic translation initiation factor 2a (eIF2α), induction of C/EBP homologous protein (CHOP), and various genes involved in protein folding and ER-associated degradation (ERAD) as surrogate markers for UPR activation. This method, albeit convenient, may be confounded by the possibility of integrating signals not directly related to stress in the ER. For example, the PERK pathway of the UPR is part of the integrated stress response that consists of three other eIF2α kinases (Ron and Walter, 2007). Activation of any of these kinases leads to eIF2α phosphorylation and induction of ATF4 and CHOP. A recent study also showed that ATF4 and CHOP can be regulated translationally in a PERK-independent manner via the TLR signaling pathways (Woo et al., 2009). Furthermore, UPR target genes such as CHOP and ER chaperones can be induced by other signaling cascades such as hormones, insulin, and cytokines/growth factors (Brewer et al., 1997; Miyata et al., 2008). Thus, downstream UPR targets alone are not best suited for accurate assessment and evaluation of UPR status, especially under physiological and disease settings.

As hyperphosphorylation of UPR sensors, IRE1α and PERK, is believed to constitute an early initiating event in UPR (Shamu and Walter, 1996; Welihinda and Kaufman, 1996), we have optimized SDS-PAGE gels incorporated with Phos-tag and Mn^{2+}, termed Phos-tag gels (Kinoshita et al., 2006), to increase the separation between the phosphorylated and nonphosphorylated forms of two UPR sensors, IRE1α and PERK (Sha et al., 2009; Yang et al., 2010). This powerful tool allows us to "visualize" and for the first time, quantitate ER stress in cells or tissues under physiological and pathological conditions. In this chapter, we describe detailed methods used to detect IRE1α and PERK activation and to quantitate the levels of ER stress.

2. Detecting UPR at the Level of UPR Sensors

2.1. Preparation of the Phos-tag gels

Dissolve Phos-tag in H_2O to a final concentration of 5 mM as instructed by the supplier (Phos-tag acrylamide AAL-107, Wako Pure Chemical Industries #304-93521). Add Phos-tag directly to the solution mix for the resolving gels to make standard 5% SDS-PAGE minigels (Table 8.1) using the Bio-Rad minigel casting systems. Final concentrations of Phos-tag in the resolving gels are 25 and 3.5 μM for IRE1α and PERK proteins, respectively. Note that no Phos-tag is in the stacking gels. For optimal results, freshly prepare gels prior to the experiments. Regular gels without Phos-tag are prepared simultaneously using the same recipe minus Phos-tag and $MnCl_2$ (Table 8.1).

2.2. Preparation of whole cell lysates

Lyse cells or frozen tissues in cold Tris-based lysis buffer (50 mM Tris pH 7.5, 150 mM NaCl, 1% Triton X-100, and 1 mM EDTA with protease inhibitors) for 15 min on ice and followed by sonication for 15 s on ice. Spin lysates at top speed for 10 min at 4 °C. Transfer the supernatant to a fresh tube and measure protein concentrations using the Bradford assay (Bio-Rad). Adjust concentrations to 0.5 ∼ 2 μg/μl using lysis buffer, add H_2O and 5× SDS sample buffer, boil for 5 min, and spin briefly. Load 15–30 μg lysates per lane. We have noted that it is preferred to prepare lysates at

Table 8.1 Recipe for 5% Phos-tag gels

Components	Resolving gels (15 ml, for two gels)		Stacking gels (5 ml, for two gels)
	IRE1α	PERK	
mQ H$_2$O (ml)	8.30	8.45	3.71
30% 29:1 acrylamide (ml)	2.5	2.5	0.667
1.5 M Tris pH 8.8 (ml)	3.75	3.75	–
1 M Tris pH 6.8 (μl)	–	–	625
10% SDS (μl)	150	150	50
10% ammonia persulfate (μl)	150	150	50
5 mM Phos-tag (μl)	75	10.5	–
10 mM MnCl$_2$ (μl)	75	10.5	–
Mix gently and well			
TEMED (μl)	15	15	8

relatively high concentrations and dilute them down similarly prior to the loading to ensure comparable salt concentration (See Section 5).

2.3. Gel running and transfer

Fill the Bio-Rad minigel running tank with running buffer (14.4 g glycine, 3 g Tris, and 1 g SDS per liter). Running conditions are 100 V for ~3 h for IRE1α and 15 mA for 15 min, followed by 5 mA for 9.5 h for PERK. Then, gently rock the Phos-tag gel in transfer buffer (14.4 g glycine and 3 g Tris with 10% methanol per liter) containing 1 mM EDTA for 10 min, followed by 10 min incubation in transfer buffer. Transfer at 90–100 V for 1.5 h in the wet transfer system, where the gel and polyvinylidene difluoride (PVDF) membrane are held within a gel hold cassette (Bio-Rad) and soaked under transfer buffer. Upon the completion of the transfer, carefully mark the PVDF membrane to indicate the protein side and then place the protein side up in a container with Tris-based buffer (TBS) containing Tween-20 (TBST; for 1 liter 1× TBST: 5.68 g NaCl, 2.4 g Tris pH 7.5, and 0.1% Tween 20).

2.4. Western blot

Incubate the PVDF membrane with the primary antibody diluted in 5% milk/TBST or 2% BSA/TBST (Table 8.2) overnight at 4 °C or 1 h at room temperature, followed by the secondary antibody. Wash extensively with TBST in-between. After a final wash in TBST, add ECL (Pierce or Amersham) or ECL$^+$ (Amersham) Western blot detecting reagents to the membrane for 1 or 5 min, respectively. Then, place PVDF membranes in a plastic

Table 8.2 Information for antibodies that work for endogenous proteins

Name (species)	Company	Catalog/clone #	Dilutions	Size^a
IRE1α	Cell signaling	3294/14C10	1:1000	125 kDa
PERK^b	Cell signaling	3192/C33E10	1:1000	140 kDa
P-Thr980 PERK^c	Cell signaling	3179/16F8	1:1000	150–170 kDa
CHOP (mouse^d)	Santa Cruz	sc7351/B3	1:500	30 kDa
XBP1	Santa Cruz	sc7160/M-186	1:1000	54, 30 kDa^e
p-eIF2α	Cell signaling	9721	1:1000	38 kDa
eIF2α	Cell signaling	9722	1:1000	38 kDa
GRP78 (goat)	Santa Cruz	sc1051/C-20	1:1000	78 kDa
HSP90	Santa Cruz	sc7947/H-114	1:5000	90 kDa

^{a} Size refers to the position of proteins on regular SDS-PAGE gels using the Precision Plus protein marker (Bio-Rad).
^{b} The size of PERK refers to non-phosphorylated PERK following λPPase treatment.
^{c} The position of p-Thr980 PERK varies depending on the activation status of PERK.
^{d} Antibody species are rabbit unless otherwise indicated.
^{a} Endogenous XBP1u protein is very hard to detect and may subject to post-translational modifications. XBP1u, 30 kDa and XBP1s, 54 kDa.

sheet cover and expose to either film or Bio-Rad ChemiDoc. Both IRE1α and PERK antibodies are very specific, even for tissues (Fig. 8.1) (Yang et al., 2010). Of note, endogenous IRE1α and PERK proteins in most cell types (e. g., HEK293T cells, mouse embryonic fibroblasts, macrophages, pancreatic acinar and β cells, hepatocytes, and adipocytes) and tissues, except skeletal muscles, are readily detectable ((Yang et al., 2010) and not shown). Skeletal muscles have very low PERK protein (Yang et al., 2010).

Strip-reprobe the IRE1α blot for HSP90 as a loading and position control. To strip, place the PVDF membrane twice in stripping solution (100 mM β-mercaptoethanol, 2% SDS, and 62.5 mM Tris pH 6.8) prewarmed at 50–55 °C for 15 min with a gentle shake. Rinse once in ddH$_2$O and twice in TBST with a gentle shake for 10 min each. The membrane is ready for reprobe.

3. QUANTITATING ER STRESS AT THE LEVEL OF UPR SENSORS

Notably, the unique pattern of IRE1α phosphorylation in the Phos-tag gel allows for a quantitative assessment of ER stress (Fig. 8.1). The percent of IRE1α being phosphorylated, as % p-IRE1α in total IRE1α, can be quantitated by the Image J or Image Lab software in the Bio-Rad ChemiDoc system, as described by the manufactures.

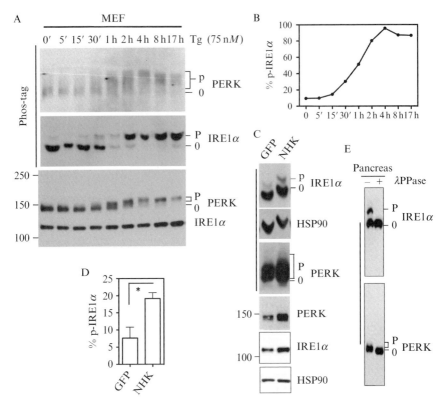

Figure 8.1 Visualization and quantitation of ER stress and UPR activation. (A) Immunoblots of IRE1α and PERK proteins in Tg-treated MEFs. (B) Quantitation of percent of phosphorylated IRE1α in total IRE1α protein in Phos-tag gels shown in A. (C) Immunoblots of IRE1α and PERK in HEK293T cells transfected with the indicated plasmids for 24 h. NHK, the unfolded form of α1-antitrypsin; GFP, negative control plasmids. (D) Quantitation of percent of phosphorylated IRE1α in total IRE1α protein in Phos-tag gels shown in C. (E) Immunoblots of IRE1α and PERK in pancreatic lysates treated with λPPase. Phos-tag gels are indicated with a bar at the left-hand side. HSP90, a loading control. "0" refers to the non- or hypophosphorylated forms of the protein, whereas "p" refers to the hyperphosphorylated forms of the protein. $\star P < 0.05$ using unpaired two-tailed Student's t-test. This data is taken or modified from Yang et al. (2010).

4. Detecting Levels of Other Common UPR Targets

Several of downstream UPR targets can be detected using Western blot (Sha et al., 2009; Yang et al., 2010), including PERK downstream targets, eIF2α, CHOP, and IRE1α downstream target XBP1s. Antibody information is listed in Table 8.2. It is important to note that, based on our experience, these downstream markers cannot be used alone to assess UPR activation under physiological and disease settings for the reasons stated in Section 1, while it is fine for the conditions using pharmacological stressors.

For many cell types and tissues, nuclear extraction is required to detect CHOP and XBP1 proteins. To this end, resuspend cells in a 6-cm dish in 200-μl ice-cold hypotonic buffer (10 mM HEPES pH 7.9, 10 mM KCl, 0.1 mM EDTA, 0.1 mM EGTA, 1 mM DTT, and protease inhibitors), followed by incubation on ice for 15 min. Add 12 μl of 10% of NP-40 to a final concentration of 0.6%. Vortex vigorously for 15 s and incubate on ice for 1 min prior to centrifugation at top speed for 1 min. Transfer the supernatant to a fresh tube as the cytosolic fraction. It is recommended that the pellet be washed once with 1-ml hypotonic buffer. Resuspend the pellet in 50-μl ice-cold high salt buffer (20 mM HEPES pH 7.9, 0.4 M NaCl, 1 mM EDTA, 1 mM EGTA, and 1 mM DTT) and vortex vigorously for 15 s every 5 min for a total of 20 min. Spin at 4 °C for 5 min and the supernatant is the nuclear fraction. For tissues, homogenize a pea-sized frozen tissue 20 times in 600 μl hypotonic buffer using dounce homogenizer with a loose pestle until no visible clear chunks. The rest of the procedure is the same as above for the cells. The quality of nuclear extraction can be checked using Western blot for nuclear proteins such as Lamin or CREB.

5. Important Tips

Phos-tag gel for IRE1α blot, not the PERK, is sensitive to salt concentrations, with different salt concentrations leading to gel curvature (Fig. 8.1C). Therefore, it is important to ensure that salt concentrations are comparable among samples. Additionally, the IRE1α blot in the Phos-tag gel is routinely reprobed with HSP90 (90 kDa vs. 110 kDa IRE1α) as a position control. For samples that are known to have very different salt concentrations, place an empty lane between them. Of note, this only affects the artistic presentation of the results, not the quantitation of percent of p-IRE1α.

For PERK, it is less important under drug-treated conditions to use Phos-tag gels, as regular gels may suffice to separate phosphorylated PERK (Fig. 8.1A). Nonetheless, addition of Phos-tag further increases the separation (Fig. 8.1A), which may be necessary under certain physiological conditions (Fig. 8.1C). Using this method, we recently showed that mild ER stress, with 10% IRE1α being phosphorylated, is detected during overnight fasting in pancreas, and is further increased by threefold upon 2-h refeeding (Yang et al., 2010). Hence, it is important to use the method described here to detect and quantitate endogenous PERK and IRE1α activation under physiological conditions.

Inclusion of positive and phosphatase-treated controls is critical to assess the extent of ER stress. Cells treated with chemical drugs (e.g., 60–300 nM Tg, 1–5 μg/ml tunicamycin, or 1 mM DTT for 2 h) should be included as positive controls. To ensure that phosphorylation accounts for the bandshift and set the baseline for nonphosphorylated PERK, cell lysates should be treated with phosphatases (Fig. 8.1E). To this end, incubate ~100 μg lysates with 0.5 μl lambda phosphatase (λPPase, New England BioLabs, NEB) in 1× PMP buffer (NEB) and 1× $MnCl_2$ at 30 °C for 30 min. Stop the reaction by the addition of 5× SDS sample buffer and boil for 5 min.

6. Concluding Remarks

The Phos-tag-based system to quantitatively assess ER stress and UPR activation has the following major advantages (Yang et al., 2010): First, dynamic ranges of PERK and IRE1α phosphorylation can be more sensitively visualized compared to regular SDS-PAGE gels; this is particularly important for physiological UPR where ER stress can be so mild that traditional methods may no longer be accurate or reliable. Second, the major breakthrough of our method lies in the unique pattern of IRE1α phosphorylation in the Phos-tag gel which allows for a *quantitative* assessment of ER stress. Finally, in comparison to using commercially available phospho-specific antibodies (e.g., P-Ser724A IRE1α and P-Thr980 PERK), our method not only provides a complete view of the overall phosphorylation status of IRE1α and PERK proteins, but also circumvents the issue of whether these specific residues are indeed phosphorylated under certain physiological conditions.

Interestingly, using this method, our data revealed that many mouse tissues and cell types display constitutive basal UPR activity, presumably to counter misfolded proteins passing through the ER (Yang et al., 2010). This observation is in line with an early report demonstrating that under physiological conditions, removal of these misfolded proteins in yeast requires coordinated action of UPR and ERAD (Travers et al., 2000). Taking it one step further, our data show that a fraction of mammalian IRE1α and PERK is constitutively active, with ~10% IRE1α being phosphorylated and activated (Yang et al., 2010). This low level of IRE1α activation may provide a plausible explanation for the inability of an earlier study to detect basal UPR in the XBP1s-GFP reporter mice (Iwawaki et al., 2004). We believe that the basal UPR activity is critical in providing quality control and maintaining ER homeostasis, as illustrated in various UPR-deficient mouse models (Harding et al., 2001; Masaki et al., 1999; Reimold et al., 2000; Scheuner et al., 2001, 2005; Yamamoto et al., 2007; Zhang et al., 2005, 2006).

As ER stress is being implicated in an increasing number of human diseases, new strategies and approaches enabling a comprehensive understanding of UPR in physiological and disease settings are urgently needed to facilitate drug design targeting UPR in conformational diseases (Kim et al., 2008; He et al., 2010). The ability to directly visualize and quantitate the early activating events is an important first step toward gaining a comprehensive view of physiological UPR. Overall, it is our hope that the method described here will aid in the diagnosis of UPR-associated diseases and improve and facilitate therapeutic strategies targeting UPR *in vivo*.

ACKNOWLEDGMENTS

We thank members of Qi laboratory for helpful discussions. This study was supported in part by Cornell startup fund, American Federation for Aging Research (RAG08061), American Diabetes Association (7-08-JF-47), and NIH R01DK082582 (to L.Q.). A patent has been filed regarding methods to assess physiological UPR.

REFERENCES

Back, S. H., Scheuner, D., Han, J., Song, B., Ribick, M., Wang, J., Gildersleeve, R. D., Pennathur, S., and Kaufman, R. J. (2009). Translation attenuation through eIF2alpha phosphorylation prevents oxidative stress and maintains the differentiated state in beta cells. *Cell Metab.* **10,** 13–26.

Basseri, S., Lhotak, S., Sharma, A. M., and Austin, R. C. (2009). The chemical chaperone 4-phenylbutyrate inhibits adipogenesis by modulating the unfolded protein response. *J. Lipid Res.* **50,** 2486–2501.

Brewer, J. W., Cleveland, J. L., and Hendershot, L. M. (1997). A pathway distinct from the mammalian unfolded protein response regulates expression of endoplasmic reticulum chaperones in non-stressed cells. *EMBO J.* **16,** 7207–7216.

Harding, H. P., Zeng, H., Zhang, Y., Jungries, R., Chung, P., Plesken, H., Sabatini, D. D., and Ron, D. (2001). Diabetes mellitus and exocrine pancreatic dysfunction in perk-/- mice reveals a role for translational control in secretory cell survival. *Mol. Cell* **7,** 1153–1163.

He, Y., Sun, S., Sha, H., Liu, Z., Yang, L., Xue, Z., Chen, H., and Qi, L. (2010). Emerging roles for XBP1, a sUPeR transcription factor. *Gene Expr.* **15,** 13–25.

Iwawaki, T., Akai, R., Kohno, K., and Miura, M. (2004). A transgenic mouse model for monitoring endoplasmic reticulum stress. *Nat. Med.* **10,** 98–102.

Kim, I., Xu, W., and Reed, J. C. (2008). Cell death and endoplasmic reticulum stress: Disease relevance and therapeutic opportunities. *Nat. Rev. Drug Discov.* **7,** 1013–1030.

Kinoshita, E., Kinoshita-Kikuta, E., Takiyama, K., and Koike, T. (2006). Phosphate-binding tag, a new tool to visualize phosphorylated proteins. *Mol. Cell. Proteomics* **5,** 749–757.

Malhotra, J. D., Miao, H., Zhang, K., Wolfson, A., Pennathur, S., Pipe, S. W., and Kaufman, R. J. (2008). Antioxidants reduce endoplasmic reticulum stress and improve protein secretion. *Proc. Natl. Acad. Sci. USA* **105,** 18525–18530.

Masaki, T., Yoshida, M., and Noguchi, S. (1999). Targeted disruption of CRE-binding factor TREB5 gene leads to cellular necrosis in cardiac myocytes at the embryonic stage. *Biochem. Biophys. Res. Commun.* **261,** 350–356.

Miyata, Y., Fukuhara, A., Matsuda, M., Komuro, R., and Shimomura, I. (2008). Insulin induces chaperone and CHOP gene expressions in adipocytes. *Biochem. Biophys. Res. Commun.* **365,** 826–832.

Reimold, A. M., Etkin, A., Clauss, I., Perkins, A., Friend, D. S., Zhang, J., Horton, H. F., Scott, A., Orkin, S. H., Byrne, M. C., Grusby, M. J., and Glimcher, L. H. (2000). An essential role in liver development for transcription factor XBP-1. *Genes Dev.* **14,** 152–157.

Ron, D., and Walter, P. (2007). Signal integration in the endoplasmic reticulum unfolded protein response. *Nat. Rev. Mol. Cell Biol.* **8,** 519–529.

Scheuner, D., Song, B., McEwen, E., Liu, C., Laybutt, R., Gillespie, P., Saunders, T., Bonner-Weir, S., and Kaufman, R. J. (2001). Translational control is required for the unfolded protein response and in vivo glucose homeostasis. *Mol. Cell* **7,** 1165–1176.

Scheuner, D., Vander Mierde, D., Song, B., Flamez, D., Creemers, J. W., Tsukamoto, K., Ribick, M., Schuit, F. C., and Kaufman, R. J. (2005). Control of mRNA translation preserves endoplasmic reticulum function in beta cells and maintains glucose homeostasis. *Nat. Med.* **11,** 757–764.

Sha, H., He, Y., Chen, H., Wang, C., Zenno, A., Shi, H., Yang, X., Zhang, X., and Qi, L. (2009). The IRE1alpha-XBP1 pathway of the unfolded protein response is required for adipogenesis. *Cell Metab.* **9,** 556–564.

Shamu, C. E., and Walter, P. (1996). Oligomerization and phosphorylation of the Ire1p kinase during intracellular signaling from the endoplasmic reticulum to the nucleus. *EMBO J.* **15,** 3028–3039.

Travers, K. J., Patil, C. K., Wodicka, L., Lockhart, D. J., Weissman, J. S., and Walter, P. (2000). Functional and genomic analyses reveal an essential coordination between the unfolded protein response and ER-associated degradation. *Cell* **101,** 249–258.

Welihinda, A. A., and Kaufman, R. J. (1996). The unfolded protein response pathway in Saccharomyces cerevisiae. Oligomerization and trans-phosphorylation of Ire1p (Ern1p) are required for kinase activation. *J. Biol. Chem.* **271,** 18181–18187.

Woo, C. W., Cui, D., Arellano, J., Dorweiler, B., Harding, H., Fitzgerald, K. A., Ron, D., and Tabas, I. (2009). Adaptive suppression of the ATF4-CHOP branch of the unfolded protein response by toll-like receptor signalling. *Nat. Cell Biol.* **11,** 1473–1480.

Yamamoto, K., Sato, T., Matsui, T., Sato, M., Okada, T., Yoshida, H., Harada, A., and Mori, K. (2007). Transcriptional induction of mammalian ER quality control proteins is mediated by single or combined action of ATF6alpha and XBP1. *Dev. Cell* **13,** 365–376.

Yang, L., Xue, Z., He, Y., Sun, S., Chen, H., and Qi, L. (2010). A Phos-tag-based method reveals the extent of physiological endoplasmic reticulum stress. *PLoS ONE* **5,** e11621.

Zhang, K., Wong, H. N., Song, B., Miller, C. N., Scheuner, D., and Kaufman, R. J. (2005). The unfolded protein response sensor IRE1alpha is required at 2 distinct steps in B cell lymphopoiesis. *J. Clin. Invest.* **115,** 268–281.

Zhang, W., Feng, D., Li, Y., Iida, K., McGrath, B., and Cavener, D. R. (2006). PERK EIF2AK3 control of pancreatic beta cell differentiation and proliferation is required for postnatal glucose homeostasis. *Cell Metab.* **4,** 491–497.

CHAPTER NINE

PI 3-Kinase Regulatory Subunits as Regulators of the Unfolded Protein Response

Jonathon N. Winnay *and* C. Ronald Kahn

Contents

1. The Unfolded Protein Response	148
2. PI 3-Kinase Regulatory Subunits as Modulators of the UPR	150
3. Assessing UPR Pathway Activation	151
3.1. PERK	152
3.2. ATF6α	153
3.3. IRE1α	153
References	157

Abstract

The endoplasmic reticulum (ER) consists of an interconnected, membranous network that is the major site for the synthesis and folding of integral membrane and secretory proteins. Within the ER lumen, protein folding is facilitated by molecular chaperones and a variety of enzymes that ensure that polypeptides obtain their appropriate, tertiary conformation (Dobson, C. M. (2004). Principles of protein folding, misfolding and aggregation. *Semin. Cell Dev. Biol.* **15,** 3–16; Ni, M., and Lee, A. S. (2007). ER chaperones in mammalian development and human diseases. *FEBS Lett.* **581,** 3641–3651.). Physiological conditions that increase protein synthesis or stimuli that disturb the processes by which proteins obtain their native conformation, create an imbalance between the protein-folding demand and capacity of the ER. This results in the accumulation of unfolded or improperly folded proteins in the ER lumen and a state of ER stress. The cellular response, referred to as the unfolded protein response (UPR), results in activation of three linked signal transduction pathways: PKR-like kinase (PERK), inositol requiring 1 α (IRE1α), and activating transcription factor 6α (ATF6α) (Ron, D., and Walter, P. (2007). Signal integration in the endoplasmic reticulum unfolded protein response. *Nat. Rev. Mol. Cell. Biol.* **8,** 519–529; Schroder, M., and Kaufman, R. (2005). ER stress and the unfolded protein response. *Mutat. Res./Fundam. Mol.*

Section on Integrative Physiology and Metabolism, Research Division, Joslin Diabetes Center, Harvard Medical School, Boston, Massachusetts, USA

Mech. Mutagen. **569,** 29–63.). Collectively, the combined actions of these signaling cascades serve to reduce ER stress through attenuation of translation to reduce protein synthesis and through activation of transcriptional programs that ultimately serve to increase ER protein-folding capacity. Recently, we and Park *et al.* have characterized a novel function for the p85α and p85β subunits as modulators of the UPR by virtue of their ability to facilitate the nuclear entry of XBP-1s following induction of ER stress (Park, S. W., Zhou, Y., Lee, J., Lu, A., Sun, C., Chung, J., Ueki, K., and Ozcan, U. (2010). Regulatory subunits of PI3K, p85alpha and p85 beta, interact with XBP1 and increase its nuclear translocation. *Nat. Med.* **16,** 429–437; Winnay, J. N., Boucher, J., Mori, M. A., Ueki, K., and Kahn, C. R. (2010). A regulatory subunit of phosphoinositide 3-kinase increases the nuclear accumulation of X-box-binding protein-1 to modulate the unfolded protein response. *Nat. Med.* **16,** 438–445.). This chapter describes the recently elucidated role for the regulatory subunits of PI 3-kinase as modulators of the UPR and provides methods to measure UPR pathway activation.

1. THE UNFOLDED PROTEIN RESPONSE

The endoplasmic reticulum (ER) is the site for the biosynthesis of lipids, sterols as well as integral and secreted proteins (Gaut and Hendershot, 1993; Wei and Hendershot, 1996). The vast majority of nascent polypeptides require folding and/or enzymatic modifications to obtain their appropriate conformation (Dobson, 2004, Ni and Lee, 2007). A variety of pathophysiological states including obesity, hyperlipidemia, nutrient deprivation, hypoxia, infection, and inflammation have been shown to disrupt ER homeostasis leading to the development of ER stress (Yoshida, 2007). In response, signaling pathways of the unfolded protein response (UPR) are activated in an attempt to resolve this condition through attenuation of translation to reduce protein load and through activation of transcriptional programs that ultimately increase protein folding capacity. This coordinated response is mediated by the combined action of three ER stress sensors: PKR-like kinase (PERK), inositol requiring 1 α (IRE1α), and activating transcription factor 6α (ATF6α) (Fig. 9.1) (Ron and Walter, 2007; Schroder and Kaufman, 2005).

PERK is a type I transmembrane protein that is composed of a luminal region that senses the presence of unfolded proteins and a cytoplasmic domain that possesses kinase activity. In the absence of ER stress, the binding of the ER chaperone BiP to the luminal domain of PERK prevents homodimerization and activation. Upon induction of ER stress, BiP is sequestered from PERK, allowing for dimerization, transphosphorylation and stimulation of kinase activity directed towards cytosolic substrates (Bertolotti *et al.*, 2000; Liu *et al.*, 2000). One such substrate is eukaryotic initiation factor 2 (eIF2α), which when phosphorylated by PERK, reduces the rate of polypeptide chain elongation, leading to a decline in mRNA

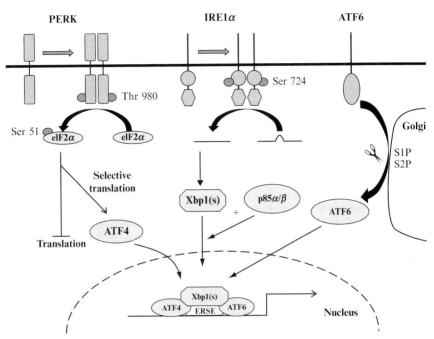

Figure 9.1 The UPR pathway. The pathway consists of three proximal sensors: PERK, IRE1α, and ATF6α. Their coordinated response serves to restore ER protein-folding capacity by inhibiting protein translation (PERK) and orchestrates a transcriptional response that ultimately increases protein-folding capacity (IRE1α and ATF6α). BiP negatively regulates these pathways by binding to the luminal domains to prevent activation. The induction of ER stress and accumulation of misfolded proteins in the ER lumen causes the release of BiP from PERK, IRE1α and ATF6α, leading to their subsequent activation. PERK dimerization leads to transphosphorylation and activation, resulting in the phosphorylation of eIF2α and a global decrease in the rate of protein translation. However, some mRNAs, such as ATF4 mRNA, are translated under these conditions, possibly by the presence of open reading frames within the 5′ untranslated region. BiP release from IRE1α permits dimerization and activation of its kinase and RNase activities to permit XBP-1 mRNA splicing, creating a translational frame-shift to produce a transcriptionally competent transcription factor (XBP-1s) that traffics to the nuclease in an p85α/β-dependent fashion. Lastly, the release of BiP from ATF6α leads to the translocation of ATF6α to the Golgi where it undergoes sequential cleavage by S1P and S2P proteases. The resultant transcription factor transits to the nuclease where it activates UPR target genes.

translation and a subsequent reduction in protein-folding demand within the ER (Harding et al., 1999; Wek and Cavener, 2007).

The second unfolded protein sensor is the ER transmembrane transcription factor, ATF6α. Accumulation of unfolded proteins in the ER lumen leads to the trafficking of full-length ATF6α (p90ATF6) to the Golgi where is undergoes sequential cleavage by site 1 (S1P) and site 2 (S2P) proteases to produce a mature, transcriptionally competent transcription factor (p50ATF6) (Shen *et al.*, 2002; Yoshida *et al.*, 2000). This transcriptionally active form of ATF6α translocates to the nucleus where it binds directly to the ERSE and cooperates with XBP-1 (X-box binding protein 1) to induce UPR target genes (Lee, 2002; Yoshida *et al.*, 2000).

The final sensor is IRE1α, which is composed of a luminal domain that serves as an unfolded protein sensor and a cytosolic portion composed of a kinase and RNase domain (Back *et al.*, 2005a; Ron and Hubbard, 2008). In a fashion similar to the activation of PERK, BiP sequestration following ER stress leads to oligomerization of IRE1α, transphosphorylation and subsequent stimulation of RNase activity (Kimata, 2004). This activity is directed toward the XBP1 pre-mRNA to generate two distinct mRNA transcripts (Fig. 9.2A). The first ORF (XBP1u) generates a transcript that encodes for a protein that is devoid of transcriptional activity. The second ORF (XBP1s) is created when IRE1α removes a 26-nucleotide (nt) intron to generate a larger protein containing both a DNA binding and transactivation domain (Back *et al.*, 2005b; Lee *et al.*, 2002). XBP1(s) translocates to the nucleus where it binds to the ER stress-response element (ERSE) in the promoters of genes involved in lipid synthesis, ER biogenesis and the UPR including BiP, ERdj4, PDI, and p58^{IPK} (Calfon *et al.*, 2002; Iwakoshi *et al.*, 2003; Lee *et al.*, 2003).

2. PI 3-Kinase Regulatory Subunits as Modulators of the UPR

Class Ia PI3Ks are a family of lipid kinases that regulate multiple cellular processes including metabolism, growth, survival, vesicular trafficking, and cytoskeletal rearrangements by virtue of their ability to generate the

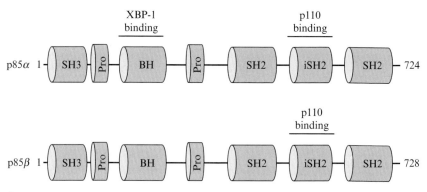

Figure 9.2 Domain structure of p85α and p85β. The NH$_2$-terminal half of p85α and p85β consists of an Src-homology 3 (SH3) domain, and two proline rich domains (Pro) surrounding the Bcl2-homology (BH) domain. The interaction of XBP-1s and p85α has been mapped to the region encompassing the BH domain. The C-terminal portion of p85α/β consists of two Src-homology 2 (SH2) domains that bind tyrosine phosphorylated YxxM motifs and an inter-SH2 (iSH2) domain that mediates binding with the catalytic subunit.

second messenger, phosphatidylinositol-3,4,5-trisphosphate (PIP_3). The enzyme exists as a heterodimer composed of a catalytic and regulatory subunit, both of which occur in several isoforms (Antonetti et al., 1996; Inukai et al., 1996, 1997; Pons et al., 1995). The p85α and closely related p85β regulatory subunits are highly homologous and share identical organization of functional domains (Fig. 9.2). The two src-homology 2 (SH2) domains bind tyrosine phosphorylated YxxM motifs with high specificity, allowing for interaction with substrates downstream of receptors possessing tyrosine kinase activity (Songyang et al., 1993). The region between these SH2 domains serves as the binding site for catalytic subunits (Holt et al., 1994). The roles of the SH3, proline rich, and BH domains of p85α and p85β are less well defined. In an attempt to identify novel, protein–protein interactions mediated by the NH_2-terminal region of p85α encompassing these domains, a bacterial two-hybrid screen was performed. One unanticipated finding of this screen revealed an interaction between the NH_2-terminal portion of p85α and XBP-1.

Additional experiments confirmed this physical association and tentatively mapped the interaction interface on p85α to the BH domain (Park et al., 2010; Winnay et al., 2010). Interestingly, overexpression of p85α or p85β led to a significant increase in the nuclear accumulation of XBP-1s, suggesting an endogenous role for p85 regulatory subunits as regulators of XBP-1s nuclear shuttling.

Not surprisingly, subsequent analysis of p85α-deficient cell lines and cells with reduced p85α expression facilitated by shRNA-mediated gene silencing showed a significant alterations in the UPR, including reduced ER stress–dependent accumulation of nuclear XBP-1s and an associated decrease in UPR target-gene induction (Holt et al., 1994; Park et al., 2010). Furthermore, mice with deletion of p85α in liver (L-Pik3r1$^{-/-}$) showed a similar attenuation of the UPR after tunicamycin administration, an effect most likely due to a reduction in the nuclear accumulation of XBP-1s (Winnay et al., 2010). Collectively, these data demonstrate an important role for p85α/β as important regulators of XBP-1s nuclear trafficking.

3. Assessing UPR Pathway Activation

In this section, we will describe specific reagents and outline protocols that will facilitate the evaluation of the PERK, ATF6α, and IRE1α activation. In addition, secondary endpoints that can be used to assess the extent of pathway activation will be described including eIF2α phosphorylation, XBP-1 splicing, and translocation as well as target-gene expression.

3.1. PERK

Based upon our experience in liver, hepatic cell lines, mouse embryonic fibroblasts (MEFs), and immortalized brown preadipocytes, the ability to detect the activation of the PERK pathway by immunoblot analysis using phospho-specific antibodies is fraught with difficulties (Winnay et al., 2010). We speculate that this may be due to differences in protein expression, degree of activation relative to other UPR pathways, or the transient nature of PERK activation in any given experimental system. In the event that PERK phosphorylation is not detected in straight lysates, immunoblots performed with anti-PERK antibodies on suitably resolved protein lysates can reveal a molecular weight shift indicative of PERK phosphorylation. Alternatively, immunoprecipitation of PERK followed by immunoblotting with phospho-PERK antibodies has been described as a reasonable approach to detect PERK phosphorylation. Lastly, the phosphorylation-state of eIF2α is often used as a surrogate to measure the extent of PERK activation. Importantly, Interpretation of these data must acknowledge that eIF2α is a substrate for a number of enzymes including PKR, GCN2 as well as PERK.

3.1.1. Preparation of cell extracts

1. Treat 80–100% confluent cells with vehicle or experimental agents.
2. Wash cells two times with ice-cold PBS and snap freeze in liquid nitrogen.
3. Thaw cells on ice until first signs of thawing are identified.
4. Add lysis buffer to obtain lysates with sufficient protein content to perform downstream analysis:
   Tris–HCl: 50 mM, pH 7.4
   NP-40: 1%
   Na-deoxycholate: 0.25%
   0.01% SDS
   NaCl: 150 mM
   EDTA: 1 mM
   Phosphatase inhibitors
   Protease inhibitors
5. Collect lysates and transfer to 1.5 ml Eppendorf tube. Rotate lysates at 4 °C for 10 min.
6. Spin lysates at 14,000 rpm for 15 min. Transfer cleared lysates to new tubes.
7. Perform protein assay to determine protein concentration.
8. Resolve protein lysates by SDS-PAGE and perform immunoblot analysis.

Required materials:

- *Cells or tissue of interest*: Liver, Huh7, MEFs, or brown preadipocytes.
- *Cell culture medium*: The above cell lines are cultured in Dulbecco's modified Eagle's medium (DMEM) supplemented with 10% FCS, 2% glutamine, penicillin, and streptomycin.
- Protease Inhibitor Cocktail (Sigma-Aldrich, St. Louis, MO)
- Phosphatase Inhibitor Cocktail 2 and 3 (Sigma-Aldrich)
- Phospho-PERK (Thr 980) (16F8) rabbit monoclonal antibody (Cell Signaling Technology, Beverly, MA)
- PERK (C33E10) rabbit monoclonal antibody (Cell Signaling Technology)
- Phospho-eIF2α (Ser 51) (119A11) rabbit monoclonal antibody (Cell Signaling Technology)
- eIF2α antibody (Cell Signaling Technology)
- BM Chemiluminescence Western Blotting Kit (Roche Applied Science, Indianapolis, IN)

3.2. ATF6α

Upon induction of ER stress, ATF6α is released from the ER and is transported to the Golgi where it undergoes cleavage by the S1P and S2P proteases. These cleavage events produce a potent transcription factor that transits to the nuclease to participate in a complex transcriptional program that ultimately serves to relieve ER stress and restore ER protein-folding capacity (Yoshida et al., 2000). Therefore, an assessment of ATF6α cleavage and nuclear accumulation serves as an accurate gauge of ATF6α pathway activation. In our experience, the preparation of nuclear extracts (see Section 3.3.2) is often required to detect the cleaved, mature ATF6α (p50ATF6α).

Required materials:

- ATF6α antibody (H-280) (Santa Cruz Biotechnology, Santa Cruz, CA)

3.3. IRE1α

The degree of IRE1α pathway activation can be assessed by performing immunoblot analysis with phospho-specific antibodies (Winnay et al., 2010). Unfortunately, the ability to detect the activation of IRE1α using this approach is dependent on the cell line and experimental context. For example, although the phosphorylation of IRE1α is easier to assess in pancreatic β-cell lines (i.e., Min6), we have experienced limited success in a variety of additional tissues and cell lines. Alternative approaches to detect IRE1α activation include evaluation of IRE1α mobility by SDS-PAGE or immunoprecipitation of IRE1α prior to immunoblot analysis with phospho-specific antibodies. Due to the difficulty of reliably detecting IRE1α phosphorylation, the IRE1α-dependent splicing of the XBP-1 mRNA is often performed as an acceptable surrogate to assess IRE1α pathway activation. In addition, the accumulation of nuclear XBP-1s is an excellent way to evaluate activation of the IRE1α pathway or to assess the impact of p85α/β on XBP-1 nuclear transport.

3.3.1. XBP-1 splicing assay

The protocol outlined in this section has been adapted from the original report that first described the IRE-1-dependent splicing of XBP-1 in response to ER stress (Calfon et al., 2002). Splicing of the XBP-1 transcript is detected by amplifying the region of the XBP-1 cDNA encompassing the 26 nt intron that is excised following induction of ER stress (Fig. 9.3A and B). To distinguish the spliced and unspliced amplicons, the PCR products can be carefully separated by agarose gel electrophoresis or digested with the PstI restriction endonuclease that digests only unspliced cDNA to produce two DNA fragments (Fig. 9.3B and C).

1. Perform reverse transcription on 1–2 μg of total RNA isolated from control and experimental conditions using random hexamers to prime the reaction.
2. Dilute the cDNA synthesis reaction 1:10–1:20 with H_2O.
3. Prepare a PCR reaction using 5.0 μl cDNA and 10 mol of mXBP-1 forward and reverse primers.
4. Perform PCR using the following conditions:

94 °C	4 min	
94 °C	10 s	(40 cycles)
66 °C	30 s	
72 °C	30 s	
72 °C	10 min	

5. Perform PstI restriction digest on 10 μl of PCR product.
6. Resolve restriction digest on 2–3% agarose gel.

In most instances, the amount of cDNA or the number of cycles of PCR amplification will need to be adjusted to work optimally in any given experimental system.

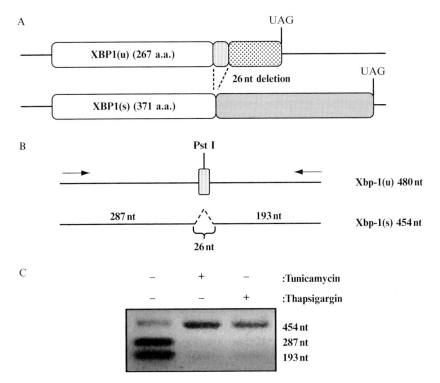

Figure 9.3 XBP-1 splicing. (A) In the absence of UPR pathway activation, the XBP-1 pre-mRNA is not spliced by IRE1α leading to translation of an 267 a.a. protein lacking a transactivation domain (XBP-1u). In contrast, IRE1α activation leads to the deletion of a 26 nt intron from the pre-mRNA to produce a translational frame-shift and production of a longer, transcriptionally competent protein product (XBP-1s). (B) Schematic representation of the region of the XBP-1u and XBP-1s cDNA's amplified with the oligonucleotides used to detect XBP-1 mRNA splicing. (C) Agarose gel electrophoresis of *Pst*I digested PCR fragments from control, tunicamycin, or thapsigargin treated mouse embryonic fibroblasts.

Required materials:

- Phospho-IRE1α (Ser 724) rabbit polyclonal antibody (Novus Biologicals, Littleton, CO)
- IRE1α (14C10) rabbit monoclonal antibody (Cell Signaling Technology)
- High Capacity cDNA Reverse Transcription Kit (Applied Biosystems, Carlsbad, CA)
- mXBP-1 forward 5′ GGCCTTGTGGTTGAGAACCAGGAG 3′, mXBP-1 reverse 5′ GAATGCCCAAAAGGATATCAGACTC 3′
- *Pst*I restriction enzyme (New England Biolabs, Ipswich, MA)
- Reagents for PCR
- Agarose Gel Electrophoresis equipment

3.3.2. XBP-1s translocation

The protocol outlined in this section allows for an assessment of XBP-1s translocation to the nucleus by performing XBP-1 immunoblot analysis on cytoplasmic and nuclear protein fractions from mammalian cell lines following induction of ER stress.

1. Perform any treatment of cells at ~90% confluency.
2. Wash two times with ice cold PBS.
3. Scrape cells into PBS containing phosphatase and protease inhibitors.
4. Centrifuge 5 min at 4000 rpm (4 °C)
5. Re-suspend cell pellet with 400 µl of buffer A containing protease inhibitors:
   10 mM HEPES (pH 7.9)
   10 mM KCl
   0.1 mM EDTA
   0.1 mM EGTA
6. Add 25 µl of 10% NP-40.
7. Vortex tube for 15 s.
8. Centrifuge for 5 min at 13,000 rpm (4 °C)
9. Collect the supernatant as the soluble, cytoplasmic protein fraction.
10. Add 50 µl of buffer B containing protease inhibitors and resuspend:
    20 mM HEPES (pH 7.9)
    400 mM NaCl
    0.1 mM EDTA
    0.1 mM EGTA
11. Rock the tube for 20 min at 4 °C.
12. Centrifuge for 15 min at 13,000 rpm (4 °C)
13. Collect the supernatant as the nuclear protein fraction.
14. Perform anti-XBP-1 immunoblots on cytoplasmic and nuclear fractions.

Required materials:

- *Cells or tissue of interest*: Liver, Huh7, MEFs, or brown preadipocytes.
- *Cell culture medium*: The above cell lines are cultured in Dulbecco's modified Eagle's medium (DMEM) supplemented with 10% FCS, 2% glutamine, penicillin, and streptomycin.
- Protease Inhibitor Cocktail (Sigma-Aldrich)
- Phosphatase Inhibitor Cocktail 2 and 3 (Sigma-Aldrich)
- Anti-XBP-1 antibody (Santa Cruz Biotechnology)
- Reagents for SDS-PAGE

REFERENCES

Antonetti, D. A., Algenstaedt, P., and Kahn, C. R. (1996). Insulin receptor substrate 1 binds two novel splice variants of the regulatory subunit of phosphatidylinositol 3-kinase in muscle and brain. *Mol. Cell. Biol.* **16,** 2195–2203.

Back, S., Schroder, M., Lee, K., Zhang, K., and Kaufman, R. (2005). ER stress signaling by regulated splicing: IRE1/HAC1/XBP1. *Methods* **35,** 395–416.

Bertolotti, A., Zhang, Y., Hendershot, L. M., Harding, H. P., and Ron, D. (2000). Dynamic interaction of BiP and ER stress transducers in the unfolded-protein response. *Nat. Cell Biol.* **2,** 326–332.

Calfon, M., Zeng, H., Urano, F., Till, J. H., Hubbard, S. R., Harding, H. P., Clark, S. G., and Ron, D. (2002). IRE1 couples endoplasmic reticulum load to secretory capacity by processing the XBP-1 mRNA. *Nature* **415,** 92–96.

Dobson, C. M. (2004). Principles of protein folding, misfolding and aggregation. *Semin. Cell Dev. Biol.* **15,** 3–16.

Gaut, J. R., and Hendershot, L. M. (1993). The modification and assembly of proteins in the endoplasmic reticulum. *Curr. Opin. Cell Biol.* **5,** 589–595.

Harding, H. P., Zhang, Y., and Ron, D. (1999). Protein translation and folding are coupled by an endoplasmic-reticulum-resident kinase. *Nature* **397,** 271–274.

Holt, K. H., Olson, L., Moye-Rowley, W. S., and Pessin, J. E. (1994). Phosphatidylinositol 3-kinase activation is mediated by high-affinity interactions between distinct domains within the p110 and p85 subunits. *Mol. Cell. Biol.* **14,** 42–49.

Inukai, K., Anai, M., van Breda, E., Hosaka, T., Katagiri, H., Funaki, M., Fukushima, Y., Ogihara, T., Yazaki, Y., Kikuchi, M., Oka, Y., and Asano, T. (1996). A novel 55-kDa regulatory subunit for phosphatidylinositol 3-kinase structurally similar to p55 PIK is generated by alternative splicing of the p85‡ gene. *J. Biol. Chem.* **271,** 5317–5320.

Inukai, K., Funaki, M., Ogihara, T., Katagiri, H., Kanda, A., Anai, M., Fukushima, Y., Hosaka, T., Suzuki, M., Shin, B. C., Takata, K., Yazaki, Y., *et al.* (1997). p85alpha gene generates three isoforms of regulatory subunit for phosphatidylinositol 3-kinase (PI 3-Kinase), p50alpha, p55alpha, and p85alpha, with different PI 3-kinase activity elevating responses to insulin. *J. Biol. Chem.* **272,** 7873–7882.

Iwakoshi, N. N., Lee, A. H., and Glimcher, L. H. (2003). The X-box binding protein-1 transcription factor is required for plasma cell differentiation and the unfolded protein response. *Immunol. Rev.* **194,** 29–38.

Kimata, Y. (2004). A role for BiP as an adjustor for the endoplasmic reticulum stress-sensing protein Ire1. *J. Cell Biol.* **167,** 445–456.

Lee, K. (2002). IRE1-mediated unconventional mRNA splicing and S2P-mediated ATF6 cleavage merge to regulate XBP1 in signaling the unfolded protein response. *Genes Dev.* **16,** 452–466.

Lee, K., Tirasophon, W., Shen, X., Michalak, M., Prywes, R., Okada, T., Yoshida, H., Mori, K., and Kaufman, R. J. (2002). IRE1-mediated unconventional mRNA splicing and S2P-mediated ATF6 cleavage merge to regulate XBP1 in signaling the unfolded protein response. *Genes Dev.* **16,** 452–466.

Lee, A. H., Iwakoshi, N. N., and Glimcher, L. H. (2003). XBP-1 regulates a subset of endoplasmic reticulum resident chaperone genes in the unfolded protein response. *Mol. Cell. Biol.* **23,** 7448–7459.

Liu, C. Y., Schroder, M., and Kaufman, R. J. (2000). Ligand-independent dimerization activates the stress response kinases IRE1 and PERK in the lumen of the endoplasmic reticulum. *J. Biol. Chem.* **275,** 24881–24885.

Ni, M., and Lee, A. S. (2007). ER chaperones in mammalian development and human diseases. *FEBS Lett.* **581,** 3641–3651.

Park, S. W., Zhou, Y., Lee, J., Lu, A., Sun, C., Chung, J., Ueki, K., and Ozcan, U. (2010). Regulatory subunits of PI3K, p85alpha and p85 beta, interact with XBP1 and increase its nuclear translocation. *Nat Med.* **16,** 429–437.

Pons, S., Asano, T., Glasheen, E., Miralpeix, M., Zhang, Y., Fisher, T. L., Myers, M. G., Sun, X. J., and White, M. F. (1995). The structure and function of p55PIK reveal a new regulatory subunit for phosphatidylinositol 3-kinase. *Mol. Cell. Biol.* **15,** 4453–4465.

Ron, D., and Hubbard, S. R. (2008). How IRE1 reacts to ER stress. *Cell* **132,** 24–26.

Ron, D., and Walter, P. (2007). Signal integration in the endoplasmic reticulum unfolded protein response. *Nat. Rev. Mol. Cell Biol.* **8,** 519–529.

Schroder, M., and Kaufman, R. (2005). ER stress and the unfolded protein response. *Mutat. Res./Fundam. Mol. Mech. Mutagen.* **569,** 29–63.

Shen, J., Chen, X., Hendershot, L., and Prywes, R. (2002). ER stress regulation of ATF6 localization by dissociation of BiP/GRP78 binding and unmasking of Golgi localization signals. *Dev. Cell* **3,** 99–111.

Songyang, Z., Shoelson, S. E., Chaudhuri, M., Gish, G. D., Pawson, T., Haser, W. G., King, F., Roberts, T., Ratnofsky, S., Lechleider, R. J., Neel, B. G., Birge, R. B., et al. (1993). SH2 domains recognize specific phosphopeptide sequences. *Cell* **72,** 767–778.

Wei, J., and Hendershot, L. M. (1996). Protein folding and assembly in the endoplasmic reticulum. *EXS* **77,** 41–55.

Wek, R. C., and Cavener, D. R. (2007). Translational control and the unfolded protein response. *Antioxid. Redox Signal.* **9,** 2357–2371.

Winnay, J. N., Boucher, J., Mori, M. A., Ueki, K., and Kahn, C. R. (2010). A regulatory subunit of phosphoinositide 3-kinase increases the nuclear accumulation of X-box-binding protein-1 to modulate the unfolded protein response. *Nat. Med.* **16,** 438–445.

Yoshida, H. (2007). ER stress and diseases. *FEBS J.* **274,** 630–658.

Yoshida, H., Okada, T., Haze, K., Yanagi, H., Yura, T., Negishi, M., and Mori, K. (2000). ATF6 activated by proteolysis binds in the presence of NF-Y (CBF) directly to the cis-acting element responsible for the mammalian unfolded protein response. *Mol. Cell. Biol.* **20,** 6755–6767.

CHAPTER TEN

The Emerging Role of Histone Deacetylases (HDACs) in UPR Regulation

Soumen Kahali, Bhaswati Sarcar, and Prakash Chinnaiyan

Contents

1. HDAC Enzymes and Cancer	159
2. HDAC Inhibitors	160
3. HDAC Inhibition and the UPR	162
3.1. Background	162
3.2. Acetylation of Grp78	163
4. Future Directions	172
References	173

Abstract

Although the function of histone deacetylases (HDACs) have primarily been associated with influencing transcription through chromatin remodeling, the capacity of these enzymes to interface with a diverse array of biologic processes by modulating a growing list of nonhistone substrates has gained recent attention. Recent investigations have demonstrated the potential of HDACs to directly regulate the unfolded protein response (UPR) through acetylation of its central regulatory protein, Grp78. Further, this appears to be an important mechanism underlying the anti-tumor activity of HDAC inhibitors. Herein, we provide a summary of the literature supporting the role HDACs play in regulating the UPR and a detailed description of methods to allow for the study of both acetylation of nonhistone proteins and UPR pathway activation following HDAC inhibition.

1. HDAC Enzymes and Cancer

Histone proteins organize DNA into regular repeating structures of chromatin. The dynamic remodeling of chromatin between relatively *open* and *closed* states plays a central role in the epigenetic regulation of gene

Department of Experimental Therapeutics and Radiation Oncology, H. Lee Moffitt Cancer Center and Research Institute, Tampa, FL, USA expression. The acetylation status of histones influence chromatin structure, which in turn, regulates gene expression (Xu et al., 2007). This potential for histones to be posttranslationally acetylated was first identified in the 1960s by Phillips (1963) and Allfrey et al. (1964). However, it was not until the mid-1990s when it was shown that hyperacetylation of core histones increase the spatial separation of DNA from histone and enhance binding of transcription factor complexes to DNA (Lee et al., 1993) and the specific enzymes involved in the reversible acetylation of histones were discovered, which include histone acetyltransferases (HATs) and histone deacetylases (HDACs). Having coined their name from their primary substrate, HATs transfer acetyl groups to the amino-terminal lysine residue in histones, which result in local expansion of chromatin and increase accessibility of regulatory proteins to DNA, whereas HDACs catalyze the removal of acetyl groups, leading to chromatin condensation and transcriptional repression. A total of 18 HDAC enzymes have since been identified and classified based on homology to yeast HDACs. Class I HDACs, which are primarily nuclear in localization, include HDAC1, 2, 3, and 8. They are related to the yeast RPD3 deacetylase and have high homology in their catalytic sites. Class II HDACs are related to yeast Hda 1 (histone deacetylase 1) and include HDAC4, 5, 6, 7, 9, and 10. Unlike Class I HDACs, Class II HDACs are primarily cytoplasmic and/or migrate between the cytoplasm and nucleus. Class III HDACs are comprised of the sirtuins, which unlike Class I and II HDACs, which are zinc-dependent enzymes, require NAD+ for their enzymatic activity. Lastly, Class IV HDAC is represented by HDAC11, which contains conserved residues in the catalytic core region shared by both class I and II enzymes (Xu et al., 2007).

2. HDAC Inhibitors

Although cancer has traditionally been considered a disease of genetic mutations and chromosomal abnormalities, a growing body of evidence suggests transcriptional control through aberrant epigenetic states also plays an important role in tumorigenesis (Bolden et al., 2006). Based on their central role in influencing chromatin structure, HDAC enzymes have been widely recognized as promising molecular targets to reverse these epigenetic alterations associated with cancer. This potential was initially observed with sodium butyrate, later identified as an HDAC inhibitor, which was found to cause morphological reversion of the transformed cell phenotype (Leder et al., 1975; Riggs et al., 1977). Since these initial findings, several other HDAC inhibitors have been identified (Richon et al., 1998; Yoshida et al., 1990) and are currently in various stages of clinical evaluation in cancer treatment (Bolden et al., 2006; Lane and Chabner, 2009). These agents can be divided into four classes based on their chemical structure (Bolden et al., 2006; Xu et al., 2007). The classes include the short-chain fatty acids (e.g., valproate, phenylbutyrate); hydroxamic acid derivatives (e.g., vorinostat (SAHA), belinostat (PXD-101), and panobinostat (LBH-589)); benzamides (e.g., entinostat (SNDX-275 or MS-275)); mocetinostat (MGCD0103); and cyclic tetrapeptides (e.g., romidepsin (FK-228)). Of these agents, vorinostat is the only compound currently approved by the Food and Drug Administration, with an indication in the treatment of refractory cutaneous T-cell lymphoma (Lane and Chabner, 2009).

It is widely perceived that a majority of these compounds are nonselective toward their HDAC substrates (excluding Class III HDACs); however, work presented by Bradner et al. suggests this to be incorrect (Bradner et al., 2010). A chemical phylogenetic analysis of class I and II HDACs as targets of a structurally diverse panel of inhibitors identified clear isoform selectivity among these compounds. Notably, few compounds were truly pan-HDAC inhibitors, with class II HDACs (excluding HDAC6) requiring several logs higher drug concentration to achieve enzyme inhibition. These findings will play an important role in our understanding of the underlying mechanism of action of this complex class of anti-cancer agents and help identify potential molecular determinants of response for individual agents.

In addition to their potential of inducing differentiation in transformed cells, HDAC inhibitors also influence a diverse array of other cellular processes regulating tumor growth, contributing toward their enthusiasm as potential anti-cancer agents. This includes induction of apoptosis, cell cycle arrest, generation of reactive oxygen species, and attenuating angiogenesis and invasion (Bolden et al., 2006). Although these potential end-consequences of HDAC inhibitors have been demonstrated, the underlying mechanism of action leading to these observed effects remains undefined. As discussed above, initial investigations focused on the potential of HDAC inhibitors to modulate chromatin structure, thereby selectively altering gene expression. However, recent investigations suggest only a small percentage of genes are actually modulated by these agents, likely secondary to a variety of processes regulating gene transcription, including methylation of histones and promoter regions in DNA (Xu et al., 2007). Further, several studies found that HDAC inhibitors induce as many genes as they repress, suggesting the mechanism of action of these agents to be significantly more complex. Along these lines, continued research focused on these enzymes has identified numerous nonhistone proteins whose functions are regulated by HDACs (Bolden et al., 2006). Based on this diverse mechanism of action, it has been suggested a more appropriate term for these enzymes are *protein* deacetylases rather than *histone* deacetylases. Interestingly, many of these substrates include transcription factors and chaperone proteins, which both play a role in regulating protein expression, thereby allowing HDAC enzymes to modulate gene expression via both histone dependent and independent mechanisms.

Of the identified nonhistone substrates of HDAC enzymes, considerable attention has focused on the chaperone protein heat shock protein 90 (Hsp90) (Bali *et al.*, 2005; Bolden *et al.*, 2006; Kovacs *et al.*, 2005; Scroggins *et al.*, 2007; Yu *et al.*, 2002). Hsp90 plays a number of important roles, including assisting protein folding, intracellular transport, maintenance and degradation of proteins, as well as facilitating cell signaling. Cancer cells are particular reliant upon these functions, making Hsp90 an attractive anti-cancer target. For example, cancer cells typically over express a number of proteins, including growth factor receptors, such as EGFR, or signal transduction proteins such as PI3K and AKT. As Hsp90 plays an important role in stabilizing these proteins, inhibition may lead to attenuation of their aberrant proliferative and survival signaling. Another important role of Hsp90 in cancer is the stabilization of mutant proteins such as v-Src, the fusion oncogene Bcr/Abl, and mutant forms of p53 that appear during cell transformation. It appears that Hsp90 can act as a "protector" of less stable proteins produced by DNA mutations (Mahalingam *et al.*, 2009).

Recent studies have demonstrated the potential for HDAC inhibitors to acetylate Hsp90, leading to inhibition of chaperone protein function. Specifically, Hsp90 acetylation following HDAC inhibition led to dissociation and degradation of Hsp90 client oncoproteins, including p53, ErbB1, ErbB2, and Raf-1 (Yu *et al.*, 2002). Further studies went on to demonstrate that the specific target was HDAC6, which was shown to bind to Hsp90 and that subsequent inhibition using siRNA attenuated binding of Hsp90 to ATP and reduced chaperone association with client oncoproteins (Bali *et al.*, 2005; Kovacs *et al.*, 2005). Scroggins *et al.* went on to identify the specific lysine residue acetylated on Hsp90, which directly influenced client protein binding (Scroggins *et al.*, 2007). Although the interaction between HDAC inhibitors and Hsp90 is of considerable interest, the relative importance of Hsp90 acetylation in HDAC inhibitor cytotoxicity is yet to be determined.

3. HDAC Inhibition and the UPR

3.1. Background

Recent studies have identified Grp78, a chaperone protein primarily localized in the endoplasmic reticulum that shares close homology with the Hsp-family proteins, as a novel nonhistone target of HDAC inhibitors (Kahali *et al.*, 2010; Rao *et al.*, 2010). Grp78 plays a central role in monitoring cell homeostasis and during stress conditions, activates the adaptive process termed the unfolded protein response (UPR). Under nonstressed conditions, Grp78 binds to its client proteins PERK, ATF6, and IRE1, preventing their signaling. However, when the ER is overloaded by newly synthesized proteins or is "stressed" by agents that cause accumulation of unfolded proteins, Grp78 binds to the unfolded proteins in the ER, freeing its client proteins (Lee et al., 2008; Schroder and Kaufman, 2005). PERK is released from its chaperone protein Grp78 to permit homodimerization and autophosphorylation, leading to activation. One of the initial targets of PERK is eukaryotic translation initiating factor 2-alpha (eIF2α). PERK phosphorylates and inactivates eIF2α, thereby globally shutting off mRNA translation and reducing the protein load on the ER. However, certain mRNAs gain a selective advantage for translation under these conditions, including the mRNA encoding activating transcription factor 4 (ATF4). The ATF4 protein is a member of the bZIP family of transcription factors, which regulate the promoters of several genes implicated in the UPR (Harding et al., 1999; Kaufman, 1999, 2004). Ire1 similarly oligomerizes in ER membranes when released by Grp78. A key downstream target of Ire1 is X box protein-1 (XBP-1) mRNA. Its endoribonuclease domain splices its mRNA, rendering it competent for translation to produce the 41 kDa XBP-1 protein, another bZIP-family transcription factor, binding to ER stress enhancer (ERSE) and unfolded protein response element (UPRE) promoters (Lin et al., 2007; Ye et al., 2000; Yoshida et al., 2001). Instead of oligomerizing in the ER as PERK and Ire1, release of ATF6 permits transport to the Golgi compartment where it is cleaved to generate the cytosolic-activated form of ATF6 that translocates to the nucleus. Among its genes regulated, it collaborates with Ire1 by inducing transcription of XBP-1 mRNA (Lin et al., 2007; Ye et al., 2000; Yoshida et al., 2001).

Thus, Grp78 regulated activation provides a direct mechanism for the three UPR transducers to sense the "stress" in the ER. Different stresses or physiologic conditions may selectively activate only one or two of the ER stress sensors; however, mechanisms permitting this selective activation of individual components of the UPR are unknown. Further, although it has been shown that a shift in binding of Grp78 from its client protein to the increased load of unfolded proteins is through an ATP-dependent competitive process (Ma and Hendershot, 2004), events leading to the respective chaperone protein dissociating from its client protein remains undefined.

3.2. Acetylation of Grp78

Recent scientific advances have provided further insight into the increasingly important role lysine acetylation plays in protein regulation. Using high-resolution mass spectrometry, 3600 modifiable lysine acetylation sites on 1750 proteins were identified. These proteins had diverse cellular functions beyond chromatin remodeling, including cell cycle regulation, slicing, nuclear transport, and actin nucleation. Of the identified proteins in this high-throughput analysis, the chaperone protein Grp78 was identified to be reversibly acetylated following HDAC inhibition (Choudhary et al., 2009). Recent publications from both our group (Kahali et al., 2010) and Rao et al. (2010) confirmed Grp78 acetylation following HDAC inhibition and went on study its underlying biologic relevance. Our initial investigations focused on the glioblastoma cell line U251, exposing cells to clinically relevant concentrations of the HDAC inhibitor vorinostat. Cell lysates were immunoprecipitated using a Grp78 specific antibody, followed by immunoblot analysis with antibodies recognizing anti-acetylated lysine residues and the client protein PERK. These studies demonstrated Grp78 acetylation at 6 and 24 h following vorinostat exposure, which was associated with the dissociation of PERK from Grp78, a key initiator in pathway activation (Fig. 10.1). Activation of PERK signaling was confirmed following exposure to both vorinostat and valproic acid, which belongs to a different class of HDAC inhibitors (Fig. 10.2A and B). Interestingly, when these studies were expanded to a panel of glioblastoma and prostate cancer cell lines, HDAC inhibitor-induced UPR activation was only identified in a subset of lines (Fig. 10.2C). We went on to confirm activation of PERK downstream signaling, including phosphorylation of eIF2α, with subsequent decrease in overall protein synthesis, along with increased expression of ATF4 and CHOP (Fig. 10.3). To determine if

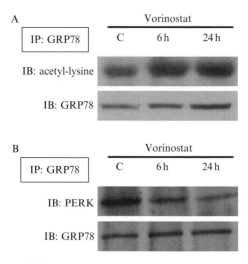

Figure 10.1 HDAC inhibition with vorinostat leads to Grp78 acetylation and dissociation with its client protein PERK. The glioblastoma cell line U251 was exposed to vorinostat (1 μM) for 6 and 24 h. Cell lysates were immunoprecipitated (IP) for Grp78 and immunoblot (IB) was performed for (A) acetyl-lysine and (B) PERK. IB for Grp78 served as loading control. Reproduced from Kahali et al. (2010), with permission.

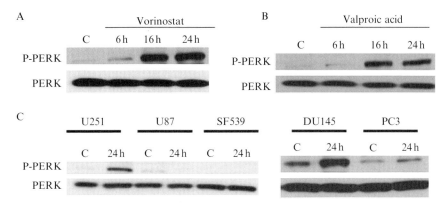

Figure 10.2 HDAC inhibitors activate the unfolded protein response (UPR). Western blot was performed on U251 cells in a time-course manner to determine if HDAC inhibitors from disparate molecular classes can activate PERK (P-PERK), a key signaling pathway associated with the UPR. The HDAC inhibitors used include (A) the hydroxamic acid vorinostat (1 μM) and (B) the short-chain fatty acid valproic acid (2.5 mM). (C) Glioblastoma (U251, U87, SF539) and prostate cancer (DU145, PC3) cell lines were treated with vorinostat (1 μM) or vehicle control for 24 h. Western blot was performed on cell lysates to evaluate for activated PERK (P-PERK). Reproduced from Kahali *et al.* (2010), with permission.

vorinostat-induced UPR activation contributed toward its anti-tumor activity, we inhibited this pathway using RNA interference of PERK. When compared with both untransfected U251 cells and those transfected with scrambled siRNA, U251 transfected with siPERK demonstrates a nearly twofold increase in clonogenic survival after vorinostat exposure, supporting the contributory role of this pathway in vorinostat-induced cytotoxicity (Fig. 10.4). Similar findings were demonstrated in breast cancer cell lines using the HDAC inhibitor panobinostat (Rao *et al.*, 2010). Both MCF-7 and MDA-MB 231 cell lines demonstrated Grp78 acetylation following 16 h panobinostat exposure. In addition, using both Western blot and confocal microscopy, HDAC6 was identified to be a specific HDAC involved with Grp78 acetylation, demonstrating its potential to both bind to and colocalize with Grp78.

Although these studies are some of the first to suggest HDACs may regulate the UPR directly through Grp78 modulation, their potential to interface with this pathway has been previously described. Baumeister *et al.* (2005) demonstrated the potential of HDAC 1 to regulate the UPR at the level of transcription. In these studies, they showed HDAC1 represses expression of Grp78 by binding to its promoter before, but not after, ER stress. Inhibition with HDAC inhibitors induced Grp78 expression leading to therapeutic resistance, which was attenuated when Grp78 was knocked down using RNA interference. In addition to directly modulating the

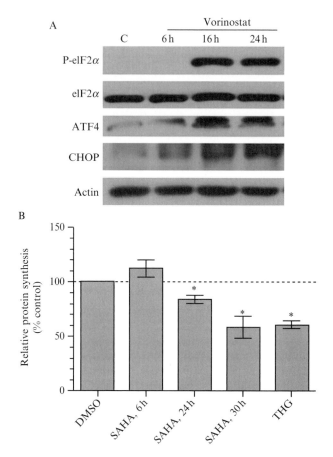

Figure 10.3 Vorinostat activates downstream signaling of the UPR. (A) Western blot performed to evaluate activated eIF2α (P-eIF2α) and translational upregulation of ATF4 and CHOP in U251 cells following vorinostat exposure. (B) Cells were treated with vorinostat (SAHA) for indicated times or thapsigargin (THG) for 24 h and exposed to methionine/cysteine-free media. Cells were then radiolabeled with [S-35]-methionine and protein lysates were collected after 40 min. Counts per minute were measured, adjusted for protein concentration and normalized to the DMSO control. (mean ± S.E.M., ⋆$p < 0.001$ vs. control by Student's t-test). Reproduced from Kahali et al. (2010), with permission.

UPR, investigators have also demonstrated the potential of HDACs to indirectly influence UPR activation. Kawaguchi et al. (2003) discovered that HDAC6, a recognized microtubule-associated deacetylase, was a crucial player in the cellular management of misfolded protein-induced stress. Specifically, they identified HDAC6 to bind both polyubiquitinated misfolded proteins and dynein motors, thereby acting to recruit misfolded protein cargo to dynein motors for transport to aggresomes. Cells deficient

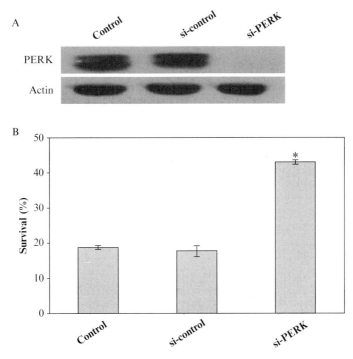

Figure 10.4 PERK signaling influences vorinostat-induced cytotoxicity. U251 cells were untransfected (control) or transfected with scrambled siRNA (si-control) and siRNA PERK (si-PERK) for 48 h. (A) Western blot was performed demonstrating successful PERK knockdown. (B) The remainder of the cells were seeded in 6-well plates, placed in the incubator for 6 h to allow to attach, treated with vorinostat (1 μM) or vehicle control for 48 h, replaced with fresh media, and then allowed to grow for 10–14 days. Percent survival was normalized to the colony forming efficiency of the untreated (vehicle control) cells. *Columns*, mean; *bars*, S.D. ★, $p < 0.01$ according to Student's *t*-test (si-PERK vs. control and si-PERK vs. si-control). Reproduced from Kahali *et al.* (2010), with permission.

in HDAC6 failed to clear misfolded protein aggregates from the cytoplasm, could not form aggresomes properly, and were hypersensitive to the accumulation of misfolded proteins.

3.2.1. Cell lines and treatment

Cell lines: U251, DU145, SF539, U87, and PC3 were obtained either from the National Cancer Institute Frederick Tumor Repository or American Type Culture Collection (ATCC).

Growth medium and condition: Cells were maintained in RPMI 1640 medium (Invitrogen, Carlsbad, CA) supplemented with either 5% (U251, DU145, or SF539) or 10% (U87 or PC3) fetal bovine serum and 5 mM glutamate. Cell cultures were maintained at 37 °C and 5% CO_2.

Other reagents

- 0.05% trypsin–EDTA solution (Invitrogen)
- 1× Ca^{2+} and Mg^{2+}-free PBS (Invitrogen)
- Vorinostat was provided by Merck Research Laboratories (Whitehouse Station, NJ) through the NCI-CTEP and was dissolved in DMSO at 1 mM and stored in aliquots at −80 °C. Vorinostat was added to the cell treatment at 1 µl final concentration.
- Valproic acid (sodium salt) was purchased from Sigma-Aldrich (St. Louis, MO) and dissolved in PBS at 100 mM, and aliquoted, stored at −20 °C.

3.2.2. Immunoprecipitation
3.2.2.1. Materials

- *Lysis buffer/washing buffer 1*: (50 mM Tris–HCl, pH 7.4, 150 mM NaCl, 1 mM EDTA, 0.5% Na-deoxycholate, 0.7 µg/ml pepstatin, 1% NP-40, and complete mini protease inhibitor cocktail (Roche Diagnostic Corp., Indianapolis, IN). The lysis buffer can be stored aliquoted in −20 °C and mixed thoroughly after thawing.
- *High salt washing buffer 2*: (50 mM Tris–HCl, pH 7.4, 500 mM NaCl, 1 mM EDTA, 0.05% sodium deoxycholate, 0.1% NP-40, and protease inhibitor cocktail.
- *Low salt washing buffer 3*: (50 mM Tris–HCl, pH 7.4, 1 mM EDTA, 0.05% sodium deoxycholate, 0.1% NP40, and protease inhibitor cocktail.
- Washing buffers 2 and 3 can be stored at 2–8 °C.
- *2× Laemmli sample buffer*: (62 mM Tris, pH 6.8, 10% glycerol, 2% sodium dodecyl sulfate (SDS), 5% β-mercaptoethanol, and 0.003% bromophenol blue)
- Protein A- and protein G-agarose beads (Roche Diagnostic Corp.)
- Tube rotator, capable of end over end rotation

3.2.2.2. Experimental procedure
Cellular extracts were prepared from approximately 5×10^6 cells in lysis buffer, incubated on ice for 15 min, and centrifuged at 10,400 rpm to remove cellular debris. Protein concentrations in the lysates were quantified using a bicinchoninic acid (BCA) protein assay kit (Thermo Fisher Scientific, Rockford, IL). Five hundred micrograms of whole-cell lysates was precleared with a 50 µl mixture of protein A-agarose and protein G-agarose beads in a tube rotator at 4 °C overnight. Fifteen micrograms of rabbit GRP78 polyclonal primary antibody (Santa Cruz Biotechnology, Santa Cruz, CA) was added to the lysate and incubated for 3 h at 4 °C. Then, 50 µl of protein A-agarose beads were added to the lysate and incubated overnight at 4 °C. Agarose–antibody–protein complexes were then pelleted by centrifugation and washed three

times with (1) ice-cold washing buffer, (2) high salt buffer, and (3) low salt buffer. After discarding the supernatant from the final wash, pelleted beads were resuspended in 40 μl of 2× Laemmli sample buffer, heated at 90 °C for 10 min, and centrifuged. Forty microliters of each sample (supernatant) was loaded onto 7.5% polyacrylamide gels, and the immunoprecipitated proteins were separated by gel electrophoresis.

3.2.3. Immunoblot analysis
3.2.3.1. Materials

- *Cell lysis buffer*: 1% Triton X-100, 1 mM phenylmethylsulfonyl fluoride, 10 μg/ml leupeptin, 1 μg/ml pepstatin-A, 2 μg/ml aprotinin, 20 mM p-nitrophenyl phosphate, 0.5 mM sodium orthovanadate, 1 mM 4-(2-aminoethyl) benzenesulfonylfluoride hydrochloride, 1 mM EDTA, 1 mg EGTA.
- The lysis buffer can be stored at 4 °C for 2 weeks or for long-term storage, 1 ml aliquots may be stored at −20 °C. One protease inhibitor cocktail tablet (Roche) was added per 10 ml lysis buffer before using.
- *TBS-Tween (TBS-T)*: TBS (20 mM Tris-base, 137 mM NaCl, pH 7.6). Required volume of Tween 20 was diluted in TBS to give a 0.1% (v/v) solution.
- *Protease inhibitor cocktail tablet*: (Roche Diagnostic Corp.)
- Phosphatase inhibitor cocktail A and B (Sigma-Aldrich)
- BCA Protein Assay Reagent (Thermo Fisher Scientific)
- Precision Plus Dual color Protein standard (Bio-Rad, Hercules, CA)

3.2.3.2. Experimental procedure For the extraction of the whole-cell lysates, cells were harvested by centrifugation at 1000 rpm for 5 min. Cell pellets were washed once with PBS, gently resuspended in 200 μl of lysis buffer, and incubated on ice with occasional vortexing for 30 min. Cell lysates were centrifuged at 12,000 rpm in a tabletop centrifuge for 15 min to remove the nuclear and cellular debris. Protein concentrations of the clarified lysate were determined by the modified BCA assay reagent, following the guidelines of the manufacturer. Proteins were resolved by sodium dodecyl sulfate-polyacrylamide gel electrophoresis (SDS-PAGE), transferred to polyvinylidene difluoride (PVDF) membranes (Millipore, Billerica, MA) using a wet transfer system (Bio-Rad). A prestained molecular marker (Bio-Rad) was loaded to the same polyacrylamide gel. Membranes were stained with 0.15% ponceau red (Sigma-Aldrich) to ensure equal loading and transfer, then blocked with 5% (w/v), dried nonfat milk in Tris-buffered saline (TBS) and incubated with primary antibody for 3 h at room temperature or overnight at 4 °C in TBS with 0.1% Tween-20 (TBST). The membranes were washed in TBST buffer three times (5 min for each wash), then incubated for 1–2 h with appropriate horseradish peroxidase-conjugated, species-specific secondary antibodies (Santa Cruz Biotechnology) at room temperature. Primary antibodies used include anti-acetyl lysine, anti-total eukaryotic initiating factor 2α (eIF2α), and anti-eIF2α Ser51 phosphospecific antibody (Cell Signaling Technology, Beverly, MA), anti-phosphospecific PERK, anti-PERK, anti-ATF4, anti-C/EBP homologous protein (CHOP), and anti-GRP78 (Santa Cruz Biotechnology), and anti-actin (Sigma-Aldrich). The membranes were washed in TBST buffer three times (10 min for each wash), and membrane bound antibody-labeled proteins were detected using chemiluminescence (Thermo Fisher Scientific) or enhanced chemiluminescence reagent kit (GE Healthcare, Piscataway, NJ) according to manufacturer recommendation. Films (MidSci, St. Louis, MO) were exposed to the membranes for varying periods of time. Reproducibility was confirmed in three individual experiments.

3.2.4. Small interfering RNA
3.2.4.1. Materials

- RPMI 1640 growth medium with 5% FBS
- 1× trypsin–EDTA solution
- 1× Ca^{2+} and Mg^{2+}-free PBS
- OPTI-MEM (Invitrogen)
- Sterile RNase free 1.5 ml microfuge tubes
- Si RNA oligos (Dharmacon, Lafayette, CO)
- 5× siRNA buffer (Dharmacon) for dilution and reconstitution of siRNA oligos
- DharmaFECT Transfection Reagent (Dharmacon). The reagent can be mixed directly with siRNA and OptiMEM, and the reagent/siRNA complex was directly added to cultured cells.
- 6-well plates, P35 and P60 dishes (Corning)
- DNase, RNase free Water (Promega, Madison, WI)
- Hemocytometer/Beckman Coulter counter
- Isotonic buffer for cell counting when using a Coulter counter

3.2.4.2. Experimental procedure The day before transfection the U251 cells were seeded at 1×10^5 cells per well in 6-well plates with fresh growth medium. The cells were allowed to reach 70% confluence on the day of transfection. Transfections can be performed in complete growth media, without the requirement for a medium change. The small interfering RNA (siRNA) constructs (siGENOME SMARTpool PERK [M-004883-03-0010], siControl nontargeting siRNA pool [D-001206-13-20], and siGENOME SMARTpool reagents) were purchased from Dharmacon. Cells were transfected with 50 nM siRNA in OPTI-MEM medium with 5% FBS according to manufacturer's protocol. In brief, 50 nM of PERK siRNA or control siRNA from 2 µM stock (i.e., 50 µl) was mixed with Opti-MEM medium (Invitrogen), to make total volume to 200 µl, and then complexed with a mixture of DharmaFECT (5–6 µl) and Opti-MEM; the total volume of the RNA-DharmaFECT complex was 200 µl. The RNA–DharmaFECT complex was then incubated for 20 min at room temperature before being added to the cells. Before transfection, the old medium was discarded, cells were washed once with 1× PBS, and 1.2 ml of fresh RPMI 1640 medium with 5% FBS was placed into each well. The final volume in each well was 2 ml. To each well RNA-DharmaFECT complex in Opti-MEM were added in a drop-wise manner. The 6-well plate was moved gently up and down and side to side to distribute and mix the complex evenly into the medium. Forty-eight hours after transfection, cells were trypsinized and plated for clonogenic survival. Efficiency of siRNA knockdown was measured using the remaining cells by Western blot.

3.2.5. Clonogenic survival

Clonogenic assay is an *in vitro* cell survival assay that evaluates all modalities of cell death based on the ability of a single cell to grow into a colony. The colony is defined to consist of at least 50 cells. Different cell lines have different plating efficiencies (PE). When untreated cells are plated as a single-cell suspension at low densities, they will grow to colonies in approximately 10–14 days. PE is the ratio of the number of colonies to the number of cells seeded. The number of colonies that arise after treatment of cells, expressed relative to the individual cells PE, is called the surviving fraction (SF).

3.2.5.1. Materials

- RPMI 1640 growth medium with 5% FBS
- 1× trypsin–EDTA solution
- 1× Ca^{2+} and Mg^{2+}-free PBS
- Snap Cap 15 ml polypropylene tube (BD Falcon) for cells dilution
- Hemocytometer/Beckman Coulter counter
- Isotonic buffer for cell counting when using a Coulter counter
- 6-well plates (Corning)
- DMSO (Sigma)
- Crystal violet (0.5% in v/v in methanol) solution for staining
- Colony counting pen

3.2.5.2. Experimental procedure Cells were seeded as single cells in 6-well plates and allowed to adhere for 6 h, treated with vorinostat (1 µM) or vehicle control (DMSO) for 48 h, then medium was replaced with drug-free medium, and cells were allowed to incubate for 10–14 days. Plates were then stained with crystal violet, and colonies consisting of 50 or more cells were manually counted. Results were normalized to the colony-forming efficiency of the vehicle control. The survival fraction was calculated based on the number of colonies formed in drug-treated cells relative to that of the untreated control.

4. Future Directions

Although the above-described studies provide insight into a potentially novel mode of action of HDAC inhibitors, several questions remain unanswered. The most fundamental of which is, what is the *specific* role HDACs play in UPR regulation? Based on the presented findings, it is tempting to speculate that acetylation may play an *active* role in client protein dissociation from Grp78 (and thereby UPR activation), rather than passive dissociation involved with competitive binding of Grp78 between its clients and misfolded proteins. Such findings would parallel the role acetylation plays in modulating client protein interactions of Hsp90. Further, the posttranslational modification of Grp78, through reversible acetylation, may allow for a unique mechanism to differentially activate selective pathways based on the specific stress the cell is experiencing. Unfortunately, a causal link between protein acetylation and its functional consequence has yet to be definitively established. In addition, the importance of the acetylation state of individual residues of Grp78 has not been defined. Therefore, a critical piece of data to support such a hypothesis involves comprehensively defining acetylation sites within the Grp78 sequence and generating point mutations to mimic acetylated/unacetylated states to establish a direct role for a specific modification with chaperone function. Such studies are currently underway to understand this process in further detail.

Beyond the limitations to our mechanistic understanding of HDAC and Grp78 interactions, the translational potential of these findings in the clinical development of HDAC inhibitors, although promising, remains unclear. As our data demonstrates UPR activation following HDAC inhibition is only present in a subset of cell lines, and that activation contributes toward their subsequent anti-tumor activity, current studies are designed to determine if UPR related proteins may serve as a biomarkers for these agents. As HDAC inhibitors are being actively investigated in many solid tumors and identifying biomarkers represents the hallmark of personalized cancer therapy, favorable findings would have important clinical implications. Further, if the above-described mechanistic studies identify a specific HDAC to be involved with UPR regulation, such findings would stimulate the development of more isotype specific inhibitors, which may provide for a broader therapeutic window.

REFERENCES

Allfrey, V. G., et al. (1964). Acetylation and methylation of histones and their possible role in the regulation of RNA synthesis. *Proc. Natl. Acad. Sci. USA* **51,** 786–794.
Bali, P., et al. (2005). Inhibition of histone deacetylase 6 acetylates and disrupts the chaperone function of heat shock protein 90: A novel basis for antileukemia activity of histone deacetylase inhibitors. *J. Biol. Chem.* **280,** 26729–26734.
Baumeister, P., et al. (2005). Endoplasmic reticulum stress induction of the Grp78/BiP promoter: Activating mechanisms mediated by YY1 and its interactive chromatin modifiers. *Mol. Cell. Biol.* **25,** 4529–4540.
Bolden, J. E., et al. (2006). Anticancer activities of histone deacetylase inhibitors. *Nat. Rev. Drug Discov.* **5,** 769–784.
Bradner, J. E., et al. (2010). Chemical phylogenetics of histone deacetylases. *Nat. Chem. Biol.* **6,** 238–243.
Choudhary, C., et al. (2009). Lysine acetylation targets protein complexes and co-regulates major cellular functions. *Science* **325,** 834–840.
Harding, H. P., et al. (1999). Protein translation and folding are coupled by an endoplasmic-reticulum-resident kinase. *Nature* **397,** 271–274.
Kahali, S., et al. (2010). Activation of the unfolded protein response contributes toward the antitumor activity of vorinostat. *Neoplasia* **12,** 80–86.
Kaufman, R. J. (1999). Stress signaling from the lumen of the endoplasmic reticulum: Coordination of gene transcriptional and translational controls. *Genes Dev.* **13,** 1211–1233.
Kaufman, R. J. (2004). Regulation of mRNA translation by protein folding in the endoplasmic reticulum. *Trends Biochem. Sci.* **29,** 152–158.
Kawaguchi, Y., et al. (2003). The deacetylase HDAC6 regulates aggresome formation and cell viability in response to misfolded protein stress. *Cell* **115,** 727–738.
Kovacs, J. J., et al. (2005). HDAC6 regulates Hsp90 acetylation and chaperone-dependent activation of glucocorticoid receptor. *Mol. Cell* **18,** 601–607.
Lane, A. A., and Chabner, B. A. (2009). Histone deacetylase inhibitors in cancer therapy. *J. Clin. Oncol.* **27,** 5459–5468.
Leder, A., et al. (1975). Differentiation of erythroleukemic cells in the presence of inhibitors of DNA synthesis. *Science* **190,** 893–894.
Lee, D. Y., et al. (1993). A positive role for histone acetylation in transcription factor access to nucleosomal DNA. *Cell* **72,** 73–84.
Lee, H. K., et al. (2008). GRP78 is overexpressed in glioblastomas and regulates glioma cell growth and apoptosis. *Neuro Oncol.* **10,** 236–243.
Lin, J. H., et al. (2007). IRE1 signaling affects cell fate during the unfolded protein response. *Science* **318,** 944–949.
Ma, Y., and Hendershot, L. M. (2004). The role of the unfolded protein response in tumour development: Friend or foe? *Nat. Rev. Cancer* **4,** 966–977.
Mahalingam, D., et al. (2009). Targeting HSP90 for cancer therapy. *Br. J. Cancer* **100,** 1523–1529.
Phillips, D. M. (1963). The presence of acetyl groups of histones. *Biochem. J.* **87,** 258–263.
Rao, R., et al. (2010). Treatment with panobinostat induces glucose-regulated protein 78 acetylation and endoplasmic reticulum stress in breast cancer cells. *Mol. Cancer Ther.* **9,** 942–952.
Richon, V. M., et al. (1998). A class of hybrid polar inducers of transformed cell differentiation inhibits histone deacetylases. *Proc. Natl. Acad. Sci. USA* **95,** 3003–3007.
Riggs, M. G., et al. (1977). n-Butyrate causes histone modification in HeLa and Friend erythroleukaemia cells. *Nature* **268,** 462–464.

Schroder, M., and Kaufman, R. J. (2005). The mammalian unfolded protein response. *Annu. Rev. Biochem.* **74,** 739–789.

Scroggins, B. T., *et al.* (2007). An acetylation site in the middle domain of Hsp90 regulates chaperone function. *Mol. Cell* **25,** 151–159.

Xu, W. S., *et al.* (2007). Histone deacetylase inhibitors: Molecular mechanisms of action. *Oncogene* **26,** 5541–5552.

Ye, J., *et al.* (2000). ER stress induces cleavage of membrane-bound ATF6 by the same proteases that process SREBPs. *Mol. Cell* **6,** 1355–1364.

Yoshida, M., *et al.* (1990). Potent and specific inhibition of mammalian histone deacetylase both in vivo and in vitro by trichostatin A. *J. Biol. Chem.* **265,** 17174–17179.

Yoshida, H., *et al.* (2001). XBP1 mRNA is induced by ATF6 and spliced by IRE1 in response to ER stress to produce a highly active transcription factor. *Cell* **107,** 881–891.

Yu, X., *et al.* (2002). Modulation of p53, ErbB1, ErbB2, and Raf-1 expression in lung cancer cells by depsipeptide FR901228. *J. Natl. Cancer Inst.* **94,** 504–513.

CHAPTER ELEVEN

Immunohistochemical Detection of Activating Transcription Factor 3, a Hub of the Cellular Adaptive–Response Network

Tsonwin Hai,[*,†,‡,§] Swati Jalgaonkar,[*,†,‡] Christopher C. Wolford,[*,†] and Xin Yin[*,†,§]

Contents

1. Introduction	176
1.1. ATF3 as a hub of the cellular adaptive–response network	176
1.2. ATF3 in unfolded protein response	177
1.3. ATF3 in disease models	179
2. An IHC Protocol for ATF3	181
2.1. Important factors for consideration in pilot experiments	181
2.2. Required materials	186
2.3. Procedures	187
Acknowledgments	192
References	192

Abstract

Activating transcription factor 3 (ATF3) gene encodes a member of the ATF family of transcription factors and is induced by various stress signals, including many of those that induce the unfolded protein response (UPR). Emerging evidence suggests that ATF3 is a hub of the cellular adaptive–response network and studies using various mouse models indicate that ATF3 plays a role in the pathogenesis of various diseases. One way to investigate the potential relevance of ATF3 to human diseases is to determine its expression in patient samples and test whether it correlates with disease progression or clinical outcomes. Due to the scarcity and preciousness of patient samples, methods that can detect ATF3 on archival tissue sections would greatly facilitate this research. In this chapter, we briefly review the roles of ATF3 in cellular

[*] Department of Molecular and Cellular Biochemistry, Ohio State University, Columbus, Ohio, USA
[†] Center for Molecular Neurobiology, Ohio State University, Columbus, Ohio, USA
[‡] Molecular, Cellular, and Developmental Biology Program, Ohio State University, Columbus, Ohio, USA
[§] Ohio State Biochemistry Program, Ohio State University, Columbus, Ohio, USA

adaptive-response and UPR, and then describe the detailed steps and tips that we developed based on general immunohistochemistry (IHC) protocols to detect ATF3 on paraffin embedded sections.

Abbreviations

ATF3	activating transcription factor 3
CREB	cyclic AMP responsive element binding protein
bZip	basic region leucine zipper
ER	endoplasmic reticulum
eIF2	eukaryotic initiation factor 2
UPR	unfolded protein response
ISR	integrated stress response
HRI	heme regulated inhibitor kinase, also called eIF2alpha kinase (EIF2AK1)
PKR	RNA-dependent protein kinase, also called EIF2AK2
PERK	PKR-like endoplasmic reticulum kinase, also called pancreatic eIF2 kinase (PEK), or EIF2AK3
GCN2	general control nonrepressible 2, also called EIF2AK4
IRE1	inositol-requiring enzyme 1
LPS	lipopolysaccharide
UV	ultraviolet
IHC	immunohistochemistry

1. Introduction

1.1. ATF3 as a hub of the cellular adaptive–response network

ATF3 is a member of the activating transcription factor/cAMP responsive element binding (CREB) protein family of transcription factors, which share the basic region-leucine zipper (bZip) DNA binding motif and bind to the consensus sequence TGACGTCA *in vitro* (for a previous review, see Hai and Hartman, 2001). The level of ATF3 mRNA is low or undetectable in normal unstressed tissues (from mice) and most cell lines, but greatly increases upon stimulation (Hai, 2006; Hai *et al.*, 1999). One striking feature of ATF3 induction is that it is neither tissue-specific nor stimulus-specific. ATF3 can be induced by a broad spectrum of stimuli and can be induced in various tissues or cell types. In fact, it is more an exception than the norm, if a stimulus does not induce ATF3. For a list of stimuli that have been shown to induce ATF3, see a previous review (Hai, 2006). Not surprisingly, many signaling pathways have been demonstrated to be involved in the induction of ATF3 by stress signals, including the JNK, p38, NFκB, PKC, and calcium signaling pathways (Hai et al., 2010). Consistently, the ATF3 promoter is packed with transcription factor binding sites, many of which are recognized by factors downstream of the signaling pathways described above. Examples include ATF, Fos/Jun, NFκB (see Hai et al., 2010 for a review), and NFAT (Wu et al., 2010). In addition to the inducibility of ATF3 expression, amino acid sequence analyses of ATF3 revealed many potential sites for modification; for example, the presence of 21 serines/threonines, one tyrosine, and 17 lysines, which together account for $\sim 20\%$ of the molecule. Although rigorous evidence for posttranslational modifications of ATF3 and their functional significance is lacking at present, this abundance of potential posttranslational modification sites is consistent with the idea that ATF3 is a target for various regulations. Taken together, ATF3 gene is induced by a variety of stress signals and signaling pathways, and its protein is a potential target for various modifications. All these indicate that ATF3 acts as an integration point for a variety of cellular controls, prompting us to put forth the idea that ATF3 functions as a "hub" of the cellular adaptive–response network that is utilized in response to the disturbance of homeostasis (more in a previous review, Hai et al., 2010).

1.2. ATF3 in unfolded protein response

Endoplasmic reticulum (ER) stress refers to the condition whereby unfolded proteins accumulate in the lumen of ER. Under this condition, the cells activate a set of responses collectively called unfolded protein response (UPR), which is composed of three signaling arms distinguished by the following ER-membrane proteins that sense the unfolded proteins in the lumen: (a) IRE1, (b) ATF6, and (c) PERK (reviews: Kaufman, 1999; Ron and Walter, 2007; Schroder and Kaufman, 2005). ATF3 is induced by the PERK pathway and plays an integral role in coordinating this pathway's function (Jiang et al., 2004). Upon activation, PERK phosphorylates eIF2α and results in the increased production of ATF4 protein, which in turn upregulates ATF3 expression (Jiang et al., 2004, and see Chapter 19). One interesting aspect of the PERK arm of UPR is that phosphorylation of eIF2α can also be achieved by other eIF2α kinases: GCN2, PKR, and HRI (for a review, see Wek et al., 2006). These kinases can be induced by many stress signals unrelated to ER stress, such as nutritional stress, double stranded RNA accumulation, UV irradiation, and heat shock (reviews, Ron and Walter, 2007; Wek et al., 2006). Thus, Ron and colleagues suggested referring to events downstream of eIF2α phosphorylation as integrated stress response (ISR) (Harding et al., 2003). Not surprisingly, all the stress signals that induce ISR also induce ATF3, presumably at least partly via the eIF2α → ATF4 pathway. However, it is possible that they also induce ATF3 expression via other signaling pathways, such as the JNK, p38, and NFκB pathways, since these pathways are known to be induced by the above stress signals and are known to be involved in the induction of ATF3. Taken together, ATF3 is induced by a variety of stress signals—including ER stress, stresses that induce ISR, and other stress signals (Fig. 11.1), again supporting the notion that ATF3 is a hub or integration point for cellular stress responses.

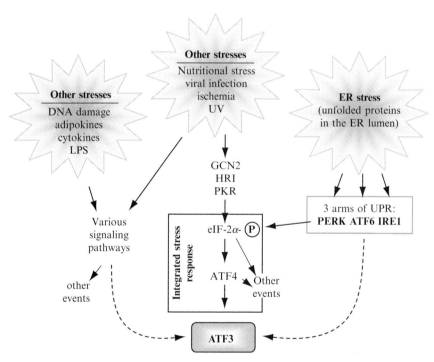

Figure 11.1 A schematic of ATF3 induction by various stress signals. See text in Section 1.2 for details. ER, endoplasmic reticulum; eIF2, eukaryotic initiation factor 2; UPR, unfolded protein response; ISR, integrated stress response; HRI, heme regulated inhibitor kinase, also called eIF2alpha kinase (EIF2AK1); PKR, RNA-dependent protein kinase, also called EIF2AK2; PERK, PKR-like endoplasmic reticulum kinase, also called pancreatic eIF2 kinase (PEK), or EIF2AK3; GCN2, general control non-repressible 2, also called EIF2AK4; IRE1, inositol-requiring enzyme 1; LPS, lipopolysaccharide; UV, ultraviolet.

1.3. ATF3 in disease models

A survey of the literature indicates that ATF3 is implicated in a diversity of diseases based on various mouse models which employed either loss- or gain-of-function approaches (for a list, see Table 11.1). One general statement summarizes these studies: ATF3 has pleiotropic effects and the consequences of its expression can be beneficial or detrimental in a context-dependent manner. As an example, ATF3 inhibits the expression of pro-inflammatory genes, such as TNFα, IL12b, and IFNγ, in immune cells (Filen *et al.*, 2010; Gilchrist *et al.*, 2006; Rosenberger *et al.*, 2008; Whitmore *et al.*, 2007). In doing so, ATF3 prevents the immune system from overreacting and thus protects the whole organism in inflammatory disease models such as septic shock, asthma, and ventilation-induced lung injury (Akram *et al.*, 2010; Gilchrist *et al.*, 2006, 2008). However, the price for this immune dampening effect is that it weakens the host defense system, making it more vulnerable to infection, such as by cytomegalovirus (Rosenberger *et al.*, 2008). This underscores the double-edge sword nature of ATF3. Another example of context dependency is the role of ATF3 in cancer development. ATF3 was demonstrated to be oncogenic in several cancer models, such as breast cancer, prostate cancer, and skin cancer (Bandyopadhyay *et al.*, 2006; Wang *et al.*, 2008; Wu *et al.*, 2010; Yin *et al.*, 2010). However, it was also demonstrated to be anti-oncogenic in a colon cancer model and a Ras-mediated transformation model (Bottone *et al.*, 2005; Lu *et al.*, 2006). Although the anti-oncogenic results are paradoxical to the above findings that ATF3 is oncogenic, they are consistent with many reports that ATF3 is pro-apoptotic and is deleterious to the tissues in transgenic mice that express ATF3 (for references, see Table 11.1 and a previous review Hai *et al.*, 2010). One potential explanation for these seemingly conflicting data is the degree of cell malignancy: ATF3 is pro-apoptotic to normal or untransformed cells, but protects the cells from death upon their malignant transformation as shown by an *in vitro* study using isogenic cell lines with varying degrees of malignancy (Yin *et al.*, 2008). However, much more work is required to understand the context-dependency of ATF3. For more description of ATF3 in different disease models, see previous reviews (Hai *et al.*, 2010; McConoughey *et al.*, 2010).

The above studies using mouse models raised an important question: does ATF3 play a role in human diseases? One approach to address this question is to examine ATF3 expression in human samples and ask whether it correlates with any disease states or clinical outcomes. Due to the scarcity and preciousness of patient samples, methods that can detect ATF3 on archival tissue sections would greatly facilitate the research. Immunohistochemistry (IHC) is a desirable assay for this purpose, because it detects proteins—the final products for many genes—and can reveal the subcellular localization of the molecules. In addition, it is generally compatible with

Table 11.1 A list of mouse models implicating ATF3 in various diseases

	Models[a]	Phenotypes	References
Cell types with ectopic expression of ATF3	**Transgenic mice** Hepatocytes Pancreatic β-cells Cardiac myocytes Basal epithelial cells	Liver dysfunction Reduced β-cell mass and defects in glucose homeostasis Conduction abnormalities and contractile dysfunction Epidermal hyperplasia, oral carcinoma, and mammary carcinoma (in biparous mice)	Allen-Jennings et al. (2001, 2002) Li et al. (2008) Okamoto et al. (2001) Wang et al. (2007, 2008)
Stress models	**ATF3 KO mice**[b] LPS-induced Septic shock Pulmonary stress [c]MCMV infection High fat diet	Hyper-inflammation and animals succumb faster Hyper-inflammation in both allergy and ventilator-induced lung injury stress models Decreased viral load Decreased ability of β-cells to function and increased glucose intolerance	Gilchrist et al. (2006) Akram et al. (2010), Gilchrist et al. (2008) Rosenberger et al. (2008) Zmuda et al. (2010)
Ectopic expression of ATF3 in cancer cells	**Injection models** Prostate, keratinocyte, melanoma, and breast cancer cells Colon cancer cells	Enhanced tumorigenicity or metastasis Reduced tumorigenicity	Bandyopadhyay et al. (2006), Ishiguro et al. (1996), Wu et al. (2010), Yin et al. (2010) Bottone et al. (2005)

Phenotypes indicate the consequences of ATF3 deficiency in the corresponding stress models or the consequences of ectopically expressing ATF3 (transgenic or injection models).

[a] Gain-of-function approach.
[b] Loss-of-function approach.
[c] ATF3 KO has a beneficial effect.

paraffin embedded sections. Below, we describe a protocol that we have developed over the years by combining and modifying different IHC protocols.

2. AN IHC PROTOCOL FOR ATF3

IHC is a finicky technique with many idiosyncratic issues that cannot be solved in a predictable manner. For each protein, it is necessary to develop a protocol that is specifically tailored for it; sometimes even with many trials and errors, it may not be possible to work out a condition that yields the most desirable quality of images. For a comprehensive review on technical aspects of IHC, see (Ramos-Vara, 2005). In Section 2.3, we detail a protocol that worked for detecting ATF3 in various tissues; however, it may need further modification for specific applications. This is because many factors affect IHC outcomes, including the batch of antibodies, the abundance of the protein of interest in the samples, and the variation in fixation time and methods. We found that even when assayed under exactly the same conditions carried out side-by-side, different tissues fixed by the same facility may yield images of different quality. Thus, it is important to carry out pilot experiments using the same tissues as desired for the real experiment. For precious samples such as tissue microarrays (TMAs), pilot experiments using TMAs may not be feasible. Obtain a few paraffin blocks of individual samples of the same tissue types and work out the conditions. If "test TMAs" (with small number of tissue cores) are available, try out the conditions on them first.

2.1. Important factors for consideration in pilot experiments

Below are several key factors that affect IHC results and should be optimized in pilot experiments.

2.1.1. Primary antibodies

Several anti-ATF3 antibodies are commercially available. We have only compared the antibodies from Atlas (# HPA001562, against amino acids 1–113) and Santa Cruz (Clone C-19, #sc-188, against the C′ terminal amino acids 163–181). They are both polyclonal antibodies generated in rabbits, but differ in their performance. In general, the Atlas antibody gives rise to better signal-to-background ratios and cleaner images. Figure 11.2 shows an example. In addition, the Atlas antibody is more consistent than the Santa Cruz antibody, providing more reproducible IHC results on both human and murine tissues. In our hands, it is also more consistent in Western blot. Some batches of the Santa Cruz antibody detect nonspecific bands, which may account for its high background (darker and less specific

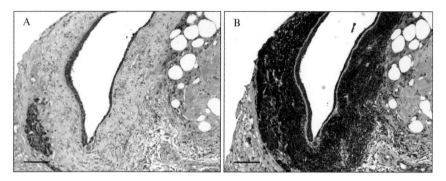

Figure 11.2 An IHC comparison of the Atlas and Santa Cruz anti-ATF3 antibodies. Human breast tumor sections were stained using optimized conditions for Atlas (1:125 dilution; A) or Santa Cruz (1:200 dilution; B) anti-ATF3 antibodies. Staining was performed with ImmPRESS anti-rabbit secondary antibody, developed using ImmPACT NovaRed substrate, and counterstained with hematoxylin. Scale bars: 200 μm.

signals compared to Atlas) and inconsistency in IHC. Due to its better performance, we recommend the Atlas antibody for IHC, despite its higher cost. For each batch of antibody, it is important to titrate the antibody for the intended application. A good starting point is 1:50, 1:125, 1:150, 1:200, and 1:300. We found that dilution in a surprisingly narrow range (such as 1:125 vs. 1:150) could make a difference in the signal-to-background ratio. After optimization, set aside the current batch of antibody for IHC use only and purchase more of the same batch (if desired). Considering the importance of primary antibodies in IHC, it is possible that the development of new antibodies recognizing different epitopes within ATF3 may further improve this assay.

2.1.2. Antigen retrieval

Fixation cross-links and modifies the protein structure, reducing the recognition of epitopes by the antibodies (Ramos-Vara, 2005). To counter this problem, various antigen retrieval methods have been developed to reverse at least some, if not all, of the cross-linking (for a review, see Ramos-Vara, 2005). Many of the retrieval methods involve a heating step, but the means of heating vary; for example, pressure cooker, steamer, microwave oven, and hot plate. In our hands, the pressure cooker method worked the best for detecting ATF3 by IHC.

2.1.3. Secondary antibody and the detection methods

The commonly used biotin–avidin system for detecting IHC signals can produce high background, especially when harsh antigen retrieval methods are used (a review, Ramos-Vara, 2005). A polymer-based system independent of avidin–biotin was developed in the 1990s (see Sabattini et al., 1998 and references therein). It consists of an inert polymer backbone conjugated with secondary antibodies and many (up to 100) molecules of reporter enzyme, such as peroxidase. Figure 11.3 shows a schematic diagram of two commercially available polymer-based systems: (a) ImmPRESS^{TM}, which uses a flexible micropolymer as the backbone, and (b) EnVison^{TM}, which uses dextran as the backbone. The polymer-based system offers two main advantages over the biotin–avidin system: first, since it is a two-step method, it is rapid and simple compared to the three-step biotin–avidin system; second, it has no background staining due to endogenous biotin or avidin (Ramos-Vara and Miller, 2006). A side-by-side comparison indicated that ImmPRESS^{TM} performed better than EnVison^{TM} in the majority of cases examined in the study (Ramos-Vara and Miller, 2006). We have used the ImmPRESS^{TM} system and found it greatly improved the IHC signals for ATF3. However, we have not compared ImmPRESS^{TM} to EnVison^{TM} side-by-side.

Many different peroxidase substrates are available for the colorimetric development of the stained tissues. We routinely use two different peroxidase substrates depending on the application: (a) DAB, which gives rise to a brown color and is slower to develop, and (b) ImmPACT NovaRed, which gives rise to a red color and is more rapid to develop. Both substrates provide excellent color contrast when counterstaining with hematoxylin; however, there are unique advantages and disadvantages of each substrate that will dictate when their use is desirable. The slower development of the DAB substrate allows for the ability to more readily discern different signal intensities within the same tissue, possibly due to tighter control of the time

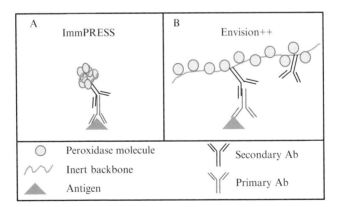

Figure 11.3 A schematic of two commercially available polymer-based systems. A flexible micropolymer backbone (ImmPRESS reagent; A) or dextran (Envision++ reagent; B) bridges a large number of peroxidase molecules to each secondary antibody. Ab, antibody. (See Color Insert.)

required to obtain the desired signal intensity. However, we have observed that signals produced by staining endogenous ATF3 using the DAB substrate are generally less sharp than those produced with the ImmPACT NovaRed substrate. Regardless of the detection method used, one needs to carry out a pilot experiment for each project in order to determine the peroxidase reaction time for optimal staining. It may be necessary to try several time points and repeat the pilot experiments a few times to find the appropriate reaction time. See Section 2.3.5 for detailed steps.

2.1.4. Signal specificity

As with any assay, it is imperative to test the specificity of the signals. Here, we present our data on ATF3 using human breast cancer samples with the following tests. (a) Depletion: We carried out depletion experiments to test the specificity. As shown in Fig. 11.4, pre-incubation of the antibody with the GST-ATF3 fusion protein completely removed the signals but pre-incubation with the same concentration of GST did not. (b) Control samples: As described in Section 1.1, ATF3 level is low or undetectable in unstressed mouse tissues. We used the breast reduction samples as a potential unstressed control tissue. In side-by-side experiments, the signal for ATF3 was greatly reduced in the breast reduction samples compared to that in the breast cancer samples (Fig. 11.5). However, this was not absolute. In most cases we could still detect a low level of ATF3, and in some cases the signal of ATF3 was quite strong (data not shown). These results are similar to that reported by MacLeod and colleagues, who analyzed ATF3 in human breast samples (Wang et al., 2008). We posit that the breast reduction samples are not un-stressed tissues, a notion supported by the abundance of fibrotic material in their stroma. (c) Tumor samples from ATF3-deficient mice: Due to the unavailability of truly unstressed human tissues as ATF3-negative controls, we resorted to ATF3 knockout (KO) mice. We crossed the ATF3

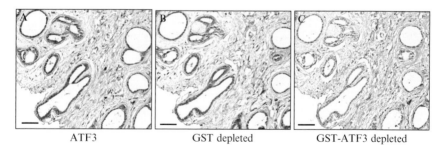

Figure 11.4 Specificity test of the Atlas anti-ATF3 antibody. Atlas anti-ATF3 antibody solution was either untreated (A), or precleared using GST (B), or GST-ATF3 (C). The resulting solutions were used to stain human breast tumor sections. Conditions were the same as those described in the Fig. 11.2. Scale bars: 100 μm.

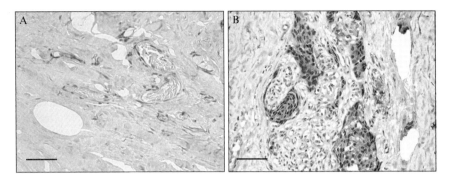

Figure 11.5 A comparison of ATF3 expression in human breast reduction (A) and breast tumor (B) samples. Tissue sections were stained using the same conditions described in the Fig. 11.2. Scale bars: 100 μm.

KO mice to transgenic mice expressing the polyoma middle T (PyMT) antigen under the control of the murine mammary tumors virus (MMTV) promoter—MMTV-PyMT mice (Davie et al., 2007; Guy et al., 1992)—to generate tumor-bearing mice in the ATF3 KO background (C57BL/6-aft3$^{-/-}$-Tg(MMTV:PyMT)). We then analyzed tumors from these mice in parallel with that from the wild type counterparts (C57BL/6-aft3$^{+/+}$-Tg (MMTV:PyMT)). As shown in Fig. 11.6, the WT tumors contained many ATF3-positive cells but the KO tumor did not, supporting the specificity of the IHC stain. We note that the optimal dilution of the primary antibody for mouse tumors may be different from that for human tumors and should be determined empirically. Two cautionary notes are relevant here. (i) Although immunoglobulin G (IgG) is a commonly used negative control for immunological assays, we found it difficult to use in this case. As described above, the ATF3 signal-to-background ratio is affected by the dilution of the primary antibody—within a surprisingly narrow range. Since the purity of the control IgG and anti-ATF3 IgG may not be the same in the commercially available preparations, it is difficult to make a fair comparison. Even at the same final protein concentration, their effective concentrations may not be the same. In experiments where a slight change in dilution makes a big difference, this is a significant issue. In addition, the subclasses (IgG$_{1-4}$) and storage (longer storage results in aggregation) of IgG affect their ability to cause background signals (see Ramos-Vara, 2005 for a review), further diminishing the utility of IgG as controls. (ii) As described in Section 1.1, the induction of ATF3 is neither tissue-specific nor stimulus-specific. Thus, in a given specimen from diseased tissues, ATF3 may be present in more than one cell type. The broad expression of ATF3 *per se* should not be interpreted as nonspecificity. Controls should be carried out for proper data interpretation, and importantly—as with all experiments—the conclusions should be tested by different assays and approaches.

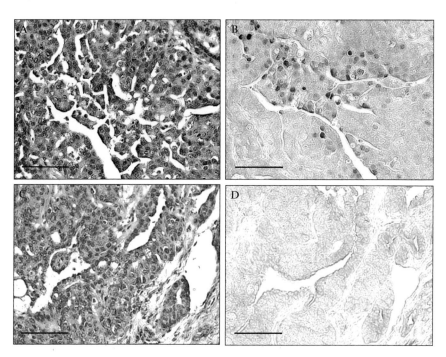

Figure 11.6 Analyses of ATF3 in tumors derived from WT or ATF3 KO mice. Serial sections of tumors from MMTV-PyMT transgenic mice in either WT (A, B) or ATF3 KO (C, D) background were stained with hematoxylin and eosin (A, C) or Atlas anti-ATF3 antibody (B, D). The slower developing DAB substrate was used in this experiment. No counterstain was applied, as it reduced the ability to discern different signal intensities. Scale bars: 100 μm. (See Color Insert.)

2.2. Required materials

Samples

- Human breast carcinoma, formalin-fixed paraffin embedded (FFPE) sections (4–5 μm)
- Human breast reduction samples, FFPE sections (4–5 μm)
- Mouse mammary tumors derived from WT mice expressing the MMTV-PyMT transgene, FFPE sections (4–5 μm)
- Mouse mammary tumors derived from ATF3 KO mice expressing the MMTV-PyMT transgene, FFPE sections (4–5 μm)

Devices

- Pressure cooker, 4 quart (Manttra Inc., #34111)
- Tissue Tek slide staining set: 12 staining dishes, rack which holds 12 staining dishes, 24-slide staining holder (Cardinal Health, #S7626-12)

- Glass baking dish (9″ × 13″)
- Metal slide staining tray (McCrone Microscopes and Accessories, #273-TRAY)
- ImmEdge hydrophobic barrier PAP pen (Vector Laboratories, #H-4000)
- Slides (Fisher Scientific, #12-550-15)
- Cover slips (Fisher Scientific, #12-548-56)
- Air vacuum device for gentle suction

Reagents and buffers

- Anti-ATF3 antibody from Atlas (# HPA001562): Polyclonal antibodies against ATF3 amino acids 1–113, made in rabbits
- Xylene (Histological grade, Fisher Scientific, #X3P-1GAL)
- Ethanol (200 proof, Fisher Scientific, BP2818-4)
- Deionized water (dH$_2$O)
- Hydrogen peroxide (Sigma, #H1009-500ML)
- *Wash buffer*: 1× Tris buffered saline with Tween-20 (TBST): 150 mM sodium chloride, 100 mM Tris–HCl, 0.1% Tween-20. Adjust pH to 7.5 before adding Tween-20.
- *Antigen unmasking solution (freshly prepared)*: 10 mM sodium citrate buffer, 0.05% Tween-20. Adjust pH to 6.0, before adding Tween-20.
- *Blocking solution:* 5% normal goat serum diluted in 1× wash buffer
- *ImmPRESS^{TM} Polymer-based detection reagent*: anti-Rabbit IgG conjugated with peroxidase (Vector, #MP-7401)
- *Substrate for peroxidase*: (a) ImmPACT NovaRed dark red peroxidase substrate (Vector, #SK-4805) or (b) DAB substrate kit for peroxidase (Vector, #SK-4100)
- Hematoxylin (Fisher, #22-220-102)
- VectaMount mounting medium (Vector, #H-5000)

Disposables

- Gloves
- Paper towels
- Kimwipes
- Plastic wrap

2.3. Procedures

For all steps below after deparaffinization, do not allow slides to dry. For all wash steps and some incubation steps below, use the slide holder, which holds multiple slides standing-up. Put it in a plastic slide staining dish filled with ∼250 ml of solution to make sure that the slides are entirely immersed (Fig. 11.7A). Although the slide holder can accommodate 24 slides, use it

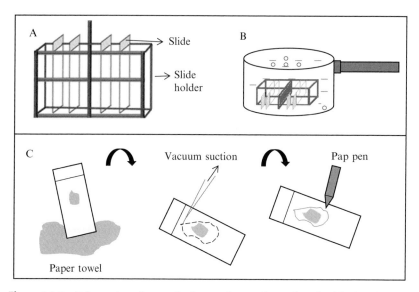

Figure 11.7 Schematics of some devices and procedures described in Section 2.3.

for maximal 12 slides with an empty slot between the adjacent slides. This prevents accidental contact between the slides. For the steps that do not use the slide holder, such as incubation with the blocking solution, primary antibody, secondary antibodies, or staining solution, the slides should be laid flat on a staining tray with the tissue-side up and placed in a moist chamber as detailed below.

2.3.1. Deparaffinization and rehydration

- Incubate the slides in a slide holder at 60 °C in an oven for 1 h to melt the paraffin. Put the slide holder in an empty slide staining dish, which keeps the slide holder steady. After this step, do not allow the slides to dry out throughout the entire protocol.
- Incubate the slides in xylene for 5–10 min; move the slides to a new container with fresh xylene for another 5–10 min.
- Incubate the slides in 100% ethanol for 5 min, move the slides to a new container with fresh 100% ethanol for a second 5 min incubation.
- Incubate the slides in 95% ethanol for 5 min, move the slides to a new container with fresh 95% ethanol for a second 5 min incubation.
- Incubate the slides one time in 75% ethanol for 5 min.
- Incubate the slides in dH_2O for 5 min, move the slides to a new container with fresh dH_2O for a second 5 min incubation.

2.3.2. Antigen retrieval using the pressure cooker method

- While hydrating the tissue sections, prepare 2 l of antigen unmasking solution and heat it in a pressure cooker until the buffer starts to boil. Do not seal the lid, since you do not want the pressure to build yet. This takes approximately 40 min, about the same time that it takes to hydrate the tissues.
- After hydration, transfer the slides to the boiling buffer. Lay the slide holder on its side as illustrated in Fig. 11.7B, so that it does not accidentally fall during incubation. Seal the lid of the pressure cooker and allow the pressure to build to its maximum (this takes about 4 min once the lid is sealed and may vary depending on the pressure cooker).
- Keep the slides at this pressure for 20 min before turning off the heat, gradually releasing the pressure, and carefully opening the lid of the pressure cooker (this takes less than 5 min and may vary depending on the pressure cooker).
- Cool the antigen unmasking solution by placing the open pressure cooker under slow-running tap water for 10 min, being careful not to dispense the water directly over top of the slides.

2.3.3. Blocking and incubation with the primary antibody

- Wash the slides in dH_2O for 5 min, three times.
- Incubate the slides in 3% hydrogen peroxide (diluted in methanol) for 10 min to block the endogenous peroxidase activity.
- Wash the slides in dH_2O for 5 min, twice.
- Incubate the slides in wash buffer one time for 5 min.
- Set up the incubation with the blocking solution by handling one slide at a time. This step should be carried out carefully but quickly to prevent the tissues from drying. Remove a slide from the buffer and damp dry it by holding it up-right with its bottom edge touching a paper towel as illustrated in Fig. 11.7C. The capillary action will drain the buffer from the slide. Then, use gentle air vacuum to dry a circle around the tissue—at a distance so as to not disturb the tissue—as illustrated in Fig. 11.7C. Use the PAP pen to trace the dry circle; this will create a heat-stable water-repellent barrier surrounding the tissue sections and allow reagents to remain on top of the sections. Note that the PAP pen will not work if the surface is wet. Lay the slides down flat on a slide staining tray with the tissue-side up. Immediately, load 100–400 µl of blocking solution over the tissue section (until the section is completely covered).
- After finishing all slides, put the tray in a moist chamber for incubation. To set up a moist chamber, put two layers of damp, but not dripping-wet,

paper towels in the bottom of a 9″ × 13″ glass baking dish, which accommodates the slide staining tray. Cover the dish with plastic wrap.
- Incubate at room temperature for 1 h.
- Remove the blocking solution by gentle air vacuum suction. Make sure to remove residual solution to avoid affecting the concentration of the primary antibody that follows.
- Load 100–400 µl of the anti-ATF3 Atlas primary antibody diluted in blocking solution to each PAP pen-enclosed area (until the section is completely covered). The optimal dilution varies for different batches of antibodies and should be determined empirically for each batch. Refer to Section 2.1.1.
- Set up a moist chamber as above and incubate the slides at 4 °C overnight.
- Remove the primary antibody, place the slides in a slide holder, and wash the slides in wash buffer for 5 min, three times.

2.3.4. Incubation with the ImmPRESS™ secondary antibody

- Remove the wash solution by gentle air vacuum suction. Make sure to remove residual solution to avoid affecting the concentration of the secondary antibody that follows.
- Load 100–400 µl of undiluted anti-rabbit Ig ImmPRESS™ reagent (until the section is completely covered).
- Set up a moist chamber and incubate the slides at room temperature for 30 min.
- Wash the slides in wash buffer for 5 min, three times.

2.3.5. Staining by peroxidase reaction

- Remove the wash solution by gentle air vacuum suction. Make sure to remove residual solution to avoid affecting the concentration of the peroxidase substrate solution that follows.
- Load 100–400 µl of undiluted peroxidase substrate solution until desired staining intensity develops. As indicated in Section 2.1.3, the reaction time should be determined beforehand by pilot experiments. However, due to potential experimental variations, close inspection should be made; it may be necessary to terminate the reaction slightly before or after the predetermined time for optimal color development. To maintain the substrate development time as close as possible among all slides, divide the samples in groups of five and carry out this step in a batch-wise manner. Add the substrate to slide 1 and note the time elapsed between adding solution to slide 1 and adding solution to slide 2. Do the same for the subsequent slides. Terminate the reaction in the same sequence with the same elapsed time. We recommend two substrate solutions: ImmPACT NovaRed and DAB substrate kit for peroxidase (see

Section 2.2, reagents). Each has its pros and cons. (i) ImmPACT NovaRed is a rapid stain and takes 2–5 min for color development, in our experience, using different tissue samples. Its color is vivid and provides a good contrast against the blue hematoxylin counter stain. However, because it develops fast, experimental variation may introduce sufficient artificial differences to affect data interpretation. (ii) DAB is a slower developing stain than ImmPACT NovaRed; in our experience the reaction time ranges from 8 to 12 min. Thus, it is less sensitive to experimental variation. However, its color is less vivid and the resulting image is not as sharp. We recommend using ImmPACT NovaRed when the quality of image is important, but using DAB when the relative signals between samples are critical.

- To terminate the reaction, immerse the slides in dH_2O. It is not necessary to remove the substrate solution by suction. Simply move the slides to the slide holder that is placed inside a slide container with ~ 250 ml dH_2O. Then, rinse the slides in tap water by changing the water in the container five times. Remove the slide holder from the container while refilling it with water to avoid dislodging the tissue sections from the slides.

2.3.6. Counterstain and coversliping

- Counterstain with freshly diluted hematoxylin (in dH_2O). The intensity of the counterstain should be faint, so that it does not interfere with data interpretation. It is best to consult the pathologist(s) who will score the IHC signals to decide the desired intensity. The precise dilution and incubation time to achieve the desired counterstain should be determined empirically and can vary depending on the batch and age of hematoxylin. A good starting point is 1:10 dilution for 1 min. After counterstaining, observe the slides closely. As soon as the desired color contrast is present, immerse the slides in dH_2O.
- Dehydrate the slides by serial incubation in series of graded alcohols: 70% (2X) → 95% (2X) → 100% (2X), for 2 min each.
- Incubate the slides in xylene for 5–10 min to dissolve the hydrophobic edge drawn around the section with the PAP pen.
- Mount by carefully placing a small drop of VectaMount permanent mounting media on the tissue section and slowly laying a cover slip on top. Start by lowering the cover slip on one side of the section; the surface tension of the mounting media will draw the cover slip close. This technique usually avoids trapping air bubbles in between the cover slip and tissues. Using Kimwipes, carefully blot away excess mounting media from the edges of the cover slip. Allow mounting media to dry by laying slides flat and incubating at room temperature overnight. Note that since the mounting media is non-aqueous, it is not necessary to further seal around the edge of the cover slip.

2.3.7. Semi-quantification of the IHC signals

If comparing signal intensity is an important component of the project, it is essential to carry out IHC on all sides simultaneously—side-by-side. It is also essential to have experienced pathologists to score the signals. Ideally, the pathologists should be unaware of—and thus unbiased by—the hypothesis, and the IHC signals should be scored by more than one pathologist.

ACKNOWLEDGMENTS

This work is supported by NIH RO1 CA118306 and DK064938 (to T. H.).

REFERENCES

Akram, A., Han, B., Masoom, H., Peng, C., Lam, E., Litvak, M., Bai, X., Shan, Y., Hai, T., Batt, J., et al. (2010). Activating Transcription Factor 3 Confers Protection against Ventilator Induced Lung Injury. *Am. J. Respir. Crit. Care Med.* **182,** 489–500.

Allen-Jennings, A. E., Hartman, M. G., Kociba, G. J., and Hai, T. (2001). The roles of ATF3 in glucose homeostasis: A transgenic mouse model with liver dysfunction and defects in endocrine pancreas. *J. Biol. Chem.* **276,** 29507–29514.

Allen-Jennings, A. E., Hartman, M. G., Kociba, G. J., and Hai, T. (2002). The roles of ATF3 in liver dysfunction and the regulation of phosphoenolpyruvate carboxykinase gene expression. *J. Biol. Chem.* **277,** 20020–20025.

Bandyopadhyay, S., Wang, Y., Zhan, R., Pai, S. K., Watabe, M., Iiizumi, M., Furuta, E., Mohinta, S., Liu, W., Hirota, S., et al. (2006). The tumor metastasis suppressor gene Drg-1 down regulates the expression of ATF3 in prostate cancer. *Cancer Res.* **66,** 11983–11990.

Bottone, F. G., Jr., Moon, Y., Kim, J. S., Alston-Mills, B., Ishibashi, M., and Eling, T. E. (2005). The anti-invasive activity of cyclooxygenase inhibitors is regulated by the transcription factor ATF3 (activating transcription factor 3). *Mol. Cancer Ther.* **4,** 693–703.

Davie, S. A., Maglione, J. E., Manner, C. K., Young, D., Cardiff, R. D., MacLeod, C. L., and Ellies, L. G. (2007). Effects of FVB/NJ and C57BL/6J strain backgrounds on mammary tumor phenotype in inducible nitric oxide synthase deficient mice. *Transgenic Res.* **16,** 193–201.

Filen, S., Ylikoski, E., Tripathi, S., West, A., Bjorkman, M., Nystrom, J., Ahlfors, H., Coffey, E., Rao, K. V., Rasool, O., and Lahesmaa, R. (2010). Activating transcription factor 3 is a positive regulator of human IFNG gene expression. *J. Immunol.* **184,** 4990–4999.

Gilchrist, M., Thorsson, V., Li, B., Rust, A. G., Korb, M., Kennedy, K., Hai, T., Bolouri, H., and Aderem, A. (2006). Systems biology approaches identify ATF3 as a negative regulator of Toll-like receptor 4. *Nature* **441,** 173–178.

Gilchrist, M., Henderson, W. R., Jr., Clark, A. E., Simmons, R. M., Ye, X., Smith, K. D., and Aderem, A. (2008). Activating transcription factor 3 is a negative regulator of allergic pulmonary inflammation. *J. Exp. Med.* **205,** 2349–2357.

Guy, C. T., Cardiff, R. D., and Muller, W. J. (1992). Induction of mammary tumors by expression of polyomavirus middle T oncogene: A transgenic mouse model for metastatic disease. *Mol. Cell. Biol.* **12,** 954–961.

Hai, T. (2006). The ATF transcription factors in cellular adaptive responses. *In* "Gene expression and Regulation," (J. Ma, ed.), pp. 322–333. Higher Education Press and Springer, Beijing, China and New York, USA.

Hai, T., and Hartman, M. G. (2001). The molecular biology and nomenclature of the activating transcription factor/cAMP responsive element binding family of transcription factors: Activating transcription factor proteins and homeostasis. *Gene* **273**, 1–11.

Hai, T., Wolfgang, C. D., Marsee, D. K., Allen, A. E., and Sivaprasad, U. (1999). ATF3 and stress responses. *Gene Expr.* **7**, 321–335.

Hai, T., Wolford, C. C., and Chang, Y. S. (2010). ATF3, a hub of the cellular adaptive-response network, in the pathogenesis of diseases: Is modulation of inflammation a unifying component? *Gene Expr.* **15**, 1–11.

Harding, H. P., Zhang, Y., Zeng, H., Novoa, I., Lu, P. D., Calfon, M., Sadri, N., Yun, C., Popko, B., Paules, R., *et al.* (2003). An integrated stress response regulates amino acid metabolism and resistance to oxidative stress. *Mol. Cell* **11**, 619–633.

Ishiguro, T., Nakajima, M., Naito, M., Muto, T., and Tsuruo, T. (1996). Identification of genes differentially expressed in B16 murine melanoma sublines with different metastatic potentials. *Cancer Res.* **56**, 875–879.

Jiang, H. Y., Wek, S. A., McGrath, B. C., Lu, D., Hai, T., Harding, H. P., Wang, X., Ron, D., Cavener, D. R., and Wek, R. C. (2004). Activating transcription factor 3 is integral to the eukaryotic initiation factor 2 kinase stress response. *Mol. Cell. Biol.* **24**, 1365–1377.

Kaufman, R. J. (1999). Stress signaling from the lumen of the endoplasmic reticulum: Coordination of gene transcriptional and translational controls. *Genes Dev.* **13**, 1211–1233.

Li, D., Yin, X., Zmuda, E. J., Wolford, C. C., Dong, X., White, M. F., and Hai, T. (2008). The repression of IRS2 gene by ATF3, a stress-inducible gene, contributes to pancreatic β-cell apoptosis. *Diabetes* **57**, 635–644.

Lu, D., Wolfgang, C. D., and Hai, T. (2006). Activating transcription factor 3, a stress-inducible gene, suppresses Ras-stimulated tumorigenesis. *J. Biol. Chem.* **281**, 10473–10481.

McConoughey, S. J., Wolford, C. C., and Hai, T. (2010). Activating transcription factor 3. *UCSD-Nat. Mol. Pages* pending.

Okamoto, Y., Chaves, A., Chen, J., Kelley, R., Jones, K., Weed, H. G., Gardner, K. L., Gangi, L., Yamaguchi, M., Klomkleaw, W., *et al.* (2001). Transgenic mice expressing ATF3 in the heart have conduction abnormalities and contractile dysfunction. *Am. J. Pathol.* **159**, 639–650.

Ramos-Vara, J. A. (2005). Technical aspects of immunohistochemistry. *Vet. Pathol.* **42**, 405–426.

Ramos-Vara, J. A., and Miller, M. A. (2006). Comparison of two polymer-based immunohistochemical detection systems: ENVISION+ and ImmPRESS. *J. Microsc.* **224**, 135–139.

Ron, D., and Walter, P. (2007). Signal integration in the endoplasmic reticulum unfolded protein response. *Nat. Rev. Mol. Cell Biol.* **8**, 519–529.

Rosenberger, C. M., Clark, A. E., Treuting, P. M., Johnson, C. D., and Aderem, A. (2008). ATF3 regulates MCMV infection in mice by modulating IFN-gamma expression in natural killer cells. *Proc. Natl Acad. Sci. USA* **105**, 2544–2549.

Sabattini, E., Bisgaard, K., Ascani, S., Poggi, S., Piccioli, M., Ceccarelli, C., Pieri, F., Fraternali-Orcioni, G., and Pileri, S. A. (1998). The EnVision++ system: A new immunohistochemical method for diagnostics and research. Critical comparison with the APAAP, ChemMate, CSA, LABC, and SABC techniques. *J. Clin. Pathol.* **51**, 506–511.

Schroder, M., and Kaufman, R. J. (2005). The mammalian unfolded protein response. *Annu. Rev. Biochem.* **74,** 739–789.

Wang, A., Arantes, S., Conti, C., McArthur, M., Aldaz, C. M., and MacLeod, M. C. (2007). Epidermal hyperplasia and oral carcinoma in mice overexpressing the transcription factor ATF3 in basal epithelial cells. *Mol. Carcinog.* **46,** 476–487.

Wang, A., Arantes, S., Yan, L., Kiguchi, K., McArthur, M. J., Sahin, A., Thames, H. D., Aldaz, C. M., and Macleod, M. C. (2008). The transcription factor ATF3 acts as an oncogene in mouse mammary tumorigenesis. *BMC Cancer* **8,** 268.

Wek, R. C., Jiang, H. Y., and Anthony, T. G. (2006). Coping with stress: eIF2 kinases and translational control. *Biochem. Soc. Trans.* **34,** 7–11.

Whitmore, M. M., Iparraguirre, A., Kubelka, L., Weninger, W., Hai, T., and Williams, B. R. (2007). Negative regulation of TLR-signaling pathways by activating transcription factor-3. *J. Immunol.* **179,** 3622–3630.

Wu, X., Nguyen, B. C., Dziunycz, P., Chang, S., Brooks, Y., Lefort, K., Hofbauer, G. F., and Dotto, G. P. (2010). Opposing roles for calcineurin and ATF3 in squamous skin cancer. *Nature* **465,** 368–372.

Yin, X., DeWille, J., and Hai, T. (2008). A potential dichotomous role of ATF3, an adaptive-response gene, in cancer development. *Oncogene* **27,** 2118–2127.

Yin, X., Wolford, C. C., McConoughey, S. J., Ramsey, S. A., Aderem, A., and Hai, T. (2010). ATF3, an adaptive-response gene, enhances TGFbeta signaling and cancer initiating cell features in breast cancer cells. *J. Cell Sci.* **123,** 3558–3565.

Zmuda, E. J., Qi, L., Zhu, M. X., Mirmira, R. G., Montminy, M. R., and Hai, T. (2010). The roles of ATF3, an adaptive-response gene, in high-fat-diet-induced diabetes and pancreatic {beta}-cell dysfunction. *Mol. Endocrinol.* 10.1210/me.2009-0463, published ahead of print, June 2.

CHAPTER TWELVE

Experimental Approaches for Elucidation of Stress-Sensing Mechanisms of the Ire1 Family Proteins

Daisuke Oikawa[*,†] and Yukio Kimata[†]

Contents

1. Introduction	196
2. Monitoring *In Vivo* Activity of Mutated Ire1 Family Proteins	199
2.1. Overview	199
2.2. Mammalian expression plasmids for differential expression of IRE1α	200
2.3. Cotransfection of the IRE1α-expression and the ERAI-reporter plasmids into the IRE1$\alpha^{-/-}$ MEF cells and the subsequent luciferase assay	200
2.4. RT-PCR amplification of the mammalian XBP1u and XBP1s mRNAs	201
2.5. Results and insights	202
3. Interaction Between BiP and the Ire1 Family Proteins	206
3.1. Overview	206
3.2. Association of BiP with Yeast Ire1	206
3.3. Association of BiP with mammalian IRE1α	207
3.4. Results and insights	208
4. Direct Interaction of Unfolded Proteins with Yeast Ire1 but not with Mammalian IRE1α	210
4.1. Overview	210
4.2. *E. coli* production and purification of luminal-domain fragments of yeast Ire1 and mammalian IRE1α	210
4.3. *In vitro* anti-aggregation assay	211

[*] Iwawaki Initiative Research Unit, Advanced Science Institute, RIKEN, Wako, Saitama, Japan
[†] Graduate School of Biological Sciences, Nara Institute of Science and Technology, Ikoma, Nara, Japan

Methods in Enzymology, Volume 490
ISSN 0076-6879, DOI: 10.1016/B978-0-12-385114-7.00012-X

© 2011 Elsevier Inc.
All rights reserved.

4.4. Results and insights	213
5. Conclusion and perspective	213
Acknowledgments	214
References	214

Abstract

Endoplasmic reticulum (ER) stress, which is often regarded as the accumulation of unfolded proteins in the ER, triggers cellular protective events including the unfolded protein response (UPR). In the yeast *S. cerevisiae*, the UPR signaling pathway starts from the ER-located transmembrane protein Ire1, the activation of which eventually leads to transcriptional induction of various genes including those encoding ER-located molecular chaperones. Mammals have two Ire1 paralogues, of which IRE1α exhibits ubiquitous tissue expression. Here, we show how we have approached study of the molecular mechanisms by which ER stress activates the Ire1 family proteins. Immunoprecipitation analyses indicated that the ER-located chaperone BiP associates with IRE1α and yeast Ire1, while ER stress dissociates these complexes. We also devised experimental systems for exogenous expression of wild-type or mutant versions of IRE1α and yeast Ire1 at appropriate levels, in order to monitor correctly their activity in evoking downstream events. An IRE1α partial deletion mutant with which BiP poorly associates showed considerable activity even under nonstress conditions, whereas a BiP-nonbinding mutant of yeast Ire1 was almost normally regulated in an ER-stress dependent manner. This finding suggests that the dissociation of BiP is the principal determinant of IRE1α's activation upon ER stress, while yeast Ire1 is largely controlled by another factor(s). Based on *in vitro* ability to inhibit aggregation of denatured proteins, we deduce that the luminal domain of yeast Ire1, but not that of IRE1α, is capable of direct interaction with unfolded proteins. Since this ability of yeast Ire1 was abolished by a mutation impairing its cellular activity, we propose that yeast Ire1 is fully activated by its direct interaction with unfolded proteins.

1. INTRODUCTION

In eukaryotic cells, secretory and cell-surface proteins are folded mostly in the endoplasmic reticulum (ER). Stress stimuli challenging the ER is cumulatively called ER stress and is often regarded as impairment of protein folding in the ER. A number of genes including those encoding ER-located chaperones are transcriptionally induced by ER stress. This phenomenon, namely the unfolded protein response (UPR), is triggered by ER-located transmembrane proteins including transmembrane endoribonuclease Ire1 (Mori, 2009; Sidrauski *et al.*, 1998). Upon ER stress, Ire1 clusters to exhibit potent endoribonuclease activity (Kimata *et al.*, 2007;

Korennykh et al., 2009). In budding yeast, this leads to cytoplasmic splicing of the *HAC1* mRNA, the product of which is translated into the transcription factor protein Hac1 which promotes the UPR (Sidrauski et al., 1998).

In higher eukaryotes, the Ire1 family is not only conserved but also expanded. Mammals and the plant Arabidopsis are reported to carry two Ire1 paralogues (Koizumi et al., 2001; Tirasophon et al., 1998; Wang et al., 1998). In mammals, the ubiquitously expressed paralogue is named as IRE1α and shows more severe knockout phenotypes than the other paralogue IRE1β (Iwawaki et al., 2009). In a similar way to the case of yeast Ire1, the endoribonuclease activity of IRE1α contributes to conversion of the XBP1u mRNA to the XBP1s mRNA, which encodes a functional transcription factor protein (Fig. 12.1A; Mori, 2009; Ron and Walter, 2007). Moreover, IRE1α digests mRNAs encoding proteins targeted to the ER (Han et al., 2009; Hollien et al., 2009; Oikawa et al., 2007a), probably in order to decrease protein load entering the ER. Metazoan cells carry another Ire1 family ER-stress sensor, namely PERK, which is a transmembrane kinase carrying an N-terminal luminal domain slightly but significantly homologous to those of IRE1α and yeast Ire1. PERK is activated by ER stress to phosphorylate an mRNA-translation component eIF2α (Harding et al., 1999). This event leads to attenuation of total protein synthesis, which should decrease protein load in the ER, together with translational induction of the transcription factor ATF4 (Ron and Walter, 2007).

These ER-stress sensors are believed to sense accumulation of unfolded or misfolded proteins in the ER, since potent ER-stressing reagents totally inhibit protein folding in the ER. When subjected to the *N*-glycosylation inhibitor tunicamycin, animal, yeast, and also plant cells evoke the UPR (Koizumi et al., 2001). Moreover, ER proteins more often undergo oxidative folding than cytosolic proteins, explaining the potent ER-stressing effect of the thiol-reducing reagent dithiothreitol (DTT). Mammalian cells also evoke the UPR when treated with thapsigargin, which is an inhibitor of the ER-located calcium pump SERCA, probably because of dysfunction of calcium-binding ER chaperones.

While the initial reports of the identification of the ER-stress sensor, yeast Ire1, were in 1993 (Cox et al., 1993; Mori et al., 1993), it has remained a long-term question as to how the ER-stress sensors actually sense ER stress. Therefore, it has also remained unclear as to whether ER accumulation of unfolded proteins *per se* actually triggers the UPR, since such a challenge to the ER may cause a secondary effect that can directly activate the ER-stress sensors. In this chapter, we describe our experimental approaches used to figure out the molecular steps by which ER stress leads to activation of the Ire1 family ER-stress sensors. The resulting

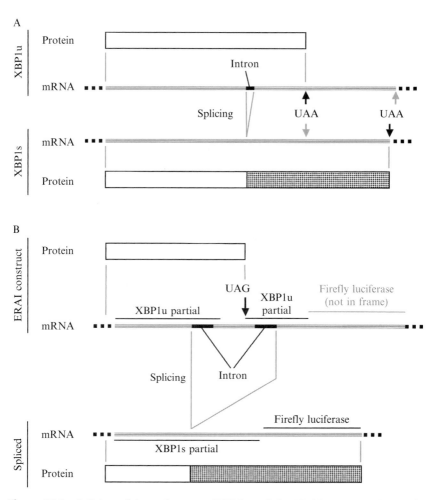

Figure 12.1 Splicing of the endogenous XBP1u and the ERAI-reporter mRNAs by IRE1α. (A) The positions of the in-frame and not-in-frame UAA termination codons are respectively marked by black and gray arrows. Since IRE1α-dependent splicing of the XBP1u mRNA causes a frame-shift, the C-terminal moiety of the XBP1s protein is completely different from that of the XBP1u protein. (B) An improved version of the ERAI reporter (Iwawaki and Akai, 2006) contains two tandem XBP1u splicing-susceptible fragments, and is carried on the mammalian expression plasmid pCAX-HA-2xXBP1ΔDBD(anATG)-LUC-F. Only when the mRNA is spliced by IRE1α, it is translated into a luciferase-containing protein.

findings cumulatively allow us to describe the stress-sensing mechanism of the Ire1 family proteins and allow us to argue that at least partly, they are actually activated directly or nearly directly by unfolded proteins.

2. Monitoring *In Vivo* Activity of Mutated Ire1 Family Proteins

2.1. Overview

Mutational analyses of the ER-stress sensors have been employed to obtain many of the important insights into their activation process upon ER stress. Since Ire1 is activated as a high-order oligomer (Kimata *et al.*, 2007; Korennykh *et al.*, 2009), it is highly possible that Ire1 or its mutants are artifactually clustered and activated when their cellular expression is excessive. In the case of the budding yeast *S. cerevisiae*, it is easy to obtain cells expressing Ire1 or its mutants at a nearly endogenous level. In our research (Kimata *et al.*, 2004, 2007; Oikawa *et al.*, 2005, 2007b), the *IRE1* gene or its mutants carrying the authentic promoter were inserted into the pRS31X-series single-copy plasmids (Sikorski and Hieter, 1989), and then used for transformation of *ire1Δ* strains. Western-blot analysis indicated that expression level of the plasmid-born Ire1 protein is indeed almost the same as that of endogenous Ire1 (Kimata *et al.*, 2007).

Since such a strategy is not applicable to mammalian cells, we generated an alternative method (Oikawa *et al.*, 2009), in order to obtain mammalian cells expressing only an exogenous version of IRE1α at an appropriate level. Immortalized mouse embryonic fibroblast (MEF) cells were obtained from a genetically engineered IRE1α$^{-/-}$ mouse (Iwawaki *et al.*, 2009). Then mammalian expression plasmids for expression of IRE1α variants from a weak promoter were transiently transfected into the IRE1α$^{-/-}$ MEF cells as described later.

In order to monitor cellular activation of the Ire1 family proteins, we prefer to check evocation of downstream molecular events. In the case of yeast Ire1, its activation is directly detectable by splicing of the *HAC1* mRNA, which is monitored by Northern blot or reverse transcription (RT)-PCR analysis of total RNA samples. Meanwhile, for more high-throughput experiments, cells carrying the UPR element-lacZ reporter gene were employed. The UPR element is a promoter element to which the Hac1 protein directly associates for transcriptional induction of its downstream genes (Kawahara *et al.*, 1998; Mori *et al.*, 1992). Therefore, evocation of the UPR was easily monitored by checking β-galactosidase activity of cells where expression of lacZ was controlled by the UPR element (Mori *et al.*, 1992).

In the case of mammalian IRE1α, Iwawaki and his colleagues generated a sophisticated reporter system to monitor its activation in cells (Iwawaki and Akai, 2006; Iwawaki *et al.*, 2004). Their ER stress-activated indicator (ERAI) is based on a fusion gene in which a splicing-susceptible fragment generated from the XBP1u cDNA was ligated to the firefly luciferase cDNA (Fig. 12.1B). Since the luciferase moiety is translated only after the IRE1α-dependent splicing of the mRNA, activation of IRE1α is easily monitored by cellular luminescence activity from luciferase. Moreover, it is also important to monitor splicing of the endogenous XBP1u mRNA by RT-PCR analysis. Because the intron sequence carried on the XBP1u mRNA is not so long (26 nucleotides), it is difficult to observe the conversion from the XBP1u to the XBP1s mRNA by Northern blot analysis.

2.2. Mammalian expression plasmids for differential expression of IRE1α

Plasmid pCAX-hIRE1α-HA was used for mammalian expression of human IRE1α under the control of the strong CAG (cytomegalovirus enhancer and chicken β-actin) promoter. This plasmid was made by insertion of the human IRE1α full-length ORF carrying three tandem C-terminal hemagglutinin (HA)-epitope sequences into the HindIII/XhoI sites of pCAX (Oikawa et al., 2007a). Plasmid pTKbasal-hIRE1α-HA was used for expression of human IRE1α at a basal level. To make this plasmid, pCAX-hIRE1α-HA was SpeI/HindIII-digested to remove the CAG promoter, and ligated with an SpeI/HindIII-digested PCR product of the basal promoter sequence of the HSV-TK (herpes simplex virus thymidine kinase) gene (the primer set 5′-AAAAAACTAGTGGCCCCGCCCAGCGTCT TGTCATTG-3′ (sense) and 5′-AAAAAGCTTCGCTGTTGACGCTGT TAAGCG-3′ (antisene) (the attached SpeI and HindIII sites are underlined)).

2.3. Cotransfection of the IRE1α-expression and the ERAI-reporter plasmids into the IRE1α$^{-/-}$ MEF cells and the subsequent luciferase assay

(1) One day before transfection, split 70–90% confluent culture at an 1:5 dilution into 24-well plates. Continue to incubate under appropriate conditions (growth medium: DMEM + 10% FBS + pen/strep; 37 °C; 5% CO_2).
(2) Make the transfection mixture using the materials supplied with Qiagen "Effecten Transfection Reagent" according to the manufacturer's instructions. For 7 wells of a 24-well plate, use 3.5 μg of plasmid mix[1], 210 μl of buffer EC, 5.6 μl of Enhancer, and 17.5 μl of Effectene Reagent.

---

[1] Plasmid mix (for 1 well); 125 ng of phRL-TK (a transfection control to express the Renilla luciferase), 12.5 ng of IRE1α-expression plasmid (pCAX-hIRE1α-HA or pTKbasal-IRE1α-HA (wild type or mutant)) and 362.5 ng of ERAI reporter (pCAX-HA-2xXBP1ΔDBD(anATG)-LUC-F).

(3) While preparing the transfection mixture, wash cells once with PBS, and add fresh growth medium (250 μl per well) to the wells.
(4) Add 1.5 ml (for 7 wells) of fresh growth medium to the transfection mixture. Immediately after mixing, add it dropwise onto the culture.
(5) 6–8 h later, wash cells twice with PBS, and add fresh growth medium to the wells.
(6) 18 h later, change to fresh growth medium with or without 2.5 μg/ml tunicamycin.
(7) 6 h later, wash cells once with PBS, which is then completely aspirated.
(8) Perform the luciferase assay using the Dual-Luciferase Reporter Assay System (Promega).
   (i) For cell lysis, add 50 μl of 1× passive lysis buffer to each well, and let stand for 15 min at 4 °C.
   (ii) Perform luciferase reaction by mixing 9 μl of the cell lysate, 45 μl of LARII solution, and 45 μl of Stop and Glo solution for each reaction.

2.4. RT-PCR amplification of the mammalian XBP1u and XBP1s mRNAs

In order to check conversion from the XBP1u to the XBP1s mRNA, total cellular RNA prepared using ISOGEN reagent (Nippon Gene) was subjected to RT-PCR.

(1) Mix 1 μg of RNA sample (in 8 μl of DEPC-treated water) with 1 μl of oligo $(dT)_{12-18}$ (0.5 μg/μl) and 1 μl of dNTPs solution (2.5 mM each).
(2) Incubate the mixture at 65 °C for 5 min and then chill on ice for 1 min.
(3) Into the mixture, add the materials supplied with Invitrogen "SuperScript II Reverse Transcriptase" (2 μl of 10× RT buffer, 4 μl of 25 mM $MgCl_2$, 2 μl of 0.1 M DTT, 1 μl of RNaseOUT, and 1 μl of SuperScript II RT (50 U/μl)).
(4) Incubate the mixture at 42 °C for 50 min and then 70 °C for 15 min. Then chill on ice.
(5) Add 1 μl of RNase H (2 U/μl) into the mixture and incubate for 20 min at 37 °C.
(6) Mix 1 μl of the RT reaction product with 1 μl of 10 μM XBP1 sense primer[2], 1 μl of 10 μM XBP1 antisense primer[2], 5 μl of 10× PCR buffer, 8 μl of dNTPs solution (2.5 mM each), 0.2 μl TaKaRa Ex-Taq (5 U/μl), and 34.8 μl water.

---

[2] For human XBP1: sense primer 5′-AGAACCAGGAGTTAAGACAGC-3′, antisense primer, 5′-AGTCAATACCGCCAGAATCC-3′; for mouse XBP1: sense primer 5′-GAACCAGGAGTTAAGAACACG-3′, antisense primer 5′-AGGCAACAGTGTCAGAGTCC-3′.

(7) *Perform PCR*: (i) 96 °C 3 min; (ii) 35 cycles of 96 °C 15 s, 55 °C 15 s, and 72 °C 1 min; (iii) 72 °C 10 min.

2.5. Results and insights

Based on insights from the deletion-scanning mutagenesis (Kimata et al., 2004), the *in vitro* partial proteolysis (Oikawa et al., 2005), and the X-ray structural determination (Credle et al., 2005), we propose to divide the luminal domain of yeast Ire1 into Subregions I to V (Fig. 12.2). We think that Subregions I and V, namely the deletable regions (Kimata et al., 2004), are loosely folded, since these regions were susceptible to *in vitro* partial proteolysis of a recombinant fragment of the Ire1 luminal domain (Oikawa et al., 2005). In contrast, the X-ray crystal structure from Subregion II to IV (Credle et al., 2005) indicated that this region forms one tightly folded domain, which we then named as the core stress-sensing region (CSSR;

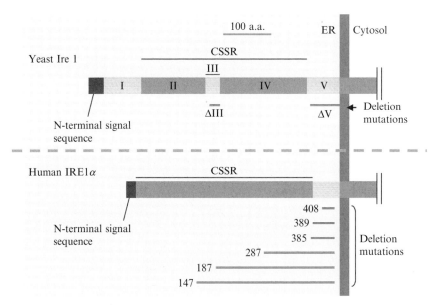

Figure 12.2 Structure and mutations of the luminal domains of yeast Ire1 and human IRE1α. Yeast Ire1 is inactivated by internal 10 a.a. deletions of Subregions II and IV but not of Subregions I, III, or V (Kimata et al., 2004). Subregions II to IV form the CSSR (Credle et al., 2005; Kimata et al., 2004). Human IRE1α also carries regions corresponding to the CSSR and probably corresponding to Subregion V. The deletion positions of yeast Ire1 mutations ΔIII (a.a. 253–272) and ΔV (a.a. 463–524) and human IRE1α mutations 408 (a.a. 409–434), 389 (a.a. 390–434), 385 (a.a. 386–434), 287 (a.a. 288–434), 187 (a.a. 188–434), and 147 (a.a. 148–434) are also indicated (Kimata et al., 2004, 2007; Oikawa et al., 2009).

Fig. 12.2). Subregion III is a loosely folded loop sticking out from the CSSR. An almost full-length deletion of Subregion V (ΔV mutation; Fig. 12.2) did not significantly alter activity of yeast Ire1, while 20-a.a. deletion of Subregion III (ΔIII mutation; Fig. 12.2) largely impaired activation of Ire1 upon ER stress (Fig. 12.3; Kimata et al., 2007). We do not think that the ΔIII mutation causes global disturbance of Ire1's structure, because its low-activity phenotype was suppressed by several second-point mutations of Ire1 (Kimata et al., 2007). Rather, as described later, this mutation is likely to impair a specific molecular event required for activation of Ire1.

As shown in Fig. 12.4A, ER stress-dependent splicing of the XBP1u mRNA was observed in IRE1α$^{+/+}$ MEF cells but not in IRE1α$^{-/-}$ MEF cells. This finding again supports the idea that IRE1α is the key factor in the splicing reaction. In our previous report (Oikawa et al., 2009) and Fig. 12.4B, the ERAI-reporter plasmid was introduced into IRE1α$^{+/+}$ MEF cells without the IRE1α-expression plasmids. As expected, this reporter was considerably induced by treatment of cells with tunicamycin. We then introduced the ERAI-reporter plasmid together with the IRE1α-expression plasmids into IRE1α$^{-/-}$ MEF cells (Fig. 12.4C; Oikawa et al., 2009). While high-level expression of IRE1α from pCAX-hIRE1α-HA caused strong and deregulated induction of the ERAI reporter, it was well

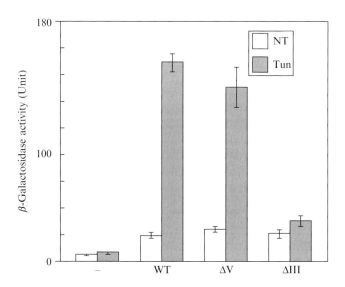

Figure 12.3 Activity of the yeast Ire1 mutants. Yeast ire1Δ cells KMY1015 (Kimata et al., 2004) carrying an UPRE-lacZ reporter plasmid (Kimata et al., 2003) were transformed with the IRE1-carrying single-copy plasmid pRS313-IRE1 or its mutant derivatives (see Fig. 12.2). The transformants were ER-stressed by 2 μg/ml tunicamycin for 4 h (Tun) or remained unstressed (NT) before assay for cellular β-galactosidase activity as performed in our previous report (Kimata et al., 2003).

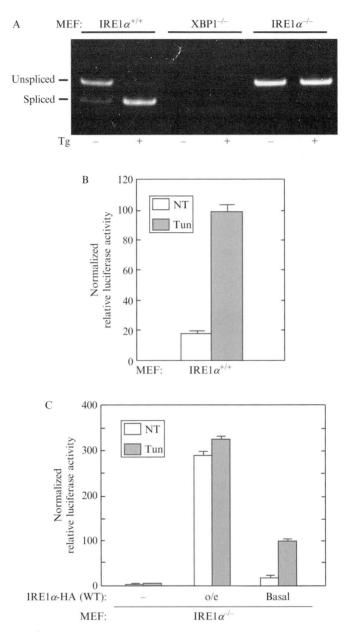

Figure 12.4 Monitoring IRE1α activation in mammalian cells. (A) MEF cells (wild-type (IRE1α$^{+/+}$), IRE1α$^{-/-}$ or XBP1$^{-/-}$) were ER-stressed with 1 μM thapsigargin (Tg) for 1 h or remained unstressed, and then analyzed by RT-PCR to amplify the XBP1 mRNAs. Positions of the XBP1u product (Unspliced) and of the XBP1s product (Spliced) on 5%-acrylamide electrophoresis are indicated. (B) After transient transfection of the ERAI-reporter plasmid set (pCAX-HA-2xXBP1ΔDBD(anATG)-LUC-F

regulated in an ER-stress dependent manner when IRE1α was moderately expressed from pTKbasal-hIRE1α-HA.

As illustrated in Fig. 12.2, human IRE1α also contains the CSSR, which shows low but significant similarity to that of yeast Ire1. Similarly to the case of yeast Ire1 (Kimata *et al.*, 2004), CSSR deletions of IRE1α (147, 187, and 287 mutations; Fig. 12.2) completely abolished activity of IRE1α (Fig. 12.5; Oikawa *et al.*, 2009). Human IRE1α also carries a juxtamembrane deletable region (Figs. 12.2 and 12.5; 385, 389, and 408 mutants; Oikawa *et al.*, 2009), and this region probably corresponds to Subregion V of yeast Ire1,

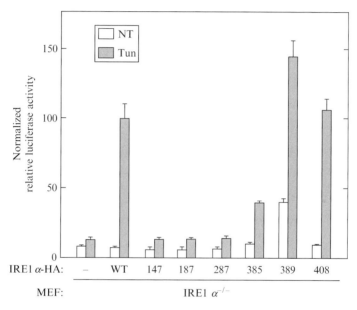

Figure 12.5 Activity of human IRE1α mutants. After transient co-transfection of the ERAI-reporter plasmid set with pTKbasal-hIRE1α-HA (WT) or its mutant derivatives (see Fig. 12.2), IRE1α$^{-/-}$ MEF cells were ER-stressed with 2.5 μg/ml tunicamycin for 6 h (Tun) or remained unstressed (NT), and then subjected to luciferase assay. Each value is normalized against that of the tunicamycin-treated "WT" sample, which is set at 100. Refer to Oikawa *et al.* (2009).

and phRL-TK), wild-type (IRE1α$^{+/+}$) MEF cells were ER-stressed by 2.5 μg/ml tunicamycin for 6 h (Tun) or remained unstressed (NT), and then subjected to luciferase assay. (C) After transient cotransfection of the ERAI-reporter plasmid set with pCAX-hIRE1α-HA (overexpression; o/e), pTKbasal-hIRE1α-HA (Basal) or no other plasmid (–), IRE1α$^{-/-}$ MEF cells were ER-stressed with 2.5 μg/ml tunicamycin (Tun) for 6 h or remained unstressed (NT). Each value is normalized against that of the tunicamycin-treated IRE1$^{+/+}$ MEF cells (B) or "Basal" sample (C), which is set at 100. Refer to Oikawa *et al.* (2009).

although evolutionary sequence conservation has not been observed. This juxtamembrane region of human IRE1α was also sensitive to *in vitro* partial proteolysis (Oikawa *et al.*, 2009), again suggesting a loosely folded structure. It is noteworthy that, unlike as is the case with yeast Ire1, an almost full-length deletion of this region, namely the 389 mutation, partly and significantly activated IRE1α even without ER stress (Fig. 12.5; Oikawa *et al.*, 2009).

3. Interaction Between BiP and the Ire1 Family Proteins

3.1. Overview

BiP is the HSP70-family molecular chaperone located in the ER lumen. Overexpression of BiP attenuates the UPR, implying negative regulation of ER-stress sensors by BiP (Bertolotti *et al.*, 2000; Kohno *et al.*, 1993). Indeed, BiP associates with endogenous IRE1α and PERK in nonstressed mammalian cells, while ER stress facilitates dissociation of this complex (Bertolotti *et al.*, 2000). As shown in this section, checking the association between BiP and mutants of the Ire1 family proteins allowed us to elucidate the physiological meaning of this interaction.

3.2. Association of BiP with Yeast Ire1

The yeast *IRE1* genes were modified to carry the triple HA tag at the C terminus of the protein (Ire1-HA; Kimata *et al.*, 2003), and cloned into the pRS423 multicopy vector plasmid (Christianson *et al.*, 1992). This is because we were unable to detect BiP association with Ire1-HA, when it was expressed from a single-copy plasmid under the control of the authentic *IRE1* promoter. In order to observe a cellular complex of BiP and Ire1-HA, yeast cells carrying this multicopy Ire1-HA plasmid, pRS423-Ire1-HA, were subjected to anti-HA immunoprecipitation as described here and in Kimata *et al.* (2003).

(1) Grow 40 ml of cells to an OD_{600} of 0.5–1.5 in SD medium.
(2) Collect the cells by centrifugation at $2000 \times g$ for 1 min.
(3) Resuspend in 200 μl of buffer A (50 mM Tris–HCl (pH 7.9), 5 mM EDTA, and 1% Triton X-100) containing 2 mM phenylmethanesulfonyl fluoride and 10 μg/ml each of pepstatin A, leupeptin, and aprotinin.
(4) Add 2 μl of Protease Inhibitor Cocktail Set III (Calbiochem).
(5) Add 200 μl of 0.5 mm acid-washed glass beads.
(6) Break the cells by vortexing six times for 30 s each at maximum speed.

(7) Puncture bottom of the tube with a 22-gauge needle and collect lysate by centrifugation (800×g for 10 s).
(8) Clarify the lysate by centrifugation at 9000×g for 10 min.
(9) Dilute 180 μl of the lysate with 800 μl of buffer B (the same composition as buffer A except containing 180 mM NaCl and 6% skim milk), and add 30 μl of 50% slurry of Protein A Sepharose 4 FF beads (GE Healthcare).
(10) Rotate for 30 min at 4 °C.
(11) Spin 800×g for 30 s.
(12) Collect supernatant and add 2 μg of mouse monoclonal anti-HA antibody 12CA5 (Roche).
(13) Rotate for 1 h at 4 °C.
(14) Spin at 9000×g for 10 min.
(15) Collect supernatant and add 50% slurry of Protein A Sepharose 4 FF beads (GE Healthcare).
(16) Rotate for 1 h at 4 °C.
(17) Wash the beads five times with buffer C (the same composition as buffer B except lacking the skim milk).
(18) Boil the beads with 7.5 μl of 4× SDS-PAGE sample buffer, 3 μl of 1 M DTT, and 4.5 μl of water for 30 s.
(19) Spin at 9000×g at room temperature for 1 min.
(20) Apply supernatant to SDS-PAGE and Western-blot analysis.

For Western-blot detection, rabbit polyclonal anti-yeast BiP (Kar2) antibody (Tokunaga *et al.*, 1992) and 12CA5 were employed.

3.3. Association of BiP with mammalian IRE1α

In order to compare BiP association with wild-type IRE1α to that with the 389 mutant, pCAX-hIRE1α-HA or its mutant derivative were transfected into HeLa cells by the calcium-phosphate method. The cells were then subjected to anti-HA immunoprecipitation (Oikawa *et al.*, 2009).

(1) One day before transfection, split 70–90% confluent culture at an 1:5 dilution into 10-cm dishes. Continue to incubate under appropriate conditions (growth medium: DMEM + 10% FBS + pen/strep; 37 °C; 5% CO_2).
(2) Replace old growth medium with 10 ml of fresh growth medium.
(3) 1 h later, add the transfection mixture[3] (1 ml for each dish) dropwise onto the culture.

---

[3] Preparation of the transfection mixture (make just before use): Mix the following materials; 20 μl of plasmid solution (pCAX-hIRE1α-HA or its mutant derivative; 1 μg/μl in TE), 30 μl of TE solution, 50 μl of 2.5 M $CaCl_2$, and 400 μl of sterilized water. Then add this mixture to 500 μl of 2× HBS with mild vortexing, and stand for 20 min at room temperature.

(4) 6 h later, wash cells twice with Hanks' solution, and add fresh growth medium to the dishes.
(5) On the next morning, change to fresh growth medium.
(6) 6 h later, change to fresh growth medium with or without 2 μM thapsigargin.
(7) 30 min later, wash cells twice with PBS, and add 1 ml of lysis buffer (20 mM HEPES (pH 7.5), 150 mM NaCl, 1 mM EDTA, 1% Triton X-100, and 1% glycerol) containing 10 μl of protease inhibitor cocktail (Sigma) to the dishes. Then stand for 15 min at 4 °C to allow cell lysis.
(8) Clarify the lysates by centrifugation at $16,000 \times g$ for 1 min at 4 °C. Small aliquots of the resulting samples are kept as "lysate" samples for the Western-blot analysis.
(9) Add 30 μl of 50% slurry of Protein G Sepharose 4 FF beads (GE Healthcare) and 1 μg of 12CA5.
(10) Rotate for 1 h at 4 °C.
(11) Wash the beads five times with lysis buffer.
(12) Boil the beads with 7.5 μl of 4× SDS-PAGE sample buffer, 3 μl of 1 M DTT, and 4.5 μl of water for 5 min.
(13) Spin at $16,000 \times g$ at room temperature for 1 min.
(14) Apply supernatant to SDS-PAGE and Western-blot analysis.

For Western-blot detection of BiP and IRE1α-HA, anti-KDEL antibody (Stressgen) and 12CA5 were respectively employed.

3.4. Results and insights

As shown in Fig. 12.6A and in our previous report (Okamura et al., 2000), BiP associates with yeast wild-type Ire1 and dissociates in response to ER stress. Our previous study indicated that activation of Ire1, together with BiP dissociation from it, was impaired in some of the yeast BiP mutant-allele strains (Kimata et al., 2003, 2004). This finding strongly supports our idea that BiP suppresses Ire1's activity. Meanwhile, we also checked association and dissociation between BiP and various Ire1 mutants and found that association of BiP with ΔV Ire1 is very weak even in the absence of ER stress (Fig. 12.6A; Kimata et al., 2004). Therefore, the BiP-association site is likely to be located on Subregion V. Since ΔV was regulated by ER stress as well as wild-type Ire1 (Fig. 12.3; Kimata et al., 2004), we propose that BiP dissociation is not sufficient for activation of yeast Ire1.

Furthermore, we were able to detect association between BiP and IRE1α in mammalian cells, together with ER stress-dependent dissociation of this complex, by using the methods described here (Fig. 12.6B; Oikawa et al., 2009). The 389 mutant IRE1α (Fig. 12.2) showed reduced association with BiP, which was, however, further decreased by ER stress (Fig. 12.6B).

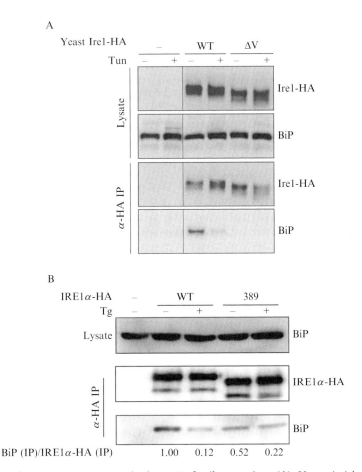

Figure 12.6 BiP association with the Ire1 family proteins. (A) Yeast *ire1Δ* cells KMY1516 (Kimata *et al.*, 2004) transformed with an wild-type (WT) Ire1-HA expression plasmid (pRS423-IRE1-HA; Kimata *et al.*, 2003) or its ΔV derivative were cultured under nonstress conditions or ER-stressed with 2 μg/ml tunicamycin (Tun) for 1 h, and then subjected to anti-HA immunoprecipitation (IP) and Western-blot analysis as performed in our previous report (Kimata *et al.*, 2004). (B) After transient transfection of pCAX-hIRE1α-HA (wild-type, WT) or its 389 mutant version, HeLa cells were ER-stressed by 2 μM thapsigargin (Tg) for 30 min or remained unstressed, and then subjected to anti-HA immunoprecipitation (IP) and Western-blot analysis (Oikawa *et al.*, 2009). The normalized ratios of the co-immunoprecipitated BiP signal to the immunoprecipitated IRE1α-HA signal, which are averages from two independent trials, are indicated.

Thus, we speculate that in the case of IRE1α, the major BiP-association site is located on the juxtamembrane deletable region in a similar way to yeast Ire1, but interaction of BiP with the CSSR may also contribute to regulation of IRE1α's activity. Since the 389 mutant was significantly activated even under nonstress conditions (Fig. 12.5), BiP association and dissociation seem to be the important determinant of IRE1α's activity (Oikawa *et al.*, 2009).

4. Direct Interaction of Unfolded Proteins with Yeast Ire1 but not with Mammalian IRE1α

4.1. Overview

Another possible regulatory scenario for the Ire1 family proteins is their direct interaction with unfolded proteins. Indeed, the X-ray structure of yeast Ire1's CSSR argues that when dimerized, it forms a groove-like structure in which unfolded proteins may be captured (Credle *et al.*, 2005). However, the groove-like structure does not seem to be suitable for capturing unfolded proteins, according to the X-ray structure of mammalian IRE1α's CSSR (Zhou *et al.*, 2006). In order to address whether the CSSRs actually have the ability to interact directly with unfolded proteins, we monitored the ability of these domains to inhibit aggregation of denatured-proteins *in vitro* (anti–aggregation assay).

4.2. *E. coli* production and purification of luminal-domain fragments of yeast Ire1 and mammalian IRE1α

Escherichia coli strain BL21 codon plus (DE3)-RIL (Strategene) was transformed with the following expression plasmids. To express maltose binding protein (MBP)-fused CSSR of yeast Ire1, pMAL-yCSSR-His and its ΔIII mutant derivative were used (Kimata *et al.*, 2007). To express unfused MBP, pMAL-His was used (Oikawa *et al.*, 2005). For expression of the luminal-domain fragments of human IRE1α, pQE-hNLD and its 389 mutant derivative were used (Oikawa *et al.*, 2009). Since all of the recombinant proteins were tagged with the polyhistidine tag, they were produced and purified as follows:

(1) Grow 400 ml of cells to an OD_{600} of 0.3 in 2× YT medium at 30 °C.
(2) Add IPTG and further incubate at 30 °C (0.3 mM (final conc.) IPTG for 1 h for the pMAL family plasmids, or 1 mM IPTG (final conc.) for 2 h for the pQE family plasmids).
(3) Collect the cells by centrifugation at 6000×*g* for 5 min.
(4) Resuspend the cells in 15 ml of *E. coli* lysis buffer (50 mM HEPES (pH 8.0), 300 mM KCl, 5 mM MgCl$_2$, 10 mM imidazole, 1% Triton X-100, 2 mM phenylmethylsulfonyl fluoride, 0.4 mg/ml benzamidine, 0.4 mg/ml pepstatin A, 0.4 mg/ml leupeptin, 0.3 mg/ml lysozyme, and 14 U/ml DNase I (TaKaRa)), and stand for 30 min on ice.

(5) Disrupt the cells by ultrasonication (15 s each, three times).
(6) Clarify the lysate by centrifugation (SRX-201 (Tomy); 8200 rpm for 10 min), and incubate with 0.5 ml of HisLink protein purification resin beads (Promega) for 12 h with rotation at 4 °C.
(7) Pack the beads into an 8-mm-diam column and wash sequentially with:
   (i) 6 ml of 50 mM HEPES (pH 8.0), 1 M KCl, 5 mM MgCl$_2$, and 0.1% Triton X-100.
   (ii) 6 ml of 50 mM HEPES (pH 8.0), 300 mM KCl, 5 mM MgCl$_2$, 0.1% Triton X-100, and 20 mM imidazole.
   (iii) 6 ml of 50 mM HEPES (pH 8.0), 300 mM KCl, 5 mM MgCl$_2$, 0.1% Triton X-100, and 40 mM imidazole.
   (iv) 3 ml of 50 mM HEPES (pH 8.0), 300 mM KCl, 5 mM MgCl$_2$, 0.1% Triton X-100, and 60 mM imidazole.
   (v) 5 ml of 50 mM HEPES (pH 8.0), 300 mM KCl, 5 mM MgCl$_2$, and 10 mM ATP.
   (vi) 3 ml of 20 mM HEPES (pH 8.0), 100 mM KCl, 5 mM MgCl$_2$, and 50% (v/v) glycerol.
(8) Elute the bead-bound protein with 50 mM HEPES (pH 8.0), 100 mM KCl, 5 mM MgCl$_2$, 200 mM imidazole and 50% (v/v) glycerol. The elution fractions (1 ml each) were analyzed by SDS-PAGE, and the fraction carrying the most sample protein was employed for the following assay.

As shown in Figs. 12.7A and C, this method allowed us to purify the sample proteins to single-band homogeneity.

4.3. *In vitro* anti-aggregation assay

As model denatured-protein substrates, luciferase (Promega) and citrate synthase (Roche) were employed. Citrate synthase was dialyzed against stock buffer (20 mM HEPES (pH 7.0), 150 mM KCl, 2 mM MgCl$_2$, and 10% (v/v) glycerol), and stored at −80 °C. For denaturing, luciferase (25 μM final concentration) and citrate synthase (50 μM final concentration) were incubated in guanidine HCl-denaturing solution (guanidine-HCl (6 M for luciferase or 4 M for citrate synthase), 20 mM HEPES (pH 7.2), 50 mM KCl, and 2 mM MgCl$_2$) for 30 min at room temperature. The denatured-protein mixtures were then diluted (50-fold dilution for luciferase or 33-fold dilution for citrate synthase) with assay buffer (20 mM HEPES (pH 7.2), 50 mM KCl, and 2 mM MgCl$_2$) in the presence or absence of the test recombinant proteins. Protein aggregation after dilution was monitored by optical absorbance at 320 nm with a spectrophotometer (DU640; Beckman Coulter) at room temperature.

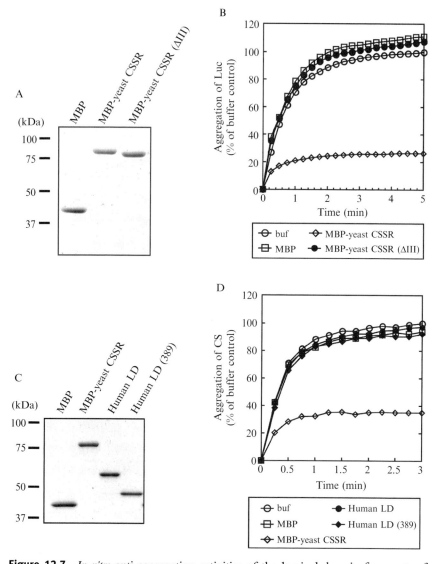

Figure 12.7 *In vitro* anti-aggregation activities of the luminal-domain fragments of yeast Ire1 or mammalian IRE1α. (A) Bacterially expressed MBP-yeast CSSR, its ΔIII mutant version or unfused MBP were purified, run on a 10% SDS-PAGE gel (1 μg protein per lane) and stained with Coomassie blue. (B) At time 0, luciferase (Luc) in guanidine HCl-denaturing solution was diluted into assay buffer containing the indicated proteins (2 μM each; buf: buffer-only). Similar results were obtained when citrate synthase was used as the denatured-protein substrate (Kimata *et al.*, 2007). (C) Bacterially expressed MBP-yeast CSSR, luminal-domain fragments of human IRE1α (human LD; wild-type or the 389 mutant) or unfused MBP were purified, run on a 2–15% SDS-PAGE gel (1 μg protein per lane) and stained with Coomassie blue. (D) At time 0, citrate synthase (CS) in guanidine HCl-denaturing solution was diluted into assay buffer containing the indicated proteins (1.5 μM each; buf: buffer-only). Similar results were obtained when luciferase was used as the denatured protein substrate (Oikawa *et al.*, 2009). In (B) and (D), turbidity of the sample mixtures is normalized against the maximal value for the "buffer-only" sample, and presented as "aggregation."

4.4. Results and insights

Our assay demonstrated anti-aggregation activity of the CSSR of yeast Ire1 (Fig. 12.7B; Kimata *et al.*, 2007), indicating that it has ability to interact with unfolded proteins. Importantly, the ΔIII mutation abolished this ability of the CSSR. Since the ΔIII Ire1 was activated only weakly in yeast cells exposed to ER stress (Fig. 12.3; Kimata *et al.*, 2007), we propose that direct interaction of unfolded proteins with the CSSR contributes to full activation of yeast Ire1.

Meanwhile, we failed to demonstrate anti-aggregation activity of the luminal domain of human IRE1α (Fig. 12.7D; Oikawa *et al.*, 2009). This seems consistent with the aforementioned observations from the X-ray structural analyses (Credle *et al.*, 2005; Zhou *et al.*, 2006), which lead us to speculate that unlike the CSSR of yeast Ire1, that of IRE1α does not interact with unfolded proteins.

5. Conclusion and perspective

Based on the overall observations described here and in our previous reports, we have proposed a model to explain how ER stress leads to activation of the Ire1 family proteins (Kimata *et al.*, 2007; Oikawa *et al.*, 2009). In the case of yeast Ire1, two molecular events, dissociation of BiP from Ire1 and direct interaction of unfolded proteins with Ire1, are required for its full activation. We think such dual regulation of Ire1 is important for the fidelity of Ire1's response, since ΔV Ire1, a BiP-nonbinding mutant, was activated by stimuli other than ER stress (Kimata *et al.*, 2004). It also should be noted that Subregion I suppresses Ire1's activity by an as yet unknown mechanism (Oikawa *et al.*, 2007b).

In the case of mammalian IRE1α, BiP association and dissociation seems to be the principal determinant of its activity (Oikawa *et al.*, 2009). Unlike yeast Ire1, IRE1α is unlikely to sense directly unfolded proteins. Moreover, IRE1α does not carry Subregion I (Fig. 12.2). Interestingly, our preliminary study showed that the luminal domain of PERK has anti-aggregation activity as well as that of yeast Ire1, suggesting direct sensing of unfolded proteins by PERK. Since downstream events evoked by IRE1α and PERK are different and often conflicting (Ron and Walter, 2007), it may be reasonable that they are regulated differently.

Considering that yeast Ire1 directly senses unfolded proteins, the notion that ER accumulation of unfolded proteins *per se* triggers the UPR seems correct at least in part. Also supporting this notion, it is possible that interaction of unfolded proteins with BiP, which is an HSP70 family chaperone, causes BiP dissociation from the Ire1 family proteins, although the actual mechanism of this BiP dissociation remains elusive. Nevertheless, some stress stimuli seem to activate the Ire1 family proteins without producing unfolded proteins, since ΔIII Ire1, with which unfolded proteins do not interact, is normally activated by ER-membrane abnormality (unpublished observation by Yukio Kimata). Furthermore, the activation mechanism of the Ire1 family proteins under various physiological conditions has not been elucidated (Rutkowski and Hegde, 2010). Thus, further studies possibly including new experimental techniques are required to fully elucidate the stress-sensing mechanism of the Ire1 family proteins.

ACKNOWLEDGMENTS

Our study concerning human IRE1α was performed together with Dr. Takao Iwawaki in his laboratory (Iwawaki Initiative Research Unit, Advanced Science Institute, RIKEN, Saitama, Japan). We greatly appreciate his scientific and financial support. Other parts of our study were performed in Nara Inst. Sci. Tech., Nara, Japan. We thank Prof. Kenji Kohno and all members of his and our laboratory for their support. Our research is supported by KAKENHI (grant numbers 22657030 and 21112516 to Y. K.; grant number 20770113 to D. O.) from MEXT or JSPS. D. O. is supported by a Research Fellowship for Young Scientists from JSPS.

REFERENCES

Bertolotti, A., Zhang, Y., Hendershot, L. M., Harding, H. P., and Ron, D. (2000). Dynamic interaction of BiP and ER stress transducers in the unfolded-protein response. *Nat. Cell Biol.* **2,** 326–332.

Christianson, T. W., Sikorski, R. S., Dante, M., Shero, J. H., and Hieter, P. (1992). Multifunctional yeast high-copy-number shuttle vectors. *Gene* **110,** 119–122.

Cox, J. S., Shamu, C. E., and Walter, P. (1993). Transcriptional induction of genes encoding endoplasmic reticulum resident proteins requires a transmembrane protein kinase. *Cell* **73,** 1197–1206.

Credle, J. J., Finer-Moore, J. S., Papa, F. R., Stroud, R. M., and Walter, P. (2005). On the mechanism of sensing unfolded protein in the endoplasmic reticulum. *Proc. Natl. Acad. Sci. USA* **102,** 18773–18784.

Han, D., Lerner, A. G., Vande Walle, L., Upton, J. P., Xu, W., Hagen, A., Backes, B. J., Oakes, S. A., and Papa, F. R. (2009). IRE1alpha kinase activation modes control alternate endoribonuclease outputs to determine divergent cell fates. *Cell* **138,** 562–575.

Harding, H. P., Zhang, Y., and Ron, D. (1999). Protein translation and folding are coupled by an endoplasmic-reticulum-resident kinase. *Nature* **397,** 271–274.

Hollien, J., Lin, J. H., Li, H., Stevens, N., Walter, P., and Weissman, J. S. (2009). Regulated Ire1-dependent decay of messenger RNAs in mammalian cells. *J. Cell Biol.* **186,** 323–331.

Iwawaki, T., Akai, R., Kohno, K., and Miura, M. (2004). A transgenic mouse model for monitoring endoplasmic reticulum stress. *Nat. Med.* **10,** 98–102.

Iwawaki, T., and Akai, R. (2006). Analysis of the XBP1 splicing mechanism using endoplasmic reticulum stress-indicators. *Biochem. Biophys. Res. Commun.* **350,** 709–715.

Iwawaki, T., Akai, R., Yamanaka, S., and Kohno, K. (2009). Function of IRE1 alpha in the placenta is essential for placental development and embryonic viability. *Proc. Natl. Acad. Sci. USA* **106,** 16657–16662.

Kawahara, T., Yanagi, H., Yura, T., and Mori, K. (1998). Unconventional splicing of HAC1/ERN4 mRNA required for the unfolded protein response. Sequence-specific and non-sequential cleavage of the splice sites. *J. Biol. Chem.* **273,** 1802–1807.

Kimata, Y., Kimata, Y. I., Shimizu, Y., Abe, H., Farcasanu, I. C., Takeuchi, M., Rose, M. D., and Kohno, K. (2003). Genetic evidence for a role of BiP/Kar2 that regulates Ire1 in response to accumulation of unfolded proteins. *Mol. Biol. Cell* **14,** 2559–2569.

Kimata, Y., Oikawa, D., Shimizu, Y., Ishiwata-Kimata, Y., and Kohno, K. (2004). A role for BiP as an adjustor for the endoplasmic reticulum stress-sensing protein Ire1. *J. Cell Biol.* **167,** 445–456.

Kimata, Y., Ishiwata-Kimata, Y., Ito, T., Hirata, A., Suzuki, T., Oikawa, D., Takeuchi, M., and Kohno, K. (2007). Two regulatory steps of ER-stress sensor Ire1 involving its cluster formation and interaction with unfolded proteins. *J. Cell Biol.* **179,** 75–86.

Kohno, K., Normington, K., Sambrook, J., Gething, M. J., and Mori, K. (1993). The promoter region of the yeast KAR2 (BiP) gene contains a regulatory domain that responds to the presence of unfolded proteins in the endoplasmic reticulum. *Mol. Cell. Biol.* **13,** 877–890.

Koizumi, N., Martinez, I. M., Kimata, Y., Kohno, K., Sano, H., and Chrispeels, M. J. (2001). Molecular characterization of two Arabidopsis Ire1 homologs, endoplasmic reticulum-located transmembrane protein kinases. *Plant Physiol.* **127,** 949–962.

Korennykh, A. V., Egea, P. F., Korostelev, A. A., Finer-Moore, J., Zhang, C., Shokat, K. M., Stroud, R. M., and Walter, P. (2009). The unfolded protein response signals through high-order assembly of Ire1. *Nature* **457,** 687–693.

Mori, K., Sant, A., Kohno, K., Normington, K., Gething, M. J., and Sambrook, J. F. (1992). A 22 bp cis-acting element is necessary and sufficient for the induction of the yeast KAR2 (BiP) gene by unfolded proteins. *EMBO J.* **11,** 2583–2593.

Mori, K., Ma, W., Gething, M. J., and Sambrook, J. (1993). A transmembrane protein with a cdc2+/CDC28-related kinase activity is required for signaling from the ER to the nucleus. *Cell* **74,** 743–756.

Mori, K. (2009). Signalling pathways in the unfolded protein response: Development from yeast to mammals. *J. Biochem.* **146,** 743–750.

Oikawa, D., Kimata, Y., Takeuchi, M., and Kohno, K. (2005). An essential dimer-forming subregion of the endoplasmic reticulum stress sensor Ire1. *Biochem. J.* **391,** 135–142.

Oikawa, D., Tokuda, M., and Iwawaki, T. (2007a). Site-specific cleavage of CD59 mRNA by endoplasmic reticulum-localized ribonuclease, IRE1. *Biochem. Biophys. Res. Commun.* **360,** 122–127.

Oikawa, D., Kimata, Y., and Kohno, K. (2007b). Self-association and BiP dissociation are not sufficient for activation of the ER stress sensor Ire1. *J. Cell Sci.* **120,** 1681–1688.

Oikawa, D., Kimata, Y., Kohno, K., and Iwawaki, T. (2009). Activation of mammalian IRE1alpha upon ER stress depends on dissociation of BiP rather than on direct interaction with unfolded proteins. *Exp. Cell Res.* **315,** 2496–2504.

Ron, D., and Walter, P. (2007). Signal integration in the endoplasmic reticulum unfolded protein response. *Nat. Rev. Mol. Cell Biol.* **8,** 519–529.

Okamura, K., Kimata, Y., Higashio, H., Tsuru, A., and Kohno, K. (2000). Dissociation of Kar2p/BiP from an ER sensory molecule, Ire1p, triggers the unfolded protein response in yeast. *Biochem. Biophys. Res. Commun.* **279,** 445–450.

Rutkowski, D. T., and Hegde, R. S. (2010). Regulation of basal cellular physiology by the homeostatic unfolded protein response. *J. Cell Biol.* **189,** 783–794.

Sidrauski, C., Chapman, R., and Walter, P. (1998). The unfolded protein response: An intracellular signalling pathway with many surprising features. *Trends Cell Biol.* **8,** 245–249.

Sikorski, R. S., and Hieter, P. (1989). A system of shuttle vectors and yeast host strains designed for efficient manipulation of DNA in *Saccharomyces cerevisiae*. *Genetics* **122,** 19–27.

Tirasophon, W., Welihinda, A. A., and Kaufman, R. J. (1998). A stress response pathway from the endoplasmic reticulum to the nucleus requires a novel bifunctional protein kinase/endoribonuclease (Ire1p) in mammalian cells. *Genes Dev.* **12,** 1812–1824.

Tokunaga, M., Kawamura, A., and Kohno, K. (1992). Purification and characterization of BiP/Kar2 protein from Saccharomyces cerevisiae. *J. Biol. Chem.* **267,** 17553–17559.

Wang, X. Z., Harding, H. P., Zhang, Y., Jolicoeur, E. M., Kuroda, M., and Ron, D. (1998). Cloning of mammalian Ire1 reveals diversity in the ER stress responses. *EMBO J.* **17,** 5708–5717.

Zhou, J., Liu, C. Y., Back, S. H., Clark, R. L., Peisach, D., Xu, Z., and Kaufman, R. J. (2006). The crystal structure of human IRE1 luminal domain reveals a conserved dimerization interface required for activation of the unfolded protein response. *Proc. Natl. Acad. Sci. USA* **103,** 14343–14348.

CHAPTER THIRTEEN

Measurement and Modification of the Expression Level of the Chaperone Protein and Signaling Regulator GRP78/BiP in Mammalian Cells

Wan-Ting Chen *and* Amy S. Lee

Contents

1. Introduction	218
2. Detection of Total GRP78	220
2.1. RT-PCR	220
2.2. Quantitative real-time PCR	220
2.3. Western blot	222
2.4. Small interfering RNA	223
3. Detection of Cytosolic GRP78 Isoform	224
3.1. RT-PCR	225
3.2. Quantitative real-time PCR	226
3.3. Comparison of RT-PCR and quantitative real-time PCR	226
3.4. Western blot	227
3.5. Small interfering RNA	227
4. Detection of Cell Surface GRP78	228
4.1. Biotinylation of cell surface proteins	228
4.2. Fluorescence-activated cell sorting (FACS) analysis	230
Acknowledgments	231
References	232

Abstract

GRP78/BiP is a major endoplasmic reticulum (ER) chaperone protein essential for protein quality control in the ER as well as a central regulator of unfolded protein response (UPR). The induction of GRP78 is well established as a marker for ER stress. Recently, mouse models targeting the *Grp78* allele indicate that GRP78 has critical roles in cancer progression, drug resistance, angiogenesis, neurological diseases, and diabetes. The discovery of a cytosolic GRP78 isoform

Department of Biochemistry and Molecular Biology, USC Norris Comprehensive Cancer Center, University of Southern California Keck School of Medicine, Los Angeles, California, USA and cell surface GRP78 adds new insights to its function beyond the ER compartment in regulating growth factor signaling and cell viability. Here, we summarize and update several approaches for the detection and quantitation of total GRP78, cytosolic GRP78 isoform, and cell surface GRP78, and the use of small interfering RNA to knockdown GRP78 expression. These techniques can be applied to culture cells as well as tissues.

1. Introduction

The endoplasmic reticulum (ER) is a cellular organelle where membrane and secretory proteins and lipids are synthesized. ER stress occurs when the amount of misfolded or unfolded proteins exceeds the folding capacity of the ER. ER stress elicits a series of adaptive pathways, termed the unfolded protein response (UPR; Rutkowski and Kaufman, 2004). GRP78/BiP is an ER chaperone required for proper folding and assembly of proteins and is also a master regulator of UPR. Upon ER stress, GRP78 is released from ER transmembrane signal sensors, ATF6, IRE1, and PERK, triggering the UPR signaling (Lee, 2005). The UPR has both survival and proapoptotic arms for eukaryotic cells to respond to the ER stress. The UPR, through upregulation of ER chaperones, transient arrest of general translation, and degradation of misfolded ER proteins, alleviates the ER stress and promotes cell survival. However, if the stress is too severe, apoptotic pathways are activated, including CHOP and caspase activation (Wang *et al.*, 2009). In addition to apoptosis, ER stress can induce autophagy, degrading cytosolic, long-lived proteins or defective organelles under starvation conditions for cell survival (Ogata *et al.*, 2006). A critical step for autophagy is the formation of the autophagosomes. Knockdown of GRP78 in the ER leads to ER expansion and inhibits autophagosome formation (Li *et al.*, 2008). This indicates that autophagy requires ER integrity, which is maintained by GRP78.

Tumor cells are subjected to ER stress due to altered glucose metabolism in the transformed cells, nutrient deprivation due to inadequate vasculature, and the tumor microenvironment being hypoxic. GRP78 is overexpressed in a wide variety of tumors and plays important roles in cancer progression, drug resistance, and angiogenesis (Lee, 2007; Wang *et al.*, 2009). The proliferative rate of glioma cells correlates with GRP78 expression levels, and downregulation of GRP78 suppresses proliferation and increases chemosensitivity (Pyrko *et al.*, 2007). Tumor-associated brain endothelial cells (TuBEC) constitutively overexpress GRP78, and chemoresistance of TuBEC can be reversed by downregulation of GRP78 (Virrey *et al.*, 2008). Knockdown of GRP78 by siRNA increases B-CLL cell apoptosis (Rosati *et al.*, 2010). On the other hand, GRP78 expression is primarily regulated at the transcriptional level (Lee, 2001). Recently, it was discovered that histone deacetylase 1 (HDAC1) suppresses basal activity of the *Grp78* promoter, and through suppression of HDAC1, HDAC inhibitors inadvertently activate GRP78 and confer resistance to the treatment (Baumeister et al., 2009).

Mouse models with genetically altered GRP78 expression are powerful systems to interrogate the requirement and function of GRP78 in cancer and other human diseases. While homozygous knockout of GRP78 results in early embryonic lethality, *Grp78* heterozygous mice are phenotypically normal (Luo et al., 2006). In transgene-driven endogenous mammary tumor, *Grp78* heterozygosity lengthens the latency period, inhibits tumor growth, and potently suppresses tumor angiogenesis while having no effect on the vasculature of normal organs (Dong et al., 2008). In an endogenous prostate cancer model, AKT activation and prostate tumorigenesis in *Pten* null prostate epithelium are partially suppressed in the heterozygous *Grp78* knockout and blocked by *Grp78* homozygous deletion in the prostate epithelium (Fu et al., 2008). Overall, these results suggest that the combination therapy of targeting GRP78 with other drugs may enhance the efficacy for cancer treatment through reducing chemoresistance or angiogenesis. Furthermore, GRP78 is also implicated in neurodegenerative and metabolism diseases. *Grp78* conditional knockout in neuron cells leads to apoptosis, decreased protein ubiquitination, and neurodegeneration (Wang et al., 2010). *Grp78* heterozygosity increases energy consumption and attenuates diet-induced obesity and insulin resistance (Ye et al., 2010).

Outside the ER lumen, GRP78 has been detected in cytosol, mitochondria, nucleus, and cell surface (Gonzalez-Gronow et al., 2009; Matsumoto and Hanawalt, 2000; Ni et al., 2009; Shin et al., 2003; Sun et al., 2006). Different functions have been ascribed to GRP78 in different subcellular locations. GRP78va, the novel isoform of GRP78, is generated by ER stress-induced alternative splicing (retention of intron 1) and translation initiation from an internal AUG codon (Ni et al., 2009). Devoid of the ER signal sequence, GRP78va is a 62 kDa cytosolic protein. Its expression is notably elevated in human leukemic cell lines and leukemia patient samples. GRP78va specifically enhances PERK signaling and has the ability to antagonize PERK inhibitor P58^{IPK} to promote cell survival in human leukemic cells under ER stress (Ni et al., 2009). GRP78va likely interacts with a variety of client proteins in the cytosol that remain to be identified.

One of the most important recent discoveries about GRP78 is that it is present on the cell surface of select cell types (Arap et al., 2004; Gonzalez-Gronow et al., 2009; Lee, 2007; Wang et al., 2009). Most well-known examples occur in cancer cells and stressed endothelial cells. ER stress promotes cell surface expression of GRP78, which is regulated in part by the KDEL retrieval mechanism (Zhang et al., 2010). In tumor cell lines, cell surface GRP78 acts as a coreceptor for binding protein ligands on the plasma membrane, transducing signals for proliferation, survival, as well as apoptosis. For example, cell surface GRP78 mediates α2-macroglobulin-induced signal transduction for survival and metastasis of prostate cancer (Gonzalez-Gronow *et al.*, 2006). GRP78 at the cell surface forms a complex with Cripto to promote tumor growth by inhibiting TGF-β signaling (Shani *et al.*, 2008). On the other hand, the interaction between cell surface GRP78 and Kringle 5 mediates the proapoptotic and antiangiogenic activity of Kringle 5 (Davidson *et al.*, 2005). Moreover, surface receptor GRP78 is required for cancer cell-specific apoptosis by extracellular Par-4 (Burikhanov *et al.*, 2009; Lee, 2009). Therefore, cell surface GRP78 can influence cell viability by binding to different ligands, and this is likely dictated by the differential expression level of these ligands in different cell contexts. Also, the preferential expression of GRP78 on the cell surface of tumor cells provides a new approach for selective targeting of tumor cells while sparing normal organs (Arap *et al.*, 2004; Jakobsen *et al.*, 2007; Kim *et al.*, 2006).

The methods for detecting total GRP78 have been discussed in a previous review (Lee, 2005). With the development of new reagents, we summarize and update the methods used for measurement of endogenous GRP78 expression in mammalian cells. In addition, we describe here special reagents and protocols developed for the identification of cytosolic GRP78 isoform and cell surface GRP78. The study of these novel forms of GRP78 outside the ER will facilitate discovery of new functions and provide more mechanistic explanations on why GRP78 plays such critical roles in health and diseases.

2. Detection of Total GRP78

2.1. RT-PCR

While traditional RNA blot provides information on both the expression level and integrity of the *Grp78* transcripts, if there is a limited amount of RNA available, RT-PCR is a good alternative. The RT-PCR protocol to determine total *Grp78* transcript has been published (Lee, 2005).

2.2. Quantitative real-time PCR

The levels of total *Grp78* transcript in different samples can be measured by quantitative real-time PCR in a highly quantitative manner.

Materials and reagents

- Trizol reagent (Invitrogen)
- PBS

- Chloroform
- Isopropanol
- 75% ethanol
- DEPC H$_2$O
- Oligo (dT) at 10 pmol/μl
- dNTP at 10 mM
- SuperScript II reverse transcriptase (Invitrogen) at 100 U/μl
- 5× RT buffer (Invitrogen)
- DTT (Invitrogen) at 0.1 M
- ddH$_2$O
- iQ SYBR Green Supermix (Bio-Rad)

Instrument

- Table-top centrifuge
- PCR machine (Bio-Rad *MyCycler*^{TM} *Thermal Cycler*)
- iCycler iQ Real-Time PCR Detection System (Bio-Rad)

Primers

- For human *Grp78*:
  Grp78 forward: 5′-CGACCTGGGGACCACCTACT-3′
  Grp78 reverse: 5′-TTGGAGGTGAGCTGGTTCTT-3′
  GAPDH forward: 5′-TCTGGTAAAGTGGATATTGTTG-3′
  GAPDH reverse: 5′-GATGGTGATGGGATTTCC-3′

2.2.1. RNA extraction

Total RNA from cell cultures or mouse tissues can be extracted by Trizol reagent (Invitrogen). To isolate RNA from the cell line, culture cells to reach 80–95% confluence. Aspirate media and wash cells with PBS. Add Trizol reagent directly into the culture dish (1 ml/10 cm^2) and then homogenize the lysate well by pipetting. Allow the samples to stand for 5 min at room temperature to permit the complete dissociation of nucleoprotein complexes. Add 0.2 ml of chloroform per milliliter of Trizol reagent used. Cap sample tubes securely. Shake tubes vigorously by inversion for 15 s and incubate them at room temperature for 10 min. Centrifuge the samples at 12,000×g for 15 min at 4 °C to separate the mixture into a lower red organic phase, an interphase, and a colorless upper aqueous phase, containing RNA. The aqueous phase is transferred to a fresh tube. Add 0.5 ml of isopropanol per milliliter of Trizol reagent to the samples and mix. Incubate samples for 10 min at room temperature and then centrifuge at 12,000×g for 10 min at 4 °C. Discard the supernatant. Wash the RNA pellet with 1 ml of 75% ethanol per 1 ml Trizol reagent used. Flick the tubes and centrifuge at $7500 \times g$ for 5 min at 4 °C. Remove the supernatant and quick-spin for 5 s. Remove the residual supernatant with a pipette. Briefly air-dry the RNA pellet for 3–5 min. Do not dry the RNA by centrifugation under vacuum. It is important not to dry the RNA pellet completely because this will decrease its solubility. Dissolve the RNA pellet in an appropriate amount of DEPC H_2O. Mix by repeated pipetting and incubate at 55–60 °C for 10–15 min. DNase I digestion is recommended before reverse transcription or Qiagen RNA extraction kit: RNeasy plus mini kit (Cat # 74134) is an alternative way to isolate RNA.

2.2.2. Reverse transcription

Incubate 1 μl oligo (dT; 10 pmol/μl), 1 μl dNTP (10 mM), and 2.5 μg RNA for a total volume of 13.5 μl in DEPC H_2O for one reaction at 65 °C for 5 min. Add 4 μl of 5× RT buffer, 2 μl of DTT (0.1 M), and 0.5 μl SuperScript II reverse transcriptase (Invitrogen) into the bottom of each tube and flick the tube. At this point, the RNA is subjected to PCR to be converted to cDNA. The first step of the PCR cycle is 42 °C incubation for 50 min–1 h and then 70 °C for 15 min.

2.2.3. Quantitative real-time PCR

cDNA samples are diluted at 1:20 and then analyzed in triplicate with the iQ SYBR Green Supermix (Bio-Rad) using iCycler iQ real-time PCR Detection System (Bio-Rad). For a single reaction setup, add 2 μl diluted cDNA in a 20-μl reaction that contains 10 μl iQ SYBR Green Supermix (Bio-Rad), 5 pmol primer mixture, and then, ddH_2O. At this point, the sample is ready for PCR amplification. The PCR parameters are 3 min incubation at 95 °C, 45 cycles of 15 s at 95 °C, 30 s at 60 °C, and 30 s at 72 °C. The relative mRNA level is calculated from primer-specific standard curves using the iCycler Data Analysis Software and normalized to GAPDH.

2.3. Western blot

The protein level of GRP78 can be easily detected in tissue culture cells or tissues by Western blot, and GRP78 is elevated in ER stress condition or after incubating with ER stress inducers. Two potent chemical inducers are thapsigargin (Tg), an inhibitor of the ER ATPase that causes Ca^{2+} efflux from the ER, and tunicamycin (Tu), which blocks N-linked glycosylation. Thus, GRP78 induction is generally served as an ER stress indicator, suggesting that the UPR is activated.

The procedure for Western blot has been described (Lee, 2005; Li et al., 2008). Currently, there are various antibodies available against GRP78.

Examples are goat polyclonal anti-GRP78 (C-20) and (N-20; Santa Cruz) that can detect GRP78 of mouse, rabbit, and human origin. There are also a rabbit polyclonal antibody (H129; Santa Cruz) against mouse, rabbit, and human GRP78 and a rabbit polyclonal antibody against human GRP78 (GeneTex, GTX102567S). In addition, mouse monoclonal anti-GRP78 (BD Biosciences) can be used to recognize GRP78 in human, mouse, and rat cells with high specificity and intensity. The updated approach for detecting protein signals is using the ECL reagent (Roche) or Supersignal chemiluminescence reagent (PIERCE) after reacting with HRP-conjugated secondary antibody. Among these antibodies, the H129 antibody also works well for immunohistochemical staining of GRP78 in tissues.

2.4. Small interfering RNA

GRP78 gene expression can be knocked down by siRNA specifically targeting the *Grp78* transcript. This is an excellent approach to manipulate GRP78 expression to explore the effects on the downstream targets.

Materials and reagents

- Lipofectamine 2000 reagent (Invitrogen)

The sequences of siRNA for human Grp78

- *Grp78* siRNA (siGrp78):
  Sense: 5'-GGAGCGCAUUGAUACUAGATT-3'
  Antisense: 5'-UCUAGUAUCAAUGCGCUCCTT-3'
- Control siRNA (siCtrl): Oligos with no matching Genebank sequences are used
  Sense: 5'-AAGGAGACGUAUAGCAACGGU-3'
  Antisense: 5'-ACCGUUGCUAUACGUCUCCUU-3'

2.4.1. siRNA transfection

The pairs of siRNA sequences are annealed and prepared as a 20 μM stock. Before transient siRNA transfection, culture cells in 12-well plates to reach 80% confluence. Lipofectamine 2000 reagent (Invitrogen) is used following the manufacturer's instruction. Harvest cells after 48–72 h of siRNA transfection. The expression levels of GRP78 are evaluated by Western blot using anti-GRP78 antibodies which are mentioned above.

3. Detection of Cytosolic GRP78 Isoform

GRP78 is encoded by a single-copy gene containing 8 exons (Fig. 13.1A). *Grp78va*, a splicing variant of *Grp78*, has an additional 112 bp intron 1 sequence which is spliced out in canonical *Grp78* transcript. It has been shown that thapsigargin (Tg), the most potent chemical inducer for *Grp78*, can also increase the level of the *Grp78va* transcript (Ni *et al.*, 2009). Both RT-PCR and quantitative real-time PCR can be used to detect cytosolic GRP78 isoform, GRP78va. Retention of intron 1

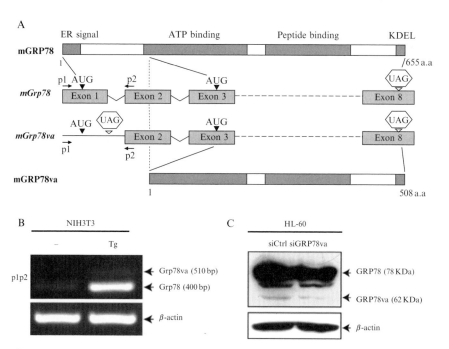

Figure 13.1 Detection of GRP78va at both mRNA and protein levels. (A) Schematic representation of GRP78 protein domains, canonical mouse *Grp78* (*mGrp78*), and alternative splicing transcript *mGrp78va*. The ER signal peptide, ATP binding domain, peptide binding domain, and KDEL motif are indicated. The numbers below indicate amino acid (a.a). The translation start codon (AUG) and stop codon (UAG) of canonical *mGrp78* are located in exon 1 and exon 8, respectively. Due to the retention of intron 1 which contains a stop codon, *mGrp78va* uses the internal AUG start site in exon 3. Arrows labeled as p1 and p2 refer to positions of the primer sets used in RT-PCR. (B) Detection of *mGrp78va* transcript in mouse cells by RT-PCR. Mouse embryo fibroblast NIH3T3 cells were nontreated (−) or treated with 300 nM thapsigargin (Tg) for 16 h, and the total RNA were extracted and subjected to RT-PCR using the p1/p2 primer sets. β-Actin served as the loading control. (C) siRNA knockdown of GRP78va in human cells. Human leukemia HL-60 cells were transfected with siCtrl or siGRP78, and Western blot was performed to determine the protein levels of GRP78 and GRP78va.

produces a UAA stop codon downstream of the initial AUG start codon in exon 1. Therefore, the translation of GRP78va starts at the next initiation site in exon 3, generating the 508 a.a. mouse GRP78va protein (in human cells, GRP78 has 654 a.a. and GRP78va has 507 a.a.) without the ER-targeting signal (Fig. 13.1A). In this section, we will also describe how to distinguish canonical GRP78 and isoform GRP78va using siRNA knockdown, followed by Western blot.

3.1. RT-PCR

By using p1/p2 primer set which spans the 5′ end of exon 1 (p1) and the exon 2 (p2; Fig. 13.1A), two clear RT-PCR bands were detected, one 400 bp (*Grp78*) and another 510 bp (*Grp78va*) band in NIH3T3 cells treated with ER stress inducer thapsigargin (Tg; Fig. 13.1B).

Materials and reagents

- Thapsigargin (Tg) at 300 nM
- dNTP at 10 mM
- ddH$_2$O
- 10× buffer
- Taq polymerase (Bioland) at 5 U/μl
- DNase I

Primers

- For mouse *Grp78* and *Grp78va*:
  p1: 5′-GCTCCGAGTCTGCTTCGTGTCT-3′
  p2: 5′-TGGACGTGAGTTGGTTCTTG-3′
- For human *Grp78* and *Grp78va* (if using human cell lines):
  p1: 5′-CAGCACAGACAGATTGACCTAT-3′
  p3: 5′-GACATCAGCACCGCACTTCTCA-3′, or
  p4: 5′-GGTGCTGATGTCCCTCTGTC-3′
  p5: 5′-CCTAACAAAAGTTCCTGAGTCCA-3′

For location of the human primer sets and details on the detection of the human *Grp78* and *Grp78va* transcripts, refer to Ni *et al.* (2009).

Instrument

- PCR machine (Bio-Rad *MyCycler*^{TM} *Thermal Cycler*)

3.1.1. RNA extraction and reverse transcription

The steps are described as above. DNase I should be used to treat RNA before RT-PCR, in order to eliminate genomic DNA contamination.

3.1.2. PCR

The cDNA generated from reverse transcription is subjected to PCR to amplify *Grp78* transcript. For each PCR reaction, add 24.4 μl ddH$_2$O, 1 μl cDNA, 3 μl of 10× buffer, 0.7 μl of dNTP (10 m*M*), 0.3 μl of the forward primer (25 pmol/μl) and the reverse primer (25 pmol/μl), respectively, and 0.3 μl of Taq polymerase for a total volume of 30 μl. The PCR cycle starts with a 3-min incubation at 94 °C, then 35 cycles of 94 °C for 30 s, 60 °C for 30 s, and 72 °C for 30 s, followed by the final incubation at 72 °C for 10 min. The PCR products are analyzed by 2% gel electrophoresis.

3.2. Quantitative real-time PCR

Quantitative real-time PCR can be used to compare the relative expression levels of total *Grp78* and *Grp78va* transcripts in different cell lines or tissues.

Materials and reagents

- iQ SYBR Green Supermix (Bio-Rad)

Instrument

- iCycler iQ Real-Time PCR Detection System (Bio-Rad)

Primers

- For human *Grp78va*:
  Grp78va forward: 5′-GGTGCTGATGTCCCTCTGTC-3′
  Grp78va reverse: 5′-TTGGAGGTGAGCTGGTTCTT-3′
  GAPDH forward: 5′-TCTGGTAAAGTGGATATTGTTG-3′
  GAPDH reverse: 5′-GATGGTGATGGGATTTCC-3′

3.2.1. RNA extraction and reverse transcription
The steps are described as above.

3.2.2. Quantitative real-time PCR
The procedures are described as above.

3.3. Comparison of RT-PCR and quantitative real-time PCR

To detect the presence or absence of specific mRNA transcripts, RT-PCR has been widely used. Since RT-PCR can identify transcript size, it is a valuable detection tool, as demonstrated by our use of RT-PCR to discover GRP78 isoform, GRP78va (Fig. 13.1B). However, RT-PCR is not considered quantitative because of the low sensitivity of ethidium bromide staining. Thus, for the PCR product to be detectable, the PCR reaction has to reach the plateau phase.

On the other hand, quantitative real-time PCR cannot provide information on the transcript size but has more reproducibility. Also, real-time PCR is more efficient than RT-PCR, because it is a single-step method of amplification and analysis. More importantly, quantitative real-time PCR requires less amount of starting material of cells or tissues and is a highly quantitative tool which is suitable for accurately measuring the relative gene expression in different samples and determining unknown sample values. For instance, quantitative real-time PCR was used to compare the relative expression levels of total *Grp78* and *Grp78va* transcripts in different cell lines or tissues (Ni *et al.*, 2009).

3.4. Western blot

The protocol for Western blot has been published (Lee, 2005). For the detection of GRP78va levels, goat polyclonal anti-GRP78 (C-20; Santa Cruz) at a dilution of 1:1000 and mouse monoclonal anti-GRP78 (BD Biosciences) at a dilution of 1:2000 can be used. Mouse monoclonal anti-β actin (Sigma) serves as the loading control (Fig. 13.1C).

3.5. Small interfering RNA

A siRNA targeting a 21-nucleotide sequence in intron 1 retained in human *Grp78va* can be used to specifically knock down the *Grp78va* transcript in human cells.

Materials and reagents

- Lipofectamine 2000 reagent (Invitrogen)
- Anti-GRP78 mouse monoclonal antibody (BD Biosciences), 1:2000
- Anti-β-actin mouse monoclonal antibody (Sigma), 1:5000

The sequences of siRNA for human Grp78va

- *Grp78va* siRNA (siGrp78va):
  Sense: 5′-AAGUGCGGUGCUGAUGUCCCU-3′;
  Antisense: 5′-AGGGACAUCAGCACCGCACUU-3′.
- Control siRNA (siCtrl):
  Sense: 5′-AAGGAGACGUAUAGCAACGGU-3′;
  Antisense: 5′-ACCGUUGCUAUACGUCUCCUU-3′.

3.5.1. siRNA transfection

This can be achieved in established cell lines; however, the transfection conditions need to be adjusted based on the efficiency of transfection on different cells. In our example, we used the human APL cell line HL-60 to perform siRNA transfection. Lipofectamine 2000 reagent (Invitrogen) is used following the manufacturer's instructions. HL-60 cells are transfected with siCtrl or siGrp78va for 72 h.

3.5.2. Determination of knockdown outcome

After siRNA transfection, the expression levels of GRP78 and GRP78va are evaluated by Western blot using anti-GRP78 monoclonal antibody (BD Biosciences), with β-actin (Sigma) as the loading control. In cells transfected with siGrp78va, the protein level of canonical GRP78 (78 kDa) was not changed as expected, whereas the level of GRP78va (62 kDa) was specifically reduced (Fig. 13.1C).

4. Detection of Cell Surface GRP78

In this section, we provide two methods to detect and analyze cell surface GRP78 in mammalian cells. Biotinylation followed by avidin pull-down is a widely used technique for detection of cell surface proteins. In addition, specific domains of GRP78 exposed on the cell surface can be detected by FACS analysis.

4.1. Biotinylation of cell surface proteins

Biotin is a small water soluble molecule, which can bind to many proteins at primary amines without alteration of their biological functions. Since biotin is charged, it cannot go through the cell membrane, and only the proteins with primary amines exposed on the cell membrane can be covalently labeled with biotin. The biotin tag can be used to purify biotin-labeled proteins by avidin, which is the natural ligand for biotin (Fig. 13.2A). Also, it has been found that under regular culture conditions, cells have very low cell surface GRP78 expression. However, ER stress induced by thapsigargin (Tg) significantly promotes cell surface localization of GRP78 (Fig. 13.2B–D; Zhang et al., 2010).

Materials and reagents

- Thapsigargin (Tg) at 300 nM
- EZ-link Sulfo-NHS-LC-Biotin (Thermoscientific) at 0.5 mg/ml
- Tris–Cl, pH 7.5

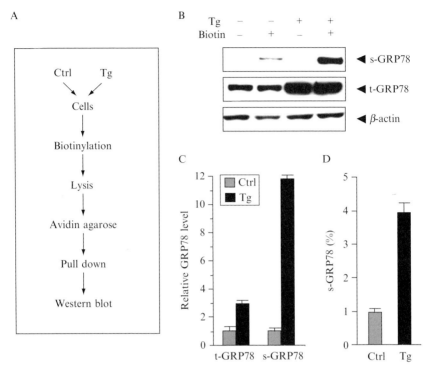

Figure 13.2 ER stress actively promotes cell surface localization of endogenous GRP78. (A) The flow chart of the biotinylation method to detect cell surface proteins in cells that are either nontreated (Ctrl) or treated with thapsigargin (Tg). (B) Western blot showing ER stress promotion of cell surface GRP78 expression. Human embryonic kidney 293T cells were either untreated (−) or treated (+) with 300 nM Tg for 16 h. The amount of lysate for measuring the total GRP78 level was 10% of the amount used for avidin pull-down. s-GRP78 and t-GRP78 represent surface and total intracellular GRP78, respectively. β-Actin served as the loading control. (C) Relative expression levels of total intracellular GRP78 and surface GRP78 in control (Ctrl) and Tg-treated cells. (D) The percentage of cell surface versus total GRP78 in control (Ctrl) and Tg-treated cells. The data from panels (C) and (D) were summarized from three to four experiments and the S.D. is shown (Zhang et al., 2010).

- Radioimmunoprecipitation assay buffer (RIPA buffer)
- Protease inhibitor (Roche Applied Science)
- Neutravidin–agarose beads (Thermoscientific)
- PBS
- 1× SDS-PAGE sample loading buffer (Sigma)
- Anti-GRP78 mouse monoclonal antibody (BD Biosciences), 1:1000
- Anti-β-actin mouse monoclonal antibody (Sigma), 1:5000

4.1.1. Cell surface protein biotinylation

Grow cells with 80–90% confluency and then treat cells with or without 300 nM thapsigargin (Tg) for 16 h. Discard the used medium and rinse the cells with cold PBS twice. Add 0.5 mg/ml EZ-link Sulfo-NHS-LC-Biotin (Thermoscientific) to cover the cell layer (1 ml/6-well plate) and then gently shake the plate at 4 °C for 30 min. After shaking, stop the biotinylation reaction by adding Tris–Cl, pH 7.5, to a final concentration of 100 mM. Discard the biotin solution, rinse the cells with ice-chilled PBS twice, and then lyse the cells with RIPA buffer supplemented with competent protease inhibitor (Roche Applied Science). Part (about 10–20%) of the cell lysate is subjected to Western blot for measuring the input total GRP78 level.

4.1.2. Avidin pull-down

Add Neutravidin–agarose beads (Thermoscientific), which trap biotinylated proteins, to the cell lysates and mix them well at 4 °C overnight. Extensively wash the Neutravidin–agarose beads five times with PBS. To elute the biotinylated proteins, boil the beads with 1× SDS-PAGE sample loading buffer (Sigma) for 5 min. The eluate together with the input are subjected to Western blot. Detect both cell surface and total GRP78 by anti-GRP78 mouse antibody (BD Biosciences) at a dilution of 1:1000. The cytosolic β-actin serves as the loading control for the total cell lysate. Its absence in the cell surface fractions serves as a control so that the cell surface protein fractions are not contaminated by cytosolic proteins.

4.2. Fluorescence-activated cell sorting (FACS) analysis

Since ER membrane is the major source for the plasma membrane and GRP78 contains both hydrophilic and hydrophobic domains, it is possible that cell surface GRP78 is derived in part from GRP78 located in the ER membrane and cycled to the cell surface (Reddy *et al.*, 2003; Zhang *et al.*, 2010). FACS is generally used to detect cell surface markers and sort cells. FACS is a good choice to test cell surface localization of GRP78, and it can also be used to investigate which domain(s) of GRP78 is exposed on the cell surface (Burikhanov *et al.*, 2009; Liu *et al.*, 2007).

Materials and reagents

- PBS
- 10% normal human serum in PBS
- Staining buffer (Dulbecco's PBS, 2% heat-inactivated fetal calf serum, 0.09% sodium azide)
- 4′,6-diamidino-2-phenylindole (DAPI, Sigma) at 1 μg/ml
- Anti-GRP78 rabbit polyclonal antibody (Novus Biologicals), 1:150

- Anti-β-actin mouse monoclonal antibody (Sigma)
- Normal rabbit IgG; normal mouse IgG
- Fluorescein isothiocyanate-conjugated antirabbit IgG (Vector Laboratories), 1:100
- Phycoerythrin-conjugated antimouse IgG (BD Biosciences), 1:40
- BD Falcon 12 × 75 polystyrene tube

Instrument

- FACS machine (BD LSR II Flow Cytometer)

4.2.1. FACS analysis

Wash the cells with cold PBS twice and aliquot 1×10^6 cells to each tube. In order to reduce nonspecific immunofluorescent background, preblock immunoglobulin Fc receptors (FcRs) on human cells by incubating the cells with 10% normal human serum in PBS for 20 min at 4 °C. Add a saturating amount (1 μg) of anti-GRP78 antibody or an immunoglobulin (Ig) isotype-matched control (normal rabbit or mouse IgG) into the cells and incubate for 40 min at 4 °C in 100 μl of staining buffer. After incubation, add 200 μl staining buffer and pellet the cells by centrifugation ($500 \times g$ for 5 min). Wash the cells with 2-ml staining buffer twice. FITC- or PE-conjugated secondary antibody (0.5 μg) is added and incubated with the cells at 4 °C for 30 min. After completely washing with staining buffer, suspend the cells in ice-cold PBS containing DAPI (1 μg/ml) and then perform FACS analysis. The cells should be protected from light throughout staining and storage. The data from FACS are analyzed by FACS-Diva (BD Biosciences) software.

4.2.2. Determination of GRP78 domains exposed on the cell surface

Since there is no permeabilization of the cells, the cell membrane is still intact; hence, we can use antibody to recognize the protein or specific domain(s) of a protein located on the cell surface. For example, by using an antibody that recognizes the middle domain of human GRP78 spanning a.a. 250–300 (Novus Biologicals) in FACS analysis, we determined that the middle domain of GRP78 is exposed on the cell surface. In contrast, β-actin, as the negative control for cell integrity, was not found on the cell surface (Zhang et al., 2010).

ACKNOWLEDGMENTS

We thank Drs. Min Ni and Yi Zhang for helpful discussions and assistance and Dr. Min Ni for contribution of figures. This work is supported by NIH grants CA027607, CA111700, and DK079999 to A.S.L.

REFERENCES

Arap, M. A., Lahdenranta, J., Mintz, P. J., Hajitou, A., Sarkis, A. S., Arap, W., and Pasqualini, R. (2004). Cell surface expression of the stress response chaperone GRP78 enables tumor targeting by circulating ligands. *Cancer Cell* **6**, 275–284.

Baumeister, P., Dong, D., Fu, Y., and Lee, A. S. (2009). Transcriptional induction of GRP78/BiP by histone deacetylase inhibitors and resistance to histone deacetylase inhibitor-induced apoptosis. *Mol. Cancer Ther.* **8**, 1086–1094.

Burikhanov, R., Zhao, Y., Goswami, A., Qiu, S., Schwarze, S. R., and Rangnekar, V. M. (2009). The tumor suppressor Par-4 activates an extrinsic pathway for apoptosis. *Cell* **138**, 377–388.

Davidson, D. J., Haskell, C., Majest, S., Kherzai, A., Egan, D. A., Walter, K. A., Schneider, A., Gubbins, E. F., Solomon, L., Chen, Z., Lesniewski, R., and Henkin, J. (2005). Kringle 5 of human plasminogen induces apoptosis of endothelial and tumor cells through surface-expressed glucose-regulated protein 78. *Cancer Res.* **65**, 4663–4672.

Dong, D., Ni, M., Li, J., Xiong, S., Ye, W., Virrey, J. J., Mao, C., Ye, R., Wang, M., Pen, L., Dubeau, L., Groshen, S., *et al.* (2008). Critical role of the stress chaperone GRP78/BiP in tumor proliferation, survival and tumor angiogenesis in transgene-induced mammary tumor development. *Cancer Res.* **68**, 498–505.

Fu, Y., Wey, S., Wang, M., Ye, R., Liao, C. P., Roy-Burman, P., and Lee, A. S. (2008). Pten null prostate tumorigenesis and AKT activation are blocked by targeted knockout of ER chaperone GRP78/BiP in prostate epithelium. *Proc. Natl. Acad. Sci. USA* **105**, 19444–19449.

Gonzalez-Gronow, M., Cuchacovich, M., Llanos, C., Urzua, C., Gawdi, G., and Pizzo, S. V. (2006). Prostate cancer cell proliferation in vitro is modulated by antibodies against glucose-regulated protein 78 isolated from patient serum. *Cancer Res.* **66**, 11424–11431.

Gonzalez-Gronow, M., Selim, M. A., Papalas, J., and Pizzo, S. V. (2009). GRP78: A multifunctional receptor on the cell surface. *Antioxid. Redox Signal.* **11**, 2299–2306.

Jakobsen, C. G., Rasmussen, N., Laenkholm, A. V., and Ditzel, H. J. (2007). Phage display derived human monoclonal antibodies isolated by binding to the surface of live primary breast cancer cells recognize GRP78. *Cancer Res.* **67**, 9507–9517.

Kim, Y., Lillo, A. M., Steiniger, S. C., Liu, Y., Ballatore, C., Anichini, A., Mortarini, R., Kaufmann, G. F., Zhou, B., Felding-Habermann, B., and Janda, K. D. (2006). Targeting heat shock proteins on cancer cells: Selection, characterization, and cell-penetrating properties of a peptidic GRP78 ligand. *Biochemistry* **45**, 9434–9444.

Lee, A. S. (2001). The glucose-regulated proteins: Stress induction and clinical applications. *Trends Biochem. Sci.* **26**, 504–510.

Lee, A. S. (2005). The ER chaperone and signaling regulator GRP78/BiP as a monitor of endoplasmic reticulum stress. *Methods* **35**, 373–381.

Lee, A. S. (2007). GRP78 induction in cancer: Therapeutic and prognostic implications. *Cancer Res.* **67**, 3496–3499.

Lee, A. S. (2009). The Par-4-GRP78 TRAIL, more twists and turns. *Cancer Biol. Ther.* **8**, 2103–2105.

Li, J., Ni, M., Lee, B., Barron, E., Hinton, D. R., and Lee, A. S. (2008). The unfolded protein response regulator GRP78/BiP is required for endoplasmic reticulum integrity and stress-induced autophagy in mammalian cells. *Cell Death Differ.* **15**, 1460–1471.

Liu, Y., Steiniger, S. C., Kim, Y., Kaufmann, G. F., Felding-Habermann, B., and Janda, K. D. (2007). Mechanistic studies of a peptidic GRP78 ligand for cancer cell-specific drug delivery. *Mol. Pharm.* **4**, 435–447.

Luo, S., Mao, C., Lee, B., and Lee, A. S. (2006). GRP78/BiP is required for cell proliferation and protecting the inner cell mass from apoptosis during early mouse embryonic development. *Mol. Cell. Biol.* **26,** 5688–5697.

Matsumoto, A., and Hanawalt, P. C. (2000). Histone H3 and heat shock protein GRP78 are selectively cross-linked to DNA by photoactivated gilvocarcin V in human fibroblasts. *Cancer Res.* **60,** 3921–3926.

Ni, M., Zhou, H., Wey, S., Baumeister, P., and Lee, A. S. (2009). Regulation of PERK signaling and leukemic cell survival by a novel cytosolic isoform of the UPR regulator GRP78/BiP. *PLoS ONE* **4,** e6868.

Ogata, M., Hino, S., Saito, A., Morikawa, K., Kondo, S., Kanemoto, S., Murakami, T., Taniguchi, M., Tanii, I., Yoshinaga, K., Shiosaka, S., Hammarback, J. A., *et al.* (2006). Autophagy is activated for cell survival after endoplasmic reticulum stress. *Mol. Cell. Biol.* **26,** 9220–9231.

Pyrko, P., Schonthal, A. H., Hofman, F. M., Chen, T. C., and Lee, A. S. (2007). The unfolded protein response regulator GRP78/BiP as a novel target for increasing chemosensitivity in malignant gliomas. *Cancer Res.* **67,** 9809–9816.

Reddy, R. K., Mao, C., Baumeister, P., Austin, R. C., Kaufman, R. J., and Lee, A. S. (2003). Endoplasmic reticulum chaperone protein GRP78 protects cells from apoptosis induced by topoisomerase inhibitors: Role of ATP binding site in suppression of caspase-7 activation. *J. Biol. Chem.* **278,** 20915–20924.

Rosati, E., Sabatini, R., Rampino, G., De Falco, F., Di Ianni, M., Falzetti, F., Fettucciari, K., Bartoli, A., Screpanti, I., and Marconi, P. (2010). Novel targets for endoplasmic reticulum stress-induced apoptosis in B-CLL. *Blood* **116,** 2713–2723.

Rutkowski, D. T., and Kaufman, R. J. (2004). A trip to the ER: Coping with stress. *Trends Cell Biol.* **14,** 20–28.

Shani, G., Fischer, W. H., Justice, N. J., Kelber, J. A., Vale, W., and Gray, P. C. (2008). GRP78 and Cripto form a complex at the cell surface and collaborate to inhibit transforming growth factor beta signaling and enhance cell growth. *Mol. Cell. Biol.* **28,** 666–677.

Shin, B. K., Wang, H., Yim, A. M., Le Naour, F., Brichory, F., Jang, J. H., Zhao, R., Puravs, E., Tra, J., Michael, C. W., Misek, D. E., and Hanash, S. M. (2003). Global profiling of the cell surface proteome of cancer cells uncovers an abundance of proteins with chaperone function. *J. Biol. Chem.* **278,** 7607–7616.

Sun, F. C., Wei, S., Li, C. W., Chang, Y. S., Chao, C. C., and Lai, Y. K. (2006). Localization of GRP78 to mitochondria under the unfolded protein response. *Biochem. J.* **396,** 31–39.

Virrey, J. J., Dong, D., Stiles, C., Patterson, J. B., Pen, L., Ni, M., Schonthal, A. H., Chen, T. C., Hofman, F. M., and Lee, A. S. (2008). Stress chaperone GRP78/BiP confers chemoresistance to tumor-associated endothelial cells. *Mol. Cancer Res.* **6,** 1268–1275.

Wang, M., Wey, S., Zhang, Y., Ye, R., and Lee, A. S. (2009). Role of the unfolded protein response regulator GRP78/BiP in development, cancer, and neurological disorders. *Antioxid. Redox Signal.* **11,** 2307–2316.

Wang, M., Ye, R., Barron, E., Baumeister, P., Mao, C., Luo, S., Fu, Y., Luo, B., Dubeau, L., Hinton, D. R., and Lee, A. S. (2010). Essential role of the unfolded protein response regulator GRP78/BiP in protection from neuronal apoptosis. *Cell Death Differ.* **17,** 488–498.

Ye, R., Jung, D. Y., Jun, J. Y., Li, J., Luo, S., Ko, H. J., Kim, J. K., and Lee, A. S. (2010). Grp78 heterozygosity promotes adaptive unfolded protein response and attenuates diet-induced obesity and insulin resistance. *Diabetes* **59,** 6–16.

Zhang, Y., Liu, R., Ni, M., Gill, P., and Lee, A. S. (2010). Cell surface relocalization of the endoplasmic reticulum chaperone and unfolded protein response regulator GRP78/BiP. *J. Biol. Chem.* **285,** 15065–15075.

CHAPTER FOURTEEN

The Endoplasmic Reticulum-Associated Degradation and Disulfide Reductase ERdj5

Ryo Ushioda *and* Kazuhiro Nagata

Contents

1. Introduction	236
2. Productive Folding of Newly Synthesized Proteins in the ER	238
2.1. ER-resident molecular chaperones: BiP and Hsp40 family proteins	238
2.2. Oxidoreductases in the ER	239
2.3. *N*-Glycosylation and lectin-like chaperones–CNX and CRT	241
3. ERAD: A Strategy for ERQC Mechanism	242
3.1. Protection system for ER stress –UPR	242
3.2. Degradation signals and its recognition	243
3.3. ERAD for nonglycosylated proteins	244
3.4. Dislocon channels for retrotranslocation	245
3.5. ERAD machinery for various substrates and ubiquitination	246
3.6. Reduction of disulfide bonds for ERAD	248
4. Conclusion	249
References	252

Abstract

The endoplasmic reticulum (ER) is an organelle where secretory or membrane proteins are correctly folded with the aid of various molecular chaperones and oxidoreductases. Only correctly folded and assembled proteins are enabled to reach their final destinations, which are called as ER quality control (ERQC) mechanisms. ER-associated degradation (ERAD) is one of the ERQC mechanisms for maintaining the ER homeostasis and facilitates the elimination of misfolded or malfolded proteins accumulated in the ER. ERAD is mainly consisting of three processes: recognition of misfolded proteins for degradation in the ER, retrotranslocation of (possibly) unfolded substrates from the ER to the cytosol through dislocation channel, and their degradation in the cytosol via

Laboratory of Molecular and Cellular Biology, Department of Molecular Biosciences, Faculty of Life Sciences, Kyoto Sangyo University, Kyoto, Japan ubiquitin-protesome system. After briefly mentioned on productive folding of nascent polypeptides in the ER, we here overview the above three processes in ERAD system by highlighting on novel ERAD factors such as EDEM and ERdj5 in mammals and yeasts.

1. INTRODUCTION

Around 30% of total proteins are synthesized in the endoplasmic reticulum (ER) as secretory and membrane proteins (Ghaemmaghami et al., 2003; Kanapin, 2003), and the ER is a major organelle of the folding of newly synthesized polypeptides. Various molecular chaperones and folding enzymes in the ER are involved in the productive folding of these nascent polypeptides. N-glycosylation-relating enzymes and oxidoreductases, including protein disulfide isomerase (PDI) and ERp57, promote the disulfide bond formation between cystein residues, which is essential for maintaining the correct ternary structure of the proteins. In spite of these well-established folding systems in the ER, misfolding of nascent polypeptides occurs systemically and/or accidentally: genetic mutations lead the proteins to the terminal misfolding, and various cellular stresses also cause the unstabilization and denaturation of the proteins. Misfolded proteins not only fail to acquire their primary function but also lead to serious cellular toxicity. In many cases, misfolded proteins expose the hydrophobic cluster of amino acids on the molecular surface, which easily leads the proteins to form aggregation via the hydrophobic interaction. Such an accumulation of misfolded protein in the ER causes the so-called ER stress.

The biosynthesis of proteins in the ER is tightly monitored by a mechanism called ER quality control (ERQC) to assure that only correctly folded and assembled proteins can reach their final destinations. When the misfolded proteins accumulated in the ER, cells exert the ERQC mechanism. The ERQC mainly consists of three pathways: (1) transient inhibition of general translation to avoid the accumulation of misfolded proteins in the ER, which is exerted by the activation of PERK (PKR [RNA-activated protein kinase]-like ER kinase), (2) renaturation of misfolded proteins by ER molecular chaperones, which is induced by the activation of ATF6 (activating transcription factor 6), and (3) the degradation of misfolded proteins by the ubiquitin–proteasome system after the substrates are transported from the ER into the cytosol. These reactions against the ER stress are called unfolded protein response (UPR). The degradation of misfolded proteins in the ER, called endoplasmic reticulum-associated degradation (ERAD), is composed mainly with three steps: (1) recognition of misfolded substrates for degradation in the ER, (2) retrotranslocation of (possibly) unfolded substrates through the

ER Quality Control System

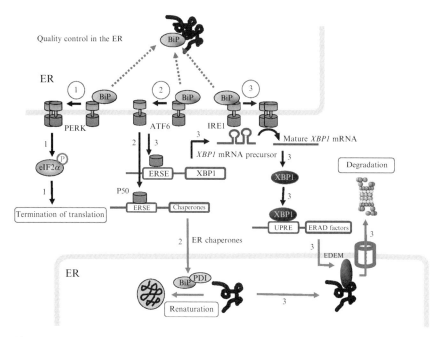

Figure 14.1 ER quality control through UPR against ER stress. When misfolded proteins accumulate in the ER, three different pathways of UPR against ER stress are booted up as follows: (1) Release of BiP from PERK activates PERK via its dimerization and activated PERK phosphorylates the translation initiation factor, eIF2, which causes the attenuation of translation. (2) Release of BiP from ATF6 activates ATF6 by cleaving off the p50 cytosolic domain from ATF6, and p50 activates the transcription of the mRNA for various molecular chaperones through binding to ERSE element. (3) p50 also promotes the transcription of the *xbp1* mRNA. When IRE1 is activated by the release of BiP, IRE1 splices the xbp1 mRNA precursor, resulting in the production of the XBP1 transcription factor, which in turn activates the transcription of several ERAD-related genes.

retrotranslocation channel (also called dislocon channel), and (3) degradation of the substrates by 26S proteasomes in the cytosol after the polyubiquitination of the substrates (Fig. 14.1).

While a lot of information over the decade promoted greatly our understanding on the ERAD mechanisms, many questions still remain for the upstream event in the ER for initiating the ERAD; for example, how are terminally misfolded proteins recognized and transferred to the ERAD pathway from the background of maturing or matured proteins, and how is the correctly folded and disulfide-bonded ternary structure unfolded before recruiting to the narrow retrotranslocation channel? Here, we overview the ERAD pathway, mainly focusing on these unsolved issues.

2. Productive Folding of Newly Synthesized Proteins in the ER

Newly synthesized polypeptides cotranslationally enter the ER through the translocon channels. N-Glycosylation of nascent polypeptides occurs cotranslationally by oligosaccharide transferase (OST), and disulfide formation between cyteine residues of polypeptides also occurs cotranslationally by ER oxidoreductases. In addition to these folding enzymes, various molecular chaperones are involved in the productive folding of nascent polypeptides.

2.1. ER-resident molecular chaperones: BiP and Hsp40 family proteins

BiP (Ig heavy-chain-binding protein) is one of the most abundant ER-resident molecular chaperones of Hsp70 family, which is conserved among all eukaryotes and is essential for cell survival. BiP is comprised of two domains: an ATPase domain and a substrate-binding domain containing a hydrophobic cleft suitable for binding to misfolded or unfolded substrates. When BiP hydrolyses bound ATP to ADP via the ATPase domain, ADP-bound BiP acquires high affinity for substrates with the conformational change to a so-called closed form where the hydrophobic cleft is closed. By binding to the substrates with high affinity, BiP prevents the misfolded proteins from aggregate formation.

Hsp40s, which possess the J domain in the structure, act as cochaperones for Hsp70 family proteins. The J domain involves His-Pro-Asp (HPD) motif required for the binding with Hsp70 family proteins including an ER homolog BiP. The binding of Hsp40 J domain to BiP stimulates the ATPase activity of BiP causing the hydrolysis of ATP to ADP (Liberek et al., 1991). GRP170 and Sil1, nucleotide exchange factors (NEFs) in the ER, exchange ADP form of BiP to ATP form, which leads the BiP to the open and accessible form to substrates. These hydrolytic cycles of BiP regulated by J domain-containing cochaperones and NEFs maintain the solubility of nascent and misfolded proteins in the ER by preventing their aggregations. BiP is also involved in the regulation of the ER stress sensors (see Section 3).

In the ER, up to 7 Hsp40s have been identified as ERdj1–7, among which ERdj1, ERdj2, ERdj4, and ERdj7 have transmembrane domains and the others are ER luminal proteins (Otero et al., 2010). ERdj3–6 have been reported to be upregulated by ER stresses, while ERdj1 and ERdj2 are not. ERdj1 and ERdj2 facilitate the translocation of newly synthesized polypeptides into the ER through the Sec61 translocon channel in corporation with BiP (Dudek et al., 2005). ERdj2 is a homolog of yeast Sec63p, which is reported as an essential factor for the posttranslational protein transport into the ER in yeast (Brodsky et al., 1995; Meyer et al., 2000). ERdj3 is identified in canine pancreatic microsomes as HEDJ (Bies et al., 2004), existing as the chaperone complex with BiP (Meunier et al., 2002). Interestingly, it is reported that ERdj3 cooperates with BiP in facilitating the dislocation of Shiga toxin from the ER lumen into the cytosol, where Shiga toxin inhibits the protein synthesis (Yu and Haslam, 2005). Shiga toxin thus takes advantage of the dislocation systems equipped in the host cells.

ERdj4 and ERdj5 are reported to enhance the ERAD of misfolded proteins (Dong et al., 2008; Ushioda et al., 2008), which is to be described later in detail (Section 3.6). ERdj6, first designated as the p58^{IPK}, was reported to negatively regulate the PERK phosphorylation (Yan et al., 2002). However, from the evidences that ERdj6 is an ER-resident protein with ER-targeting signal and that ERdj6 binds to BiP through the tetratricopeptide repeat domain and works with BiP as a cochaperone (Petrova et al., 2008; Rutkowski et al., 2007), ERdj6 is most probably involved in the productive folding of newly synthesized proteins in the ER lumen. Recent mass spectrometric analysis with canine pancreatic microsomes revealed the presence of further Hsp40 family protein ERdj7, which possesses the transmembrane domain and luminal domain (Zahedi et al., 2009), while its function is still unknown. As described above, Hsp40 family proteins in the ER cooperatively regulate and cooperate with the wide spectrum of BiP functions in the ER.

2.2. Oxidoreductases in the ER

The disulfide bonds between two cysteine residues serve to stabilize the ternary structure of the protein when they are intramolecular disulfides, and make a quaternary structure when intermolecular disulfides are formed between two subunits or proteins. PDI and other approximately 20 proteins residing in the ER lumen belong to the thioredoxin super family. Thioredoxin super family members possess the thioredoxin-like domain consisting of four β sheets and three α helices. The active site in the thioredoxin-like domain lies at Cys-X-X-Cys (CXXC) motif on N-terminal α helix. The pair of cystein residues in CXXC motif is responsible for oxidoreductase activity; CXXC motifs accept electrons from the substrates when it oxidizes the substrates and transfer the electrons to the substrates when it reduces the substrates. Thioredoxin family proteins can rearrange the disulfide bonds to the different pair of cysteine residues, which is called isomerization, and the isomerization of disulfides is important for the correct folding to the mature form of proteins.

PDI is one of the most abundant oxidoreductase in the ER possessing four thioredoxin like domains, designated as a, b, b′, and a′ domains (Darby et al., 1996; Freedman et al., 1998; Givol et al., 1964; Kemmink et al., 1997). The a and a′ domains contain CGHC motif on the α helix, both of which are active as the oxidoreductase. However, b and b′ domains do not contain CXXC motif and inactive in the oxidoreductase activity.

The oxidative reaction of the substrate with reduced thiol pairs is catalyzed by the PDI with oxidized CXXC motif where the two cysteines of CXXC are disulfide bonded (Schwaller et al., 2003; Walker and Gilbert, 1997). PDI reduced by catalyzing this oxidative reaction is again oxidized by Ero1, which is a flavin-containing protein in the ER (Cabibbo et al., 2000; Pagani et al., 2000). A number of potential substrates of PDI have been identified by cross-linking strategy, followed by immunoprecipitation (Ashworth et al., 1999; Klappa et al., 1998; Miyaishi et al., 1998; Roth, 1987; Vandenbroeck et al., 2006), which revealed that PDI is involved in the maturation of several proteins, including procollagen and thyroglobulin, via its oxidative activity (Di Jeso et al., 2005; Forster and Freedman, 1984; Koivu and Myllyla, 1987). Moreover, PDI, P5, and other ER-folding factors assemble into the large complex, which serve as a folding machinery for the productive-folding pathway in the ER of living cells (Meunier et al., 2002).

ERp57 is also an abundant ER-resident PDI family protein. The most distinct feature of ERp57 is the interaction with ER lectin-like chaperone calnexin (CNX) and calreticulin (CRT) (Frickel et al., 2004; Pollock et al., 2004; Russell et al., 2004) (see Section 2.3). Glycosylated substrates are concomitantly folded and Oxidised/isomerised by CNX/CRT and ERp57, respectively (Molinari and Helenius, 1999; Oliver et al., 1997). Actually, the molecular maturation of major histocompatibility complex (MHC) class I heavy chain and CD1δ were delayed when cells were treated with ERp57 siRNA or with castanospermine, an inhibitor for glucosidase I and II, suggesting the inevitable role of the ERp57 and CNX/CRT complex in the productive folding of newly synthesized proteins (Kang and Cresswell, 2002; Zhang et al., 2006).

ERp72, also an abundant ER oxidoreductase, is reported to be involved in the productive folding of various substrates (Cotterill et al., 2005; Menon et al., 2007; Meunier et al., 2002; Schaiff et al., 1992; Sorensen et al., 2006). Interestingly, ERp72 and PDI have opposing functions in retention and retrotranslocation of cholera toxin (CT), a toxic agent that crosses the ER membrane to reach the cytosol during intoxication (Forster et al., 2006). PDI promotes the unfolding of CT subunits for the dislocation from ER to cytosol, while ERp72 prevents the dislocation by stabilizing the compactly folded conformation of CT. Such opposing functions were also observed for the retrotranslocation, followed by cytosolic degradation via ubiquitin–proteasome system for endogenous proteins.

As the oxidoreductase reaction is intrinsically the cascade reaction in terms of the transfer of electrons, the reaction might form a network or a cascade among various oxidoreductases and substrates. However, very few information is available now for more than 20 ER oxidoreductases.

2.3. *N*-Glycosylation and lectin-like chaperones–CNX and CRT

N-Glycan consisting of Man5GlcNAc2 is generated as a precursor bound with dolichol phosphate on the cytosolic side of rough ER and flip over across the membrane into the lumen of the ER. After maturing to the core glycan, which consists of $Glc_3Man_9GlcNAc_2$, *N*-glycan is transferred *en bloc* onto the asparagine residues in the consensus sequence (Asn-X-Ser/Thr) of the cotranslationally entering polypeptides by the OST (Caramelo *et al.*, 2004; Kornfeld and Kornfeld, 1985). Mammalian OST is a complex consisting of several proteins, including Ribophorin I, Ribophorin II, STT, and so on, among which STT subunit is important for catalytic activity of OST (Yan and Lennarz, 2005). Once the *N*-glycosylation is completed, glucoses and mannoses on the *N*-glycans are then trimmed by ER glucosidases and mannosidases during the productive folding in the ER.

Both CNX and CRT have spheral β sandwich domain and the long-arm structure with two β strands (Schrag *et al.*, 2001). The β sandwich domain, a lectin domain, generates the high affinity ($K_d = 10^{-5}$–10^{-6} M) for the binding with monoglucose form *N*-glycan. Long-arm structure provides a binding site for ERp57. While CNX is a transmembrane protein which possesses the basic amino acid residues as a retention signal in the ER membrane, CRT is an ER luminal protein with KDEL sequence at the C-terminus as an ER retention signal. CNX and CRT bind to Ca^{2+} with high affinity, and partially contribute for the retention of Ca^{2+} as a Ca^{2+} store in the ER. CRT knockout mice died in the fetal life due to the fatal cardiac defect (Mesaeli *et al.*, 1999). Surprisingly, CNX is not essential for mouse survival because the knockout of CNX in mouse was carried to full term, although 50% died within 48 h after birth and showed motor disorders, associated with a dramatic loss of large myelinated nerve fibers (Denzel *et al.*, 2002).

CNX and CRT recognize the monoglucose form of *N*-glycans on the substrates via lectin domain. The monoglucose form of *N*-glycans is generated either by the trimming of glucose with ER glucosidase I and II, or by the reglucosylation of the $Man_9GlcNAc_2$ form of *N*-glycans by UDP-glucose: glycoprotein glucosyltransferase (UGGT). CNX and CRT accelerate the folding of the monoglucose form of glycoproteins as molecular chaperones, and the trimming of the final glucose by glucosidase II causes the release of the substrates from the CNX. If the released substrates are correctly folded, they are secreted through the central secretory pathway. However, the substrates have not been fully or correctly folded yet, they are reglucosylated by UGGT and become the target of CNX and CRT for the next folding attempt, suggesting that UGGT recognizes the exposure of hydrophobic region on proteins like molecular chaperones. These CNX/CRT cycles are important not only for productive folding of the glycoproteins but also for the quality control for the misfolded glycoproteins.

3. ERAD: A Strategy for ERQC Mechanism

3.1. Protection system for ER stress –UPR

ER stress is characterized as the accumulation of misfolded proteins in the ER, and once ER stress occurs, cells respond to such adverse circumstances by triggering the UPR. UPR is triggered by the expression of genetically mutated proteins in the ER, modification of newly synthesized proteins by the treatment with glycosylation inhibitors such as tunicamycin, and disturbance of ER homeostasis such as Ca ion and redox environment. In some cases, overloaded expression of normal secretory proteins even causes the ER stress. Actually, ER stress occurs consistently in the pancreas where a lot of various digestive enzymes are actively synthesized in and secreted through the ER.

UPR which is conserved from yeast to human is a representative strategy for dealing with the adverse situation caused in the ER such as the accumulation of unfolded/misfolded proteins (Harding *et al.*, 2002; Mori, 2000; Patil and Walter, 2001; Schroder and Kaufman, 2005). When misfolded proteins accumulate in the ER, eukaryotic cells convey its information to the nucleus and activate the transcription for target genes. Most target genes of UPR are engaged in the protection mechanism by releasing the adverse effects caused by the ER stress. For example, the induction of molecular chaperons and folding enzymes in ER promotes the refolding of misfolded proteins and prevents the cells from toxic aggregation formation.

In the mammalian cells, three ER membrane proteins, PERK, ATF6, and IRE1, work as sensors for ER stress. Once the misfolded proteins are accumulated in the ER, ER molecular chaperones, including BiP, are recruited to the misfolded proteins to prevent the aggregate formation. In the usual condition, BiP is bound to these sensor proteins and maintains those as inactive state. However, when ER stress occurs, the release of BiP from those sensor proteins activates themselves by different ways.

When BiP is liberated from PERK by ER stress, PERK is activated via the dimerization and phosphorylates the translation initiation factor eIF2a, causing the termination of translation. When the activation of ATF6 after ER stress is achieved by the intramembranous cleavage of ATF6, released p50 cytosolic fragment of ATF6 is transported to the nucleus and activate the transcription of a set of genes, including ER molecular chaperones and oxidoreductases. Thus, translation attenuation by PERK pathway keeps the folding capacity in the ER by suppressing the further synthesis of misfolding-prone proteins, and activation of the synthesis of ER molecular chaperones and oxidoreductases by ATF6 pathway promotes the refolding and/or renaturation of misfolded proteins in the ER. However, if these two strategies are not sufficient enough to circumvent the adverse conditions caused by ER stress, cells respond to it by the third strategy.

The third sensor on ER membrane is IRE1 (Inositol requirement I) (Iwawaki et al., 2001; Tirasophon et al., 1998; Wang et al., 1998), the activation of which is performed via the self-oligomerization, followed by the phosphorylation of cytosolic RNase domain. After the splicing of *xbp1* (x-box binding protein 1) mRNA precursor, the transcription of which is activated by the transcription factor p50 ATF6 after ER stress, XBP1 activates the transcription of genes harboring a cis-acting element, unfolded protein response element (UPRE), at the promoter region(Calfon et al., 2002; Yoshida et al., 2001). UPRE-harboring genes encode various ERAD components, including HRD1, EDEM and Derlin-2, and Derlin-3, and these factors facilitate the disposal of terminally misfolded proteins from the ER (Calfon et al., 2002; Oda et al., 2006).

3.2. Degradation signals and its recognition

How are the proteins to be discarded through ERAD pathway discriminated from the folding intermediates, and what is the signal for degradation? Precise dictation of degradation signal for ERAD is particularly important because of the massive capacity of protein synthesis in the ER. In the case of N-glycosylated proteins, N-glycans act as the degradation signal and are recognized by lectin-like proteins in the ER.

After removal of the third glucose from core N-linked oligosaccharide by glucosidase II, the substrates are released from CNX/CRT. If the substrates acquire the properly folded conformation, the substrates are transported to the Golgi through vesicular transport (Section 2.3). However, if the substrates are terminally misfolded in spite of repeated folding attempts through CNX/CRT cycles, mannose trimming from the middle branch of $Man_9GlcNAc_2$ by ER α1,2-mannosidase I (ER ManI) occurs and provides a degradation signal which consists of $Man_8GlcNAc_2$ isomer B (Man8B). Accessibility of UGGT to the Man8B form of glycoprotein is much lowered compared to that of Man9 form, suggesting that rechallenge for folding by reglucosylation and by entering the CNX/CRT cycle is very limited after mannose trimming.

When the mannose trimming was impaired in the ER ManI-deficient yeast mutant, the ERAD for N-glycosylated substrates was considerably delayed (Jakob et al., 1998), which suggests that the Man8B form serves as the degradation signal for glycoprotein ERAD. This finding anticipated the existence of some lectin-like molecule(s) serving the recognition mechanism for sorting the substrates into degradation pathway after mannose trimming. EDEM (ER-degradation enhancing α-mannosidase-like protein), which is induced by ER stress, was discovered as a putative Man8B lectin (Hosokawa et al., 2001; Jakob et al., 2001; Nakatsukasa et al., 2001), which received the terminally misfolded substrates from CNX (Molinari et al., 2003; Oda et al., 2003). In mammals, EDEM family is comprised of three members, EDEM1, -2, and -3, all of which share a conserved α-mannosidase-like domain with ER mannosidase I and a collection of Golgi-resident enzymes (Clerc et al., 2009; Hirao et al., 2006; Hosokawa et al., 2009; Mast et al., 2005; Olivari et al., 2005, 2006). These EDEM family proteins accelerate ERAD of N-glycosylated substrates in the mannose trimming-dependent manner.

EDEM1 was reported to have modest mannosidase activity, which trims off mannose from A branch and C branch in N-linked oligosaccharides (Hosokawa et al., 2009; Olivari et al., 2006). However, this enzymatic activity in EDEM1 is not essential for promoting ERAD. EDEM2 was reported to possess no enzymatic activity in spite of its ERAD-enhancing activity (Mast et al., 2005; Olivari et al., 2005). In contrast, EDEM3 has a mannose-trimming activity which is required for the activation of glycoprotein ERAD (Clerc et al., 2009; Hirao et al., 2006). These results provide an insight into the functional redundancy among EDEM families, while the mannosidase and the lectin activities of EDEM should be more clearly demonstrated by in vitro assay using purified recombinant EDEM proteins, which has not succeeded at present because of the difficulty in expression/purification of EDEM.

Another lectin-like molecule, Yos9p (yeast osteosarcoma 9), was also reported to be involved in the activation of glycoprotein ERAD by genome-wide screening in yeast (Buschhorn et al., 2004). Yos9p contains a mannose-6-phosphate homolog (MRH) domain which is a putative lectin domain. Yos9p bound with misfolded substrates bearing the Man8B and $Man_5GlcNAc_2$ (Man5) oligosaccharides and accelerated the elimination of these forms of glycoproteins from the ER (Bhamidipati et al., 2005; Kim et al., 2005; Szathmary et al., 2005). In vitro study by frontal affinity chromatography showed that Yos9 binds to N-glycan exposing a terminal α1,6-linked mannose, such as $Man_7GlcNAc_2$ (Man7) or Man5 (Quan et al., 2008), while another report showed that Yos9 could recognize misfolded substrates independently of sugar chain (Bhamidipati et al., 2005). These results suggest that Yos9 may recognize both the specific glycan structures as lectin and the protein portion like chaperones.

Recently, the mammalian homologs of Yos9, OS9, and XTP3B were reported to be involved in the activation of ERAD (Christianson et al., 2008; Hosokawa et al., 2008). OS9 has MRH domain and interacts with GRP94 (94 kDa glucose-regulated protein), an ER-resident Hsp90 family protein (Christianson et al., 2008). In mammalian cells, OS9 promotes the transfer of α1-antitrypsin variant NHK to HRD1 complex which is a member comprising ERAD machinery (Section 3.5).

3.3. ERAD for nonglycosylated proteins

In the case of N-glycosylated substrates, N-glycan and its trimming provide important signals for disposal of misfolded proteins from the ER. What is the signal for degradation of nonglycosylated substrates? Although only few information on the mechanistic analysis on the quality control of nonglycosylated proteins is available, the involvement of BiP in the recognition of misfolded substrates through the exposed hydrophobic surfaces was reported (Blond-Elguindi et al., 1993). As described in Section 2.1, BiP binding and releasing for the substrates are ATP-dependent. This cycle promotes the productive folding and prevents the aggregate formation. In the yeast, Kar2p (yeast BiP) was shown to keep the solubility of substrates and to be engaged in ERAD. Disruption of Jem1p and Scj1, members of Hsp40 family in ER, caused the defect in ERAD, suggesting that the regulation of Kar2 by Jem1p and Scj1p as cochaperones of Kar2 is necessary for the enhancement of ERAD (Fewell et al., 2001; Nishikawa et al., 2001). These results suggest that BiP (Kar2p) might be one of the key factors for recognition and recruitment of terminally misfolded substrates to ERAD pathway.

3.4. Dislocon channels for retrotranslocation

Retrotranslocation of the substrates from the ER to the cytosol through dislocation channel in the ER membrane is an essential process for ERAD. Newly synthesized polypeptides are cotranslationally inserted into the ER through Sec61 channel (Kostova and Wolf, 2003; Matlack et al., 1998). Sec61 translocon channel consists of Sec61α, Sec61β, and Sec61γ as a hetero trimmer, among which Sec61β and Sec61γ are single pass transmembrane proteins, and Sec61α is a multispanning transmembrane protein (Johnson and van Waes, 1999). This complex interacts with TRAM (translocation-associated membrane protein), which is a multispanning transmembrane protein forming the core structure of the translocon channel.

The Sec61 channel is also proposed to compose the dislocon channel for ERAD based on some experimental evidences; Sec61 channel was shown to be involved in the retrotranslocation of MHC class I (Wiertz, 1996) and subunit of CT (Schmitz et al., 2000), and the mutation of a component of Sec61 channel caused the delay of the degradation of soluble and membrane substrates (Pilon et al., 1997; Plemper et al., 1997). If the Sec61 channel also works as the dislocon channel for ERAD, we wonder how on earth the Sec61 channel distinguishes its function for the retrotranslocation from that for the translocation.

The Derlin family proteins are another candidate for the components of the retrotranslocation channel. It is to be noted that the Derlin family proteins are included in the ERAD machinery (Section 3.5) and are involved in ERAD (Katiyar et al., 2005; Lilley and Ploegh, 2004, 2005; Oda et al., 2006; Okuda-Shimizu and Hendershot, 2007; Schulze et al., 2005; Wahlman et al., 2007; Ye et al., 2004, 2005; Younger et al., 2006).

3.5. ERAD machinery for various substrates and ubiquitination

The expression of a variety of ERAD components is upregulated via the IRE1–XBP1 pathway of the UPR. In general, ERAD complexes contain E3 ubiquitin ligases, Derlin family proteins, and other accessory proteins to recruit the substrates to 26S proteasome in the cytosol. Three different pathways of ERAD have been proposed in yeast as ERAD-L, ERAD-C, and ERAD-M in response to where the substrates are misfolded (Carvalho et al., 2006; Denic et al., 2006; Gauss et al., 2006). ERAD-L targets the substrates misfolded in the ER lumen to the ERAD pathway, where Hrd1p forms E3 ubiquitin ligase complex with Hrd3p (yeast homolog of mammalian SEL1), and this core complex then interacts with Yos9p (yeast OS9), Kar2p (yeast BiP), Der1p (yeast Derlin1 family), Usa1p, and Ubx2p. Yos9p and Kar2p are assumed to recognize the ERAD substrate for recruiting it to the disposal pathway, and Der1p to compose the dislocon channel (Section 3.4) (Lilley and Ploegh, 2004; Ye et al., 2004) (Fig. 14.2A). UBL-domain-containing Usa1p connects Hrd1p/Hrd3p with Der1p, and ubiquitin regulatory X (UBX) domain-containing Ubx2p recruits 26S proteasome to the cytosolic side of the Hrd1–Hrd3 complex via Cdc48p. Cdc48p, a AAA^+ ATPase, is believed to withdraw the substrates from the ER lumen to the cytosol in an ATPase-dependent manner. During the ER stress, these structural components assemble as one complex and fulfill the efficient successive functions including dislocation, ubiquitination, and degradation by proteasome.

However, transmembrane substrates containing misfolded portions in the cytosolic and membrane domains are degraded via ERAD-C and ERAD-M pathways, respectively. In ERAD-C complex, ubiquitin ligase Doa10p interacts with Ubx2p instead of Hrd1p/Hrd3p in ERAD-L, and Doa10p–Ubx2p complex interacts with proteasome via Cdc48p (Fig. 14.2B). ERAD-M contains the Hrd1p/Hrd3p E3 ubiquitin ligase complex similar to ERAD-L (Carvalho et al., 2006) (Fig. 14.2A). Thus, the ERAD machinery on ER membrane exchanges the components to deal with the variety of substrates.

In mammalian cells, HRD1 (mammalian homolog of yeast Hrd1p) and SEL1L (mammalian homolog of Hrd3p) form the E3 ubiquitin ligase complex on ER membrane, which also contains Derlin-2 and Derlin-3 serving in the recruitment of p97, a mammalian homolog of Cdc48p, to the HRD1/SEL1L complex (Fig. 14.2C). Additionally, Derlin-2 and Derlin-3 link EDEM to p97 and 26S proteasome (Oda et al., 2006). Another mammalian homolog of yeast Hrd1p, gp78, also interacts with p97 and E3 ubiquitin ligase RMA, and promotes the ERAD of mutant CFTR (cystic fibrosis transmembrane conductance regulator) (Morito et al., 2008). OS9 and XTP3B are the mammalian homolog of yeast Yos9p, associated with HRD1–SEL1 complex, and involved in the recognition

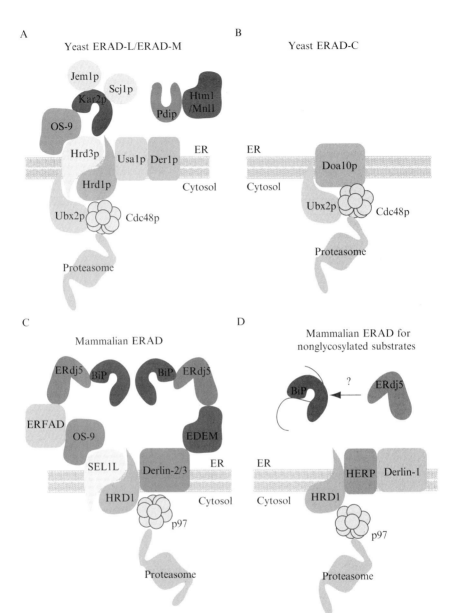

Figure 14.2 ERAD machinery around the dislocon channels (A, B) ERAD complexes in yeast. (A) Hrd1p–Hrd3p complex for ERAD-L and ERAD-M. (B) Doa10p complex for ERAD-C. Yos9p recognizes N-glycans on misfolded proteins, which are then ubiquitinated by Hrd1p in the cytosol and degraded by the proteasome. (C, D) Mammalian ERAD complexes. (C) HRD1–SEL1 complex. OS-9 (and possibly, XTP3-B) recognizes N-glycans on misfolded proteins, which are then ubiquitinated by HRD1. EDEM also recognizes N-glycans on misfolded proteins and recruits them to Derlin-2/3 and p97. ERdj5 interacts with EDEM and/or BiP and promotes dislocation of misfolded proteins by cleaving their disulfide bonds. (D) HRD1–HERP complex for nonglycosylated proteins. BiP recognizes misfolded nonglycoproteins and recruits them to the HERP complex that contains p97 and Derlin-1. It is possible that ERdj5 or other Hsp40 family regulates the dislocation process.

and recruitment of degradation substrate (Christianson *et al.*, 2008; Hosokawa *et al.*, 2008). Derlin-1, the other Derlin family protein, associates with HERP, which is a ubiquitin-like membrane protein induced by ER stress, and is essential for the dislocation and degradation of human MHC class I heavy chain (Lilley and Ploegh, 2004; Ye *et al.*, 2004) (Fig. 14.2D). Interestingly, this HERP complex was shown to interact with nonglycosylated proteins following BiP binding (Okuda-Shimizu and Hendershot, 2007), and thus HERP complex may specifically facilitate the degradation of misfolded nonglycosylated proteins.

3.6. Reduction of disulfide bonds for ERAD

Newly synthesized polypeptides entering the ER are cotranslationally *N*-glycosylated by OGT and oxidized by ER-resident oxidoreductases, including Ero1 and PDI. In some cases, proteins make oligomes through the disulfide bonds and/or by hydrophobic interactions, which are assumed to be difficult in being retrotranslocated through the narrow dislocon channel. In yeast, Kar2p (yeast BiP) and Pdi p (yeast PDI) are reported to keep the solubility of misfolded substrates and prevent the formation of oligomer or aggregates until dislocation (Gillece *et al.*, 1999; Lumb and Bulleid, 2002; Molinari *et al.*, 2002; Tsai and Rapoport, 2002). The ternary structure stabilized by intramolecular disulfide bonds may also interfere with the efficient dislocation of the substrates through the dislocation channel.

Because the redox state in the ER is more oxidative compared with that in the cytosol, the environment is suitable for the formation of disulfide bonds which stabilize the ternary structures of proteins and assemble the subunits of proteins to make oligomers. Such disulfide bond formation in the oxidative environment may be disadvantageous for dislocation of the substrates through dislocon channel, which was exemplified by the fact that treatment of the cells with oxidizing reagent inhibited the ERAD and that reducing reagent dithiothreitol facilitated the ERAD of intermolecular disulfide-containing proteins (Hosokawa *et al.*, 2006; Mancini *et al.*, 2000; Tortorella *et al.*, 1998). These results suggest the importance of reductive cleavage of intra- or intermolecular disulfide bonds in spite of the oxidative environment of the ER.

Recently, we have identified an ER-resident oxidoreductase ERdj5 as an EDEM-binding protein (Ushioda *et al.*, 2008). ERdj5 is upregulated by UPR and has a J domain at the N-terminus and four thioredoxin-like domains with active CXXC motifs. N-terminal J domain serves as a platform interacting with BiP and accelerates the ATPase activity of BiP (Cunnea *et al.*, 2003; Hosoda *et al.*, 2003). In an *in vitro* assay, we detected the reducing activity of purified ERdj5 in the ER redox condition. We used model substrates for showing the involvement of ERdj5 in the ERAD, NHK variant of α1-antitrypsin (NHK) and the immunoglobulin M joint chain

(J chain), both of which have the intermolecular disulfide bonds making dimmers and multimers, respectively. When overexpressed in mammalian cells, ERdj5s accelerated the ERAD of these substrates by facilitating the cleavage of disulfide bonds. When cells were treated with kifunensine, an ER mannosidase inhibitor, ERdj5 could not promote the degradation of glycosylated substrates NHK, suggesting that ERdj5 cooperates with EDEM, because the inhibition of mannose trimming by kifunensine prevents the substrates to be recognized by EDEM.

By incorporating the new factor ERdj5 as the reductase, we can draw the sequential pathway in the ERAD process as follows: (1) terminally misfolded glycoproteins are transferred from CNX to EDEM after the trimming of mannose residues from Man9GlcNAc2 core oligosaccharide to the M8B form of N-glycans, (2) EDEM binds to the fourth thioredoxin domain of ERdj5 (Ushioda et al., unpublished observation) and presents the substrates to ERdj5, (3) ERdj5 reduces the disulfide bonds of the substrates, making an extended form of polypeptides, (4) the ATP form of BiP binds to the J domain of ERdj5 and accepts the substrates from ERdj5, (5) the J domain of ERdj5 accelerates the ATPase activity of BiP and converts BiP from the ATP-bound form to the ADP-bound form, and (6) BiP in the ADP form binds the substrates more tightly and presumably pulls out the substrates from the ERdj5–EDEM complex and recruits the substrates to the dislocon channel (Fig. 14.3).

ERdj5 is conserved in mammals, vertebrates, and nematodes, but not in yeast. In yeast, Htm1p/Mnl1p (yeast homolog of EDEM) interacts specifically with the Pdi1p (yeast homolog of PDI). This interaction has been shown to be required for the acceleration of ERAD by Htm1p/Mnl1, suggesting that Pdi1p might functionally substitute ERdj5 as the reductase in yeast (Sakoh-Nakatogawa et al., 2009).

More recently, it has been shown that ERFAD, which binds with FAD and NADPH, interacts with ERdj5, SEL1, and OS-9 and promotes the ERAD (Riemer et al., 2009). Although the significance of these interactions is still unclear, it suggests that ERdj5-reducing activity may be one of the major players for the ERAD.

4. Conclusion

Following recent studies, the details of the ERAD systems have been revealed by degrees. As the ER is the major organelle of the protein synthesis and folding, degradation should be closely monitored by several ways. Particularly, it is of importance to distinguish the substrates for disposal from folding intermediates. Trimming of N-oligosaccharides provides not only the signal for the productive folding which is recognized by

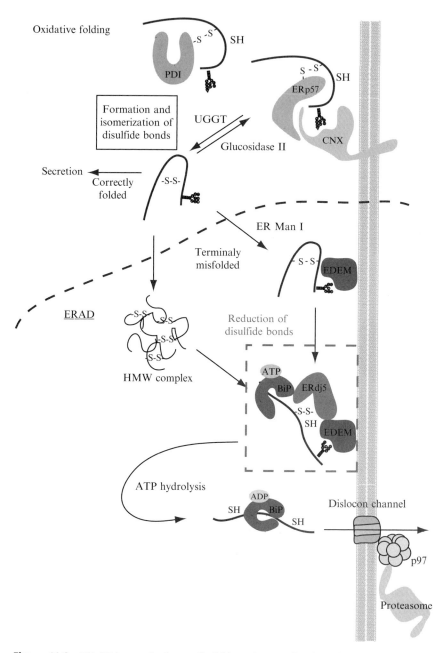

Figure 14.3 ERdj5 is required as a disulfide reductase for degradation of misfolded proteins in ER. Oxidative folding and glycosylation of nascent polypeptides are promoted by ER-resident oxidoreductases, molecular chaperones, OST, and glucosidases. After trimming of glucose and mannose residues, terminally misfolded proteins with the

lectin-like chaperones CNX and CRT, but also the degradation signals which is recognized by lectin-like ERAD components such as EDEM and Yos9p. BiP is also assumed to be a player for the ERAD of nonglycosylated substrates by keeping the solubility for efficient dislocation. Upregulation of several ERAD components by the UPR, particularly via IRE1–XBP1 pathway, makes the elimination of misfolded proteins accumulated under ER stress more efficient to accommodate such an adverse situation.

The oxidative environment of the ER is suitable for efficient disulfide bond formation for productive folding of newly synthesized polypeptides. However, we underscore here the importance of reduction of preformed inter- and intramolecular disulfide bonds even in such an oxidative ER environment to facilitate the ERAD. In this regard, ERdj5, which was identified for the first time as the ER-resident disulfide reductase, plays a fundamental role in accelerating the ERAD by cleaving the disulfide bonds and making the extended form of polypeptides, which is beneficial for efficient retrotranslocation through the dislocon channel. It is worth noting that ERdj5 makes a complex with EDEM as a monitor for degradation and BiP as a chaperone possessing an ability to keep the substrates in an unfolded state. Thus, the supramolecular complex comprising EDEM, ERdj5, and BiP may have a central role in the ERAD pathway.

A lot of questions still remain unaddressed. How is the mechanism of substrate transfer in the case of nonglycosylated proteins? Is ERdj5 also necessary for nonglycosylated proteins? How and from what molecules does ERdj5 acquire the reductive forces in the oxidative ER environment? Are other ER oxidoreductases also involved in reducing the disulfides of misfolded proteins for accelerating ERAD? More generally, there are more than 20 PDI-like oxidoreductases in the ER. However, information on the interactions of these proteins is not available, except for a few proteins. Proteome analysis may be necessary for answering this question. All these questions are to be addressed in the future to enable a better understanding of the mechanism of quality control in the ER.

Man8B form of N-linked oligosaccharide are transferred from calnexin to EDEM. ERdj5 bound to EDEM then reduces the disulfide bonds of the misfolded proteins, resulting in dissociation of the covalent multimeric subunits. ERdj5 binds to the ATP-bound form of BiP through its DnaJ domain and activates ATP hydrolysis of BiP, resulting in the dissociation of BiP from ERdj5. In turn, the ADP-bound form of BiP binds substrates strongly, presumably pulls out the substrates from the EDEM–ERdj5 complex, and holds them in a dislocation-competent state until they are transferred to the retrotranslocation channel.

REFERENCES

Ashworth, J. L., Kelly, V., Wilson, R., Shuttleworth, C. A., and Kielty, C. M. (1999). Fibrillin assembly: Dimer formation mediated by amino-terminal sequences. *J. Cell Sci.* **112,** 3549–3558.
Bhamidipati, A., Denic, V., Quan, E. M., and Weissman, J. S. (2005). Exploration of the topological requirements of ERAD identifies Yos9p as a lectin sensor of misfolded glycoproteins in the ER lumen. *Mol. Cell* **19,** 741–751.
Bies, C., Blum, R., Dudek, J., Nastainczyk, W., Oberhauser, S., Jung, M., and Zimmermann, R. (2004). Characterization of pancreatic ERj3p, a homolog of yeast DnaJ-like protein Scj1p. *Biol. Chem.* **385,** 389–395.
Blond-Elguindi, S., Cwirla, S. E., Dower, W. J., Lipshutz, R. J., Sprang, S. R., Sambrook, J. F., and Gething, M. J. (1993). Affinity panning of a library of peptides displayed on bacteriophages reveals the binding specificity of BiP. *Cell* **75,** 717–728.
Brodsky, J. L., Goeckeler, J., and Schekman, R. (1995). BiP and Sec63p are required for both co- and posttranslational protein translocation into the yeast endoplasmic reticulum. *Proc. Natl. Acad. Sci. USA* **92,** 9643–9646.
Buschhorn, B. A., Kostova, Z., Medicherla, B., and Wolf, D. H. (2004). A genome-wide screen identifies Yos9p as essential for ER-associated degradation of glycoproteins. *FEBS Lett.* **577,** 422–426.
Cabibbo, A., Pagani, M., Fabbri, M., Rocchi, M., Farmery, M. R., Bulleid, N. J., and Sitia, R. (2000). ERO1-L, a human protein that favors disulfide bond formation in the endoplasmic reticulum. *J. Biol. Chem.* **275,** 4827–4833.
Calfon, M., Zeng, H., Urano, F., Till, J. H., Hubbard, S. R., Harding, H. P., Clark, S. G., and Ron, D. (2002). IRE1 couples endoplasmic reticulum load to secretory capacity by processing the XBP-1 mRNA. *Nature* **415,** 92–96.
Caramelo, J. J., Castro, O. A., De Prat-Gay, G., and Parodi, A. J. (2004). The endoplasmic reticulum glucosyltransferase recognizes nearly native glycoprotein folding intermediates. *J. Biol. Chem.* **279,** 46280–46285.
Carvalho, P., Goder, V., and Rapoport, T. A. (2006). Distinct ubiquitin–ligase complexes define convergent pathways for the degradation of ER proteins. *Cell* **126,** 361–373.
Christianson, J. C., Shaler, T. A., Tyler, R. E., and Kopito, R. R. (2008). OS-9 and GRP94 deliver mutant α1-antitrypsin to the Hrd1/SEL1L ubiquitin ligase complex for ERAD. *Nat. Cell Biol.* **10,** 272–282.
Clerc, S., Hirsch, C., Oggier, D. M., Deprez, P., Jakob, C., Sommer, T., and Aebi, M. (2009). Htm1 protein generates the N-glycan signal for glycoprotein degradation in the endoplasmic reticulum. *J. Cell Biol.* **184,** 159–172.
Cotterill, S. L., Jackson, G. C., Leighton, M. P., Wagener, R., Mäkitie, O., Cole, W. G., and Briggs, M. D. (2005). Multiple epiphyseal dysplasia mutations in MATN3 cause misfolding of the A-domain and prevent secretion of mutant matrilin-3. *Hum. Mutat.* **26,** 557–565.
Cunnea, P. M., Miranda-Vizuete, A., Bertoli, G., Simmen, T., Damdimopoulos, A. E., Hermann, S., Leinonen, S., Huikko, M. P., Gustafsson, J. A., Sitia, R., and Spyrou, G. (2003). ERdj5, an endoplasmic reticulum (ER)-resident protein containing DnaJ and thioredoxin domains, is expressed in secretory cells or following ER stress. *J. Biol. Chem.* **278,** 1059–1066.
Darby, N. J., Kemmink, J., and Creighton, T. E. (1996). Identifying and characterizing a structural domain of protein disulfide isomerase. *Biochemistry* **35,** 10517–10528.
Denic, V., Quan, E. M., and Weissman, J. S. (2006). A luminal surveillance complex that selects misfolded glycoproteins for ER-associated degradation. *Cell* **126,** 349–359.

Denzel, A., Molinari, M., Trigueros, C., Martin, J. E., Velmurgan, S., Brown, S., Stamp, G., and Owen, M. J. (2002). Early postnatal death and motor disorders in mice congenitally deficient in calnexin expression. *Mol. Cell. Biol.* **22,** 7398–7404.

Di Jeso, B., Park, Y. N., Ulianich, L., Treglia, A. S., Urbanas, M. L., High, S., and Arvan, P. (2005). Mixed-disulfide folding intermediates between thyroglobulin and endoplasmic reticulum resident oxidoreductases ERp57 and protein bisulfide isomerase. *Mol. Cell. Biol.* **25,** 9793–9805.

Dong, M., Bridges, J. P., Apsley, K., Xu, Y., and Weaver, T. E. (2008). ERdj4 and ERdj5 are required for endoplasmic reticulum-associated protein degradation of misfolded surfactant protein C. *Mol. Biol. Cell* **19,** 2620–2630.

Dudek, J., Greiner, M., Mller, A., Hendershot, L. M., Kopsch, K., Nastainczyk, W., and Zimmermann, R. (2005). ERj1p has a basic role in protein biogenesis at the endoplasmic reticulum. *Nat. Struct. Mol. Biol.* **12,** 1008–1014.

Fewell, S. W., Travers, K. J., Weissman, J. S., and Brodsky, J. L. (2001). The action of molecular chaperones in the early secretory pathway. *Annu. Rev. Genet.* **35,** 149–191.

Forster, S. J., and Freedman, R. B. (1984). Catalysis by protein disulphide-isomerase of the assembly of trimeric procollagen from procollagen polypeptide chains. *Biosci. Rep.* **4,** 223–229.

Forster, M. L., Sivick, K., Park, Y. N., Arvan, P., Lencer, W. I., and Tsai, B. (2006). Protein disulfide isomerase-like proteins play opposing roles during retrotranslocation. *J. Cell Biol.* **173,** 853–859.

Freedman, R. B., Gane, P. J., Hawkins, H. C., Hlodan, R., McLaughlin, S. H., and Parry, J. W. L. (1998). Experimental and theoretical analyses of the domain architecture of mammalian protein disulphide-isomerase. *Biol. Chem.* **379,** 321–328.

Frickel, E. M., Frei, P., Bouvier, M., Stafford, W. F., Helenius, A., Glockshuber, R., and Ellgaard, L. (2004). ERp57 is a multifunctional thiol-disulfide oxidoreductase. *J. Biol. Chem.* **279,** 18277–18287.

Gauss, R., Jarosch, E., Sommer, T., and Hirsch, C. (2006). A complex of Yos9p and the HRD ligase integrates endoplasmic reticulum quality control into the degradation machinery. *Nat. Cell Biol.* **8,** 849–854.

Ghaemmaghami, S., Huh, W. K., Bower, K., Howson, R. W., Belle, A., Dephoure, N., O'Shea, E. K., and Weissman, J. S. (2003). Global analysis of protein expression in yeast. *Nature* **425,** 737–741.

Gillece, P., Luz, J. M., Lennarz, W. J., De La Cruz, F. J., and Romisch, K. (1999). Export of a cysteine-free misfolded secretory protein from the endoplasmic reticulum for degradation requires interaction with protein disulfide isomerase. *J. Cell Biol.* **147,** 1443–1456.

Givol, D., Goldberger, R. F., and Anfinsen, C. B. (1964). Oxidation and disulfide interchange in the reactivation of reduced ribonuclease. *J. Biol. Chem.* **239,** 3114–3116.

Harding, H. P., Calfon, M., Urano, F., Novoa, I., and Ron, D. (2002). Transcriptional and translational control in the Mammalian unfolded protein response. *Annu. Rev. Cell Dev. Biol.* **18,** 575–599.

Hirao, K., Natsuka, Y., Tamura, T., Wada, I., Morito, D., Natsuka, S., Romero, P., Sleno, B., Tremblay, L. O., Herscovics, A., Nagata, K., and Hosokawa, N. (2006). EDEM3, a soluble EDEM homolog, enhances glycoprotein endoplasmic reticulum-associated degradation and mannose trimming. *J. Biol. Chem.* **281,** 9650–9658.

Hosoda, A., Kimata, Y., Tsuru, A., and Kohno, K. (2003). JPDI, a novel endoplasmic reticulum-resident protein containing both a BiP-interacting J-domain and thioredoxin-like motifs. *J. Biol. Chem.* **278,** 2669–2676.

Hosokawa, N., Wada, I., Hasegawa, K., Yorihuzi, T., Tremblay, L. O., Herscovics, A., and Nagata, K. (2001). A novel ER alpha-mannosidase-like protein accelerates ER-associated degradation. *EMBO Rep.* **2,** 415–422.

Hosokawa, N., Wada, I., Natsuka, Y., and Nagata, K. (2006). EDEM accelerates ERAD by preventing aberrant dimer formation of misfolded α1-antitrypsin. *Genes Cells* **11**, 465–476.

Hosokawa, N., Wada, I., Nagasawa, K., Moriyama, T., Okawa, K., and Nagata, K. (2008). Human XTP3-B forms an endoplasmic reticulum quality control scaffold with the HRD1-SEL1L ubiquitin ligase complex and BiP. *J. Biol. Chem.* **283**, 20914–20924.

Hosokawa, N., Kamiya, Y., Kamiya, D., Kato, K., and Nagata, K. (2009). Human OS-9, a lectin required for glycoprotein endoplasmic reticulum-associated degradation, recognizes mannose-trimmed N-glycans. *J. Biol. Chem.* **284**, 17061–17068.

Iwawaki, T., Hosoda, A., Okuda, T., Kamigori, Y., Nomura-Furuwatari, C., Kimata, Y., Tsuru, A., and Kohno, K. (2001). Translational control by the ER transmembrane kinase/ribonuclease IRE1 under ER stress. *Nat. Cell Biol.* **3**, 158–164.

Jakob, C. A., Burda, P., Roth, J., and Aebi, M. (1998). Degradation of misfolded endoplasmic reticulum glycoproteins in *Saccharomyces cerevisiae* is determined by a specific oligosaccharide structure. *J. Cell Biol.* **142**, 1223–1233.

Jakob, C. A., Bodmer, D., Spirig, U., Battig, P., Marcil, A., Dignard, D., Bergeron, J. J. M., Thomas, D. Y., and Aebi, M. (2001). Htm1p, a mannosidase-like protein, is involved in glycoprotein degradation in yeast. *EMBO Rep.* **2**, 423–430.

Johnson, A. E., and van Waes, M. A. (1999). The translocon: A dynamic gateway at the ER membrane. *Annu. Rev. Cell Dev. Biol.* **15**, 799–842.

Kanapin, A. (2003). Mouse proteome analysis. *Genome Res.* **13**, 1335–1344.

Kang, S. J., and Cresswell, P. (2002). Calnexin, calreticulin, and ERp57 cooperate in disulfide bond formation in human CD1d heavy chain. *J. Biol. Chem.* **277**, 44838–44844.

Katiyar, S., Joshi, S., and Lennarz, W. J. (2005). The retrotranslocation protein Derlin-1 binds peptide:N-glycanase to the endoplasmic reticulum. *Mol. Biol. Cell* **16**, 4584–4594.

Kemmink, J., Darby, N. J., Dijkstrat, K., Nilges, M., and Creighton, T. E. (1997). The folding catalyst protein disulfide isomerase is constructed of active and inactive thioredoxin modules. *Curr. Biol.* **7**, 239–245.

Kim, W., Spear, E. D., and Ng, D. T. W. (2005). Yos9p detects and targets misfolded glycoproteins for ER-associated degradation. *Mol. Cell* **19**, 753–764.

Klappa, P., Ruddock, L. W., Darby, N. J., and Freedman, R. B. (1998). The b' domain provides the principal peptide-binding site of protein disulfide isomerase but all domains contribute to binding of misfolded proteins. *EMBO J.* **17**, 927–935.

Koivu, J., and Myllyla, R. (1987). Interchain disulfide bond formation in types I and II procollagen. Evidence for a protein disulfide isomerase catalyzing bond formation. *J. Biol. Chem.* **262**, 6159–6164.

Kornfeld, R., and Kornfeld, S. (1985). Assembly of asparagine-linked oligosaccharides. *Annu. Rev. Biochem.* **54**, 631–664.

Kostova, Z., and Wolf, D. H. (2003). For whom the bell tolls: Protein quality control of the endoplasmic reticulum and the ubiquitin-proteasome connection. *EMBO J.* **22**, 2309–2317.

Liberek, K., Marszalek, J., Ang, D., Georgopoulos, C., and Zylicz, M. (1991). *Escherichia coli* DnaJ and GrpE heat shock proteins jointly stimulate ATPase activity of DnaK. *Proc. Natl. Acad. Sci. USA* **88**, 2874–2878.

Lilley, B. N., and Ploegh, H. L. (2004). A membrane protein required for dislocation of misfolded proteins from the ER. *Nature* **429**, 834–840.

Lilley, B. N., and Ploegh, H. L. (2005). Multiprotein complexes that link dislocation, ubiquitination, and extraction of misfolded proteins from the endoplasmic reticulum membrane. *Proc. Natl. Acad. Sci. USA* **102**, 14296–14301.

Lumb, R. A., and Bulleid, N. J. (2002). Is protein disulfide isomerase a redox-dependent molecular chaperone? *EMBO J.* **21**, 6763–6770.

Mancini, R., Fagioli, C., Fra, A. M., Maggioni, C., and Sitia, R. (2000). Degradation of unassembled soluble Ig subunits by cytosolic proteasomes: Evidence that retrotranslocation and degradation are coupled events. *FASEB J.* **14,** 769–778.

Mast, S. W., Diekman, K., Karaveg, K., Davis, A., Sifers, R. N., and Moremen, K. W. (2005). Human EDEM2, a novel homolog of family 47 glycosidases, is involved in ER-associated degradation of glycoproteins. *Glycobiology* **15,** 421–436.

Matlack, K. E., Mothes, W., and Rapoport, T. A. (1998). Protein translocation: Tunnel vision. *Cell* **92,** 381–390.

Menon, S., Lee, J., Abplanalp, W. A., Yoo, S. E., Agui, T., Furudate, S. I., Kim, P. S., and Arvan, P. (2007). Oxidoreductase interactions include a role for ERp72 engagement with mutant thyroglobulin from the rdw/rdw rat dwarf. *J. Biol. Chem.* **282,** 6183–6191.

Mesaeli, N., Nakamura, K., Zvaritch, E., Dickie, P., Dziak, E., Krause, K. H., Opas, M., MacLennan, D. H., and Michalak, M. (1999). Calreticulin is essential for cardiac development. *J. Cell Biol.* **144,** 857–868.

Meunier, L., Usherwood, Y. K., Tae Chung, K., and Hendershot, L. M. (2002). A subset of chaperones and folding enzymes form multiprotein complexes in endoplasmic reticulum to bind nascent proteins. *Mol. Biol. Cell* **13,** 4456–4469.

Meyer, H. A., Grau, H., Kraft, R., Kostka, S., Prehn, S., Kalies, K. U., and Hartmann, E. (2000). Mammalian Sec61 is associated with Sec62 and Sec63. *J. Biol. Chem.* **275,** 14550–14557.

Miyaishi, O., Kozaki, K. I., Iida, K. I., Isobe, K. I., Hashizume, Y., and Saga, S. (1998). Elevated expression of PDI family proteins during differentiation of mouse F9 teratocarcinoma cells. *J. Cell. Biochem.* **68,** 436–445.

Molinari, M., and Helenius, A. (1999). Glycoproteins form mixed disulphides with oxidoreductases during folding in living cells. *Nature* **402,** 90–93.

Molinari, M., Galli, C., Piccaluga, V., Pieren, M., and Paganetti, P. (2002). Sequential assistance of molecular chaperones and transient formation of covalent complexes during protein degradation from the ER. *J. Cell Biol.* **158,** 247–257.

Molinari, M., Calanca, V., Galli, C., Lucca, P., and Paganetti, P. (2003). Role of EDEM in the release of misfolded glycoproteins from the calnexin cycle. *Science* **299,** 1397–1400.

Mori, K. (2000). Tripartite management of unfolded proteins in the endoplasmic reticulum. *Cell* **101,** 451–454.

Morito, D., Hirao, K., Oda, Y., Hosokawa, N., Tokunaga, F., Cyr, D. M., Tanaka, K., Iwai, K., and Nagata, K. (2008). Gp78 cooperates with RMA1 in endoplasmic reticulum-associated degradation of CFTRΔF508. *Mol. Biol. Cell* **19,** 1328–1336.

Nakatsukasa, K., Nishikawa, S. I., Hosokawa, N., Nagata, K., and Endo, T. (2001). Mnl1p, an α-mannosidase-like protein in yeast *Saccharomyces cerevisiae*, is required for endoplasmic reticulum-associated degradation of glycoproteins. *J. Biol. Chem.* **276,** 8635–8638.

Nishikawa, S. I., Fewell, S. W., Kato, Y., Brodsky, J. L., and Endo, T. (2001). Molecular chaperones in the yeast endoplasmic reticulum maintain the solubility of proteins for retrotranslocation and degradation. *J. Cell Biol.* **153,** 1061–1069.

Oda, Y., Hosokawa, N., Wada, I., and Nagata, K. (2003). EDEM as an acceptor of terminally misfolded glycoproteins released from calnexin. *Science* **299,** 1394–1397.

Oda, Y., Okada, T., Yoshida, H., Kaufman, R. J., Nagata, K., and Mori, K. (2006). Derlin-2 and Derlin-3 are regulated by the mammalian unfolded protein response and are required for ER-associated degradation. *J. Cell Biol.* **172,** 383–393.

Okuda-Shimizu, Y., and Hendershot, L. M. (2007). Characterization of an ERAD pathway for nonglycosylated BiP substrates, which require Herp. *Mol. Cell* **28,** 544–554.

Olivari, S., Galli, C., Alanen, H., Ruddock, L., and Molinari, M. (2005). A novel stress-induced EDEM variant regulating endoplasmic reticulum-associated glycoprotein degradation. *J. Biol. Chem.* **280,** 2424–2428.

Olivari, S., Cali, T., Salo, K. E. H., Paganetti, P., Ruddock, L. W., and Molinari, M. (2006). EDEM1 regulates ER-associated degradation by accelerating de-mannosylation of folding-defective polypeptides and by inhibiting their covalent aggregation. *Biochem. Biophys. Res. Commun.* **349**, 1278–1284.

Oliver, J. D., Van Der Wal, F. J., Bulleid, N. J., and High, S. (1997). Interaction of the thiol-dependent reductase ERp57 with nascent glycoproteins. *Science* **275**, 86–88.

Otero, J. H., Lizak, B., and Hendershot, L. M. (2010). Life and death of a BiP substrate. *Semin. Cell Dev. Biol.* **21**, 472–478.

Pagani, M., Fabbri, M., Benedetti, C., Fassio, A., Pilati, S., Bulleid, N. J., Cabibbo, A., and Sitia, R. (2000). Endoplasmic reticulum oxidoreductin 1-Lβ (ERO1-Lβ), a human gene induced in the course of the unfolded protein response. *J. Biol. Chem.* **275**, 23685–23692.

Patil, C., and Walter, P. (2001). Intracellular signaling from the endoplasmic reticulum to the nucleus: The unfolded protein response in yeast and mammals. *Curr. Opin. Cell Biol.* **13**, 349–355.

Petrova, K., Oyadomari, S., Hendershot, L. M., and Ron, D. (2008). Regulated association of misfolded endoplasmic reticulum lumenal proteins with P58/DNAJc3. *EMBO J.* **27**, 2862–2872.

Pilon, M., Schekman, R., and Romisch, K. (1997). Sec61p mediates export of a misfolded secretory protein from the endoplasmic reticulum to the cytosol for degradation. *EMBO J.* **16**, 4540–4548.

Plemper, R. K., Bohmler, S., Bordallo, J., Sommer, T., and Wolf, D. H. (1997). Mutant analysis links the translocon and BiP to retrograde protein transport for ER degradation. *Nature* **388**, 891–895.

Pollock, S., Kozlov, G., Pelletier, M. F., Trempe, J. F., Jansen, G., Sitnikov, D., Bergeron, J. J. M., Gehring, K., Ekiel, I., and Thomas, D. Y. (2004). Specific interaction of ERp57 and calnexin determined by NMR spectroscopy and an ER two-hybrid system. *EMBO J.* **23**, 1020–1029.

Quan, E. M., Kamiya, Y., Kamiya, D., Denic, V., Weibezahn, J., Kato, K., and Weissman, J. S. (2008). Defining the glycan destruction signal for endoplasmic reticulum-associated degradation. *Mol. Cell* **32**, 870–877.

Riemer, J., Appenzeller-Herzog, C., Johansson, L., Bodenmiller, B., Hartmann-Petersen, R., and Ellgaard, L. (2009). A luminal flavoprotein in endoplasmic reticulum-associated degradation. *Proc. Natl. Acad. Sci. USA* **106**, 14831–14836.

Roth, R. A. (1987). In vivo cross-linking of protein disulfide isomerase to immunoglobulins. *Biochemistry* **26**, 4179–4182.

Russell, S. J., Ruddock, L. W., Salo, K. E. H., Oliver, J. D., Roebuck, Q. P., Llewellyn, D. H., Roderick, H. L., Koivunen, P., Myllyharju, J., and High, S. (2004). The primary substrate binding site in the b' domain of ERp57 is adapted for endoplasmic reticulum lectin association. *J. Biol. Chem.* **279**, 18861–18869.

Rutkowski, D. T., Kang, S. W., Goodman, A. G., Garrison, J. L., Taunton, J., Katze, M. G., Kaufman, R. J., and Hegde, R. S. (2007). The role of p58IPK in protecting the stressed endoplasmic reticulum. *Mol. Biol. Cell* **18**, 3681–3691.

Sakoh-Nakatogawa, M., Nishikawa, S. I., and Endo, T. (2009). Roles of protein-disulfide isomerase-mediated disulfide bond formation of yeast Mnl1p in endoplasmic reticulum-associated degradation. *J. Biol. Chem.* **284**, 11815–11825.

Schaiff, W. T., Hruska, K. A., Jr., McCourt, D. W., Green, M., and Schwartz, B. D. (1992). HLA-DR associates with specific stress proteins and is retained in the endoplasmic reticulum in invariant chain negative cells. *J. Exp. Med.* **176**, 657–666.

Schmitz, A., Herrgen, H., Winkeler, A., and Herzog, V. (2000). Cholera toxin is exported from microsomes by the Sec61p complex. *J. Cell Biol.* **148**, 1203–1212.

Schrag, J. D., Bergeron, J. J., Li, Y., Borisova, S., Hahn, M., Thomas, D. Y., and Cygler, M. (2001). The structure of calnexin, an ER chaperone involved in quality control of protein folding. *Mol. Cell* **8**, 633–644.

Schroder, M., and Kaufman, R. J. (2005). ER stress and the unfolded protein response. *Mutat. Res.* **569**, 29–63.

Schulze, A., Standera, S., Buerger, E., Kikkert, M., Van Voorden, S., Wiertz, E., Koning, F., Kloetzel, P. M., and Seeger, M. (2005). The ubiquitin-domain protein HERP forms a complex with components of the endoplasmic reticulum associated degradation pathway. *J. Mol. Biol.* **354**, 1021–1027.

Schwaller, M., Wilkinson, B., and Gilbert, H. F. (2003). Reduction-reoxidation cycles contribute to catalysis of disulfide isomerization by protein-disulfide isomerase. *J. Biol. Chem.* **278**, 7154–7159.

Sorensen, S., Ranheim, T., Bakken, K. S., Leren, T. P., and Kulseth, M. A. (2006). Retention of mutant low density lipoprotein receptor in endoplasmic reticulum (ER) leads to ER stress. *J. Biol. Chem.* **281**, 468–476.

Szathmary, R., Bielmann, R., Nita-Lazar, M., Burda, P., and Jakob, C. A. (2005). Yos9 protein is essential for degradation of misfolded glycoproteins and may function as lectin in ERAD. *Mol. Cell* **19**, 765–775.

Tirasophon, W., Welihinda, A. A., and Kaufman, R. J. (1998). A stress response pathway from the endoplasmic reticulum to the nucleus requires a novel bifunctional protein kinase/endoribonuclease (Ire1p) in mammalian cells. *Genes Dev.* **12**, 1812–1824.

Tortorella, D., Story, C. M., Huppa, J. B., Wiertz, E. J. H. J., Jones, T. R., and Ploegh, H. L. (1998). Dislocation of type I membrane proteins from the er to the cytosol is sensitive to changes in redox potential. *J. Cell Biol.* **142**, 365–376.

Tsai, B., and Rapoport, T. A. (2002). Unfolded cholera toxin is transferred to the ER membrane and released from protein disulfide isomerase upon oxidation by Ero1. *J. Cell Biol.* **159**, 207–215.

Ushioda, R., Hoseki, J., Araki, K., Jansen, G., Thomas, D. Y., and Nagata, K. (2008). ERdj5 is required as a disulfide reductase for degradation of misfolded proteins in the ER. *Science* **321**, 569–572.

Vandenbroeck, K., Martens, E., and Alloza, I. (2006). Multi-chaperone complexes regulate the folding of interferon-γ in the endoplasmic reticulum. *Cytokine* **33**, 264–273.

Wahlman, J., DeMartino, G. N., Skach, W. R., Bulleid, N. J., Brodsky, J. L., and Johnson, A. (2007). Real-time fluorescence detection of ERAD substrate retrotranslocation in a mammalian in vitro system. *Cell* **129**, 943–955.

Walker, K. W., and Gilbert, H. F. (1997). Scanning and escape during protein-disulfide isomerase-assisted protein folding. *J. Biol. Chem.* **272**, 8845–8848.

Wang, X. Z., Harding, H. P., Zhang, Y., Jolicoeur, E. M., Kuroda, M., and Ron, D. (1998). Cloning of mammalian Ire1 reveals diversity in the ER stress responses. *EMBO J.* **17**, 5708–5717.

Wiertz, E. J. (1996). The human cytomegalovirus us11 gene product dislocates MHC class I heavy chains from the endoplasmic reticulum to the cytosol. *Cell* **84**, 769–779.

Yan, A., and Lennarz, W. J. (2005). Unraveling the mechanism of protein N-glycosylation. *J. Biol. Chem.* **280**, 3121–3124.

Yan, W., Frank, C. L., Korth, M. J., Sopher, B. L., Novoa, I., Ron, D., and Katze, M. G. (2002). Control of PERK eIF2α kinase activity by the endoplasmic reticulum stress-induced molecular chaperone P58IPK. *Proc. Natl. Acad. Sci. USA* **99**, 15920–15925.

Ye, Y., Shibata, Y., Yun, C., Ron, D., and Rapoport, T. A. (2004). A membrane protein complex mediates retro-translocation from the ER lumen into the cytosol. *Nature* **429**, 841–847.

Ye, Y., Shibata, Y., Kikkert, M., Van Voorden, S., Wiertz, E., and Rapoport, T. A. (2005). Recruitment of the p97 ATPase and ubiquitin ligases to the site of retrotranslocation at the endoplasmic reticulum membrane. *Proc. Natl. Acad. Sci. USA* **102,** 14132–14138.

Yoshida, H., Matsui, T., Yamamoto, A., Okada, T., and Mori, K. (2001). XBP1 mRNA is induced by ATF6 and spliced by IRE1 in response to ER stress to produce a highly active transcription factor. *Cell* **107,** 881–891.

Younger, J. M., Chen, L., Ren, H. Y., Rosser, M. F. N., Turnbull, E. L., Fan, C. Y., Patterson, C., and Cyr, D. M. (2006). Sequential quality-control checkpoints triage misfolded cystic fibrosis transmembrane conductance regulator. *Cell* **126,** 571–582.

Yu, M., and Haslam, D. B. (2005). Shiga toxin is transported from the endoplasmic reticulum following interaction with the luminal chaperone HEDJ/ERdj3. *Infect. Immun.* **73,** 2524–2532.

Zahedi, R. P., Vlzing, C., Schmitt, A., Frien, M., Jung, M., Dudek, J., Wortelkamp, S., Sickmann, A., and Zimmermann, R. (2009). Analysis of the membrane proteome of canine pancreatic rough microsomes identifies a novel Hsp40, termed ERj7. *Proteomics* **9,** 3463–3473.

Zhang, Y., Baig, E., and Williams, D. B. (2006). Functions of ERp57 in the folding and assembly of major histocompatibility complex class I molecules. *J. Biol. Chem.* **281,** 14622–14631.

CHAPTER FIFTEEN

Structural Insight into the Protective Role of P58(IPK) during Unfolded Protein Response

Jiahui Tao *and* Bingdong Sha

Contents

1. Endoplasmic Reticulum Stress and Unfolded Protein Response	260
2. P58(IPK) Might be a Dual-Function Protein	261
3. Knocking Out P58(IPK) in Mouse Models	261
4. P58(IPK) is an ER-Resident Hsp40	262
5. Crystal Structure of Mouse P58(IPK) TPR Domain	263
6. P58(IPK) Functions as a Molecular Chaperone to Bind Unfolded Proteins Using Subdomain I	264
7. Structural Basis for P58(IPK) J Domain–BiP Interaction	267
8. Working Model of P58(IPK) During UPR	267
9. Future Research	267
References	268

Abstract

P58(IPK) has been identified as an ER molecular chaperone to maintain protein-folding homeostasis. P58(IPK) expression can be significantly upregulated during unfolded protein responses (UPR), and it may play important roles in suppressing the ER protein aggregations. To investigate the mechanism how P58(IPK) functions to promote protein folding within ER, we have determined the crystal structure of P58(IPK) TPR domain at 2.5 Å resolution. P58(IPK) contains nine TPR motifs and a C-terminal J domain within its primary sequence. The crystal structure of P58(IPK) revealed three subdomains (I, II, and III) with similar folds and each domain contains three TPR motifs. Our data also showed that P58(IPK) acts as a molecular chaperone by interacting with the unfolded proteins such as luciferase, rhodanese, and insulin. The P58(IPK) structure reveals a conserved hydrophobic patch located in subdomain I that may be involved in binding the misfolded polypeptides. We have proposed a working model for P58(IPK) to act together with Bip to prevent protein aggregations and promote protein foldings within ER.

Department of Cell Biology, University of Alabama at Birmingham, Birmingham, Alabama, USA

1. Endoplasmic Reticulum Stress and Unfolded Protein Response

In eukaryotic cells, endoplasmic reticulum (ER) is a membranous organelle responsible for biogenesis of secreted and membrane proteins. About one-third of the newly translated proteins are transported into the ER lumen (Zhang and Kaufman, 2008), where they are folded, modified, and assembled with the help of ER-resident chaperones. An imbalance between the unfolded protein influx into ER and the protein-folding capacity of ER results in accumulation of misfolded protein in the ER lumen, a condition called ER stress (Ron and Walter, 2007). To protect stressed cells and ensure the fidelity of protein folding, eukaryotic cells have developed a signaling transduction pathway named unfolded protein response (UPR). Three major ER stress sensors have been identified, IRE1 (Cox et al., 1993; Mori et al., 1993), PERK (Harding et al., 1999), and ATF6 (Haze et al., 1999), which sense the protein-folding status in the ER lumen and regulate UPR-related genes expression.

Three different mechanisms are employed to cope with ER stress: (1) global protein synthesis is downregulated to reduce the protein-folding burden on the ER lumen (Harding et al., 1999); (2) ER-resident chaperones are upregulated to increase the protein-folding capacity of ER (Kozutsumi et al., 1988); and (3) misfolded protein accumulated in the ER lumen is redirected to the degradation pathway (Travers et al., 2000). If the protein-folding homeostasis is not reestablished, cell apoptosis will be initiated to avoid further damage to other cells (Scheuner et al., 2006).

BiP, also known as GRP78, is one of the major ER-resident chaperones, which is upregulated during ER stress (Hendershot, 2004; Lee, 2001). BiP prevents misfolded protein aggregation by binding to the exposed hydrophobic residues (Flynn et al., 1991). BiP also promotes protein folding through a polypeptide binding-releasing cycle at the cost of ATP. Depletion of cellular ATP will inhibit the folding of secretory proteins (Gething, 1999). The protein-folding activity of BiP is regulated by other proteins, for example, Hsp40 family members (P58(IPK)) (Petrova et al., 2008) and nucleotide exchange factors (Sil1) (Tyson and Stirling, 2000).

Previous studies suggested that P58(IPK) serves as a negative regulator of PKR and PERK signaling in cytosol (Melville et al., 1999; Tang et al., 1999; Yan et al., 2002b). Recent work indicated that P58(IPK) functions as a molecular chaperone as well as a regulator of BiP in the presence of ER stress (Petrova et al., 2008; Rutkowski et al., 2007; Tao et al., 2010). With the P58(IPK) TPR domain structure determined (Tao et al., 2008, 2010), our understanding of this dual-function protein has been advanced.

2. P58(IPK) MIGHT BE A DUAL-FUNCTION PROTEIN

P58(IPK), also referred to as DnaJC3, is first identified as a cytosolic protein (Melville et al., 1997) and a negative regulator of PKR (double-stranded RNA-dependent protein kinase) (Polyak et al., 1996; Tan et al., 1998). PKR is a part of the host defensive system of the eukaryotic cells against RNA virus (Katze, 1995). Upon RNA virus infection, the presence of viral double-stranded RNA activates PKR, which downregulates global protein synthesis by catalyzing the phosphorylation of the α-subunit of eukaryotic initiation factor 2 (eIF2α) (Galabru and Hovanessian, 1987). However, influenza virus evolves a new strategy to inactivate PKR. Upon viral infection, cytosolic protein P58(IPK) is released from its own inhibitor, Hdj1 (Melville et al., 1997), and inhibits the kinase activity of PKR, which will finally affect the phosphorylation status of eIF2α (Lee et al., 1990, 1992, 1994). With the help of the host protein P58(IPK), influenza virus can restore global protein synthesis, which is crucial for the production of massive viral protein.

Recent studies indicate that P58(IPK) functions as an ER-resident molecular chaperone (Rutkowski et al., 2007; Tao et al., 2010) and a regulator of BiP (Petrova et al., 2008). Studies in mouse model and cultured cells indicate that P58(IPK) is upregulated by UPR (Petrova et al., 2008; Yan et al., 2002a). In contrast to previous studies, P58(IPK) is translocated into ER lumen instead of staying in the cytosol (Petrova et al., 2008; Rutkowski et al., 2007). In cultured cells, the global protein synthesis is not affected by knocking out P58(IPK) (Rutkowski et al., 2007). However, the protein-folding capacity of the ER is reduced in the absence of P58 (IPK). P58(IPK) associates with unfolded protein in the ER lumen, suggesting its role as a molecule chaperone in stressed ER. Overexpression of P58 (IPK) in stressed cells stimulates the maturation of secreted protein (Rutkowski et al., 2007). The C-terminal J domain of P58(IPK) is capable of stimulating ATP hydrolysis by BiP, supporting that P58(IPK) is a regulator of BiP, the ER luminal Hsp70 (Petrova et al., 2008).

3. KNOCKING OUT P58(IPK) IN MOUSE MODELS

Knockout of P58(IPK) results in upregulation of apoptosis-related genes and elevated pancreatic β-cell failure. Though β-cells undergo apoptosis in P58(IPK)$^{-/-}$ mice, the knockout mice only develop moderate diabetic phenotype for the reason that there are still adequate viable β-cells that will respond to glucose and produce insulin (Ladiges et al., 2005). Expression of human carbonic anhydrase IV (hCAIV) mutant proteins in kidneys caused ER stress and progressive renal injury in male transgenic mice. Haploinsufficiency of P58(IPK) exacerbates the kidney cell apoptosis caused by accumulation of hCAIV mutants (Datta et al., 2010). Knockout of Sil1, a nucleotide exchange factor of BiP, results in ER stress in Purkinje cells. Depletion of P58(IPK) ameliorates ER stress and neurodegeneration in Sil1$^{-/-}$ mice, suggesting that P58(IPK) is a regulator of ATP/ADP cycling of BiP (Zhao et al., 2010).

4. P58(IPK) IS AN ER-RESIDENT HSP40

P58(IPK) is composed of an N-terminal ER-targeting signal, a TPR domain, and a C-terminal J domain. Deletion of the 26-residue N-terminal ER-targeting signal blocks the translocation of P58(IPK) into ER lumen, suggesting that ER-targeting signal is required for P58(IPK) ER localization. Sucrose gradient velocity sedimentation identifies endogenous P58 (IPK) as a component of ER lumen under normal and stressed conditions (Rutkowski et al., 2007). The ER-targeting signal is cleaved after the translocation, suggesting that the signal peptide is not required for P58 (IPK)'s function in ER lumen.

The 43-kDa core TPR domain is composed of nine tandemly organized TPR (tetratricopeptide repeat) motifs, 34-residue helix-turn-helix structure (Tao et al., 2010). Of the nine TPR motifs, TPR6 is proposed to be responsible for interaction with PKR kinase domain (Tang et al., 1996). Mutant P58(IPK) protein with TPR6 deleted (P58IPKΔTPR6) is unable to inhibit PKR kinase activity in vitro and in vivo. TPR7 is binding site P52^{rIPK}, a repressor of P58(IPK) (Gale et al., 2002). P58(IPK) mutant with TPR7 deleted inhibits PKR kinase activity more effectively, suggesting that TPR7 is a negative regulatory element in P58(IPK) (Gale et al., 1998). P58(IPK) TPR domain preferentially bind unfolded protein to folded protein in vitro and in vivo. Previous studies suggested that TPR1–TPR3 are responsible for P58(IPK) self-association (Gale et al., 1996; Melville et al., 1997, 1999). The crystal structure solved recently supports that TPR1–TPR3 are the binding site of unfolded protein, which forms the structural basis for P58(IPK)'s unfolded protein-binding activity in ER lumen (Tao et al., 2010).

The C-terminal J domain of P58(IPK) is the hallmark of Hsp40 family member (Kampinga and Craig, 2010). P58(IPK) associates with BiP ATPase domain in ER lumen via J domain (Petrova et al., 2008). As the J domain in other Hsp40 proteins, P58(IPK) J domain is capable of stimulating ATP hydrolysis by BiP. Mutation of the highly conserved HPD motif in P58 (IPK) J domain will completely abolish the interaction between P58(IPK) and BiP (Petrova et al., 2008; Rutkowski et al., 2007). BiP binding promotes dissociation of unfolded protein from P58(IPK) TPR domain, suggesting that the association of unfolded protein with P58(IPK) is regulated by BiP.

5. CRYSTAL STRUCTURE OF MOUSE P58(IPK) TPR DOMAIN

Crystal structure of the core TPR domain (residue 35–393) is determined at 2.5 Å resolution by our group. The high-resolution structure enables us to understand the function of P58(IPK) better. As predicted before, nine TPR motifs are present in P58(IPK) TPR domain structure (Fig. 15.1). Three structurally similar subdomains are identified, which are composed of TPR1–TPR3, TPR4–TPR6, and TPR7–TPR9, respectively (Fig. 15.1). The three subdomains are organized in a head-to-tail fashion, which provides P58(IPK) TPR domain an elongated molecular shape with the length of about 100 Å. Moreover, the three subdomains adopt very similar protein folds in the crystal structure. Previous structural studies have shown that TPR motifs can utilize a groove formed by concave surface to interact with their protein ligands. In P58(IPK) TPR domain

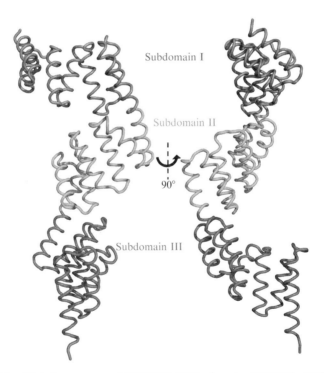

Figure 15.1 Global structure of P58(IPK) TPR domain P58(IPK). TPR domain (residue 35–393) contains nine TPR motifs. TPR1–TPR3, TPR4–TPR6, and TPR7–TPR9 comprise subdomain I, II, and III, respectively. Three subdomains are linked by two long helixes.

structure, a large groove can be identified within each subdomain I, II, and III, which we propose to be the binding site for the binding partners of P58(IPK).

6. P58(IPK) FUNCTIONS AS A MOLECULAR CHAPERONE TO BIND UNFOLDED PROTEINS USING SUBDOMAIN I

The surface potential drawing of P58(IPK) suggests that the groove located in subdomain I is hydrophobic, while the grooves of subdomain II and III are negatively charged (Fig. 15.2). Mapping of hydrophobic residues

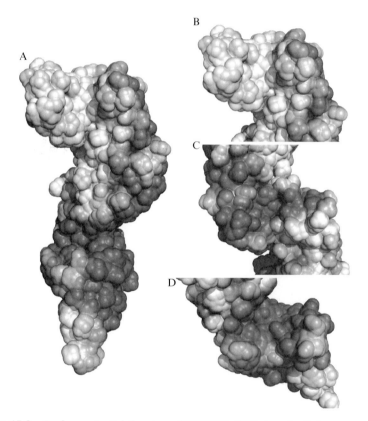

Figure 15.2 Surface potential drawing of P58(IPK) TPR domain. Solvent-accessible surface of P58(IPK) TPR domain is colored according to their potential, red for negative, white for neutral, and blue for positive (Fig. 15.2A). The groove of subdomain I is neutral (Fig. 15.2B), while grooves of subdomain II and III are negatively charged (Fig. 15.2C,D). (See Color Insert.)

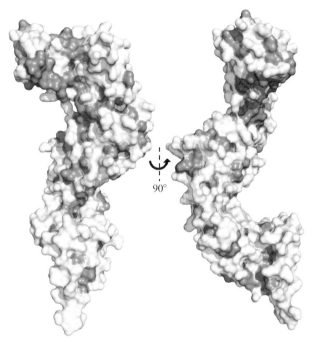

Figure 15.3 Mapping of hydrophobic residues on P58(IPK) TPR domain. Hydrophobic residues are colored green and hydrophilic residues are colored white. Subdomain I groove region is more hydrophobic than that of subdomain II and III, indicating that the groove of subdomain I might be the potential binding site for unfolded protein. The orientation of P58(IPK) in this figure is similar as that in Fig. 15.1. (For interpretation of the references to color in this figure legend, the reader is referred to the Web version of this chapter.)

on the surface of P58(IPK) TPR domain indicates that subdomain I groove is highly hydrophobic which makes it an ideal binding site for the exposed hydrophobic residues of the unfolded protein (Fig. 15.3). Several key residues in the subdomain I groove have been identified, including Y71, Y75, R76, and F104. These residues are 100% conserved among several model species and form a hydrophobic patch in the groove of subdomain I (Fig. 15.4). Mutation of each residue to Ala significantly reduces the binding affinity between P58(IPK) TPR domain and unfolded model protein luciferase (Fig. 15.5). These mutants also exhibit compromised binding ability for unfolded rhodanese and insulin (data not shown). Based on current available data, we hypothesize that subdomain I of P58(IPK) is responsible for unfolded protein capture during UPR. It is likely that P58(IPK) may interact with the unfolded protein using hydrophobic interactions to protect the unfolded protein from aggregations.

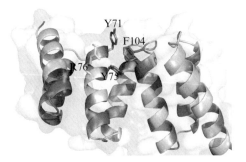

Figure 15.4 Mapping of key residues in subdomain I groove Y71, Y75, R76, and F104 are located in the center of subdomain I groove. These four residues are completely conserved among mouse, human, frog, fruit fly, and worm, suggesting their importance for the function of P58(IPK). Mutation of any of these residues will disrupt the interaction between P58(IPK) domain and unfolded protein (luciferase, rhodanese, insulin).

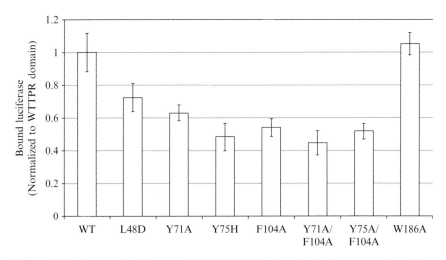

Figure 15.5 Interaction between P58(IPK) TPR domain (or its mutants) and the unfolded luciferase. Denatured luciferase is added into 96-well plates coated with wild-type or mutant P58(IPK) TPR domain. The bound luciferase is quantified by antiluciferase antibody. Mutation of any residue in the hydrophobic patch in subdomain I groove (Y71, Y75, F104, L48) will significantly reduce the binding affinity of TPR–luciferase interaction. Double mutation will exacerbate the situation. W186A mutation, which is located at the junction between subdomain I and II, does not affect the binding.

7. Structural Basis for P58(IPK) J Domain–BiP Interaction

The crystal or solution structure of P58(IPK) J domain has not been described yet. Studies of Hsp40 family members suggest that the J domain of different Hsp40 members might be similar in structure. The HPD motif, key site for BiP binding, is also present in the middle of P58(IPK) J domain. Thus, it is reasonable to speculate that P58(IPK) interacts with BiP ATPase domain in a similar way as auxilin J domain and Hsc70 ATPase domain (Jiang et al., 2007).

8. Working Model of P58(IPK) During UPR

Based on the crystal structure of P58(IPK) TPR domain and related biochemical studies, a working model of P58(IPK) can be established. Once accumulation of misfolded protein in the ER lumen is detected by ER stress sensors (IRE1, PERK, ATF6), the stress signal is transmitted through the ER membrane, and the ER molecular chaperones, including P58(IPK), may be upregulated. P58(IPK) is guided into the ER lumen by its N-terminal ER-targeting signal. In the stressed cells, P58(IPK) captures unfolded protein by use of the hydrophobic patch in the groove of subdomain I. The binding between P58(IPK) and unfolded protein prevents unfolded protein from forming aggregations. The P58(IPK) J domain then recruits BiP and forms a transient complex which is composed of BiP, P58(IPK), and unfolded protein. BiP binding may initiate the dissociation of unfolded protein from P58(IPK). In the meantime, P58(IPK) J domain will stimulate ATP hydrolysis by BiP ATPase domain. Once the handover is complete, P58(IPK) will dissociate from the BiP-unfolded protein complex and enter the next round of unfolded protein capture (Fig. 15.6).

9. Future Research

Determining the complex crystal structure for P58(IPK)/PKR is crucial for us to understand the role of P58(IPK) in regulating PKR kinase activity. In addition, the handover process of unfolded protein from P58(IPK) to BiP is not understood completely. It will be of great interest to elucidate the mechanism how P58(IPK) interacts with the unfolded protein and transport it to BiP at the molecular level.

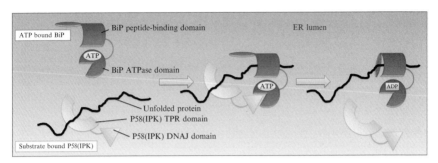

Figure 15.6 Working model of P58(IPK). P58(IPK) TPR domain captures the unfolded protein in the ER lumen. In the presence of ATP, P58(IPK) and BiP associate with each other via the J domain and the ATPase domain, respectively. The unfolded protein substrate will be handed over from P58(IPK) to BiP. The hydrolyzation of ATP to ADP results in a conformational change in BiP, which forms the structural basis for (i) release of P58(IPK) from BiP; (ii) stabilized binding of unfolded protein substrate to BiP peptide-binding domain.

REFERENCES

Cox, J. S., Shamu, C. E., and Walter, P. (1993). Transcriptional induction of genes encoding endoplasmic reticulum resident proteins requires a transmembrane protein kinase. *Cell* **73** (6), 1197–1206.

Datta, R., Shah, G. N., Rubbelke, T. S., Waheed, A., Rauchman, M., Goodman, A. G., Katze, M. G., and Sly, W. S. (2010). Progressive renal injury from transgenic expression of human carbonic anhydrase IV folding mutants is enhanced by deficiency of p58IPK. *Proc. Natl. Acad. Sci. USA* **107**(14), 6448–6452.

Flynn, G. C., Pohl, J., Flocco, M. T., and Rothman, J. E. (1991). Peptide-binding specificity of the molecular chaperone BiP. *Nature* **353**(6346), 726–730.

Galabru, J., and Hovanessian, A. (1987). Autophosphorylation of the protein kinase dependent on double-stranded RNA. *J. Biol. Chem.* **262**(32), 15538–15544.

Gale, M., Jr., Tan, S. L., Wambach, M., and Katze, M. G. (1996). Interaction of the interferon-induced PKR protein kinase with inhibitory proteins P58IPK and vaccinia virus K3L is mediated by unique domains: Implications for kinase regulation. *Mol. Cell. Biol.* **16**(8), 4172–4181.

Gale, M., Jr., Blakely, C. M., Hopkins, D. A., Melville, M. W., Wambach, M., Romano, P. R., and Katze, M. G. (1998). Regulation of interferon-induced protein kinase PKR: Modulation of P58IPK inhibitory function by a novel protein, P52rIPK. *Mol. Cell. Biol.* **18**(2), 859–871.

Gale, M., Jr., Blakely, C. M., Darveau, A., Romano, P. R., Korth, M. J., and Katze, M. G. (2002). P52rIPK regulates the molecular cochaperone P58IPK to mediate control of the RNA-dependent protein kinase in response to cytoplasmic stress. *Biochemistry* **41**(39), 11878–11887.

Gething, M. J. (1999). Role and regulation of the ER chaperone BiP. *Semin. Cell Dev. Biol.* **10**(5), 465–472.

Harding, H. P., Zhang, Y., and Ron, D. (1999). Protein translation and folding are coupled by an endoplasmic-reticulum-resident kinase. *Nature* **397**(6716), 271–274.

Haze, K., Yoshida, H., Yanagi, H., Yura, T., and Mori, K. (1999). Mammalian transcription factor ATF6 is synthesized as a transmembrane protein and activated by proteolysis in response to endoplasmic reticulum stress. *Mol. Biol. Cell* **10**(11), 3787–3799.

Hendershot, L. M. (2004). The ER function BiP is a master regulator of ER function. *Mt. Sinai J. Med.* **71**(5), 289–297.

Jiang, J., Maes, E. G., Taylor, A. B., Wang, L., Hinck, A. P., Lafer, E. M., and Sousa, R. (2007). Structural basis of J cochaperone binding and regulation of Hsp70. *Mol. Cell* **28**(3), 422–433.

Kampinga, H. H., and Craig, E. A. (2010). The HSP70 chaperone machinery: J proteins as drivers of functional specificity. *Nat. Rev. Mol. Cell Biol.* **11**(8), 579–592.

Katze, M. G. (1995). Regulation of the interferon-induced PKR: Can viruses cope? *Trends Microbiol.* **3**(2), 75–78.

Kozutsumi, Y., Segal, M., Normington, K., Gething, M. J., and Sambrook, J. (1988). The presence of malfolded proteins in the endoplasmic reticulum signals the induction of glucose-regulated proteins. *Nature* **332**(6163), 462–464.

Ladiges, W. C., Knoblaugh, S. E., Morton, J. F., Korth, M. J., Sopher, B. L., Baskin, C. R., MacAuley, A., Goodman, A. G., LeBoeuf, R. C., and Katze, M. G. (2005). Pancreatic beta-cell failure and diabetes in mice with a deletion mutation of the endoplasmic reticulum molecular chaperone gene P58IPK. *Diabetes* **54**(4), 1074–1081.

Lee, A. S. (2001). The glucose-regulated proteins: Stress induction and clinical applications. *Trends Biochem. Sci.* **26**(8), 504–510.

Lee, T. G., Tomita, J., Hovanessian, A. G., and Katze, M. G. (1990). Purification and partial characterization of a cellular inhibitor of the interferon-induced protein kinase of Mr 68,000 from influenza virus-infected cells. *Proc. Natl. Acad. Sci. USA* **87**(16), 6208–6212.

Lee, T. G., Tomita, J., Hovanessian, A. G., and Katze, M. G. (1992). Characterization and regulation of the 58,000-dalton cellular inhibitor of the interferon-induced, dsRNA-activated protein kinase. *J. Biol. Chem.* **267**(20), 14238–14243.

Lee, T. G., Tang, N., Thompson, S., Miller, J., and Katze, M. G. (1994). The 58, 000-dalton cellular inhibitor of the interferon-induced double-stranded RNA-activated protein kinase (PKR) is a member of the tetratricopeptide repeat family of proteins. *Mol. Cell. Biol.* **14**(4), 2331–2342.

Melville, M. W., Hansen, W. J., Freeman, B. C., Welch, W. J., and Katze, M. G. (1997). The molecular chaperone hsp40 regulates the activity of P58IPK, the cellular inhibitor of PKR. *Proc. Natl. Acad. Sci. USA* **94**(1), 97–102.

Melville, M. W., Tan, S. L., Wambach, M., Song, J., Morimoto, R. I., and Katze, M. G. (1999). The cellular inhibitor of the PKR protein kinase, P58(IPK), is an influenza virus-activated co-chaperone that modulates heat shock protein 70 activity. *J. Biol. Chem.* **274**(6), 3797–3803.

Mori, K., Ma, W., Gething, M. J., and Sambrook, J. (1993). A transmembrane protein with a cdc2+/CDC28-related kinase activity is required for signaling from the ER to the nucleus. *Cell* **74**(4), 743–756.

Petrova, K., Oyadomari, S., Hendershot, L. M., and Ron, D. (2008). Regulated association of misfolded endoplasmic reticulum lumenal proteins with P58/DNAJc3. *EMBO J.* **27**(21), 2862–2872.

Polyak, S. J., Tang, N., Wambach, M., Barber, G. N., and Katze, M. G. (1996). The P58 cellular inhibitor complexes with the interferon-induced, double-stranded RNA-dependent protein kinase, PKR, to regulate its autophosphorylation and activity. *J. Biol. Chem.* **271**(3), 1702–1707.

Ron, D., and Walter, P. (2007). Signal integration in the endoplasmic reticulum unfolded protein response. *Nat. Rev. Mol. Cell Biol.* **8**(7), 519–529.

Rutkowski, D. T., Kang, S. W., Goodman, A. G., Garrison, J. L., Taunton, J., Katze, M. G., Kaufman, R. J., and Hegde, R. S. (2007). The role of p58IPK in protecting the stressed endoplasmic reticulum. *Mol. Biol. Cell* **18**(9), 3681–3691.

Scheuner, D., Patel, R., Wang, F., Lee, K., Kumar, K., Wu, J., Nilsson, A., Karin, M., and Kaufman, R. J. (2006). Double-stranded RNA-dependent protein kinase phosphorylation of the alpha-subunit of eukaryotic translation initiation factor 2 mediates apoptosis. *J. Biol. Chem.* **281**(30), 21458–21468.

Tan, S. L., Gale, M. J., Jr., and Katze, M. G. (1998). Double-stranded RNA-independent dimerization of interferon-induced protein kinase PKR and inhibition of dimerization by the cellular P58IPK inhibitor. *Mol. Cell. Biol.* **18**(5), 2431–2443.

Tang, N. M., Ho, C. Y., and Katze, M. G. (1996). The 58-kDa cellular inhibitor of the double stranded RNA-dependent protein kinase requires the tetratricopeptide repeat 6 and DnaJ motifs to stimulate protein synthesis in vivo. *J. Biol. Chem.* **271**(45), 28660–28666.

Tang, N. M., Korth, M. J., Gale, M., Jr., Wambach, M., Der, S. D., Bandyopadhyay, S. K., Williams, B. R., and Katze, M. G. (1999). Inhibition of double-stranded RNA- and tumor necrosis factor alpha-mediated apoptosis by tetratricopeptide repeat protein and cochaperone P58(IPK). *Mol. Cell. Biol.* **19**(7), 4757–4765.

Tao, J., Wu, Y., Ron, D., and Sha, B. (2008). Preliminary X-ray crystallographic studies of mouse UPR responsive protein P58(IPK) TPR fragment. *Acta Crystallogr. F Struct. Biol. Cryst. Commun.* **64**(Pt 2), 108–110.

Tao, J., Petrova, K., Ron, D., and Sha, B. (2010). Crystal structure of P58(IPK) TPR fragment reveals the mechanism for its molecular chaperone activity in UPR. *J. Mol. Biol.* **397**(5), 1307–1315.

Travers, K. J., Patil, C. K., Wodicka, L., Lockhart, D. J., Weissman, J. S., and Walter, P. (2000). Functional and genomic analyses reveal an essential coordination between the unfolded protein response and ER-associated degradation. *Cell* **101**(3), 249–258.

Tyson, J. R., and Stirling, C. J. (2000). LHS1 and SIL1 provide a lumenal function that is essential for protein translocation into the endoplasmic reticulum. *EMBO J.* **19**(23), 6440–6452.

Yan, W., Frank, C. L., Korth, M. J., Sopher, B. L., Novoa, I., Ron, D., and Katze, M. G. (2002a). Control of PERK eIF2alpha kinase activity by the endoplasmic reticulum stress-induced molecular chaperone P58IPK. *Proc. Natl. Acad. Sci. USA* **99**(25), 15920–15925.

Yan, W., Gale, M. J., Jr., Tan, S. L., and Katze, M. G. (2002b). Inactivation of the PKR protein kinase and stimulation of mRNA translation by the cellular co-chaperone P58 (IPK) does not require J domain function. *Biochemistry* **41**(15), 4938–4945.

Zhang, K., and Kaufman, R. J. (2008). From endoplasmic-reticulum stress to the inflammatory response. *Nature* **454**(7203), 455–462.

Zhao, L., Rosales, C., Seburn, K., Ron, D., and Ackerman, S. L. (2010). Alteration of the unfolded protein response modifies neurodegeneration in a mouse model of Marinesco-Sjogren syndrome. *Hum. Mol. Genet.* **19**(1), 25–35.

CHAPTER SIXTEEN

Principles of IRE1 Modulation Using Chemical Tools

Kenneth P. K. Lee[*,†] *and* Frank Sicheri[*,†]

Contents

1. Introduction: The Unfolded Protein Response	272
2. Structural Biology of IRE1–XBP1 Signaling	276
2.1. Mechanism of misfolded protein detection	277
2.2. Mechanism of effector activation	278
3. Chemical Approaches to Modulate IRE1 Function	282
3.1. Chemical genetic modulation of IRE1 using 1NM-PP1	283
3.2. Kinase-inhibitors mimick ADP to stimulate IRE1 effector dimerization and activate IRE1 nuclease function	283
3.3. The unusual mode of IRE1 activation by Quercetin	284
3.4. Focal points of IRE1 control	286
3.5. Principles of IRE1 chemical modulation	287
References	292

Abstract

Perturbations that derail the proper folding and assembly of proteins in the endoplasmic reticulum (ER) lead to misfolded protein accrual in the ER—a toxic condition known as ER stress. The unfolded protein response (UPR) is a signaling system evolved to detect and rectify ER stress. IRE1 is the most ancient member of the ER stress transducers and is conserved in all eukaryotes. In response to ER stress, IRE1 activates a UPR-dedicated transcription factor called X-box binding protein 1 (XBP1) in metazoans (or HAC1 in yeast) to bolster the productive capacity of the ER and purge misfolded proteins from the ER. To activate XBP1/HAC1, IRE1 cleaves XBP1/HAC1 mRNA twice to eliminate an inhibitory intron using a dormant nuclease function in its cytoplasmic effector region (IRE1^{cyto}). Recent structural, molecular, and chemical biological approaches have greatly advanced our molecular understanding of how IRE1 transduces ER stress. Here we highlight a sampling of these advances with a bias toward structure and the insights they provide. We also propose a set of

[*] Program in Systems Biology, Samuel Lunenfeld Research Institute, Mount Sinai Hospital, Toronto, Ontario, Canada
[†] Department of Molecular and Medical Genetics, University of Toronto, Toronto, Ontario, Canada

Methods in Enzymology, Volume 490
ISSN 0076-6879, DOI: 10.1016/B978-0-12-385114-7.00016-7

© 2011 Elsevier Inc.
All rights reserved.

principles for IRE1 chemical modulation that may assist in the development of tools to better understand how IRE1 function contributes to health and disease and perhaps ultimately the development of new methods of therapeutic intervention.

1. INTRODUCTION: THE UNFOLDED PROTEIN RESPONSE

Many environmental insults (e.g., ischemic hypoxia) and developmental programs (e.g., plasma cell differentiation) can diminish or overwhelm the biosynthetic capacity of the endoplastic reticulum (ER), respectively. As a consequence, ER client proteins that fail to be correctly modified, folded, or assembled (collectively termed misfolded or unfolded proteins) cluster nonspecifically and accumulate inside the ER. This is harmful to the cell. Cells subjected to conditions leading to the buildup of misfolded proteins in the ER are said to experience ER stress.

If left unchecked, ER stress can lead to the disruption of vital cellular functions and ultimately trigger cell death (Fribley *et al.*, 2009; Kim *et al.*, 2006, 2008; Szegezdi *et al.*, 2006). Perhaps not surprisingly, ER stress has been linked to wide-ranging human diseases in recent years and its relation to development and pathophysiology is a subject of intense scrutiny (Boot-Handford and Briggs, 2010; Kaufman, 2002; Lin *et al.*, 2008; McGuckin *et al.*, 2010; Naidoo, 2009; Ogawa *et al.*, 2007; Scheuner and Kaufman, 2008; van der Kallen *et al.*, 2009).

A complex system of homeostatic signaling pathways, collectively termed the unfolded protein response (UPR), has evolved to elicit adaptive measures that allow eukaryotes to adjust the productive capacity of the ER according to its demand (Bernales *et al.*, 2006; Hetz and Glimcher, 2009; Kaufman and Cao, 2010; Malhotra and Kaufman, 2007; Mori, 2009; Ron and Walter, 2007). In response to ER stress, the UPR initiates mechanisms that remove misfolded proteins from the ER and amplify its biosynthetic capacity to restore homeostasis. The ability to trigger cell suicide via apoptosis (or programmed cell death), if homeostasis cannot be reestablished, is also built into the UPR signaling circuitry. In this respect, the UPR system acts as an arbitrator of cell fate (life or death) under conditions of heightened ER stress.

At least three signaling pathways compose the UPR in all metazoans studied to date. These pathways are mediated by three ER stress transducers: *A*ctivating *T*ranscription *F*actor 6 (ATF6), *P*KR-like *ER K*inase (PERK), and *I*nositol *RE*quiring protein *1* (IRE1; Fig. 16.1). ATF6, PERK, and IRE1 detect misfolded protein buildup in the ER lumen and communicate this information to the nucleus by activating downstream transcription

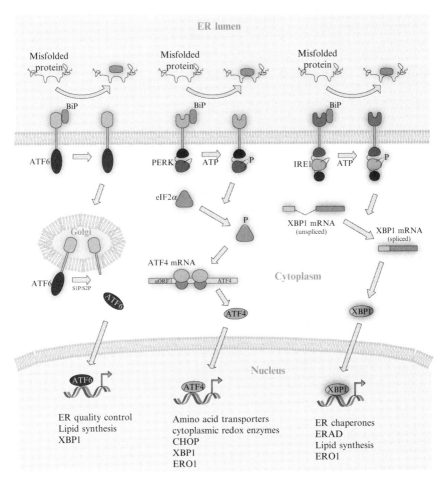

Figure 16.1 The unfolded protein response. *Signaling by ATF6*. ATF6 is a transmembrane protein resident in the ER membrane under normal conditions. Binding of the Hsp70 family ER chaperone, BiP, to the luminal region of ATF6 is thought to retain ATF6 in the ER. Under conditions of elevated ER stress, BiP dissociates from ATF6, resulting in the translocation of ATF6 to the Golgi apparatus. Cleavage of ATF6 by site 1 protease (S1P) and site 2 protease (S2P) releases the cytoplasmic effector portion of ATF6 from the lipid bilayer. The liberated ATF6 effector functions as a potent transcription factor that is imported into the nucleus to upregulate a subset of UPR targets genes, including those involved in ER quality control and lipid synthesis, as well as the ER stress responsive transcription factor XBP1. *Signaling by PERK*. ER stress is thought to trigger the release of BiP from the ER luminal region of PERK, which is similar in sequence to the luminal ER stress sensing region of IRE1. Dissociation of BiP from PERK stimulates the oligomerization of PERK in the plane of the ER membrane. This facilitates *trans*-autophosphorylation of PERK within its activation segment, potentiating PERK kinase activity. PERK is known to phosphorylate the α subunit of eukaryotic translation initiation factor-2 (eIF2α) on Ser51. Phosphorylated eIF2α suppresses global translation by inhibiting the guanine nucleotide exchange factor eIF2B, which services

factors (Bernales *et al.*, 2006; Hetz and Glimcher, 2009; Kaufman and Cao, 2010; Malhotra and Kaufman, 2007; Mori, 2009; Ron and Walter, 2007). Once active, these transcription factors travel into the nucleus and execute transcriptional programs that enhance the production capability of the ER by restructuring this organelle.

Many receptors on the cell surface communicate extracellular signals to the nucleus by modulating transcription factor phosphorylation. Contrary to this expectation, ATF6, PERK, and IRE1 mediate intracellular signaling by controlling distinct steps in the biogenesis of their cognate transcription factor proteins.

ATF6 is a transmembrane ER stress sensor whose cognate effector transcription factor comprises its cytoplasmic portion. In the presence of ER stress, ATF6 transits to the Golgi where membrane proteases site 1 protease (S1P) and site 2 protease (S2P) release the ATF6 transcription factor region from the membrane. The liberated ATF6 fragment is a basic leucine-zipper (bZIP) family transcription factor, which then migrates into the nucleus to upregulate genes involved in ER-associated degradation (ERAD), lipid synthesis, and protein folding in the ER.

PERK is a receptor kinase constitutively resident in the ER membrane. Misfolded proteins activate the cytoplasmic kinase domain of PERK, which site-specifically phosphorylates the translation initiation factor eIF2α on residue Ser51. This results in global shutdown of translation, which is thought to alleviate the ER synthetic load. Paradoxically, eIF2α-Ser51 phosphorylation selectively enhances the translation of mRNAs encoding

---

the translation initiation factor eIF2. Depressed global protein translation alleviates the biosynthetic load on the ER, thereby reducing ER stress. Interestingly, global translational inhibition exerted by eIF2α phosphorylation also promotes the selective translation of stress responsive transcripts encoding the transcription factor activating transcription factor 4 (ATF4). ATF4 upregulates amino acid transporters, genes that counteract oxidative stress, as well as ER oxidase 1 (ERO1), which promotes disulphide bond formation in the ER. In addition, ATF4 upregulates the C/EBP-homologous protein (CHOP), which is a transcription factor linked to stress-induced programmed cell death. *Signaling by IRE1*. Disengagement of BiP from the ER luminal region of IRE1 triggered by ER stress is believed to induce the oligomerization of IRE1 in the ER. *Trans*-autophosphorylation of IRE1 within cytoplasmic kinase domain follows the clustering of IRE1 molecules in the plane of the ER membrane. IRE1 autophosphorylation elicits an otherwise quiescent nuclease activity encoded by the cytoplasmic effector region of IRE1. Nuclease active IRE1 catalyzes the nonconventional excision of a small intron from transcripts encoding the ER stress-responsive transcription factor X-box binding protein-1 (XBP1). The resultant exons are joined by an RNA ligase activity in the cytoplasm. Spliced XBP1 messages are translated to generate active XBP1 protein, which functions as a potent transcription factor that promotes the expression of genes that aid protein folding in the ER (including ER-resident chaperones and ERO1), lipid synthesis as well as the removal of misfolded protein from the ER by a process known as ER-associated degradation (ERAD).

ATF4. ATF4 protein upregulates ER-resident chaperones, foldases, oxidases as well as a proapoptotic transcription factor called CCAAT-enhancer-binding *HO*mologous *P*rotein (CHOP).

IRE1 is the most ancient transducer of ER stress in the UPR system. Of the three branches of the metazoan UPR, only the IRE1 branch is conserved from yeast to humans. Like PERK, IRE1 is an ER-stress responsive receptor kinase embedded in the ER membrane. However, ER stress also triggers a latent nuclease activity present in the cytoplasmic effector region of IRE1. Upon stimulation, IRE1 nuclease activity executes the first step of a nonconventional splicing reaction that eliminates an intron from mRNAs encoding the bZIP transcription factor X-box binding protein 1 (XBP1). Translation of spliced XBP1 transcripts generates active XBP1 protein, which promotes cytoprotective mechanisms including ER/Golgi expansion, clearance of misfolded ER proteins, and folding of nascent polypeptides in the ER.

In addition to the prosurvival IRE1–XBP1 pathway, IRE1 also interfaces with the apoptosis-signaling machinery. Active IRE1 protein binds the cytoplasmic molecular scaffold tumor necrosis associated factor 2 (TRAF2; Urano *et al.*, 2000), which recruits apoptosis-signaling regulating kinase 1 (ASK1; Nishitoh *et al.*, 2002). ASK1 in turn relays ER stress signals to the c-Jun N-terminal kinase (JNK) pathway (Urano *et al.*, 2000). IRE1–TRAF2 complexes also potentiate apoptosis by the recruitment and activation of caspase-12 (Yoneda *et al.*, 2001).

More recently, IRE1 was implicated in nonspecific mRNA decay induced by ER stress. This phenomenon is associated with ER stress-dependent apoptosis (Han *et al.*, 2009; Hollien and Weissman, 2006; Hollien *et al.*, 2009). The destabilized transcripts include mRNAs predicted to localize to the ER as well as many cytosolic mRNAs. Both the kinase and site-specific nuclease activities of IRE1 are required for the reported nondiscriminatory transcript destruction. How precisely IRE1 ribonuclease activity is harnessed to destroy RNA rather than medicate splicing remains to be determined. Regardless, it is now well-appreciated that IRE1 determines cell fate by adjusting the balance between its prosurvival and proapoptotic signaling outputs according to ER stress severity (Lin *et al.*, 2007, 2008, 2009).

The study of the UPR has revealed many novel paradigms of intracellular signaling. Further examination of UPR signaling mechanisms promises to reveal additional, and perhaps unexpected, features and interest in this area is burgeoning. Here we attempt to illustrate in broad strokes a portrait of the current state-of-the-art in IRE1 structural biology. We distill from this body of knowledge a set of focal points of IRE1 regulation and propose strategies to manipulate signaling output from IRE1 by targeting these focal points using small molecules.

2. Structural Biology of IRE1–XBP1 Signaling

IRE1 is an ~120 kDa (~1115 aa) type 1 monotopic membrane protein in the ER. Humans encode two paralogous IRE1 proteins (IRE1α and IRE1β) whereas yeast encodes a single IRE1 protein (IRE1p). The IRE1 polypeptide is comprised of an N-terminal ER luminal region, followed by a hydrophobic membrane-spanning segment, which is in turn followed by a cytoplasmic effector region. The luminal region of IRE1 composes a novel protein fold that confers misfolded protein sensitivity to IRE1 (Credle et al., 2005; Zhou et al., 2006). This region is functionally exchangeable with the luminal portion of PERK despite weak sequence similarity between IRE1 and PERK in this region (Liu et al., 2000). The IRE1 cytoplasmic effector region constitutes a kinase–nuclease dual-enzyme constructed from a eukaryotic protein kinase domain fused rigidly at its C-terminus to an alpha-helical subdomain termed the kinase-extension nuclease (KEN) domain, which contains the IRE1 nuclease active site (Lee et al., 2008). This fused kinase-KEN domain module is also present in the antiviral protein RNase L, which destroys cellular and viral RNAs to curtail viral infections.

Like many receptor kinases on the cell surface, intracellular signaling from IRE1 is triggered by stimulus-induced self-association in its resident membrane compartment (Shamu and Walter, 1996). When ER stress is below an activating threshold, the luminal sensor region of quiescent IRE1 molecules associates with an ER-resident Hsp70-family molecular chaperone called BiP. Elevated levels of ER stress releases BiP from IRE1. In parallel, IRE1 molecules aggregate in the plane of the ER membrane. These events correlate with the potentiation of latent autokinase and nuclease activities encoded in the cytoplasmic dual-enzyme of IRE1.

Once its nuclease function is active, IRE1 makes two site-specific incisions in transcripts encoding the transcription factor XBP1 (or HAC1 in yeast; Calfon et al., 2002; Cox and Walter, 1996; Gonzalez et al., 1999; Lee et al., 2002; Sidrauski and Walter, 1997). These cuts occur after the third nucleotide in the 7-base loops of RNA stem loops found twice in XBP1/HAC1 mRNA; the 7-base loops have the consensus sequence "CNGNNG" where N is any nucleotide. The outcome is the excision of a 26-nt intron from XBP1 mRNA. The corresponding intron in HAC1 mRNA is ~200 nt in length. The resultant exons are joined together by an unidentified RNA ligase activity in metazoans or by tRNA ligase RLG1 in yeast (Sidrauski et al., 1996).

Translation of spliced XBP1/HAC1 mRNA produces active XBP1/HAC1 protein. XBP1 and HAC1 are potent transcription factors that dramatically alter the transcriptional landscape of the cell. Up to 3% of all genes in the yeast genome, among which are genes involved in ERAD and ER expansion, are modulated by the IRE1–HAC1 signaling pathway (Travers *et al.*, 2000). Remodeling of gene expression by the IRE1–XBP1 pathway in humans is believed to be similar, if not broader, in scope.

2.1. Mechanism of misfolded protein detection

How the IRE1 luminal-domain senses ER stress is an important fundamental question in IRE1 structural biology. Structural and functional characterization of the luminal domain of yeast IRE1p (yLD; Credle *et al.*, 2005) and human IRE1α (hLD; Zhou *et al.*, 2006) have led to two distinct models that rationalize the induction of IRE1 oligomerization and activation by misfolded proteins. In the "direct engagement" model, binding of misfolded proteins to the yLD mediates its polymerization, resulting in activation of IRE1 cytoplasmic functions. In the "negative regulation" model, the intrinsic ability of the hLD to form tight dimers is opposed by BiP binding while titration of BiP by misfolded proteins permits IRE1 self-association and activation.

The yLD was crystallized as a dimeric assembly that further polymerizes into (infinite) helical filaments. In yeast, access to the polymeric configuration of yLD dimers was critical for the propagation of ER stress signals by IRE1p. As the yLD was exclusively monomeric in solution (even at high concentrations), direct binding of misfolded proteins to the yLD was hypothesized to mediate the formation of yLD helical polymers. Indeed, a deep hydrophobic furrow large enough to accommodate extended oligopeptides spans the yLD dimer. Disruption of the yLD hydrophobic furrow crippled ER stress signaling by IRE1p in yeast, consistent with a potential role of this putative peptide binding site in misfolded protein detection, at least in yeast. A more recent observation that yLD can suppress the aggregation of denatured proteins *in vitro* lends further support to the "direct engagement" model (Kimata *et al.*, 2007).

The universality of the "direct engagement" model was challenged by findings from structural and functional analysis of the hLD. The hLD composes a similar three-dimensional fold as the yLD and adopts a yLD-like dimer arrangement to generate a hydrophobic groove traversing the hLD dimer. However, unlike its yeast counterpart, hLD exists as a constitutive dimer in solution that does not self-assemble into helical polymers during crystallization. Access to the hLD dimer configuration was indispensible for UPR signaling by human IRE1α *in vivo*. Although the luminal domain of human IRE1a has a strong propensity to dimerize, this observation alone does not invalidate the "direct engagement" model (i.e., unfolded proteins may still bind across the dimer composed hydrophobic groove to more strongly promote dimerization of the luminal-domain). However, the dimer-spanning groove in the hLD crystal structure is too narrow to engage hydrophobic peptides and residues proposed to engage incoming hydrophobic peptides are either buried or charged. In addition, hLD did not suppress denatured protein aggregation *in vitro* (Oikawa *et al.*, 2009). Neither have hLD and yLD structures been solved in complex with synthetic hydrophobic peptides to date. In light of the incompliance of the human IRE1α system with the "direct recognition" model, the "negative regulation" model was proposed to explain the responsiveness of human IRE1α to ER stress. The preeminent repressive role of BiP in the "negative regulation" model is further supported by a recent study demonstrating that the association of BiP to the yLD sets the activating threshold of IRE1 in response to ER stress (Pincus *et al.*, 2010).

It is possible that the divergences between the "direct engagement" and "negative regulation" models may reflect evolutionary adaptations to the different types of ER stress encountered by yeast and metazoans. Future effort in this area may resolve the controversy over the precise mechanism by which IRE1 senses misfolded proteins in the ER and lead to a unified model of ER stress detection.

2.2. Mechanism of effector activation

Dimerization/polymerization of IRE1 molecules in the plane of the ER membrane induced by ER stress coincides with *trans*-autophosphorylation of its cytoplasmic effector through it autokinase activity. Both events are required for activation of nuclease output from wild-type IRE1 and subsequent nonconventional splicing of XBP1/HAC1 mRNA *in vivo* (Aragon *et al.*, 2009; Mori *et al.*, 1993; Shamu and Walter, 1996). The requirement for autophosphorylation *in vivo* can be uncoupled from IRE1 nuclease activation via the enforced occupation of an analog-sensitized version of the nucleotide-binding pocket of the IRE1 kinase domain using the ATP-competitive kinase inhibitor 1NM-PP1 (Han *et al.*, 2009; Papa *et al.*, 2003). This result had the connotation that binding of an adenine nucleotide to the IRE1 kinase domain may also be an important step in the activation of nuclease output from wild-type IRE1 protein *in vivo*.

The crystal structure of the yeast IRE1 effector module complexed to ADP (ADP-IRE1^{cyto}), coupled with mutagenic analysis, provided a first view of the location and architecture of the IRE1 nuclease active site and explained how this machinery catalyzes cleavage of phosphodiester bonds in RNA (Lee *et al.*, 2008). Additional features in the ADP-IRE1^{cyto} crystal structure also hinted at how IRE1 nuclease activity may be linked to IRE1^{cyto} dimerization, autophosphorylation, and nucleotide binding (Lee *et al.*, 2008).

ADP-IRE1^{cyto} forms a symmetric dimer of two elongated trilobal molecules arranged in a parallel back-to-back configuration. Each IRE1^{cyto} subunit is constructed from a canonical kinase domain cemented at its C-terminus to a novel alpha-helical subdomain formed by the C-terminal kinase-extension region of IRE1. The location of the IRE1 nuclease active site was traced to a large conserved concave surface assembled via dimerization of the novel alpha-helical domain. Through comparative structural analysis, the functions of the four invariant catalytic residues in the IRE1 nuclease active site were determined by analogy to the tRNA endonuclease active site. In light of these findings, the novel alpha-helical subdomain formed by the IRE1 C-terminal kinase extension was dubbed the KEN domain. Additional biochemical and genetic analysis established that IRE1 must access the symmetric dimer configuration imposed by back-to-back association of two IRE1 kinase domains observed in the ADP-IRE1^{cyto} crystal structure to effect RNA cleavage. Biophysical measurements further demonstrate that dimerization of IRE1^{cyto} in solution requires ADP binding and that ADP binds preferentially to autophosphorylated IRE1^{cyto} but not nonphosphorylated IRE1^{cyto}.

On the basis of these results, the following "effector dimer" model of IRE1 activation was proposed: ER stress stimulates multimerization of IRE1 in the ER membrane, which triggers *trans*-autophosphorylation of the IRE1 kinase domain; this in turn facilitates unimpeded ADP/ATP binding to the nucleotide-binding cleft, which in turn potentiates back-to-back dimerization of the effector region; finally, back-to-back dimerization of IRE1 assembles the IRE1 nuclease active site, switching on the latent nuclease function of IRE1.

The nuclease active site and dimer interfaces of yeast and human IRE1 (and indeed RNase L) are conserved. Thus, the mechanisms of RNA cleavage and modulation of nuclease activity by dimerization revealed by the ADP-IRE1^{cyto} crystal structure are likely shared by IRE1 and RNase L proteins in both fungal and metazoan species. With the catalytic mechanism of RNA cleavage by IRE1 now understood, elucidation of the basis for sequence-specific RNA substrate recognition by IRE1 awaits the structure of the IRE1 effector complexed to an RNA hairpin substrate.

A recent crystal structure of the yeast IRE1 effector bound to the kinase inhibitor APJ29 offers the interesting new perspective that IRE1 nuclease activity may be controlled via regulated polymerization of IRE1 effector dimers into helical filaments on the surface of the ER membrane (Korennykh *et al.*, 2009). The IRE1 cytoplasmic effector construct used in the Korennykh study, Ire1KR32, formed large but soluble oligomeric aggregates at low concentration but precipitated at high concentration. The APJ29–Ire1KR32 complex crystallized as a helical stack of seven dimers that are nearly identical to the back-to-back dimer observed in the ADP-IRE1^{cyto} crystal structure (14 subunits total in the helical oligomer). This helical assembly can, at least in principle, be extended infinitely by appending additional APJ29–Ire1KR32 dimers to the filament termini. Disruption of intermolecular contacts that perpetuate helical stacking of APJ29–Ire1KR32 dimers suppressed nuclease output from Ire1KR32 as well as the formation of high-order oligomeric structures by Ire1KR32 in solution. In addition, Ire1KR32 (encompassing residues 641–1115 of yeast IRE1) exhibits a near 100-fold higher nuclease output compared to an N-terminal truncation mutant Ire1KR (encompassing residues 673–1115), which does not oligomerize/polymerize. Further, oligomerization/polymerization of nonphosphorylated kinase-inactive Ire1KR32(D878A) at high concentrations resulted in nuclease output that is an $\sim$100-fold higher than a phosphorylated but nonpolymerizing construct, Ire1KR24.

Based on these observations, a revised model of IRE1 nuclease activation was proposed. In this "effector oligomer/polymer" model, aggregation of the luminal domain of IRE1 promotes oligomerization/polymerization of nonphosphorylated IRE1 effectors, activating IRE1 nuclease function. Organization of the cytoplasmic effector region into helical filaments, presumably preceded by effector dimerization, positions the activation segmentation for *trans*-autophosphorylation. Activation segment phosphorylation and nucleotide binding to the kinase domain reinforce the helical oligomer/polymer and further enhance nuclease output.

Since repeating units of back-to-back dimers compose the helical oligomer/polymer assembly of APY29–Ire1KR32, the "effector oligomer/polymer" model complements the "effector dimer" model by bringing an additional layer of complexity (i.e., polymerization of back-to-back dimers) to our understanding of IRE1 nuclease modulation. The "effector oligomer/polymer" model provides an innovative framework to rationalize the cytoplasmic events accompanying the potentiation of IRE1 nuclease activity. Like all good models, it raises many interesting new questions worthy of further pursuit.

Effector region oligomerization/polymerization was proposed to be the dominant activating transition of IRE1 nuclease output as the kinase-inactive variant of Ire1KR32(D828A) can still oligomerize/polymerize and cleave RNA robustly *in vitro*. As the D828A mutation crippled UPR signaling from IRE1(D828A) *in vivo* (Mori et al., 1993), it is interesting to contemplate whether effector oligomerization/polymerization alone is sufficient to stimulate IRE1 nuclease activity *in vivo*. Additional experiments *in vivo* may supply the missing information to address this issue.

Oligomerization/polymerization of Ire1KR32 is completely dependent on eight amino acids "EKKRKRG" at its N-terminus. As the entire N-terminus is disordered in the APY29–Ire1KR32 crystal structure, how the sequence "EKKRKRG" induces effector oligomerization/polymerization is currently unknown and should be the focus of future efforts.

The helical assembly of back-to-back dimers in the APY29–Ire1KR32 crystal structure represents a potential arrangement in which *trans*-autophosphorylation of the IRE1 cytosolic effector takes place. Interestingly, *trans*-autophosphorylation can also be explained by an alternate face-to-face arrangement of back-to-back dimers of IRE1^{cyto} that was captured in two different crystal forms (Fig. 16.2; Lee et al., 2008; Wiseman et al., 2010).

Chemical Modulation of IRE1 Function

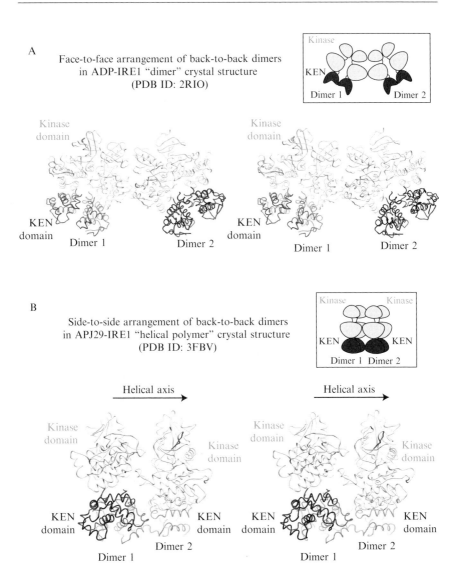

Figure 16.2 Both face-to-face and side-to-side arrangements can explain *trans*-autophosphorylation. (A) Face-to-face arrangement of back-to-back dimers of the IRE1 effector module. Stereo view of a higher order arrangement of back-to-back dimers observed in the crystal structures of IRE1^{cyto} bound to ADP (PDB ID: 2RIO, spacegroup $P6_5$) and the kinase inhibitors Cdk1/2 inhibitor III (PDB ID: 3LJ1, spacegroup $P3_212$) and Jak inhibitor 1 (PDB ID: 3LJ2, spacegroup $P3_212$). This face-to-face arrangement places the kinase domain activation segment of one IRE1^{cyto} subunit in close proximity to the kinase active site of the adjacent back-to-back IRE1^{cyto} dimer. The polypeptide backbone of IRE1^{cyto} is shown as a tube. The kinase domain is shown in white and the KEN domain is shown in black. Ligands are omitted for clarity. Inset shows a cartoon representation of the face-to-face arrangement of back-to-back dimers.

Further studies are needed to differentiate between the helical or face-to-face arrangement as the operational assembly during *trans*-autophosphorylation of the IRE1 effector region *in vivo*.

Finally, it is interesting to note that although dimerization interfaces in both luminal and cytoplasmic regions of IRE1 are conserved in yeast and metazoans, surfaces mediating polymerization in yeast IRE1 crystal structures are highly divergent. Neither is the N-terminal oligomer/polymer-inducing sequence in Ire1KR32 conserved in metazoan IRE1 proteins. Interestingly, a recent study reports that a fragment of the human IRE1α effector region, which includes a portion of the juxtamembrane region (potentially containing an oligomer/polymer-inducing signal), also formed higher order oligomers *in vitro* (Li *et al.*, 2010). However, owing to the lack of conserved oligomer/polymer-inducing features in human IRE1α, the basis of human IRE1α effector oligomer/polymer formation cannot be readily extrapolated from the yeast IRE1 effector oligomer/polymer structure. Furthermore, the requirement for effector oligomerization/polymerization in human IRE1α signaling *in vivo* is still unknown. More experiments in metazoan systems will be key to understanding the general relevance of effector oligomerization/polymerization in the IRE1 signaling process.

3. Chemical Approaches to Modulate IRE1 Function

Chronic ER stress has been attributed to the development of diseases associated with secretory tissue dysfunction. Proapoptotic signaling via IRE1 kinase activity is thought to contribute in part to the harmful effects of ER stress in diseases such as diabetes. Conversely, prosurvival effects emanating from IRE1 nuclease function is suspected to play a role in the development of cancers such as multiple myeloma. Given the potential involvement of IRE1 signaling in disease development, the value of developing novel chemical tools capable of selectively modulating proapoptotic or prosurvival signaling outputs from IRE1 is self-evident. With the

---

(B) Side-to-side arrangement of back-to-back dimers in the helical polymer form of the IRE1 effector module. Stereo view of two back-to-back dimers observed in the APJ29–Ire1KR32 helical polymer (PDB ID: 3FBV). This side-by-side arrangement of back-to-back Ire1^{cyto} dimers also positions the kinase domain activation segment close to the kinase active site of the adjacent back-to-back Ire1^{cyto} dimer. The same representation scheme in (A) is used. Inset shows a cartoon representation of the side-to-side arrangement of dimers. The helical axis of the helical polymer in the APJ29–Ire1KR32 crystal structure is parallel to the plane of the page and its direction is indicated by a black arrow.

multistep processes governing IRE1 function now understood in greater detail, new opportunities emerge to artificially control IRE1 signaling. Drug discovery in this area holds great potential for the identification of new IRE1 modulating small molecules. These molecules may be used as tools to study the relationship between IRE1 function and disease, and possibly as lead compounds for the development of novel therapeutics. In the following section, we will briefly examine chemical strategies employed in recent years to modulate signaling output from IRE1. We will then proceed to synthesize, from structural and biochemical data covered in this and previous sections, a set of focal points of control that impinge on signaling output from IRE1. We then turn our attention to a collection of potential strategies to modulate IRE1 function using chemical tools.

3.1. Chemical genetic modulation of IRE1 using 1NM-PP1

The first attempt to directly modulate signaling output from IRE1 using pharmacological tools was conducted in the yeast system using a chemical genetics approach (Papa et al., 2003). To selectively target the kinase domain of IRE1 in yeast using the ATP-competitive kinase inhibitor 1NM-PP1, the bulky L745 residue at the "gatekeeper" position in the adenine-binding pocket of the kinase active site was mutated to glycine. The L745G mutation was predicted to sensitize IRE1 toward 1NM-PP1 inhibition by granting access to a hydrophobic pocket in the kinase domain N-lobe, with the expectation that kinase activity would be preserved. However, the L745G mutation silenced IRE1 kinase output and severely crippled HAC1 mRNA processing by IRE1. Surprisingly, 1NM-PP1 rescued UPR signaling through IRE1. This result can be rationalized with reference to the ADP-IRE1^{cyto} crystal structure. Given the ATP-competitive nature of 1NM-PP1 in other kinase systems, it is reasonable to predict that 1NM-PP1 and ADP engage the nucleotide-binding pocket of the kinase active sites of wild-type IRE1 and IRE1(L745G) utilizing similar binding modes. By mimicking ADP binding to the nucleotide-binding pocket, 1NM-PP1 may stabilize the "closed" interlobe configuration of the IRE1 kinase domain conducive to the back-to-back dimer arrangement observed in the ADP-IRE1^{cyto} crystal structure. Thus, by promoting back-to-back dimerization, 1NM-PP1 stimulates IRE1 nuclease activity.

3.2. Kinase-inhibitors mimick ADP to stimulate IRE1 effector dimerization and activate IRE1 nuclease function

Recent findings corroborate the notion that ATP-competitive kinase inhibitors can stimulate IRE1 nuclease function by mimicking the binding of adenine nucleotides to the IRE1 kinase active site. A series of synthetic kinase inhibitors known to suppress the activity of certain kinases in an ATP-competitive mode (APY29, APY24, sunitinib, Cdk1/2 inhibitor III, and JAK1 inhibitor 2) have been shown to potentiate the *in vitro* nuclease output of phosphorylated yeast IRE1 effector fragments (Korennykh *et al.*, 2009; Wiseman *et al.*, 2010).

APY29, Cdk1/2 inhibitor III, and JAK inhibitor 1 in complex with two variants of the yeast IRE1 cytoplasmic effector have been captured in two distinct crystal forms. The APY29–Ire1KR32 complex crystallizes as helical stacks of dimers that are nearly indistinguishable from ADP-IRE1^{cyto} back-to-back dimers (discussed in Section 2.2). IRE1^{cyto} in complex with Cdk1/2 inhibitor III or JAK inhibitor 1 crystallized in a crystal form distinct from ADP-IRE1^{cyto} and APY29–Ire1KR32 but, nevertheless, adopted a back-to-back dimer arrangement also nearly identical to that observed in the ADP-IRE1^{cyto} crystal structure. In all three kinase inhibitor-IRE1 crystal structures, the binding mode of the small-molecule ligand is highly analogous to that of ADP. The region occupied by all three chemical modulators in the IRE1 kinase active site overlap strongly with ADP in the ADP-IRE1^{cyto} crystal structure (Fig. 16.3). Also similar to ADP, all three synthetic kinase inhibitors form hydrogen-bonding interactions with the peptide-backbone of the flexible hinge region connecting the N- and C-lobes of the kinase domain. However, unlike ADP, APY29, Cdk1/2 inhibitor III, and JAK inhibitor 1 all engage the IRE1 kinase active site in a magnesium-independent manner (Korennykh *et al.*, 2009; Wiseman *et al.*, 2010).

To summarize, ADP, APY29, Cdk1/2 inhibitor III, and JAK inhibitor 1 all bound the IRE1 cytoplasmic effector in a manner compatible with the formation of back-to-back dimers, which correlates with stimulation of IRE1 nuclease activity. In the case of APY29, back-to-back dimer formation induced by APY29 engagement of the kinase active site facilitates propagation of the rod-shaped oligomeric/polymeric IRE1 assembly. These observations, in aggregate, support the prediction that chemical modulators that mimick the binding of ADP to the kinase active site of IRE1 will likely stabilize the "closed" conformation of the kinase domain and promote back-to-back dimer formation, thereby stimulating IRE1 nuclease activity.

3.3. The unusual mode of IRE1 activation by Quercetin

In an effort to discover new small-molecule modulators of IRE1 nuclease output, quercetin was discovered as a powerful stimulator of IRE1 nuclease function among a panel of known ATP-competitive small-molecule kinase-inhibitors (Wiseman *et al.*, 2010). To understand the mechanism of action of quercetin, IRE1^{cyto} was cocrystallized with ADP and quercetin. The IRE1^{cyto}-ADP-quercetin crystal structure revealed a novel ligand-binding pocket (designated the Q-site) at the KEN domain interface in

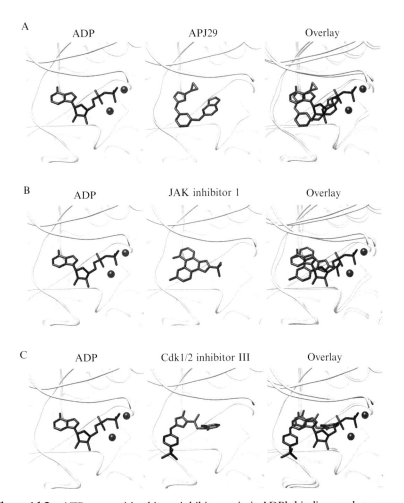

Figure 16.3 ATP-competitive kinase inhibitors mimic ADP's binding mode to engage the IRE1 kinase active site. (A–C) Crystal structures of the IRE1 kinase-nuclease effector module in complex with ATP-competitive kinase inhibitors that positively modulate IRE1 nuclease output: (A) APJ29 (PDB ID: 3FBV), (B) Jak inhibitor 1 (PDB ID: 3LJ1), (C) Cdk1/2 inhibitor III (PDB ID: 3LJ2). Close-up views of the nucleotide-binding pocket in the IRE1 kinase active sites are shown. The structure of the ADP-IRE1^{cyto} complex (PDB ID: 2RIO) is shown alone and overlaid onto the IRE1-kinase inhibitor structures for comparison. In all panels, the polypeptide backbone is shown as a gray tube. Ligands are shown in ball-and-stick representation and are colored black.

back-to-back IRE1^{cyto} dimers, in which a symmetric dimer of quercetin molecules was nestled. *In vitro* and *in vivo* mutational analysis confirmed that quercetin functions by binding the Q-site to promote IRE1 dimerization, which stimulates IRE1 nuclease activity. The observation that two symmetrically arranged molecules of quercetin in direct physical contact simultaneously engage the Q-site provides further impetus to exploit physical drug–drug interactions in designing small-molecules targeting protein–protein interaction surfaces (Shokat, 2010). The finding that ligand engagement at the Q-site can modulate IRE1 signaling output hints at the existence of endogenous cytoplasmic ligands that target the novel Q-site and function in concert with ER stress signals to modulate IRE1 function.

The unusual mechanism of action (i.e., engagement of the Q-site) of quercetin (a nonspecific ATP-competitive kinase inhibitor) in promoting IRE1 nuclease function contrasts that of three other ATP-competitive kinase inhibitors APY29, JAK inhibitor I, and Cdk1/2 inhibitor III, which stimulate IRE1 nuclease activity through engagement of the nucleotide-binding site in the kinase domain.

Quercetin stimulated the nuclease activity of nonphosphorylated IRE1^{cyto} by accessing the Q-site. Intriguingly, ADP inhibited the stimulatory effect of quercetin when bound to nonphosphorylated IRE1^{cyto}. This unexpected inhibitory behavior of ADP contrasts its potentiating effect on quercetin-dependent nuclease activity from the phosphorylated form of IRE1^{cyto}. This finding highlights a previously unappreciated role of IRE1 phosphorylation in modifying the functional output of nucleotide binding to the kinase domain. More work is needed to establish the physiological relevance of this phenomenon.

3.4. Focal points of IRE1 control

By integrating insights garnered from structural studies of yeast and human IRE1, it is possible to identify features in the mechanics of IRE1 regulation where fungal and metazoan systems intersect. These features represent focal points of control that may be accessed using small molecules to modulate specific aspects of IRE1 function.

IRE1 luminal region dimerization is prominently featured in both the "direct recognition" model and the "negative regulation" model of misfolded protein detection (discussed in Section 2.1). Both models predict that luminal-domain dimerization constitutes a critical element in the control of IRE1 signaling.

Dimerization of the IRE1 cytoplasmic effector region is also a theme common to both the "effector dimer" model and the "effector oligomer/polymer" model. Both models place back-to-back dimerization of the IRE1 effector region at the center of IRE1 nuclease activation.

Connected to effector dimerization is the issue of ligand engagement at the IRE1 kinase active site. In all IRE1 effector region crystal structures determined to date, occupation of the IRE1 kinase active site by small molecules that mime ADP's mode of binding stabilized the "closed" interlobe configuration of the IRE1 kinase domain (discussed in Section 3.2).

This productive "closed" kinase domain configuration facilitates access to the back-to-back dimer arrangement of the IRE1 effector region, which is essential to IRE1 nuclease activation. Enforced occupancy of the IRE1 kinase active site by 1NM-PP1 was critical to IRE1-mediated signaling in both yeast and mammalian cells (Han et al., 2009; Papa et al., 2003). Thus, stabilization of the "closed" kinase domain configuration by ligand engagement at the kinase active site is a key milestone during activation of IRE1 nuclease activity.

The identification of the Q-site opens a new dimension to manipulate dimerization of the IRE1 effector region pharmacologically. There are currently no known native ligands of the Q-site and the occupancy of this novel ligand-binding pocket is not essential for ER stress signaling by IRE1 (Wiseman et al., 2010). However, quercetin binding at the Q-site was sufficient to potently stimulate the nuclease activity of IRE1 by promoting back-to-back dimerization of the IRE1 effector region. In addition, the Q-site is conserved from yeast to humans. Thus, the Q-site represents an alternate (general) access point to control IRE1 effector dimerization using small molecules.

The "effector dimer" model and "effector oligomer/polymer" model prophesize distinct requirements for autophosphorylation in the potentiation of IRE1 nuclease function. Notwithstanding, kinase-deficit variants of IRE1 are severely impaired for UPR signaling in both fungal and mammalian systems (Mori et al., 1993; Tirasophon et al., 2000). Thus, it is clear that autophosphorylation of wild-type IRE1 protein is a critical step in wild-type IRE1 nuclease activation *in vivo*.

3.5. Principles of IRE1 chemical modulation

Prosurvival and proapoptotic signals emanating from IRE1 have been attributed to its nuclease and kinase enzymatic outputs, respectively. Given its connection to disease there is great interest to develop compounds capable of finessing signaling outputs from IRE1 to therapeutic effect.

The task of designing strategies to manipulate IRE1 function chemically is greatly facilitated by information available on IRE1 structure and regulation. Regulated dimerization of the IRE1 luminal and effector regions are hallmarks of IRE1 activation in all eukaryotic systems studied to date. With the benefit of this insight, one can propose that using small molecules to modulate the self-association steps in the IRE1 activation process may be a plausible approach to control IRE1 signaling.

Controlling protein–protein interactions using small molecules has traditionally been challenging owing to the paucity of "druggable" features at protein–protein interaction surfaces. Fortunately, multiple ligand-binding sites directing self-association are already built into the IRE1 effector region. The nucleotide-binding site in the kinase domain and the Q-site at the KEN domain dimer interface are entry points that allow ligand-based control of IRE1 effector dimerization.

As discussed, ATP-competitive compounds that engage the IRE1 kinase active site by mimicking the binding mode of ADP potentiate IRE1 nuclease activity by stabilizing the "closed" configuration of the IRE1 kinase domain (Fig. 16.4A and B). As stabilization of the "closed" kinase domain configuration is conducive to back-to-back dimerization of the IRE1 effector region, we refer to positive modulators of IRE1 nuclease function that engage the kinase active site as "kinase-directed dimer promoters."

Given the positive modulatory effect of the "closed" kinase domain configuration on IRE1 nuclease activity, we infer that kinase-directed compounds that do not support the "closed" kinase domain configuration would suppress IRE1 nuclease output. Indeed, we have identified a number of compounds that inhibit IRE1 nuclease activity by targeting the IRE1 kinase active site (data not shown). We predict that kinase active site targeting compounds that stabilize "overopen," "overclosed," or other interlobe arrangements deviating from the "closed" configuration, will suppress IRE1 nuclease activity by prohibiting back-to-back dimerization of the IRE1 effector module (Fig. 16.4C). We refer to kinase-directed negative modulators of IRE1 nuclease output that inhibit back-to-back dimerization of the IRE1 effector as "kinase-directed dimer breakers."

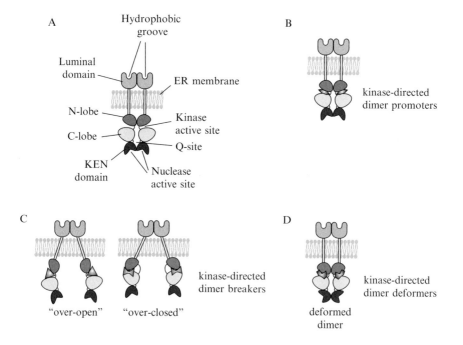

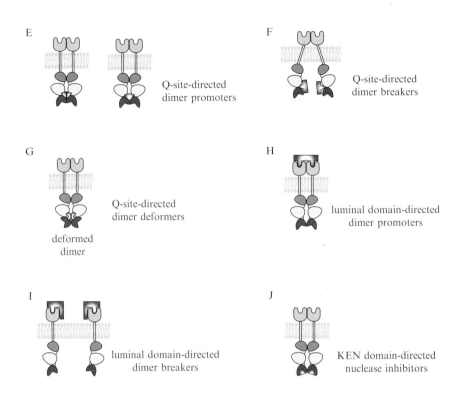

Figure 16.4 A blueprint for modulating IRE1 using chemical tools. (A) Schematic of an IRE1 back-to-back dimer in the plane of the ER membrane. Both the luminal domain and the cytoplasmic effector module are shown in their dimerized forms. The orientation of the N- and C-lobes in the IRE1 kinase domain as shown represents a "closed" configuration that facilitates formation of the back-to-back dimer arrangement observed in all IRE1 effector crystal structures determined to date. (B–J) Predicted modes of action of different classes of IRE1 chemical modulators (see text for details). (B) Kinase-directed dimer promoters, (C) Kinase-directed dimer breakers, (D) Kinase-directed dimer deformers, (E) Q-site-directed dimer promoters, (F) Q-site-directed dimer breakers, (G) Q-site-directed dimer deformers, (H) Luminal domain-directed dimer promoters, (I) Luminal domain-directed dimer breakers, (J) KEN domain-directed nuclease inhibitors.

Alternatively, it is possible that kinase-directed negative modulators of IRE1 nuclease function act by stabilizing a distorted dimer that does not support RNA cleavage (Fig. 16.4D). We refer to this last class of negative modulators targeting the kinase active site as "kinase-directed dimer deformers."

We note that all kinase-directed IRE1 modulators have the double duty of being allosteric modulators of IRE1 nuclease activity and orthosteric inhibitors of IRE1 kinase output, owing to their ATP-competitive nature. Hence it may be possible to selectively inhibit IRE1 kinase output while preserving IRE1 nuclease function using "kinase-directed dimer promoters." Alternatively, it may be possible to simultaneously suppress both kinase and nuclease outputs from IRE1 using "kinase-directed dimer breakers" or "kinase-directed dimer deformers."

The kinase active site provides opportunities for the discovery of allosteric modulators that exert indirect control over IRE1 effector dimerization. In contrast, the Q-site, by virtue of its location at the interface of paired KEN domains, is an excellent candidate for the development of chemical modulators that directly stimulate or suppress IRE1 effector dimerization.

In the IRE1^{cyto}-ADP-quercetin crystal structure, two quercetin molecules interact symmetrically to mediate an intricate network of hydrogen bonding and hydrophobic interactions across paired KEN domains at the Q-site. This network of interactions promotes back-to-back dimerization of the IRE1 effector, thereby enhancing IRE1 nuclease activity. We predict that small molecules that mimic the binding mode of quercetin to the Q-site will function as positive modulators of IRE1 nuclease output. We refer to this class of IRE1 modulators as "Q-site-directed dimer promoters" (Fig. 16.4E).

"Q-site-directed dimer promoters" can potentially exploit the symmetric drug–drug interaction utilized by quercetin to engage the twofold symmetric Q-site, as previously noted (Shokat, 2010). In principle, "Q-site-directed dimer promoters" can also be ligands in which symmetric drug–drug interactions are enforced (e.g., by cross-linking a pair of quercetin molecules) to reduce the entropic cost of Q-site engagement and thereby enhance the stability of the resultant IRE1 effector back-to-back dimer. Alternatively, "Q-site-directed dimer promoters" may be asymmetric molecules that bind across the Q-site to stabilize the back-to-back dimer arrangement.

We also predict that Q-site binding compounds can potentially suppress IRE1 nuclease output by destabilizing the back-to-back dimer arrangement of the IRE1 effector region. We refer to this class of negative modulators of IRE1 nuclease function as "Q-site-directed dimer breakers" (Fig. 16.4F). Alternatively, we predict that allosteric inhibition of IRE1 nuclease activity may be achieved by Q-site binders that induce a distorted version of the IRE1 effector dimer that is unfavorable toward RNA cleavage by IRE1. We refer to ligands of the Q-site that deform the back-to-back IRE1 effector dimer as "Q-site-directed dimer deformers" (Fig. 16.4G).

As the Q-site is removed from the kinase active site, one would expect that IRE1 modulators targeting the Q-site will not impinge on IRE1 kinase output. Hence, Q-site-directed compounds can be used as selective allosteric modulators of IRE1 nuclease activity. However, it is worth nothing that

ADP inhibits quercetin-dependent nuclease output from nonphosphorylated IRE1$^{\text{cyto(D797A)}}$ (Wiseman et al., 2010). This inhibitory effect is contrary to the stimulatory effect of ADP on quercetin activated phosphorylated wild-type IRE1^{cyto}. As a corollary, phosphorylated and nonphosphorylated pools of IRE1 protein may respond differently to Q-site targeting compounds. This can have desired or undesired consequences and is a potential consideration in drug design efforts aimed at targeting the IRE1 Q-site. More work is needed to understand the basis of this phenomenon.

We note that although no ligands that target the putative peptide groove in the IRE1 luminal domain have been identified to date, this potential ligand-binding site is hydrophobic, canyon-like in form and strategically positioned across the luminal-domain dimer. These attributes make the dimer-spanning groove in the IRE1 luminal domain an attractive target for the discovery of compounds that directly modulate luminal-domain dimerization.

It is conceivable that small molecules that bind across the dimer-spanning groove will stabilize the luminal-domain dimer. We refer to this group of compounds as "luminal-domain-directed dimer promoters" (Fig. 16.4H). We also propose that it may be possible to identify small-molecule inhibitors of luminal-domain dimerization that bind to the half-site that compose the hydrophobic groove in a luminal domain. These luminal-domain dimerization inhibitors can potentially prohibit luminal-domain self-association by steric hindrance or inducing conformational changes in the luminal domain that disfavor dimerization. We refer to this group of compounds as "luminal-domain-directed dimer breakers" (Fig. 16.4I).

As self-association of the IRE1 luminal-domain stimulates both the kinase and nuclease activities of cytoplasmic effector, we expect that "luminal-domain-directed dimer promoters" will be positive allosteric modulators of both IRE1 kinase and nuclease outputs. Similarly, we expect "luminal-domain-directed dimer breakers" will be negative allosteric modulators of both cytoplasmic enzyme activities of IRE1.

Lastly, the most direct route to selectively inhibit nuclease output from IRE1 would be through the use of small molecules that directly target the IRE1 nuclease active site. The "pocket-like" nature of the IRE1 nuclease active site in the KEN domain (observed in both effector dimer and effector polymer crystal forms (Korennykh et al., 2009; Wiseman et al., 2010) indicates that this site is also an attractive target for drug discovery. We refer to IRE1 nuclease inhibitors that bind to the IRE1 nuclease active site in the KEN domain as "KEN domain-directed nuclease inhibitors" (Fig. 16.4J).

Many of aforementioned classes of IRE1 modulators have yet to be identified but the mode of action of these classes compounds can be predicted based on our current understanding of mechanistic details the

IRE1 activation process. Our goal here is to provide a blueprint for chemical modulation of IRE1 function by highlighting potential avenues to access specific aspects of IRE1 signaling pharmacologically. This may be advantageous in various disease states. For example, it may be desirable to selectively inhibit proapoptotic kinase activity from IRE1 while preserving prosurvival signals from IRE1 nuclease output in pancreatic beta-cells in patients with Type-II diabetes. For this purpose, drugs that function as "kinase-directed dimer promoters" can be considered. It may also be desirable to selectively suppress prosurvival signaling from IRE1 nuclease output while maintaining prop-apoptotic IRE1 kinase function in multiple myeloma patients. This may be accomplished through the use of "Q-site-directed dimer-breakers", for instance.

It is hoped that the set of principles of IRE1 chemical modulation outlined here will help advance the development of new tools to better understand how IRE1 function contributes to health and disease and perhaps ultimately the development of new methods of therapeutic intervention.

REFERENCES

Aragon, T., van Anken, E., Pincus, D., Serafimova, I. M., Korennykh, A. V., Rubio, C. A., and Walter, P. (2009). Messenger RNA targeting to endoplasmic reticulum stress signalling sites. *Nature* **457,** 736–740.

Bernales, S., Papa, F. R., and Walter, P. (2006). Intracellular signaling by the unfolded protein response. *Annu. Rev. Cell Dev. Biol.* **22,** 487–508.

Boot-Handford, R. P., and Briggs, M. D. (2010). The unfolded protein response and its relevance to connective tissue diseases. *Cell Tissue Res.* **339,** 197–211.

Calfon, M., Zeng, H., Urano, F., Till, J. H., Hubbard, S. R., Harding, H. P., Clark, S. G., and Ron, D. (2002). IRE1 couples endoplasmic reticulum load to secretory capacity by processing the XBP-1 mRNA. *Nature* **415,** 92–96.

Cox, J. S., and Walter, P. (1996). A novel mechanism for regulating activity of a transcription factor that controls the unfolded protein response. *Cell* **87,** 391–404.

Credle, J. J., Finer-Moore, J. S., Papa, F. R., Stroud, R. M., and Walter, P. (2005). On the mechanism of sensing unfolded protein in the endoplasmic reticulum. *Proc. Natl. Acad. Sci. USA* **102,** 18773–18784.

Fribley, A., Zhang, K., and Kaufman, R. J. (2009). Regulation of apoptosis by the unfolded protein response. *Methods Mol. Biol.* **559,** 191–204.

Gonzalez, T. N., Sidrauski, C., Dorfler, S., and Walter, P. (1999). Mechanism of non-spliceosomal mRNA splicing in the unfolded protein response pathway. *EMBO J.* **18,** 3119–3132.

Han, D., Lerner, A. G., Vande Walle, L., Upton, J. P., Xu, W., Hagen, A., Backes, B. J., Oakes, S. A., and Papa, F. R. (2009). IRE1alpha kinase activation modes control alternate endoribonuclease outputs to determine divergent cell fates. *Cell* **138,** 562–575.

Hetz, C., and Glimcher, L. H. (2009). Fine-tuning of the unfolded protein response: Assembling the IRE1alpha interactome. *Mol. Cell* **35,** 551–561.

Hollien, J., and Weissman, J. S. (2006). Decay of endoplasmic reticulum-localized mRNAs during the unfolded protein response. *Science* **313,** 104–107.

Hollien, J., Lin, J. H., Li, H., Stevens, N., Walter, P., and Weissman, J. S. (2009). Regulated Ire1-dependent decay of messenger RNAs in mammalian cells. *J. Cell Biol.* **186**, 323–331.

Kaufman, R. J. (2002). Orchestrating the unfolded protein response in health and disease. *J. Clin. Invest.* **110**, 1389–1398.

Kaufman, R. J., and Cao, S. (2010). Inositol-requiring 1/X-box-binding protein 1 is a regulatory hub that links endoplasmic reticulum homeostasis with innate immunity and metabolism. *EMBO Mol. Med.* **2**(6), 189–192.

Kim, R., Emi, M., Tanabe, K., and Murakami, S. (2006). Role of the unfolded protein response in cell death. *Apoptosis* **11**, 5–13.

Kim, I., Xu, W., and Reed, J. C. (2008). Cell death and endoplasmic reticulum stress: Disease relevance and therapeutic opportunities. *Nat. Rev. Drug Discov.* **7**, 1013–1030.

Kimata, Y., Ishiwata-Kimata, Y., Ito, T., Hirata, A., Suzuki, T., Oikawa, D., Takeuchi, M., and Kohno, K. (2007). Two regulatory steps of ER-stress sensor Ire1 involving its cluster formation and interaction with unfolded proteins. *J. Cell Biol.* **179**, 75–86.

Korennykh, A. V., Egea, P. F., Korostelev, A. A., Finer-Moore, J., Zhang, C., Shokat, K. M., Stroud, R. M., and Walter, P. (2009). The unfolded protein response signals through high-order assembly of Ire1. *Nature* **457**, 687–693.

Lee, K., Tirasophon, W., Shen, X., Michalak, M., Prywes, R., Okada, T., Yoshida, H., Mori, K., and Kaufman, R. J. (2002). IRE1-mediated unconventional mRNA splicing and S2P-mediated ATF6 cleavage merge to regulate XBP1 in signaling the unfolded protein response. *Genes Dev.* **16**, 452–466.

Lee, K. P., Dey, M., Neculai, D., Cao, C., Dever, T. E., and Sicheri, F. (2008). Structure of the dual enzyme Ire1 reveals the basis for catalysis and regulation in nonconventional RNA splicing. *Cell* **132**, 89–100.

Li, H., Korennykh, A. V., Behrman, S. L., and Walter, P. (2010). Mammalian endoplasmic reticulum stress sensor IRE1 signals by dynamic clustering. *Proc. Natl. Acad. Sci. USA* **107**(37), 16113–16118.

Lin, J. H., Li, H., Yasumura, D., Cohen, H. R., Zhang, C., Panning, B., Shokat, K. M., Lavail, M. M., and Walter, P. (2007). IRE1 signaling affects cell fate during the unfolded protein response. *Science* **318**, 944–949.

Lin, J. H., Walter, P., and Yen, T. S. (2008). Endoplasmic reticulum stress in disease pathogenesis. *Annu. Rev. Pathol.* **3**, 399–425.

Lin, J. H., Li, H., Zhang, Y., Ron, D., and Walter, P. (2009). Divergent effects of PERK and IRE1 signaling on cell viability. *PLoS ONE* **4**, e4170.

Liu, C. Y., Schroder, M., and Kaufman, R. J. (2000). Ligand-independent dimerization activates the stress response kinases IRE1 and PERK in the lumen of the endoplasmic reticulum. *J. Biol. Chem.* **275**, 24881–24885.

Malhotra, J. D., and Kaufman, R. J. (2007). The endoplasmic reticulum and the unfolded protein response. *Semin. Cell Dev. Biol.* **18**, 716–731.

McGuckin, M. A., Eri, R. D., Das, I., Lourie, R., and Florin, T. H. (2010). ER stress and the unfolded protein response in intestinal inflammation. *Am. J. Physiol. Gastrointest. Liver Physiol.* **298**, G820–G832.

Mori, K. (2009). Signalling pathways in the unfolded protein response: Development from yeast to mammals. *J. Biochem.* **146**, 743–750.

Mori, K., Ma, W., Gething, M. J., and Sambrook, J. (1993). A transmembrane protein with a cdc2+/CDC28−related kinase activity is required for signaling from the ER to the nucleus. *Cell* **74**, 743–756.

Naidoo, N. (2009). Cellular stress/the unfolded protein response: Relevance to sleep and sleep disorders. *Sleep Med. Rev.* **13**, 195–204.

Nishitoh, H., Matsuzawa, A., Tobiume, K., Saegusa, K., Takeda, K., Inoue, K., Hori, S., Kakizuka, A., and Ichijo, H. (2002). ASK1 is essential for endoplasmic reticulum stress-

induced neuronal cell death triggered by expanded polyglutamine repeats. *Genes Dev.* **16,** 1345–1355.
Ogawa, S., Kitao, Y., and Hori, O. (2007). Ischemia-induced neuronal cell death and stress response. *Antioxid. Redox Signal.* **9,** 573–587.
Oikawa, D., Kimata, Y., Kohno, K., and Iwawaki, T. (2009). Activation of mammalian IRE1alpha upon ER stress depends on dissociation of BiP rather than on direct interaction with unfolded proteins. *Exp. Cell Res.* **315,** 2496–2504.
Papa, F. R., Zhang, C., Shokat, K., and Walter, P. (2003). Bypassing a kinase activity with an ATP-competitive drug. *Science* **302,** 1533–1537.
Pincus, D., Chevalier, M. W., Aragon, T., van Anken, E., Vidal, S. E., El-Samad, H., and Walter, P. (2010). BiP binding to the ER-stress sensor Ire1 tunes the homeostatic behavior of the unfolded protein response. *PLoS Biol.* 8e1000415.
Ron, D., and Walter, P. (2007). Signal integration in the endoplasmic reticulum unfolded protein response. *Nat. Rev. Mol. Cell Biol.* **8,** 519–529.
Scheuner, D., and Kaufman, R. J. (2008). The unfolded protein response: A pathway that links insulin demand with beta-cell failure and diabetes. *Endocr. Rev.* **29,** 317–333.
Shamu, C. E., and Walter, P. (1996). Oligomerization and phosphorylation of the Ire1p kinase during intracellular signaling from the endoplasmic reticulum to the nucleus. *EMBO J.* **15,** 3028–3039.
Shokat, K. M. (2010). A drug–drug interaction crystallizes a new entry point into the UPR. *Mol. Cell* **38,** 161–163.
Sidrauski, C., and Walter, P. (1997). The transmembrane kinase Ire1p is a site-specific endonuclease that initiates mRNA splicing in the unfolded protein response. *Cell* **90,** 1031–1039.
Sidrauski, C., Cox, J. S., and Walter, P. (1996). tRNA ligase is required for regulated mRNA splicing in the unfolded protein response. *Cell* **87,** 405–413.
Szegezdi, E., Logue, S. E., Gorman, A. M., and Samali, A. (2006). Mediators of endoplasmic reticulum stress-induced apoptosis. *EMBO Rep.* **7,** 880–885.
Tirasophon, W., Lee, K., Callaghan, B., Welihinda, A., and Kaufman, R. J. (2000). The endoribonuclease activity of mammalian IRE1 autoregulates its mRNA and is required for the unfolded protein response. *Genes Dev.* **14,** 2725–2736.
Travers, K. J., Patil, C. K., Wodicka, L., Lockhart, D. J., Weissman, J. S., and Walter, P. (2000). Functional and genomic analyses reveal an essential coordination between the unfolded protein response and ER-associated degradation. *Cell* **101,** 249–258.
Urano, F., Wang, X., Bertolotti, A., Zhang, Y., Chung, P., Harding, H. P., and Ron, D. (2000). Coupling of stress in the ER to activation of JNK protein kinases by transmembrane protein kinase IRE1. *Science* **287,** 664–666.
van der Kallen, C. J., van Greevenbroek, M. M., Stehouwer, C. D., and Schalkwijk, C. G. (2009). Endoplasmic reticulum stress-induced apoptosis in the development of diabetes: Is there a role for adipose tissue and liver? *Apoptosis* **14,** 1424–1434.
Wiseman, R. L., Zhang, Y., Lee, K. P., Harding, H. P., Haynes, C. M., Price, J., Sicheri, F., and Ron, D. (2010). Flavonol activation defines an unanticipated ligand-binding site in the kinase-RNase domain of IRE1. *Mol. Cell* **38,** 291–304.
Yoneda, T., Imaizumi, K., Oono, K., Yui, D., Gomi, F., Katayama, T., and Tohyama, M. (2001). Activation of caspase-12, an endoplastic reticulum (ER) resident caspase, through tumor necrosis factor receptor-associated factor 2-dependent mechanism in response to the ER stress. *J. Biol. Chem.* **276,** 13935–13940.
Zhou, J., Liu, C. Y., Back, S. H., Clark, R. L., Peisach, D., Xu, Z., and Kaufman, R. J. (2006). The crystal structure of human IRE1 luminal domain reveals a conserved dimerization interface required for activation of the unfolded protein response. *Proc. Natl. Acad. Sci. USA* **103,** 14343–14348.

CHAPTER SEVENTEEN

Methods to Study Stromal-Cell Derived Factor 2 in the Context of ER Stress and the Unfolded Protein Response in *Arabidopsis thaliana*

Andrea Schott *and* Sabine Strahl

Contents

1. Introduction	296
2. *Arabidopsis sdf2* T-DNA Insertion Lines	297
2.1. Genotyping *sdf2* mutants	298
3. Inducing and Monitoring ER Stress in *Arabidopsis sdf2* Mutants	299
3.1. Plant growth and application of ER stress-inducing agents	300
3.2. Monitoring UPR induction	301
3.3. ER stress-induced phenotypes	302
4. Analyzing the ER Stress-Induced Expression Pattern of *SDF2* in *Arabidopsis*	304
4.1. Transgenic *Arabidopsis* lines expressing an *SDF2* promoter GUS reporter fusion	304
4.2. qRT-PCR	307
4.3. SDF2-specific antibodies	307
5. The Subcellular Localization of SDF2	309
5.1. Transient expression of SDF2 fusions in tobacco leaves	309
5.2. Analyzing localization of SDF2-HA by sucrose density gradient centrifugation	310
5.3. Analyzing localization of SDF2–roGFP by confocal microscopy	312
6. Purification of Recombinant SDF2 Protein from *E. coli*	313
6.1. GST-SDF2 fusion construct	313
6.2. Production of the GST-SDF2 fusion protein in *E. coli*	314
6.3. Affinity purification of GST-SDF2	315
6.4. Removal of the GST-tag	316
Acknowledgments	316
References	317

Department of Cell Chemistry, Centre for Organismal Studies (COS), University of Heidelberg, Heidelberg, Germany

Abstract

The accumulation of misfolded or unfolded polypeptides in the endoplasmic reticulum (ER) provokes ER stress and triggers protective signaling pathways termed the unfolded protein response (UPR). Stromal cell-derived factor 2 (SDF2)-type proteins are conserved throughout the animal and plant kingdoms. Upon UPR activation transcription of *SDF2*-type genes is significantly enhanced in metazoan and plants, suggesting an evolutionarily conserved role. However, the precise molecular function of SDF2-type proteins still needs to be established. Most eukaryotes have two SDF2 homologous, whereas the model plant *Arabidopsis thaliana* has a single SDF2, thus representing an ideal model system to study the functional role of SDF2-type proteins.

This chapter provides techniques to study SDF2 in the context of ER stress in *Arabidopsis*. We describe available *sdf2* mutants, and methods to evaluate ER stress sensitivity of seedlings. Further, we summarize tools and methods that are helpful to monitor UPR induction in general (e.g., SDF2 promoter–reporter fusion constructs and SDF2-specific antibodies). In Section 6, we provide protocols for the expression and purification of recombinant SDF2 protein that can be used for further biochemical studies.

1. Introduction

The most prominent phenomenon induced by the unfolded protein response (UPR) is the transcriptional induction of different endoplasmic reticulum (ER) chaperone systems (reviewed in Urade, 2009). One of the major ER chaperone systems depends on the HSP70 family member immunoglobulin heavy chain-binding protein (BiP) (Bertolotti et al., 2000). In mammalian cells, BiP has been demonstrated to associate with the molecular chaperones glucose-regulated protein 94 (GRP94), calcium-binding protein 1 (CaBP1), protein disulfide isomerase (PDI), ER Hsp40-like co-chaperone (ERdj3), cyclophilin B, ER protein 72 kDa (ERp72), GRP170, UDP-glucosyltransferase, and stromal cell-derived factor 2-like1 (SDF2-L1) to form a large ER-localized complex of chaperones and folding enzymes (Bies et al., 2004; Meunier et al., 2002). In contrast to most of these factors, the molecular function of SDF2-type proteins is largely unknown.

SDF2-type proteins have been initially identified in mouse and human (Fukuda et al., 2001; Hamada et al., 1996). They are highly conserved throughout the animal and plant kingdoms. In the genomes of vertebrates and some land plants (e.g., *Oryza sativa* and *Vitis vinifera*) two SDF2-type proteins are encoded (Nekrasov et al., 2009; Schott et al., 2010). Transcription of *SDF2*-type genes is significantly upregulated in metazoans and plants at the onset of the UPR, suggesting an evolutionarily conserved function of SDF2-type proteins (Adachi et al., 2008; Fukuda et al., 2001; Kamauchi et al., 2005; Martinez and Chrispeels, 2003; Schott et al., 2010).

The principle mechanisms of ER protein quality control and the UPR are conserved between metazoans and plants (Urade, 2009; Vitale and Boston, 2008). As the model plant *Arabidopsis thaliana* has only a single *SDF2* gene, it provides an ideal model to study SDF2-type proteins. The *Arabidopsis SDF2* gene (At2g25110) is composed of six exons and five introns (accession number: NM_128068). *Arabidopsis* SDF2 shares an overall protein sequence similarity of 55% and 52% with its murine/human homologs SDF2 and SDF2-L1, respectively. It is an ER-resident protein, whose abundance is increased in response to ER stress conditions (Schott *et al.*, 2010).

Typically, SDF2-type proteins contain three copies of a highly conserved MIR motif, which is named after three of the proteins in which it occurs (protein O-*m*annosyltransferases, *i*nositol-1,4,5-triphosphate and *r*yanodine receptors; Ponting, 2000). The 3D crystal structure of *Arabidopsis* SDF2 at 1.95 Å resolution (protein database entry 3MAL) revealed a typical β-trefoil fold with a pseudo threefold symmetry made up by structural repeats that correspond to the three MIR motifs covering amino acid residues 34–88 (MIR1), 96–151 (MIR2), and 154–208 (MIR3) (Radzimanowski *et al.*, 2010; Schott *et al.*, 2010). Analyses of the surface characteristics identified several putative interaction sites for protein as well as carbohydrate ligands, suggesting that SDF2 might interact with ER glycoproteins (Schott *et al.*, 2010). This is further supported by the fact that SDF2 and its interacting partner ERdj3b are required for the maturation and secretion of the *N*-glycosylated surface-exposed leucine-rich receptor kinase EFR in *Arabidopsis* (Nekrasov *et al.*, 2009).

Arabidopsis sdf2 mutant seedlings are particularly sensitive toward ER stress conditions and show severe developmental and morphological defects upon treatment with ER stress-inducing agents (Schott *et al.*, 2010). In these *Arabidopsis sdf2* mutants, the EFR receptor is retained in the ER lumen and degraded via the ER-associated protein degradation pathway (Nekrasov *et al.*, 2009). These data indicate that SDF2 is involved in the folding and quality control of ER glycoproteins, albeit the precise molecular function of SDF2 still needs to be established.

In this chapter, we summarize methods and tools for the detailed analysis of SDF2 in the context of ER stress and the UPR in the model organism *Arabidopsis*.

2. *ARABIDOPSIS SDF2* T-DNA INSERTION LINES

To study the functional role of *SDF2 in vivo*, *Arabidopsis* T-DNA insertion mutants are a valuable resource. Various stock collections are available to obtain mutant lines (for links see The Arabidopsis Information

Resource (TAIR); www.arabidopsis.org). So far, three *Arabidopsis sdf2* T-DNA insertion lines have been characterized and can be ordered from the Nottingham Arabidopsis stock center (NASC). The knockout mutants *sdf2-2* (SALK_141321; Nekrasov *et al.*, 2009; Schott *et al.*, 2010) and *sdf2-5* (WiscDsLox293–296invl23; Schott *et al.*, 2010) contain T-DNA integrations in exon 2 and exon 5, respectively, of the *SDF2* coding sequence. Line *sdf2-6* (SALK_077481) containing a T-DNA integration in the *SDF2* promoter (at position bp −294 from the ATG start codon), represents a knockdown line with significantly reduced *SDF2* transcript and protein levels (Schott *et al.*, 2010).

2.1. Genotyping *sdf2* mutants

Arabidopsis sdf2 mutant lines and Columbia-0 wild-type plants (as a control) are grown on soil under standard conditions and genomic DNA is prepared according to Edwards *et al.* (1991). General methods for growth and handling *Arabidopsis* are described in detail in Arabidopsis–A Laboratory Manual edited by Weigel and Glazebrook (2002).

Materials

Edwards-buffer (200 mM Tris-HCl, pH 8.0, 25 mM EDTA, 250 mM NaCl, 0.5% SDS), plastic pestles, 70% ethanol, isopropanol, TE buffer (10 mM Tris-HCl, pH 7.5, 1 mM EDTA).

Procedure

1. Plant tissue (5-50 mg) is ground in a microfuge tube with a disposable plastic pestle.
2. 400 μl of Edwards-buffer is added and samples are briefly vortexed. Samples are kept at room temperature until all are processed.
3. Samples are centrifuged for 5 min at 20,000 × g.
4. 300 μl of the supernatant are transferred to a fresh tube, mixed with 300 μl of isopropanol, and incubated for 5 min at room temperature.
5. Samples are centrifuged for 10 min at 20,000 × g; the supernatant is removed and discarded.
6. The DNA-pellet is washed with 700 μl of 70% ethanol and dried.
7. The DNA pellet is resuspended in 25-50 μl of TE buffer or sterile water and samples are incubated for 3 min at 95 °C.
8. 1 μl of genomic DNA is used as template in a 25-μl PCR reaction.

Two separate PCR reactions are carried out. The wild-type *SDF2* allele is detected with gene-specific primers flanking the T-DNA integration site (Table 17.1; left primer LP and right primer RP). The mutant allele is detected using a primer specific for the left border sequence of the T-DNA

Table 17.1 Genotyping of *sdf2* mutants: primer sequences; and sizes of expected PCR products[a]

Line	Primer sequence		PCR product (bp)
	LP 5'–3'	RP 5'–3'	WT allele
sdf2-2	cacaggtcacatatactcgc	atcatctcctcctccgtcttc	946
sdf2-5	gttaaacctgtgcctgggacaac	ccaaaacaagatatattcgtgcc	1007
sdf2-6	attgctgaccactgccggatc	ttcacaatttatgccggatctagtacc	874
	BP 5'–3'	RP 5'–3'	Mutant allele
sdf2-2	attttgccgatttcggaac	atcatctcctcctccgtcttc	612
sdf2-5	aatcgccttgcagcacatcc	ccaaaacaagatatattcgtgcc	533
sdf2-6	attttgccgatttcggaac	ttcacaatttatgccggatctagtacc	~440

[a] This research was originally published in Schott *et al.* (2010). Arabidopsis stromal-derived factor 2 (SDF2) is a crucial target of the unfolded protein response in the endoplasmic reticulum. © The American society of Biochemistry and Molecular Biology.

(border primer–BP) and a gene-specific primer RP (for primer sequences and product sizes see Table 17.1). Standard PCR reactions are performed using DreamTaq™ DNA polymerase (Fermentas) according to the manufacturer's instruction.

3. Inducing and Monitoring ER Stress in *Arabidopsis sdf2* Mutants

The most frequently applied agents to provoke ER stress and induce the UPR in *Arabidopsis* are the reducing agent dithiothreitol (DTT); and tunicamycin, an inhibitor of protein *N*-glycosylation. Both chemicals interfere with protein folding and maturation in the ER, however, through different mechanisms. Tunicamycin inhibits *N*-acetylglucosamine phosphotransferase, which catalyzes the initial step of *N*-glycan biosynthesis, and prevents *N*-glycosylation of newly synthesized proteins in the ER (Elbein, 1984). In contrast, DTT affects protein folding by interfering with disulfide bond formation. Both agents trigger the UPR in *Arabidopsis* and as a result induce the expression of characteristic UPR target genes such as the Hsp70 chaperones *BIP1-3*, *PDI*, calnexin (*CNX*), calreticulin (*CRT*) (Kamauchi *et al.*, 2005; Martinez and Chrispeels, 2003) as well as *SDF2* (Schott *et al.*, 2010). In *Arabidopsis*, ER stress particularly affects seedling development with most notable changes in root morphology (Schott *et al.*, 2010; Watanabe and Lam, 2008). As determinants for an increased sensitivity of *Arabidopsis sdf2* mutants toward ER stress, expansion of green cotyledons, root elongation, and chlorophyll content can be compared with wild-type seedlings. The following section describes methods to provoke and monitor UPR in *Arabidopsis*, and to characterize ER stress-induced phenotypes in *sdf2* mutants.

3.1. Plant growth and application of ER stress-inducing agents

The impact of ER stress on growth and development of *Arabidopsis* is analyzed best in seedlings. To grow *Arabidopsis* plantlets under sterile culture conditions, seeds have to be surface sterilized (Weigel and Glazebrook, 2002). Thereto, seeds are incubated for 10 min in 70% ethanol/0.1% Triton X-100 with occasional agitation. After a short centrifugation to sediment the seeds, the supernatant is removed. Seeds are then incubated for 10 min in 2.5% hypochloride/0.1% Triton X-100 with mild agitation. After removal of the hypochloride solution, seeds are washed three times with sterile water and kept therein. In order to enhance coordinated germination, seeds are stratified for 2 days at 4 °C (Weigel and Glazebrook, 2002). After stratification, seeds are sown on MS agar plates containing 1× Murashige and Skoog (MS) salts (Duchefa), 1% sucrose, 10 mM 2-(N-morpholino)ethanesulfonic acid (MES) pH 5.7, and 0.8% agar (Duchefa). MS plates are sealed with a gas-permeable tape such as surgical tap (Micropore) and incubated in a plant growth chamber at 21 °C under long-day conditions with 16 h light and 8 h dark. The above described conditions are defined as standard growth conditions.

Stock solutions

DTT: 1,4-dithiothreitol (Carl Roth); 1 M stock solution in Millipore water; freshly prepared and filter-sterilized.

Tunicamycin: tunicamycin from *Streptomyces* sp. (Sigma); 5 mg/ml stock solution in DMSO; stored at 4 °C up to 6 months.

To cause ER stress conditions during seed germination and seedling growth on solid medium, either 1-2.5 mM DTT or 0.01-0.05 μg/ml tunicamycin are added to MS plates after autoclaving. It is advisable to determine the concentration range for individual tunicamycin batches, since variations have been observed. Cool MS plates down to approximately 50 °C before adding-on DTT or tunicamycin. We recommend preparing these plates on the day of use to obtain standardized conditions. Seedlings are cultivated as detailed above.

To transiently induce ER stress in 4- to 7-day-old *Arabidopsis* seedlings, seeds are germinated on MS medium under standard conditions. At the desired growth stage, seedlings are removed from the plates under sterile conditions using a forceps. Seedlings are placed in sterile 6-well culture plates (Greiner) containing MS medium (20–50 seedlings per well) and incubated for 24 h under standard conditions. The seedlings are then transferred under sterile conditions, to fresh MS medium containing either 1–5 mM DTT or 0.1–5 μg/ml tunicamycin. Seedlings are further incubated under standard growth conditions for 24–72 h (DTT treatment) or for 5–24 h (tunicamycin treatment).

3.2. Monitoring UPR induction

To control effectiveness of DTT and tunicamycin treatments, transcript levels of UPR marker genes such as *CRT2*, *BIP1/2*, and *BIP3* (Kamauchi et al., 2005; Koizumi, 1996; Noh et al., 2003) can be monitored by semiquantitative reverse transcription polymerase chain reaction (sqRT-PCR). Thereto, RNA is extracted from *Arabidopsis* wild-type and *sdf2* mutant lines and reverse transcribed as detailed in Section 4.2. Semiquantitative RT-PCR is performed using DreamTaq™ DNA polymerase (Fermentas) according to the manufacturer's instructions. PCR reactions (20 μl total volume) contain cDNA derived from 20 ng RNA, 200 nM of forward and reverse primers, and 400 μM of each dNTP. *BIP1/2*, *BIP3*, *CRT2*, and *PP2A* transcripts are amplified using gene-specific primers. PCR reactions are carried out with an initial denaturation step at 95 °C for 3 min, followed by cycles of 95 °C for 30 s, 59 °C for 30 s, and 72 °C for 15 s. Primer sequences, size of PCR products and cycle numbers used are summarized in Table 17.2. The number of cycles used for amplification with each primer pair was adjusted to be in the linear range.

Table 17.2 Semiquantitative PCR of UPR target genes: primer sequences; sizes of expected PCR products; PCR conditions[a]

Gene	Forward primer 5'–3'	Reverse primer 5'–3'	PCR product (bp)	Cycle no.
BIP1,2	ggtgacactcacttgggaggtga	ctcacattcccttcggagctta	135	32
BIP3	cacggttccagcgtatttcaat	ataagctatggcagcacccgtt	118	40
CRT2	cctgcggaatctgatgctgaa	gagcggtggcgtctttctca	130	31
PP2A	gcagtatcgcttctcgctccagta	tgttctccacaaccgcttggtc	166	31

[a] This research was originally published in Schott et al. (2010). Arabidopsis stromal-derived factor 2 (SDF2) is a crucial target of the unfolded protein response in the endoplasmic reticulum. © The American society of Biochemistry and Molecular Biology.

3.3. ER stress-induced phenotypes

3.3.1. General seedling development and root morphology

For the analysis of seedling phenotypes under ER stress conditions, seeds are sown on MS plates with or without effectors (DTT or tunicamycin) and grown as detailed above. As DTT and tunicamycin batches may vary in quality, it is advisable to test various concentrations ranging from 0.5 to 2.5 mM for DTT and 0.01 to 0.05 μg/ml for tunicamycin. In order to grow wild-type and mutant lines under identical conditions, a defined number of seeds is placed within segments on the same plate (seed number should not exceed ∼250 per 90 mm Petri dish) (see Fig. 17.1A). For each treatment, three to four replicates are analyzed.

To quantify general growth defects of developing seedlings, plates are incubated in horizontal position for up to 20 days. On different days, the percentage of seedlings that generate green expanded cotyledons is determined as a measure for proper seedling development. When all seedlings grown under nonstress conditions uniformly show green expanded cotyledons (usually around day 6), the developmental index is calculated for each line and each treatment. Thereto, the percentage of seedlings that developed green cotyledons under ER stress conditions is divided by the percentage of seedlings that developed green cotyledons under control conditions (MS). For each replicate, the developmental index is determined, and mean values and standard deviations are calculated.

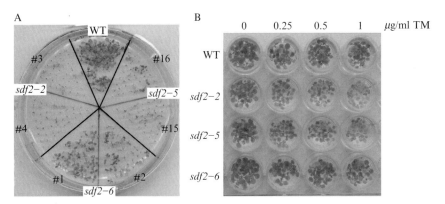

Figure 17.1 ER stress sensitivity of *sdf2* mutants. (A) Growth phenotypes of *sdf2* mutant seedlings upon DTT treatment. Seedlings were grown on MS medium containing 2.5 mM DTT for 18 days. For each mutant line, seeds from two different individuals were analyzed. (B) Phenotypes of *sdf2* mutant seedlings upon tunicamycin (TM) treatment. Six-day-old seedlings were incubated in MS medium supplemented with the indicated amounts of TM for 4 days. (This research was originally published in Schott *et al.*, 2010. Arabidopsis stromal-derived factor 2 (SDF2) is a crucial target of the unfolded protein response in the endoplasmic reticulum. © The American society of Biochemistry and Molecular Biology.)

To evaluate root growth and morphology, plates are incubated in horizontal position for 6 days. After that, seedlings are carefully removed from the agar plates using forceps not to damage the roots. For the documentation of root size and morphology, a total of at least 10 seedlings is placed in series on an MS plate. Thereby, hypocotyls are aligned and roots are fully expanded.

To make sure that the observed phenotypes are specific for ER stress rather than a general stress response, abiotic stresses such as osmotic and salt stress may be applied. Various detailed protocols are available (Bolle, 2009).

3.3.2. Quantification of chlorophyll content

Seedlings are grown on MS plates for 6 days as described in Section 3.1. Then, seedlings are carefully transferred from the plates to liquid medium under sterile conditions using forceps. Thereto, the wells of a 24-well culture plate (Greiner) are filled with medium containing different concentrations of tunicamycin (0.5, 0.75, and 1 μg/ml are recommended) or an equal volume of DMSO as mock treatment. The seedlings are incubated for 3–4 days under standard growth conditions until a phenotype is visible. Tunicamycin treatment results in bleaching of leaf margins or entire leaves in contrast to mock treated seedlings (see Fig. 17.1B). After this period, phenotypes of the seedlings are documented and chlorophyll a and b contents are determined.

Thereto, residual liquid is removed from the seedlings. After determination of the fresh weight, seedlings are transferred to a 2-ml tube and frozen in liquid nitrogen. The frozen plant material is ground using a ball mill (Retsch) and 1 ml of 80% acetone (v/v) is added to extract chlorophylls. After vigorous vortexing for 1 min, samples are centrifuged at $16,000 \times g$ for 2 min at 4 °C. The supernatant is taken off and collected in a 2-ml tube. The remaining pellet is extracted once again with 1 ml of 80% acetone. Supernatants are combined and the adsorption is measured at 646, 663, and 470 nm using a spectrophotometer. Total amounts of chlorophyll a and b (μg/ml plant extract) are calculated according to the equations $C_a = 12.21 A_{663} - 2.81 A_{646}$ and $C_b = 20.13 A_{646} - 5.03 A_{663}$, respectively (Lichtenthaler and Wellburn, 1985). The amount of chlorophyll a and b is figured up and subsequently related to the seedlings fresh weight.

Arabidopsis sdf2 mutant plants do not show obvious phenotypes under normal growth conditions (Schott *et al.*, 2010). However, using the methods described here we could demonstrate that *sdf2* mutant seedlings are specifically sensitive toward ER stress, induced by the accumulation of unfolded proteins (Schott *et al.*, 2010). Seeds of *sdf2* null-mutants germinate but the majority of seedlings fail to develop green expanded cotyledons in the presence of prolonged ER stress conditions (Fig. 17.1). Primary roots are extremely short and formation of root hairs and lateral roots is suppressed. Compared to wild-type, chlorophyll a and b contents are decreased by about 30–50%. Characterization of *sdf2–6* knockdown seedlings revealed that already minor amounts of SDF2 protein allow seedlings to gradually overcome ER stress (Fig. 17.1).

4. Analyzing the ER Stress-Induced Expression Pattern of *SDF2* in *Arabidopsis*

Expression of the *SDF2* gene is induced in response to ER stress (Kamauchi *et al.*, 2005; Martinez and Chrispeels, 2003; Schott *et al.*, 2010) allowing the usage of *SDF2* as a reporter gene to monitor UPR induction. Further, characterization of the spatial and temporal expression pattern of *SDF2* in wild-type plants and changes in transcript levels and/or expression pattern due to ER stress conditions defines tissues and organs that are likely to be affected in *sdf2* mutants, thus facilitating phenotypic analyses. Information about the expression pattern of *Arabidopsis SDF2* can be obtained by various methods. Distribution of *SDF2* transcripts can be analyzed by *SDF2* promoter β-glucuronidase (GUS) reporter fusions as well as quantitative RT-PCR (qRT-PCR). However, since the amount of mRNA transcripts does not necessarily correlate with protein levels, SDF2 specific antibodies can be used to quantify the SDF2 protein.

4.1. Transgenic *Arabidopsis* lines expressing an *SDF2* promoter GUS reporter fusion

To analyze tissue- and stage-specific gene expression *in planta*, the *Escherichia coli uid*A gene, which encodes the enzyme GUS is most frequently applied (Jefferson *et al.*, 1986). Transgenic *Arabidopsis* lines expressing the *uidA* gene under transcriptional control of the *SDF2* promoter are available (Schott *et al.*, 2010). These plants can be assayed for GUS activity by staining with the chromogenic substrate 5-bromo-4-chloro-3-indolyl β-D-glucuronide (X-Gluc). Sites of GUS enzymatic activity are visible due to the formation of an indigo-blue precipitate after cleavage of X-Gluc and reflect *SDF2* promoter activity.

Generally, *ProSDF2:GUS* fusions can be applied as reporter to measure UPR induction in *Arabidopsis*. Thus, in the following section we summarize methods necessary to analyze transgenic *Arabidopsis ProSDF2:GUS* lines.

4.1.1. Generation of *SDF2* promoter GUS reporter plants

For expression of the *uidA* gene under transcriptional control of the *SDF2* promoter, the binary plasmid pAS64 has been generated (Schott *et al.*, 2010). Plasmid pAS64 is based on the GATEWAY-compatible binary vector pBGWFS7 (Karimi *et al.*, 2002) and contains 2.47 kb of the 5′

upstream promoter sequence of *SDF2*. It confers spectinomycin and glufosinate ammonium- (BASTA) resistance to bacteria and plants, respectively.

Arabidopsis plants can be stably transformed with pAS64 using the *Agrobacterium tumefaciens*-mediated gene transfer (Hellens *et al.*, 2000). We recommend *A. tumefaciens* strain C58C1Rif^R, a rifampicin-resistant strain containing Ti plasmid pGV2260 (carbenicillin^R) (Deblaere *et al.*, 1985). Preparation of electro-competent *A. tumefaciens* cells and transformation protocols are described in Weigel and Glazebrook (2002). *Arabidopsis* plants are transformed using the floral dip method (Clough and Bent, 1998). Using this method, transformation efficiencies of 0.1–1% can be achieved.

4.1.2. Histochemical analysis of *SDF2* promoter GUS reporter plants

Materials

500 mM sodium phosphate buffer, pH 7.2, 10% Triton X-100, 100 mM potassium ferrocyanide (Sigma; should be stored at 4 °C in the dark), 100 mM potassium ferricyanide (Sigma; should be stored at 4 °C in the dark), 100 mM X-Gluc (Duchefa) in *N,N*-dimethylformamide (DMF) (should be stored at -20 °C in the dark), 90% acetone, ethanol (20%, 35%, 50%, and 70%), FAA fixative (50% ethanol, 10% glacial acetic acid, 5% formaldehyde), vacuum pump, stereo microscope MZ FLIII (Leica)

Procedure

In situ GUS staining is performed according to the protocol in Weigel and Glazebrook (2002).
1. The desired material (e.g., entire seedlings, leaves, flowers) is collected from transgenic plants expressing *ProSDF2:GUS* and placed in 90% acetone on ice.
2. When all samples have been harvested, they are incubated in 90% acetone for 20 min at room temperature.
3. During the incubation time, the staining buffer (50 mM sodium phosphate buffer, pH 7.2, 0.2% Triton X-100, 2 mM potassium ferrocyanide, 2 mM potassium ferricyanide) is freshly prepared and cooled to 4 °C.
4. Samples are washed twice with freshly prepared cold staining buffer on ice.
5. The staining buffer is replaced by staining solution (staining buffer with X-Gluc added to a final concentration of 2 mM). Samples are infiltrated under vacuum for 20 min on ice. This step is repeated to ensure proper infiltration of the sample material.
6. Samples are incubated at 37 °C for 1 h, followed by 25 °C overnight.

7. Staining solution is removed and samples are incubated in 20% ethanol at room temperature for 30 min. This step is repeated using 35% and 50% ethanol.
8. Samples are incubated in FAA fixative for 30 min at room temperature. The fixative is removed and samples are stored in 70% ethanol.
9. The GUS staining patterns are recorded using a Leica MZ FLIII stereo microscope.

Using the procedures described several transgenic *Arabidopsis ProSDF2: GUS* lines were created (*ProSDF2:GUS* #4, #13, #16, #26, and #27). GUS staining of transgenic *ProSDF2:GUS* lines #4 and #27 established that *SDF2* expression is prominent in zones with elevated levels of protein biosynthesis such as fast growing tissues, differentiating cells and meristematic regions (Schott *et al.*, 2010). In 5- to 7-day-old seedlings, GUS staining is mainly observed in the elongation zones of shoots and primary roots (Fig. 17.2). In response to prolonged ER stress, *ProSDF2:GUS* reporter lines exhibit stress-induced growth reduction and strongly elevated expression of GUS in roots (Fig. 17.2A). In response to transient ER stress, *ProSDF2*:GUS expression is induced especially in primary and lateral roots (Fig. 17.2B).

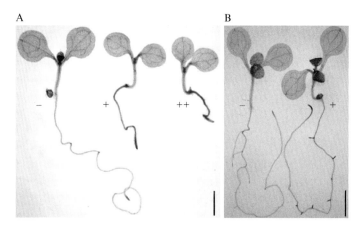

Figure 17.2 Analysis of *SDF2* expression. (A, B) *SDF2* promoter activity detected in GUS reporter plants. Transgenic plants harboring the *ProSDF2:GUS* reporter were stained for GUS activity to monitor sites of *SDF2* expression. (A) Seven-day-old seedlings either grown on MS plates (−) or MS medium supplemented with 2 mM (+) or 2.5 mM DTT (++). Bar: 1 mm. (B) Seven-day-old seedlings grown overnight in liquid MS medium were treated for 24 h with 1 mM DTT (+). Mock-treated seedlings were grown in liquid MS medium (−). Bar: 2 mm. (This research was originally published in Schott *et al*. 2010. Arabidopsis stromal-derived factor 2 (SDF2) is a crucial target of the unfolded protein response in the endoplasmic reticulum. © The American society of Biochemistry and Molecular Biology.) (See Color Insert.)

4.2. qRT-PCR

Its high sensitivity and accuracy qualify qRT-PCR as a superior method to quantify *SDF2* transcript levels, especially in plant tissues with low RNA yields or for the detection of minor variations of mRNA levels.

To quantify *SDF2* expression in *Arabidopsis* seedlings, total RNA is extracted from five 6-day-old seedlings using the RNeasy plant mini kit (Qiagen) following the manufacturer's instructions. DNA contaminations are removed by DNase digest using RQ1 RNase-free DNase (Promega). The quality of RNA is verified by gel electrophoresis on a denaturing agarose gel and RNA concentrations are determined by measuring absorbance at 260 nm. RNA samples can be stored in aliquots at $-80\,^\circ$C.

Synthesis of cDNA from 500 ng of total *Arabidopsis* RNA is performed using the iScript Select cDNA Synthesis Kit (Bio-Rad) with oligo(dT) primers. Quantitative qRT-PCR is carried out using the iCycler iQ real-time PCR detection system (Bio-Rad) and iQ Sybergreen Supermix (Bio-Rad) in a total reaction volume of 20 µl. Each reaction contains cDNA derived from 10 ng *Arabidopsis* total RNA and 100 nM gene-specific primers (100 nM). For the quantification of *SDF2* mRNA levels, forward primer 1057 (5'-tcctatatcagggaacttagaggttagctg-3') and reverse primer 1058 (5'-cggactctttggtcctgtttcca-3') are used. As an internal control, mRNA of protein phosphatase 2A (*PP2A*) (At1g13320) is quantified using the forward primer 1014 (5'-gcagtatcgcttctcgctccagta-3') and reverse primer 1015 (5'-tgttctccacaaccgcttggtc-3') (Czechowski *et al.*, 2005).

35 PCR cycles are carried out at 98 $^\circ$C for 10 s, 59 $^\circ$C for 20 s, and 72 $^\circ$C for 15 s. Serial dilutions of template cDNAs are conducted to determine primer efficiencies for all primer combinations. Each reaction is performed in triplicate. Data are analyzed using the relative standard curve method and the level of *SDF2* transcripts are normalized to reference transcripts of *PP2A*.

qPCR revealed an at least threefold induction of *SDF2* in *Arabidopsis* seedlings after DTT treatment (Schott *et al.*, 2010).

4.3. SDF2-specific antibodies

Specific antibodies are available that allow detection and quantification of the SDF2 protein in *Arabidopsis* (Schott *et al.*, 2010). These antibodies have successfully been used for Western blot analysis (see below) as well as immuno-localization of SDF2 protein (Schott *et al.*, 2010).

4.3.1. Affinity purified antibodies directed against SDF2

Polyclonal antibodies directed against SDF2 were generated in rabbits using recombinant glutathione *S*-transferase (GST)-SDF2 (SDF2 amino acid residues 24–218) fusion protein purified from *E. coli* (Schott *et al.*, 2010). The fusion protein was purified as described in Section 6 omitting cleavage of the GST affinity tag. Immunization of rabbits and affinity purification of SDF2 antibodies were performed at Pineda Antikörper Service (Berlin, Germany). SDF2-specific antibodies were enriched from immune sera at day 210 and 250 after immunization by affinity-purification using SDF2 (residues Ser24 to Lys218) polypeptide coupled to cyanogen bromide-activated sepharose.

Affinity-purified SDF2 antibodies are applied in 1:500 dilutions for Western blot analysis (see below), and in 1:20 dilutions for immuno-localization (Schott et al., 2010).

4.3.2. Preparation of microsomal membranes from *Arabidopsis* seedlings

Materials

Extraction buffer (50 mM HEPES, pH 7.5, 5 mM MgCl$_2$, 10 mM KCl); protease inhibitors: leupeptin, aprotinin, *trans*-epoxysuccinyl-L-leucylamido-(4-guanido)-butane (E-64), pepstatin and *o*-phenanthroline. All protease inhibitors were purchased from Sigma. Sonopuls GM70 sonifier (Bandelin), mortar and pestle, DC Protein Assay (Bio-Rad).

Procedure

1. Five-day-old *Arabidopsis* seedlings (30–150) are homogenized in extraction buffer (including protease inhibitors 2 μg/ml leupeptin, 2 μg/ml aprotinin, 1 μg/ml E-64, 0.7 μg/ml pepstatin, and 1 mM *o*-phenanthroline) using mortar and pestle.
2. The homogenate is centrifuged at 2000×g for 10 min at 4 °C to remove cell debris.
3. The supernatant is transferred to a fresh tube and centrifuged at 20,000 × g for 20 min at 4 °C.
4. The resulting supernatant is sonified twice with 60% maximum intensity for 20 s on ice using a Sonopuls sonifier.
5. The suspension is then subjected to ultra-centrifugation at 100,000×g for 1 h at 4 °C.
6. The resulting supernatant containing soluble microsomal contents is removed.
7. The resulting microsomal membrane pellet is suspended in an applicable volume (10–50 μl) of extraction buffer. Protein concentrations of the different factions are determined using the DC Protein Assay (Bio-Rad).

4.3.3. Western blot analysis of microsomal membranes

Equivalent amounts of protein (20–50 μg) are separated on 12% polyacrylamide-gels and transferred to nitrocellulose membranes (GE Healthcare). Transfer efficiency to nitrocellulose membranes is routinely monitored by

Ponceau S staining. Nitrocellulose membranes are blocked with 5% skim milk in 1× Tris-buffered saline (20 mM Tris, pH 7.4, 137 mM NaCl, 3 mM KCl) containing 0.1% Tween-20 (TBS-T). Affinity purified polyclonal rabbit SDF2 antibodies (1:500), polyclonal rabbit VHA subunit c antibodies (1:2000) and secondary antibody-HRP-conjugates (anti-mouse and anti-rabbit, 1:5000; Sigma) used for immuno-decoration are diluted in 5% skim milk/TBS-T. Antibody incubations are carried out for 1 h at room temperature or overnight at 4 °C. After each incubation step, membranes are washed three times with TBS-T. Protein-antibody complexes are visualized by enhanced chemiluminescence using the Pierce SuperSignal® West Pico substrates (Thermo Fischer).

Western Blot analysis revealed that *Arabidopsis* SDF2 (the apparent molecular weight is 21 kDa) is an ER membrane-associated protein. Affinity-purified SDF2 antibodies show cross-reactivity toward an unknown 30 kDa protein present in microsomal membrane fractions. The 30 kDa protein is not related to SDF2 and can therefore be used as loading control. Analyses of microsomal membranes confirmed that SDF2 protein abundance is increased in response to ER stress conditions (Schott *et al.*, 2010).

5. The Subcellular Localization of SDF2

SDF2-type proteins possess N-terminal signal peptides that direct their translocation into the ER lumen. Mammalian SDF2-type proteins are further characterized by C-terminal ER retention motifs (Fukuda *et al.*, 2001; Hamada *et al.*, 1996). In contrast plant-SDF2-type proteins lack typical ER retention motifs. To demonstrate ER-localization of SDF2 *in planta*, heterologous expression in tobacco leaf epidermal cells provides an excellent system. Thereto, we fused SDF2 to an epitope tag allowing detection in microsomal subfractions after separation on a sucrose density gradient. Alternatively, SDF2 fused to a redox-sensitive GFP2 (roGFP) was localized by fluorescent microscopy.

5.1. Transient expression of SDF2 fusions in tobacco leaves

5.1.1. Hemagglutinin (HA) epitope-tagged SDF2 expression construct

The binary plasmid pAS76 (Schott *et al.*, 2010) carries the *SDF2* coding sequence (bp +1 to +654), fused to three copies of the HA epitop (*SDF2: HA*), under transcriptional control of the CaMV 35S promoter. It further contains a kanamycin resistance marker for the selection in bacteria and plants.

5.1.2. SDF2–roGFP reporter construct

The binary plasmid pAS75 carrying *SDF2:roGFP* under transcriptional control of the enhanced CaMV 35S promoter and a bacterial kanamycin-resistance marker is available.

To create pAS75, the *SDF2* coding sequence (bp +1 to +654) was amplified using primers 970 5′-gagtctagaatggctttaggattcttctgtc-3′ (*Xba*I restriction site is underlined) and 971 5′-tctgtcgaccttgctgctctcattaaggg-3′ (*Sal*I restriction site is underlined) and cloned into the binary vector pVKH18En6-SEC22:roGFP (Brach *et al.*, 2009) linearized with *Xba*I/*Sal*I. Thereby, the *SEC22* coding sequence was replaced by the *SDF2* gene.

5.1.3. Infiltration of tobacco leaves

Materials

Six- to eight-week-old *Nicotiana benthamiana* or *Nicotiana tabacum* plants grown on soil (Weigel and Glazebrook, 2002), YEB medium, infiltration buffer (10 mM MES, pH 5.6, 150 μM acetosyringone (Sigma))

Procedure

1. *A. tumefaciens* cells, transformed with the respective binary vector, are grown overnight at 28 °C in 30 ml of YEB medium with 100 μg/ml rifampicin, 50 μg/ml carbenicillin, and 50 μg/ml spectinomycin until stationary phase.
2. Cultures are centrifuged at 3000×g for 10 min at room temperature.
3. The cell pellet is suspended in 10 ml of infiltration buffer and incubated for 2 h at 28 °C with gentle agitation. Acetosyringone is used to promote infection.
4. Cell suspensions are adjusted to OD_{600} 1 with infiltration buffer and used for pressure infiltration into the lower epidermis of *N. benthamiana* or *N. tabacum* leaves using a disposable plastic syringe (without needle). For later recovery, infiltrated leaf areas were tagged using a marker pen. Further analyses were carried out 48 h after infiltration.

5.2. Analyzing localization of SDF2-HA by sucrose density gradient centrifugation

Microsomal membranes derived from different organelles have characteristic densities and can be separated by isopycnic centrifugation on a linear sucrose density gradient. The association of ribosomes with rough ER membranes is Mg^{2+}-dependent. Removal of Mg^{2+} results in the dissociation of ribosomes from the ER membranes and a shift from higher to lower densities in the sucrose gradient (Pryme, 1986). Therefore, sucrose gradients prepared with or without Mg^{2+} are used to distinguish ER-derived membranes from those of other cellular compartments.

Materials

Ultracentrifuge L8-M (Beckmann), swing-out rotor, TH64-1 (Sorvall), 14 ml ultracentrifugation tubes (Sorvall), Brix Scale Refractometer (Reichert GmbH), extraction buffer A (50 mM HEPES, pH 7.5, 5 mM MgCl$_2$, 10 mM KCl, 0.5% polyvinylpyrrolidone 40 (Sigma), 10% sucrose), extraction buffer B (50 mM HEPES, pH 7.5, 5 mM EDTA 10 mM KCl, 0.5% polyvinylpyrrolidone 40, 10% sucrose), gradient buffer A (50 mM HEPES, pH 7.5, 5 mM MgCl$_2$, 10 mM KCl, 0.5% polyvinylpyrrolidone 40), gradient buffer B (50 mM HEPES, pH 7.5, 5 mM EDTA, 10 mM KCl, 0.5% polyvinylpyrrolidone 40). All buffers contain the following amounts of protease inhibitors: 2 µg/mL leupeptin, 2 µg/mL aprotinin, 1 µg/mL E-64, 0.7 µg/mL pepstatin, and 1 mM *o*-phenanthroline.

Procedure

For the separation by sucrose density gradient centrifugation, microsomal membranes are isolated from three infiltrated tobacco leaves as described in Section 4.3.2 except that extraction buffer A or B, is used for protein extraction. Further, sonification of membranes is omitted.

1. 10 ml each of 20%, 26%, 32%, 38%, 44%, and 50% (w/w) sucrose solutions is prepared in the respective gradient buffer A or B. Sugar concentrations are measured and adjusted with a Reichert Brix Scale Refractometer.
2. In order to obtain linear sucrose gradients (20–50%, w/w), 1.7 ml of each sucrose solution is pipetted sequentially into a 14-ml ultracentrifuge tube starting with 50% sucrose solution at the bottom. The tubes are then incubated for 4–5 h at 4 °C.
3. Microsomal membranes, suspended in 500 µl of the appropriate extraction buffer are layered on top of the respective linear sucrose gradient.
4. Centrifugation of the gradient is carried out at 100,000×g overnight at 4 °C in a swing out rotor.
5. Immediately after centrifugation, 800-µl fractions are collected from the top. For each gradient fraction, sucrose and protein concentrations are determined.
6. To recover proteins from each fraction, 200 µl of 100% trichloroacetic acid (TCA) are added and samples are incubated for 1 h at 4 °C. The precipitate is pelleted at 20,000×g for 30 min at 4 °C.
7. Protein pellets are washed twice with 800-µl of ice-cold acetone and dried using a vacuum concentrator (Bachuber).

8. Protein pellets are suspended in 50 μl of SDS sample buffer and incubated for 5 min at 37 °C. 5μl of each fraction are used for Western blot analysis. SDF2-HA is detected using HA-directed antibodies (Covance Research Products Inc.) diluted 1:5000. To identify the distribution of organelle specific membranes in gradient fractions, ER marker proteins calnexin and calreticulin are detected using antibodies described by Pimpl et al. (2000). Antibodies directed against plasma membrane H^+-ATPase (PMA) are reported in Morsomme et al. (1998).

Following this protocol, ER localization of SDF2-HA was confirmed (Schott et al., 2010). SDF2-HA cofractionates with the ER marker proteins calnexin and calreticulin. Upon depletion of Mg^{2+} ions SDF2-HA follows the density shift observed for ER marker proteins.

5.3. Analyzing localization of SDF2–roGFP by confocal microscopy

Recombinant fusions with green fluorescent protein (GFP) are a common tool to analyze the subcellular distribution of proteins in living cells. The use of roGFP further allows identification of cellular compartments not only based on morphology but also on the redox-state of the respective compartment (Meyer and Brach, 2009; Meyer et al., 2007). When roGFP is excited at 405 and 488 nm, different fluorescent intensities are observed due to the formation of an intramolecular disulfide bond. Thus, the redox-state of roGFP can be determined by ratiometric analysis. The localization of roGFP in the cytoplasm, results in a reduced probe and in low 405/488 nm fluorescent ratios (Brach et al., 2009). In contrast, high 405/488 nm ratios are observed if roGFP is targeted to the oxidizing environment of the ER lumen (Brach et al., 2009).

Materials

Confocal microscope with excitation wavelength 405 and 488 nm (e.g., Zeiss LSM510 META laser scanning microscope), glass slides, overslips, insulation tape, plant material expressing SDF2–roGFP, and roGFP marker proteins.

Procedure

1. SDF2–roGFP, ER localized roGFP–HDEL (Meyer et al., 2007), and cytosolic roGFP (Meyer et al., 2007) are transiently expressed in tobacco leaves as described above.
2. Infiltrated leaf areas (~0.5 cm × 0.5 cm) are excised omitting the larger veins and mounted in water on glass slides. To avoid squashing of the samples, insulation tape is used to create a spacer between slide and coverslip.

3. Samples are screened for green fluorescence in a Zeiss LSM510 META laser scanning microscope (Carl Zeiss) using the epifluorescence mode at 490 nm excitation. To carry out ratiometric imaging, cells showing medium to strong fluorescence should be chosen.
4. Excitation of roGFP was performed at 405 and 488 nm in multitrack mode, which allowed parallel recording of the two different signals corresponding to both laser wavelengths. To guarantee correct ratiometric analysis, images are taken in line mode with laser switching after each line. Emission is collected using a band-pass filter at 505–530 nm.
5. Ratiometric analyses of the images are carried out with a custom-written MATLAB script which is available on request from A.J. Meyer (Brach *et al.*, 2009). Briefly, after background subtraction, the 405 nm image is divided by the 488 nm image. A Gaussian fit is applied to the ratio value distribution of the calculated ratio image in order to determine the mean ratio value. The ratio images are illustrated using a color-map (Brach *et al.*, 2009).

For a more detailed protocol concerning the use of the confocal laser scanning microscope in order to create images for the ratiometric analyses, we refer the reader to Meyer and Brach (2009).

The SDF2–roGFP fusion protein reveals a typical ER pattern, staining the nuclear envelope and the peripheral ER network. In addition, SDF2–roGFP features a high 405/488 nm fluorescent ratio and thus an oxidized state (Fig. 17.3). Staining pattern and redox-state observed for SDF2–roGFP are comparable with the ER marker protein roGFP–HDEL.

6. Purification of Recombinant SDF2 Protein from *E. coli*

Heterologous expression and purification of recombinant SDF2 protein from *E. coli* enables further detailed biochemical and structural characterization. For affinity purification, SDF2 is expressed without its N-terminal signal peptide as GST-fusion protein. A PreScission protease cleavage site located between the N-terminal GST-tag and SDF2 allows the removal of the GST moiety after the initial affinity purification step. SDF2, purified according to the following procedures, has been successfully used for crystallization and structure determination (Radzimanowski *et al.*, 2010; Schott *et al.*, (2010).

6.1. GST-SDF2 fusion construct

Plasmid pAS62 encoding an in frame fusion of a GST-tag followed by a PreScission protease cleavage site at the N-terminus of SDF2 (amino acid residues 24–218) is available (Radzimanowski *et al.*, 2010). To create this

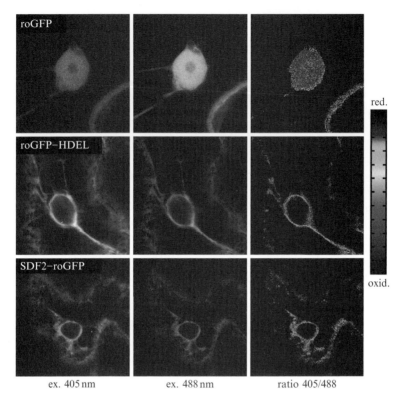

Figure 17.3 Ratiometric imaging of SDF2–roGFP. SDF2–roGFP fusion indicates ER localization of SDF2. SDF2–roGFP, the ER-localized roGFP–HDEL (Meyer et al., 2007), and cytosolic roGFP (Meyer et al., 2007) were transiently expressed in N. benthamiana leaf epidermal cells under control of the CaMV 35S promoter. Images were taken by confocal laser scanning microscopy with excitation at 405 and 488 nm as indicated. Ratio images 405/488 nm were calculated and converted into pseudocolors. The color scale shows the pseudocolor coding for reduced (blue) and oxidized roGFP (red). (See Color Insert.)

plasmid, the SDF2 coding sequence (bp $+70$ to $+657$) was cloned in the vector pGEX-6P-1 (GE Healthcare).

6.2. Production of the GST-SDF2 fusion protein in *E. coli*

Materials

E. coli strain BL21 (DE3) (Stratagene), LB medium, isopropyl-β-D-thiogalactopyranoside (IPTG, Applichem), 1× TBS (20 mM Tris-HCl, pH 7.4, 137 mM NaCl, 3 mM KCl)

Procedure

1. After transformation of *E. coli* BL21 cells with pAS62, a single ampicillin-resistant colony is used to inoculate 20 ml of LB medium containing 100 μg/ml ampicillin for overnight at 37 °C.
2. The next day, the preculture is diluted 1:100 into 1 l of LB medium without antibiotics and incubated with shaking at 37 °C for 3 h.
3. The culture is cooled down to room temperature. Protein production is induced by adding 100 mM IPTG to a final concentration of 0.4 mM and the culture is further incubated overnight at 18 °C with shaking.
4. Cells are harvested by centrifugation at 3500 × g for 30 min and the cell pellet is washed with 200 ml of 1× TBS. At this point, cells can be directly used for the purification procedure or stored at −80 °C after freezing in liquid nitrogen.

6.3. Affinity purification of GST-SDF2

Materials

M-110L Microfluidizer (Microfluidics), GSTtrap™ HP columns (GE Healthcare), reduced glutathione (GSH, Applichem), DTT (Carl Roth). GSH and DTT are always freshly added to the buffers. Lysis buffer (50 mM Tris-HCl, pH 8.0, 300 mM NaCl, 0.1 mM EDTA, 1 mM phenylmethylsulphonyl fluoride (PMSF), 1 mM benzamidine (BA))

Procedure

1. Frozen cell pellets are thawed on ice, while freshly prepared pellets are incubated on ice until further processing. The bacterial pellet is resuspended in 30 ml of lysis buffer containing protease inhibitors and mechanically ruptured using an M-110L Microfluidizer.
2. The cell lysate is cleared by centrifugation at 23,000 × g for 30 min at 4 °C. To avoid clogging of affinity columns, it is advisable to remove remaining particles from the supernatant by passing it through a 0.45-μm filter.
3. Before applying the cell lysate, a 1-ml GSTrap™ HP column is equilibrated with 5 column volumes (CV) of lysis buffer.
4. The cell lysate is loaded onto the GSH affinity column and the column is washed with 10–15 CV of lysis buffer.
5. The bound GST fusion protein is finally eluted from the column using 5 CV of lysis buffer containing freshly added 20 mM glutathione (GSH) and 1 mM DTT.

6.4. Removal of the GST-tag

Materials

PreScission Protease (GE Healthcare): PreScission protease is a genetically engineered form of the rhinoviral 3C cystein protease fused to GST (Walker *et al.*, 1994). This offers the advantage, to remove the cleaved GST-tag together with the PreScission protease in one purification step. The protease is highly active at low temperatures.

GSTtrap^{TM} HP columns (GE Healthcare), Vivaspin 6 MWCO 10,000 concentrator (Vivascience), HiLoad 16/60 Superdex 200 pg (GE Healthcare), and chromatography system UPC10 (GE Healthcare). PreScission protease cleavage buffer (50 mM Tris-HCl, pH 7.5, 150 mM NaCl, 1 mM EDTA, 1 mM DTT); size exclusion buffer (10 mM Tris-HCl, pH 7.5, 150 mM NaCl).

Procedure

1. To remove GSH, the eluate of the GSTrap^{TM} HP column is dialyzed against PreScission protease cleavage buffer overnight.
2. The GST-SDF2 fusion protein in treated with PreScission protease (GE Healthcare) (1 U/100 µg fusion protein) for 10 h at 4 °C according to the manufacturer's instructions.
3. The cleaved and the PreScission protease are removed in one step by loading the protein solution on two sequentially connected 1 ml GSTrap^{TM} HP columns. The flow through containing the SDF2 protein in PreScission protease cleavage buffer is collected.
4. If downstream applications such as protein crystallization require high amounts of pure recombinant protein, the protein can be further purified by size exclusion chromatography. Thereto, the protein solution is concentrated using Vivaspin 6 concentrator (MWCO 10,000) and purified on a Superdex 200/16–60 size exclusion column (GE Healthcare) equilibrated in 10 mM Tris-HCl (pH 7.5), 150 mM NaCl using the chromatography system UPC10 (GE Healthcare).

Upon elution, the protein-containing fractions are pooled and concentrated as described before. Finally, protein concentration is determined and the purity is analyzed by SDS-PAGE using 12% polyacrylamide gels and Coomassie-staining. It is recommended not to freeze purified SDF2. Instead, the pure protein can be stored for short-time periods at 4 °C (Fig. 17.4).

ACKNOWLEDGMENTS

We are very grateful to A. Meyer and S. Soyk for assistance with the SDF2–roGFP analysis. We thank G. Hinz and P. Pimpl for technical advice and S. Keller for excellent technical

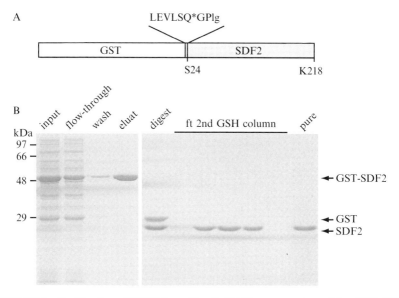

Figure 17.4 Purification of SDF2 from E. coli. (A) Representation of the GST–SDF2 fusion protein. SDF2 without its N-terminal signal peptide is fused to the C-terminus of GST. The PreScission protease recognition sequence is specified with uppercase letters. The cleavage site is marked with an asterisk. Additional amino acids introduced due to the cloning strategy are indicated in lowercase letters. (B) Purification of SDF2. Protein samples of each purification step were separated on a 12% polyacrylamide gel and visualized by Coomassie staining. Input, flow-through, wash fraction, and eluate of the affinity purification of GST–SDF2, cleavage of the GST-tag (digest), flow through (ft) of the second affinity column, and pure SDF2 after size exclusion chromatography are shown.

assistance. We are very grateful to R. Serrano and his group members for the generous help in analyzing stress-induced growth phenotypes. We especially thank I. Sinning, S. Ravaud, J. Radzimanowski for very effective cooperation. We thank M. Buettner for critical reading of the manuscript. This work was partially supported by the Deutsche Forschungs Gemeinschaft (SFB638). S. Strahl is member of CellNetworks—Cluster of Excellence (EXC81).

REFERENCES

Adachi, Y., Yamamoto, K., Okada, T., Yoshida, H., Harada, A., and Mori, K. (2008). ATF6 is a transcription factor specializing in the regulation of quality control proteins in the endoplasmic reticulum. Cell Struct. Funct. **33,** 75–89.

Bertolotti, A., Zhang, Y., Hendershot, L. M., Harding, H. P., and Ron, D. (2000). Dynamic interaction of BiP and ER stress transducers in the unfolded-protein response. Nat. Cell Biol. **2,** 326–332.

Bies, C., Blum, R., Dudek, J., Nastainczyk, W., Oberhauser, S., Jung, M., and Zimmermann, R. (2004). Characterization of pancreatic ERj3p, a homolog of yeast DnaJ-like protein Scj1p. Biol. Chem. **385,** 389–395.

Bolle, C. (2009). Phenotyping of abiotic responses and hormone treatments in Arabidopsis. *Methods Mol. Biol.* **479,** 35–59.

Brach, T., Soyk, S., Muller, C., Hinz, G., Hell, R., Brandizzi, F., and Meyer, A. J. (2009). Non-invasive topology analysis of membrane proteins in the secretory pathway. *Plant J.* **57,** 534–541.

Clough, S. J., and Bent, A. F. (1998). Floral dip: A simplified method for Agrobacterium-mediated transformation of *Arabidopsis thaliana. Plant J.* **16,** 735–743.

Czechowski, T., Stitt, M., Altmann, T., Udvardi, M. K., and Scheible, W. R. (2005). Genome-wide identification and testing of superior reference genes for transcript normalization in Arabidopsis. *Plant Physiol.* **139,** 5–17.

Deblaere, R., Bytebier, B., De Greve, H., Deboeck, F., Schell, J., Van Montagu, M., and Leemans, J. (1985). Efficient octopine Ti plasmid-derived vectors for Agrobacterium-mediated gene transfer to plants. *Nucleic Acids Res.* **13,** 4777–4788.

Edwards, K., Johnstone, C., and Thompson, C. (1991). A simple and rapid method for the preparation of plant genomic DNA for PCR analysis. *Nucleic Acids Res.* **19,** 1349.

Elbein, A. D. (1984). Inhibitors of the biosynthesis and processing of N-linked oligosaccharides. *CRC Crit. Rev. Biochem.* **16,** 21–49.

Fukuda, S., Sumii, M., Masuda, Y., Takahashi, M., Koike, N., Teishima, J., Yasumoto, H., Itamoto, T., Asahara, T., Dohi, K., and Kamiya, K. (2001). Murine and human SDF2L1 is an endoplasmic reticulum stress-inducible gene and encodes a new member of the Pmt/rt protein family. *Biochem. Biophys. Res. Commun.* **280,** 407–414.

Hamada, T., Tashiro, K., Tada, H., Inazawa, J., Shirozu, M., Shibahara, K., Nakamura, T., Martina, N., Nakano, T., and Honjo, T. (1996). Isolation and characterization of a novel secretory protein, stromal cell-derived factor-2 (SDF-2) using the signal sequence trap method. *Gene* **176,** 211–214.

Hellens, R., Mullineaux, P., and Klee, H. (2000). Technical focus: A guide to agrobacterium binary Ti vectors. *Trends Plant Sci.* **5,** 446–451.

Jefferson, R. A., Burgess, S. M., and Hirsh, D. (1986). beta-Glucuronidase from *Escherichia coli* as a gene-fusion marker. *Proc. Natl. Acad. Sci. USA* **83,** 8447–8451.

Kamauchi, S., Nakatani, H., Nakano, C., and Urade, R. (2005). Gene expression in response to endoplasmic reticulum stress in *Arabidopsis thaliana. FEBS J.* **272,** 3461–3476.

Karimi, M., Inze, D., and Depicker, A. (2002). GATEWAY vectors for Agrobacterium-mediated plant transformation. *Trends Plant Sci.* **7,** 193–195.

Koizumi, N. (1996). Isolation and responses to stress of a gene that encodes a luminal binding protein in *Arabidopsis thaliana. Plant Cell Physiol.* **37,** 862–865.

Lichtenthaler, H. K., and Wellburn, A. R. (1985). Determination of total carotenoids and chlorophylls a and b of leaf extracts in different solvents. *Biochem. Soc. Trans.* **11,** 591–592.

Martinez, I. M., and Chrispeels, M. J. (2003). Genomic analysis of the unfolded protein response in Arabidopsis shows its connection to important cellular processes. *Plant Cell* **15,** 561–576.

Meunier, L., Usherwood, Y. K., Chung, K. T., and Hendershot, L. M. (2002). A subset of chaperones and folding enzymes form multiprotein complexes in endoplasmic reticulum to bind nascent proteins. *Mol. Biol. Cell* **13,** 4456–4469.

Meyer, A. J., and Brach, T. (2009). Dynamic redox measurements with redox-sensitive GFP in plants by confocal laser scanning microscopy. *Methods Mol. Biol.* **479,** 93–107.

Meyer, A. J., Brach, T., Marty, L., Kreye, S., Rouhier, N., Jacquot, J. P., and Hell, R. (2007). Redox-sensitive GFP in *Arabidopsis thaliana* is a quantitative biosensor for the redox potential of the cellular glutathione redox buffer. *Plant J.* **52,** 973–986.

Morsomme, P., Dambly, S., Maudoux, O., and Boutry, M. (1998). Single point mutations distributed in 10 soluble and membrane regions of the *Nicotiana plumbaginifolia* plasma membrane PMA2 H+-ATPase activate the enzyme and modify the structure of the C-terminal region. *J. Biol. Chem.* **273,** 34837–34842.

Nekrasov, V., Li, J., Batoux, M., Roux, M., Chu, Z. H., Lacombe, S., Rougon, A., Bittel, P., Kiss-Papp, M., Chinchilla, D., van Esse, H. P., Jorda, L., et al. (2009). Control of the pattern-recognition receptor EFR by an ER protein complex in plant immunity. *EMBO J.* **28,** 3428–3438.

Noh, S. J., Kwon, C. S., Oh, D. H., Moon, J. S., and Chung, W. I. (2003). Expression of an evolutionarily distinct novel BiP gene during the unfolded protein response in *Arabidopsis thaliana. Gene* **311,** 81–91.

Pimpl, P., Movafeghi, A., Coughlan, S., Denecke, J., Hillmer, S., and Robinson, D. G. (2000). In situ localization and in vitro induction of plant COPI-coated vesicles. *Plant Cell* **12,** 2219–2236.

Ponting, C. P. (2000). Novel repeats in ryanodine and IP3 receptors and protein O-mannosyltransferases. *Trends Biochem. Sci.* **25,** 48–50.

Pryme, I. F. (1986). Compartmentation of the rough endoplasmic reticulum. *Mol. Cell. Biochem.* **71,** 3–18.

Radzimanowski, J., Ravaud, S., Schott, A., Strahl, S., and Sinning, I. (2010). Cloning, recombinant production, crystallization and preliminary X-ray diffraction analysis of SDF2-like protein from *Arabidopsis thaliana. Acta Crystallogr. F Struct. Biol. Cryst. Commun.* **66,** 12–14.

Schott, A., Ravaud, S., Keller, S., Radzimanowski, J., Viotti, C., Hillmer, S., Sinning, I., and Strahl, S. (2010). Arabidopsis stromal-derived Factor2 (SDF2) is a crucial target of the unfolded protein response in the endoplasmic reticulum. *J. Biol. Chem.* **285,** 18113–18121.

Urade, R. (2009). The endoplasmic reticulum stress signaling pathways in plants. *Biofactors* **35,** 326–331.

Vitale, A., and Boston, R. S. (2008). Endoplasmic reticulum quality control and the unfolded protein response: Insights from plants. *Traffic* **9,** 1581–1588.

Walker, P. A., Leong, L. E., Ng, P. W., Tan, S. H., Waller, S., Murphy, D., and Porter, A. G. (1994). Efficient and rapid affinity purification of proteins using recombinant fusion proteases. *Biotechnology (NY)* **12,** 601–605.

Watanabe, N., and Lam, E. (2008). BAX inhibitor-1 modulates endoplasmic reticulum stress-mediated programmed cell death in Arabidopsis. *J. Biol. Chem.* **283,** 3200–3210.

Weigel, D., and Glazebrook, J. (2002). Arabidopsis: A Laboratory Manual. Cold Spring Harbor Laboratory Press, Cold Spring Harbor, NY.

CHAPTER EIGHTEEN

NITROSATIVE STRESS-INDUCED S-GLUTATHIONYLATION OF PROTEIN DISULFIDE ISOMERASE

Joachim D. Uys, Ying Xiong, *and* Danyelle M. Townsend

Contents

1. Introduction	322
2. Identification and Confirmation of S-Glutathionylated Proteins in Cells	324
2.1. Materials	325
2.2. Methods	325
3. Identification of Target Cysteine Residues	327
3.1. Materials	327
3.2. Methods	328
4. Characterization of Structural and Functional Consequences of S-Glutathionylated PDI	329
4.1. Materials	329
4.2. Methods	329
5. Summary	331
Acknowledgments	331
References	331

Abstract

Oxidative and nitrosative stress result in the accumulation of reactive oxygen and nitrogen species (ROS/RNS) which trigger redox-mediated signaling cascades through posttranslational modifications on cysteine residues, including S-nitrosylation (P-SNO) and S-glutathionylation (P-SSG). Protein disulfide isomerase (PDI) is the most abundant chaperone in the endoplasmic reticulum and facilitates protein folding via oxidoreductase activity. Prolonged or acute nitrosative stress blunts the activity of PDI through the formation of PDI–SNO and PDI–SSG. The functional implication is that reduced activity for the period of time leads to an accumulation of misfolded or unfolded proteins and activation of the unfolded protein response. Redox regulation of PDI and downstream

Department of Pharmaceutical and Biomedical Sciences, Medical University of South Carolina, Charleston, South Carolina, USA signaling events provides an integration point for the functional determination of cell survival pathways. Herein, we describe the methodologies to globally identify S-glutathionylated targets of ROS/RNS; validate and identify the specific cysteine targets and characterize the structural and functional consequences.

1. INTRODUCTION

Oxidation and reduction (redox) reactions play an essential role in numerous cell-signaling cascades including those associated with proliferation, apoptosis, and inflammatory responses. Oxidative and nitrosative stress results from the imbalance between the production of oxidants and their removal by antioxidants, leading to the accumulation of reactive oxygen and nitrogen species (ROS/RNS). Elevated levels of nitric oxide (NO) provide the primary source of RNS. NO is an endogenous, diffusible, transcellular messenger shown to participate in survival and death pathways (Moncada, 1997) and can alter protein function directly through posttranslational modifications (nitration or S-nitrosylation) or indirectly through interactions with oxygen, superoxide, thiols, and heavy metals, the products of which cause protein S-glutathionylation. Altered NO homeostasis can lead to the release of RNS and ROS, each of which have been implicated in a number of human pathologies, including neurodegenerative disorders, cystic fibrosis, cancer, and aging (Ilic et al., 1999; Tieu et al., 2003; Townsend and Tew, 2003).

Glutathione (GSH), a tripeptide of cysteine, glutamic acid, and glycine, represents one of the most prevalent and important antioxidant buffers in the cell. The ratio of GSH (reduced) and its disulfide, GSSG (oxidized), contributes to the redox potential of the cell and thereby, is key in defining redox homeostasis. Oxidative or nitrosative stress induced by physiological or pathological conditions leads to a decreased ratio of GSH/GSSG. As a consequence, reduced cysteine residues (-SH) that have a low pK_a are reactive and can become oxidized into protein sulfenic acids (P-SOH), nitrosylated (P-SNO), or S-glutathionylated (P-SSG). These proteins are referred to as redox sensors and the corresponding posttranslational modifications lead to structural and functional changes that govern signal transduction pathways.

Unlike the cytosol where GSH/GSSG $\sim 100{:}1$, the endoplasmic reticulum (ER) favors an oxidizing environment ($\sim 3{:}1$) which facilitates protein folding. This unique ER environment also provides a platform to sense oxidative and nitrosative stress. Stress upon the ER results in the accumulation of misfolded proteins, leading to activation of the unfolded protein response (UPR). The UPR has three primary functions: (1) restore normal function of the cell by halting protein translation, (2) activate the signaling pathways that lead to increased production of molecular chaperones involved in protein folding, and (3) trigger the degradation of terminally misfolded proteins (Townsend, 2007).

The ER contains several key chaperones that catalytically mediate protein folding and prevent aggregation of proteins as they undergo maturation. The most abundant chaperone in the lumen of the ER is protein disulfide isomerase (PDI). It is not surprising that PDI acts as a redox sensor and can be S-nitrosylated (Uehara et al., 2006) and/or S-glutathionylated (Townsend et al., 2006, 2009b) following nitrosative stress (Fig. 18.1). PDI is organized into five domains (a, b, b′, a′, and c) and the C-terminal KDEL sequence retains it to the ER. The crystal structure of yeast PDI suggests that the four thioredoxin domains (a, b, b′, a′) form a twisted U shape with the catalytic domains (a and a′) facing each other and an internal hydrophobic surface that interacts with misfolded proteins (Tian et al., 2006). The cysteine residues within the a and a′ domains (catalytic) are targets for redox regulation (Fig. 18.1). The functional implication is that PDI has a reduced activity for the period of time that the cysteine residue is S-glutathionylated, resulting in accumulated unfolded/misfolded proteins hence triggering activation of the UPR. The coordination of PDI and

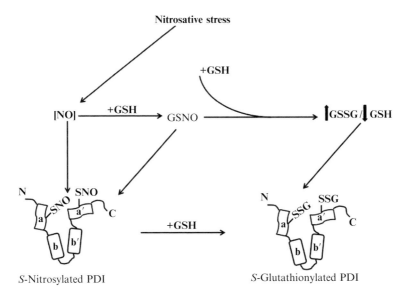

Figure 18.1 Nitrosative stress induced S-glutathionylation. Nitrosative stress leads to an alteration in the ratio of GSH/GSSG and formation of reactive nitrogen species. PDI can be posttranslationally modified to form S-nitrosylated or S-glutathionylated proteins. The cellular consequence leads to an accumulation of unfolded proteins and activation of the UPR.

downstream signaling events provides an integration point for the functional determination of cell survival pathways. Here, we describe the methodology for the detection, confirmation, and validation of the functional consequences of S-glutathionylation of PDI.

2. IDENTIFICATION AND CONFIRMATION OF S-GLUTATHIONYLATED PROTEINS IN CELLS

S-Glutathionylation is an important posttranslational modification that governs redox-mediated signaling events. Identification and validation of target cysteine residues have been limited by detection methods that are not compatible with reducing agents. A two-pronged approach was utilized to identify and confirm that PDI is S-glutathionylated (Townsend et al., 2006, 2009b). First, using two-dimension isoelectric focusing under nonreducing conditions PDI was isolated and identified by mass spectrometry as a target for S-glutathionylation (Townsend et al., 2006). Second, S-glutathionylation of PDI was confirmed via immunoprecipitation of PDI and Western blot analysis for P-SSG. This general approach is outlined in Fig. 18.2 and has been used *in vitro* and *in vivo* to identify S-glutathionylated proteins following both oxidative and nitrosative stress (Findlay et al., 2006; Townsend et al., 2006, 2008, 2009a,b). *In situ* detection of S-glutathionylated proteins has

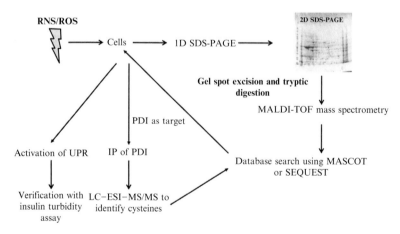

Figure 18.2 Identification, validation, and characterization of PDI-SSG and its functional consequences. A general scheme for identifying S-glutathionylated proteins by MALDI-TOF showed that PDI is a target for redox regulation. Subsequent confirmation in cells via immunoprecipitation validated the modification. The cysteine targets were mapped using LC–MS–ESI from recombinant protein and in cells. The structural and functional consequences were evaluated both *in vitro* and in cells.

been recently reported (Aesif et al.) and collectively these tools can define the role of PDI in multiple pathologies associated with misfolded proteins.

2.1. Materials

PABA/NO; GSH; 20 mM PBS; protein A/G Plus agarose; IgG; anti-PDI (Affinity BioReagents, Golden, CO); anti-SSG (Virogen, Watertown, MA); lysis buffer (20 mM Tris–HCl, pH 7.5, 15 mM NaCl, 1 mM EDTA, 1 mM EGTA, 1% Triton X-100, 2.5 mM sodium pyrophosphate, and 1 mM β-glycerophosphate with freshly added protease and phosphatase inhibitors, 5 mM NaF and 1 mM Na$_3$VO$_4$); nonreducing loading sample buffer; 10% SDS-PAGE gel 100 μg whole cell protein; 2-D lysis buffer (8 M urea, 2 M thiourea, 4% CHAPS, 20 mM Spermine, 1 mM PMSF); rehydration buffer (8 M urea, 2% CHAPS, 0.5% IPG buffer); equilibration buffer (6 M urea, 30% glycerol, 2% SDS, 50 mM Tris–HCl, pH 8.8).

2.2. Methods

2.2.1. Induction of S-glutathionylation in cells

Prior to identifying specific protein targets of S-glutathionylation, a dose- and time-curve is performed using the anti-glutathionylation (anti-SSG) antibody as the final readout. For the time course, $\sim 5 \times 10^5$ HL60 cells per treatment group are seeded and treated with 25 μM PABA/NO (IC$_{50}$) for 0–8 h with 30-min to 1-h increments. Wash cells prior to harvest with PBS and resuspend the pellets in lysis buffer and incubate on ice for 30 min. Sonicate lysates for 10 s and centrifuge for 30 min at 10,000×g at 4 °C. Determine the protein concentration with the Bradford reagent using IgG as a standard. Do not freeze cell lysate. Add 50 μg cell lysate in nonreducing sample loading buffer and heat to $\sim$95 °C for 10 min. Separate the proteins under nonreducing conditions on 10% SDS-PAGE gels. Transfer proteins onto a nitrocellulose membrane (Bio-Rad, Hercules, CA). Block nonspecific binding with blocking buffer (20 mM Tris–HCl, pH 7.5, 150 mM NaCl, 0.1% Tween 20, 1 μM protease inhibitors, 5 mM NaF, and 1 mM Na$_3$VO$_4$) containing 10% nonfat dried milk for 1 h. Incubate the membrane with anti-SSG (1:100) overnight at 4 °C. Wash the membrane 3× with PBS for 15 min and incubate with secondary antibody conjugated to horseradish peroxidase for 1 h. Remove the secondary antibody with 3× washes using PBS. Develop the blots with enhanced chemiluminescence detection reagents and scan using a transilluminator. Strip the membrane and reprobe with anti-actin. Calculate the relative abundance of P-SSG to actin for each treatment group.

For the dose-curve, treat $\sim 5 \times 10^5$ cells with 0–50 μM PABA/NO for the time at which peak P-SSG levels were observed and perform immunoblot analysis as described above. To identify specific proteins, it is necessary to run two-dimensional gels and excise bands using the optimal dose/time regime determined in this series of experiments.

2.2.2. Identification of S-glutathionylated proteins via 2-D SDS-PAGE and MALDI-TOF mass spectrometry

From the dose- and time-curve, treat the cells with an IC_{50} dose for the time to reach maximal global S-glutathionylation. For SKOV3 or HL60 cells, treat cells with diluent or 25 µM PABA/NO for 1 h. Wash cells prior to harvest with PBS and resuspend the pellets in lysis buffer and incubate on ice for 30 min. Sonicate lysates for 10 s and centrifuge for 30 min at $10,000 \times g$ at 4 °C. Determine the protein concentration with the Bradford reagent using IgG as a standard. (Two gels per treatment group are needed for Coomassie staining and Western blot.)

For the first dimension, 200 µg of whole cell protein will be resuspended in 2-D-lysis buffer. Prepare the pH gradient strips according to the manufacturers' suggestion in rehydration buffer without DTT. The suspension will be divided equally and run on immobilized pH gradients covering exponentially the pH range from 3 to 10 as follows: rehydration for 12 h; 50 µA/strip at 500 V for 1 h, 1000 V for 1 h, and then 8000 V for $\sim$3 h. Equilibrate according to manufacturers' directions but do not include DTT in EQ Buffer 1 or P-SSG modifications will be lost.

For the second dimension, run the immobilized pH gradient strip on a 10% polyacrylamide gel for $\sim$15 min at 10 mA/gel then increase to $\sim$25 mA/gel for 4 h or until the band reaches the bottom of the gel. One gel will be stained using Coomassie brilliant blue R250 or G250 which is mass spectrometry compatible. The other gel will be transferred and blotted for P-SSG. Place the Coomassie-stained gel in HPLC grade water until the Western blot for P-SSG is complete. Place the developed film for P-SSG onto a light box and align the gel. Excise the band(s) of interest with a clean scalpel and transfer to a microcentrifuge tube. Following trypsin digestion, peptide mass analysis can be performed by MALD-TOF mass spectrometry and the protein identification can be assessed using software from the National Center for Biotechnology Information protein database.

2.2.3. Confirmation of S-glutathionylated PDI

Pre-clear 800 µg cell lysate from control and treated cells with protein A/G plus agarose. Precipitate PDI with the anti-PDI antibody (1:100) overnight at 4 °C. Spin and wash the precipitate three times with lysis buffer. Resuspend the precipitated PDI in nonreducing sample loading buffer and heat to $\sim$95 °C for 10 min. Separate the proteins under nonreducing conditions on 10% SDS-PAGE gels. Transfer proteins onto a nitrocellulose membrane (Bio-Rad). Block nonspecific binding with blocking buffer

(20 mM Tris–HCl, pH 7.5, 150 mM NaCl, 0.1% Tween 20, 1 μM protease inhibitors, 5 mM NaF, and 1 mM Na$_3$VO$_4$) containing 10% nonfat dried milk for 1 h. Incubate the membrane with anti-SSG (1:100) overnight at 4 °C. Wash the membrane 3× with PBS for 15 min and incubate with secondary antibody conjugated to horseradish peroxidase for 1 h. Remove the secondary antibody with 3× washes using PBS. Develop the blots with enhanced chemiluminescence detection reagents and scan using a transilluminator. Strip the membrane and reprobe with anti-PDI.

3. Identification of Target Cysteine Residues

Specificity is critical in ascribing any posttranslational modification. Unlike phosphorylation, there is no sequence signature that is a hallmark for S-glutathionylation. However, reduced cysteine residues (-SH) that have a low pK_a (vicinal or close to basic amino acid residues) are reactive and can become oxidized into protein sulfenic acids (P-SOH), nitrosylated (P-SNO), or S-glutathionylated (P-SSG). As such, it is important to search the database for each of these modifications concurrently. Using the biotin switch assay, Uehara *et al.* (2006) reported that PDI is S-nitrosylated in neuronal cells and tissue from PD and AD patients. Oxidation or S-glutathionylation was not investigated in these samples. Both P-SNO and P-SSG of PDI lead to diminished isomerase activity and activation of the UPR. Mass spectrometry analysis of PDI following nitrosative stress showed that both active site cysteines are S-glutathionylated; however, P-SNO was not detected, suggesting that P-SNO residues are subject to rapid conversion to a P-SSG product (Townsend *et al.*, 2009b). The following technique utilizes recombinant PDI–SSG as a positive control in the mass spectrometry analysis of PDI in cells following nitrosative stress.

3.1. Materials

Recombinant PDI; PABA/NO; GSH; 20 mM PBS; control and PABA/NO-treated cell pellets; protein A/G Plus agarose; IgG; anti-PDI (Affinity BioReagents); anti-SSG (Virogen); lysis buffer (20 mM Tris–HCl, pH 7.5, 15 mM NaCl, 1 mM EDTA, 1 mM EGTA, 1% Triton X-100, 2.5 mM sodium pyrophosphate, and 1 mM β-glycerophosphate with freshly added protease and phosphatase inhibitors, 5 mM NaF and 1 mM Na$_3$VO$_4$); nonreducing loading sample buffer; 10% SDS-PAGE gel; LysC. Biospin-size exclusion micro-spin columns (Bio-Rad).

3.2. Methods

3.2.1. In vitro S-glutathionylation of PDI

Incubate 2 mg/mL PDI (>95% homogeneous) in 20 mM phosphate buffer (pH 7.4) for 1 h at room temperature as follows: (a) control; (b) 25 µM PABA/NO and 1 mM GSH; (c) 25 µM PABA/NO; and (d) 1 mM GSH. Excess PABA/NO and GSH will be eliminated through Biospin-6 Size exclusion micro-spin columns (Bio-Rad) with 20 mM PBS. Five microliters of native and PDI-SSG will be used for immunoblot analysis and the remaining will be digested with LysC and evaluated by mass spectrometry.

3.2.2. S-glutathionylation of PDI in cells

Treat cells with 25 µM PABA/NO for 4 h. Wash cells prior to harvest with PBS and resuspend the pellets in lysis buffer and incubate on ice for 30 min. Sonicate lysates for 10 s and centrifuge for 30 min at 10,000×g at 4 °C. Determine the protein concentration with the Bradford reagent using IgG as a standard. Immunoprecipitate PDI as described in Section 2.2. Divide the sample equally and load on two SDS-PAGE gels; one for transfer and immunoblot confirmation as described in Section 2.2 and the other for staining and mass spectral analysis.

3.2.3. Mass spectrometry identification of target cysteine residues

Stain the proteins using Coomassie brilliant blue R250 or G250 which is mass spectrometry compatible. Place the gel in HPLC grade water for a few hours. Place the gel onto a light box to excise the band of interest with a clean scalpel, transfer to a microcentrifuge tube, and spin down. The gel will be destained by adding 100 µl of 100 mM ammonium bicarbonate/acetonitrile (1:1, v/v) and vortexing. Depending on the stain intensity, this will require occasional vortexing for ∼30 min. Remove and add 500 µl acetonitrile and incubate at room temperature until the gel pieces are clear and shrunk. Remove acetonitrile prior to digesting the protein. To digest, cover the gel piece with LysC buffer at 4 °C for 30 min. It is important to keep the gel plugs wet during the enzymatic cleavage. Incubate the tubes at 37 °C overnight in an air circulation thermostat. To extract the peptide digestion products, add 100 µl of extraction buffer (1:2, 5% formic acid/acetonitrile) to each tube and shake for 15 min at 37 °C. Separate the digested fragments (20 µl) by HPLC using a C18 RP column. Elute the peptides during a 30-min gradient from 2% to 70% with 0.2% formic acid using a 180 nL/min flow rate. The eluted peptides can be detected on an ion trap mass spectrometer operated in data acquisition mode with dynamic exclusion enable. Dynamic exclusion will prevent duplication of MS/MS experiments of the same precursor ions over an elution window of 3 min and thus allow lower abundance peptides to be sequenced and more complete sequence coverage. Analysis of the LysC digested protein with GPS Explorer software using Mascot showed that two fragments were detected with a molecular mass [+305.6] compatible with a single S-glutathionylation in the PABA/NO-treated samples that was not present in untreated. The S-glutathionylated fragments (43–57 and 387–401) correspond to the active sites in the a and a′ domain, and each contain a CXXC motif. Both active sites of PDI are altered and thereby inhibit protein function.

4. Characterization of Structural and Functional Consequences of S-Glutathionylated PDI

Activation of the UPR is triggered by the accumulation of misfolded or unfolded proteins. PDI is a key player in protein folding and as such, its dysregulation can lead to activation of the UPR. The structural and functional consequences of S-glutathionylation of PDI can be assessed directly using recombinant protein. The functional consequences can be assessed indirectly in cells and tissues by detection of activation of the UPR concurrent with S-glutathionylation. Methods to detect activation of the UPR are reviewed in this issue. Here, we describe *in vitro* assays for assessing structural (circular dichroism (CD) and intrinsic fluorescence) and functional aspects (insulin turbidity assay) of native and modified PDI (Holmgren, 1979; Lundstrom and Holmgren, 1990).

4.1. Materials

Native or S-glutathionylated PDI (1 mg/ml in sodium phosphate buffer); 10 mg/ml insulin in 50 mM Tris–HCl, pH 7.5; 100 mM DTT; 100 mM sodium EDTA, pH 7.0.

4.2. Methods

4.2.1. Determination of the structural consequences of P-SSG

The effect of PDI-SSG on secondary structure can be examined by spectroscopic analysis (Townsend *et al.*, 2009b). *In vitro* S-glutathionylation of PDI is carried out as described in Section 3.2 and the excess GSH and PABA/NO removed by running the samples through Biospin columns. Add 1 ml PDI, PDI-SSG or diluent to a 10 × 10 × 40 mm quartz cuvette. Record protein tryptophan fluorescence on an F 2500 spectrofluorometer (Hitachi) where excitation and emission slits are 2.5 and 5.0 nm, respectively. The excitation wavelength at 295 nm will minimize the effect of protein tyrosine and phenylalanine residues. Subtract background (diluent) spectra from the final emission of the protein ($N = 3$). Remove the sample and use for CD.

CD measurements can be carried out on a 202 AVIV Associates (Lakewood, NJ) using a semi-micro quartz rectangular $1 \times 10 \times 40$ mm cuvette. Maintain samples at 22 °C using a Pelletier element. Record spectra while scanning in the far-ultraviolet region (190–260 nm), with bandwidth of 1.0 nm, step size of 1.0 nm, integration time of 30 s three times for buffer control, PDI, and PDI-SSG. The output of the CD spectrometer ($N = 3$) will be reported according to the protein concentration, amino acid content, and cuvette thickness into molecular ellipticity units (degrees/cm^2/dmol).

4.2.2. Determination of isomerase activity

The insulin turbidity assay is based on the ability of PDI to reduce the disulfide bonds within insulin, resulting in the precipitation of an insoluble β-chain which can be followed at 650 nm (Holmgren, 1979; Lundstrom and Holmgren, 1990) (Fig. 18.3). DTT has been shown to reduce the disulfide bonds completely and is used as a positive control. Prepare a

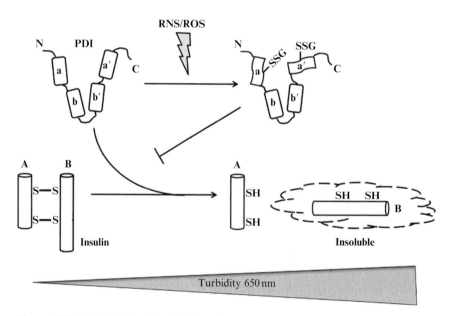

Figure 18.3 PDI-SSG activity is blunted as measured by the insulin turbidity assay. PDI breaks the two disulfide bonds between two insulin chains (A and B) that results in precipitation of the B chain. The turbidity is measured as an increase in the absorbance at 650 nm. RNS/ROS induces S-glutathionylation of PDI in the a and a$'$ domains, changes the protein conformation, and abolishes its enzyme activity.

5 ml reaction mix (4.2 ml sodium phosphate buffer, 130 µl EDTA; 670 µl insulin). Prepare native and S-glutathionylated PDI to a final dilution of 3.6 µg PDI. Positive control (100 mM DTT); negative control (no PDI or DTT). In a 96-well plate, add the following per well (in triplicate): 75 µl reaction mix; 23.5 µl sodium phosphate buffer; and 1 µl DTT. Equilibrate the plate to 25 °C and read at A_{650}. Add 1.5 µL PDI (native or test) or 1.5 µL DTT (positive control) or diluent (negative control) and read every 5 min for 1 h. Plot the absorbance/minute after subtracting from the negative control.

5. SUMMARY

Redox regulation of proteins is emerging as an important aspect to signaling cascades. Identification and characterization of S-glutathionylated moieties will be critical to ascribing structural and functional consequences. The methodologies described herein can be used to identify and validate S-glutathionylated proteins in different cells and tissues following oxidative or nitrosative stress.

ACKNOWLEDGMENTS

The research relevant to this publication was supported by the National Cancer Institute grants CA08660 and CA117259 and NIH R56 ES017453.

REFERENCES

Aesif, S. W., Janssen-Heininger, Y. M., and Reynaert, N. L. (2010). Protocols for the detection of S-glutathionylated and S-nitrosylated proteins in situ. *Methods Enzymol.* **474**, 289–296.

Findlay, V. J., Townsend, D. M., Morris, T. E., Fraser, J. P., He, L., and Tew, K. D. (2006). A novel role for human sulfiredoxin in the reversal of glutathionylation. *Cancer Res.* **66**, 6800–6806.

Holmgren, A. (1979). Thioredoxin catalyzes the reduction of insulin disulfides by dithiothreitol and dihydrolipoamide. *J. Biol. Chem.* **254**, 9627–9632.

Ilic, T. V., Jovanovic, M., Jovicic, A., and Tomovic, M. (1999). Oxidative stress indicators are elevated in de novo Parkinson's disease patients. *Funct. Neurol.* **14**, 141–147.

Lundstrom, J., and Holmgren, A. (1990). Protein disulfide-isomerase is a substrate for thioredoxin reductase and has thioredoxin-like activity. *J. Biol. Chem.* **265**, 9114–9120.

Moncada, S. (1997). Nitric oxide in the vasculature: Physiology and pathophysiology. *Ann. NY Acad. Sci.* **811**, 60–67. discussion 67–69.

Tian, G., Xiang, S., Noiva, R., Lennarz, W. J., and Schindelin, H. (2006). The crystal structure of yeast protein disulfide isomerase suggests cooperativity between its active sites. *Cell* **124**, 61–73.

Tieu, K., Ischiropoulos, H., and Przedborski, S. (2003). Nitric oxide and reactive oxygen species in Parkinson's disease. *IUBMB Life* **55**, 329–335.

Townsend, D. M. (2007). S-glutathionylation: Indicator of cell stress and regulator of the unfolded protein response. *Mol. Interv.* **7,** 313–324.
Townsend, D., and Tew, K. (2003). Cancer drugs, genetic variation and the glutathione-S-transferase gene family. *Am. J. Pharmacogenomics* **3,** 157–172.
Townsend, D. M., Findlay, V. J., Fazilev, F., Ogle, M., Fraser, J., Saavedra, J. E., Ji, X., Keefer, L. K., and Tew, K. D. (2006). A glutathione S-transferase pi-activated prodrug causes kinase activation concurrent with S-glutathionylation of proteins. *Mol. Pharmacol.* **69,** 501–508.
Townsend, D. M., He, L., Hutchens, S., Garrett, T. E., Pazoles, C. J., and Tew, K. D. (2008). NOV-002, a glutathione disulfide mimetic, as a modulator of cellular redox balance. *Cancer Res.* **68,** 2870–2877.
Townsend, D. M., Manevich, Y., He, L., Hutchens, S., Pazoles, C. J., and Tew, K. D. (2009a). Novel role for glutathione S-transferase pi. Regulator of protein S-glutathionylation following oxidative and nitrosative stress. *J. Biol. Chem.* **284,** 436–445.
Townsend, D. M., Manevich, Y., He, L., Xiong, Y., Bowers, R. R., Jr., Hutchens, S., and Tew, K. D. (2009b). Nitrosative stress-induced s-glutathionylation of protein disulfide isomerase leads to activation of the unfolded protein response. *Cancer Res.* **69,** 7626–7634.
Uehara, T., Nakamura, T., Yao, D., Shi, Z. Q., Gu, Z., Ma, Y., Masliah, E., Nomura, Y., and Lipton, S. A. (2006). S-nitrosylated protein-disulphide isomerase links protein misfolding to neurodegeneration. *Nature* **441,** 513–517.

CHAPTER NINETEEN

Methods for Analyzing eIF2 Kinases and Translational Control in the Unfolded Protein Response

Brian F. Teske, Thomas D. Baird, *and* Ronald C. Wek

Contents

1. Introduction to Translational Control in the UPR	334
1.1. The eIF2 kinase PERK is central for translational control and the UPR	334
1.2. Phosphorylation of eIF2α is not unique to ER stress	336
1.3. Key strategies and methodologies for analysis of translation control in the UPR	337
2. Measuring Activation of the eIF2 Kinase Pathway During ER Stress	338
2.1. Strategies for measuring induction of eIF2α~P and the ISR	338
2.2. Measurement of eIF2α~P and the ISR by immunoblot analyses	339
2.3. Determining whether the observed induction of eIF2α~P and the ISR results from ER stress	341
2.4. Measuring the phosphorylation status of PERK during cellular stress	343
2.5. Detection of UPR activation through PERK *in vitro* kinase assay	344
3. Investigating General and Gene-Specific Translation Control	345
3.1. Measuring protein synthesis by [^{35}S] methionine/cysteine incorporation	345
3.2. Polysome profiling by sucrose gradient centrifugation	346
3.3. Isolation and characterization of RNA in sucrose gradients	349
3.4. *ATF4* 5′UTR luciferase assay to study preferential translation	351
Acknowledgments	353
References	353

Department of Biochemistry and Molecular Biology, Indiana University School of Medicine, Indianapolis, Indiana, USA

Abstract

Endoplasmic reticulum (ER) stress induces a program of translational and transcriptional regulation, designated the unfolded protein response (UPR), that collectively remedies stress damage and restores ER homeostasis. The protein kinase PERK facilitates the translational control arm of the UPR by phosphorylation of eIF2, a translation initiation factor that combines with GTP to escort initiator Met-tRNA$_i^{Met}$ to the ribosomal machinery during the initiation of protein synthesis. Phosphorylation of the alpha subunit of eIF2 on serine-51 inhibits global translation initiation, which reduces the influx of nascent polypeptides into the overloaded ER. eIF2 phosphorylation also facilitates the preferential translation of stress-related mRNAs, such as *ATF4* which in turn activates the transcription of UPR genes. In this chapter, we present experimental strategies and methods for establishing and characterizing global and gene-specific translation control induced by eIF2 phosphorylation (eIF2α~P) during ER stress. These methods include assays for the detection of eIF2α~P and its target genes. We also discuss strategies to address whether a given ER stress condition triggers eIF2α~P through PERK, as opposed to other stress conditions activating alternative members of the eIF2 kinase family. Additionally, experimental descriptions are provided for detecting and quantifying a repression in global translation initiation, and identifying stress-induced preferential translation, such as that described for *ATF4*. Together, these experimental descriptions will provide a useful molecular "toolkit" to study each feature of the translational control processes invoked during ER stress.

1. INTRODUCTION TO TRANSLATIONAL CONTROL IN THE UPR

1.1. The eIF2 kinase PERK is central for translational control and the UPR

The endoplasmic reticulum (ER) is a cellular organelle that is a site for calcium storage, lipid biosynthesis, and entry, folding and assembly of proteins destined for the secretory pathway. Calcium dysregulation, oxidative damage, and accumulation of misfolded proteins can trigger ER stress and induce a regulatory network involving both transcriptional regulation and translational control mechanisms that are jointly referred to as the unfolded protein response (UPR) (Harding et al., 2003; Marciniak and Ron, 2006; Ron and Walter, 2007; Schroder and Kaufman, 2005; Wek and Cavener, 2007). The UPR enhances the expression of genes involved in protein folding, processing, and transport, as well as clearance of terminally malfolded proteins through the ER-associated degradation (ERAD) system, which can collectively alleviate the underlying stress and return the ER to homeostasis.

PKR-like endoplasmic reticulum kinase (PERK, also known as PEK/EIF2AK3) is a Type I ER transmembrane protein that mediates the translational control arm of the UPR through phosphorylation of eIF2, a translation initiation factor which combines with GTP and escorts initiator Met-tRNA$_i^{Met}$ to the ribosomal machinery during the initiation of protein synthesis (Fig. 19.1). The precise mechanisms by which PERK recognizes ER stress is not fully understood, but it is proposed that the ER chaperone GRP78/BiP binds the amino-terminal portion of PERK located in the ER lumen, repressing the eIF2 kinase (Bertolotti *et al.*, 2000; Ma *et al.*, 2002; Marciniak and Ron, 2006; Schroder and Kaufman, 2005). Perturbations in

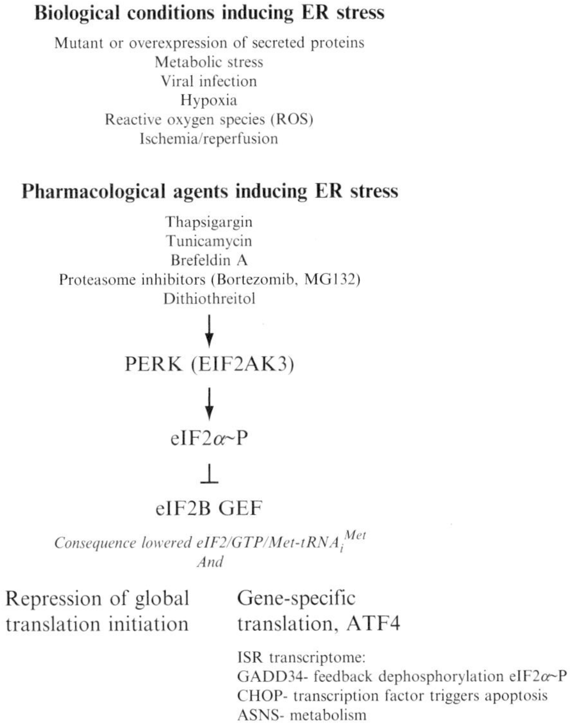

Biological conditions inducing ER stress

Mutant or overexpression of secreted proteins
Metabolic stress
Viral infection
Hypoxia
Reactive oxygen species (ROS)
Ischemia/reperfusion

Pharmacological agents inducing ER stress

Thapsigargin
Tunicamycin
Brefeldin A
Proteasome inhibitors (Bortezomib, MG132)
Dithiothreitol

↓

PERK (EIF2AK3)

↓

eIF2α~P

⊥

eIF2B GEF

Consequence lowered eIF2/GTP/Met-tRNA$_i^{Met}$
And

| Repression of global translation initiation | Gene-specific translation, ATF4 |

ISR transcriptome:
GADD34- feedback dephosphorylation eIF2α~P
CHOP- transcription factor triggers apoptosis
ASNS- metabolism

Figure 19.1 Different biological and pharmacological conditions induce ER stress, eliciting eIF2α~P and subsequent translational control. Many different biological conditions or pharamacological agents can disrupt the ER and induce eIF2α~P by PERK. Phosphorylation of eIF2α changes this translation initiation factor from a substrate to a competitive inhibitor of the guanine nucleotide exchange factor, eIF2B. The resulting reduction in eIF2-GTP levels represses global translation initiation by preventing the eIF2 delivery of Met-tRNA$_i^{Met}$ to the translation apparatus. Phosphorylation of eIF2α also preferentially enhances translation of stress-related mRNAs, such as that encoding *ATF4*. ATF4 is a transcription activator of the integrated stress response, featuring *GADD34*, *CHOP*, and *ASNS*.

the ER that can affect protein folding are suggested to lead to an accumulation of malfolded protein which recruit GRP78, thus releasing the ER chaperone from the regulatory portion of PERK. Release of the repressing GRP78 is then suggested to facilitate PERK dimerization, resulting in conformational changes which culminate in PERK autophosphorylation and activation of the cytoplasmic protein kinase domain. Enhanced PERK phosphorylation of the α subunit of eIF2 (eIF2α∼P) at serine-51 blocks the function of eIF2B, a guanine nucleotide exchange factor that recycles eIF2-GDP to eIF2-GTP (Fig. 19.1). The resulting lowered levels of eIF2-GTP blocks binding and delivery of the initiator Met-tRNA$_i^{Met}$ to the ribosomal machinery, thus significantly reducing global protein synthesis and diminishing the influx of nascent polypeptides into the overloaded ER secretory pathway. Accompanying this repression of global translation, eIF2α∼P selectively enhances the translation of specific stress-related mRNAs, such as that encoding *ATF4* (Fig. 19.1) (Harding *et al.*, 2000; Lu *et al.*, 2004; Vattem and Wek, 2004). The ATF4 transcription factor functions in conjunction with additional ER stress regulators, ATF6 and IRE1, to induce the transcriptional component of the UPR (Marciniak and Ron, 2006; Ron and Walter, 2007; Schroder and Kaufman, 2005). Therefore, eIF2α∼P by PERK directs both global and gene-specific translational control, and the preferential translation of key stress-related mRNAs is also central for eIF2α∼P regulation of the transcriptional arm of the UPR.

1.2. Phosphorylation of eIF2α is not unique to ER stress

In addition to the UPR signaling network, translation control is directed by other eIF2 kinases that each respond to different stress conditions (Sonenberg and Hinnebusch, 2009; Wek and Cavener, 2007; Wek *et al.*, 2006). In mammalian cells, these additional eIF2 kinases include GCN2 (EIF2AK4), which is activated in response to nutrient deprivation and UV irradiation, PKR (EIF2AK2) that participates in an antiviral defense mechanism triggered by interferon, and HRI (EIF2AK1), which is activated by heme deprivation and oxidative stress in erythroid tissues. Analogous to that described for PERK and ER stress, activation of each of these eIF2 kinases abruptly blocks translation initiation by reducing the recycling of eIF2-GDP to eIF2-GTP. For example, in response to starvation for amino acids, eIF2α∼P by GCN2 serves to conserve nutrients (Anthony *et al.*, 2004; Hinnebusch, 2005; Wek *et al.*, 2006) while eIF2α∼P by HRI insures that globin synthesis is coupled to heme availability (Chen, 2007). In each case, eIF2α∼P also leads to preferential translation of *ATF4*, a master transcriptional regulator of stress-related genes involved in metabolism, nutrient uptake, protein folding and assembly, the redox status of cells, and apoptosis (Harding *et al.*, 2003; Marciniak and Ron, 2006; Schroder and Kaufman, 2005). The idea that ATF4 is a targeted downstream factor that integrates signaling from PERK and other eIF2 kinases has led to the eIF2α~P/ATF4 pathway being designated as the integrated stress response (ISR) (Harding et al., 2003).

It should be emphasized that these additional eIF2 kinases are not thought to function in connection with the other UPR regulators, ATF6 and IRE1. For example, GCN2 functions in conjunction with the mTOR-signaling pathway in response to starvation for amino acids (Cherkasova and Hinnebusch, 2003; Cherkasova et al., 2010; Staschke et al., 2010). Therefore, it is important to experimentally establish that eIF2α~P and translational control identified in response to a given stress condition are part of the UPR, as opposed to other stress pathways that activate the ISR. It is also important to note that while the ISR serves essential adaptive functions in response to ER stress and other stress arrangements, perturbations in or unabated induction of these stress responses can contribute to morbidity. The processes by which the ISR can adversely affect cells is not well understood, but central to this process is the ATF4-target gene *CHOP/ GADD153*, a transcriptional regulator for which extended expression during stress can trigger apoptosis (Marciniak et al., 2004; McCullough et al., 2001; Puthalakath et al., 2007; Ron and Walter, 2007; Rutkowski et al., 2006) (Fig. 19.1).

1.3. Key strategies and methodologies for analysis of translation control in the UPR

This chapter will focus on experimental strategies and methods for establishing and characterizing global and gene-specific translation control induced by eIF2α~P during the UPR. These strategies and methods are divided into Sections 2 and 3.

Section 2 will describe the experimental methods for establishing whether the stress condition induces eIF2α~P and expression of key downstream target genes. It is critical when designing experiments that eIF2α~P be measured following a time course for a given stress condition and following different degrees of the stress treatment. For example, if a particular pharmacological agent is added to cultured cells, a range of doses and exposure times should be analyzed. This is important because the different components of the ISR occur stepwise and each can be transitory. With the initiation of ER stress, eIF2α~P occurs rapidly, leading to repression of global translation initiation coincident with preferential translation of *ATF4*. Accumulating ATF4 enhances the transcription of a collection of genes subject to the ISR. Included among the ATF4-targeted genes is *GADD34*, a regulatory subunit of Type 1 protein phosphatase (PP1c) that facilitates dephosphorylation of eIF2α~P (Brush et al., 2003; Ma and Hendershot, 2003; Marciniak and Ron, 2006; Novoa et al., 2001, 2003) (Fig. 19.1). Thus there is a feedback mechanism that allows for resumption of translation once the ISR transcriptome is produced. In this way, isolated measurements of eIF2α~P during the course of a stress response may provide an incomplete picture, with little evidence of the translational control processes at later time points, while the ISR transcriptome may be fully implemented. Section 2 will also detail the methods and approaches for establishing whether eIF2α~P induced in response to a given stress condition is truly a result of ER stress, or is rather a consequence of another stress condition that can induce the ISR, but is independent of the ER. It is problematic measuring the precise stress signals perturbing the ER, and instead ER stress has typically been measured by determining the changes in activation of the ER stress sensors, IRE1, ATF6, and PERK, and expression of their downstream targets. Again it is important to consider both the timing and dose of the stress treatment, as ER stress may be secondary, occurring much later as a consequence of larger stress damage to cells.

Section 3 describes methods for measuring global translation and gene-specific translation, with an emphasis on delineating whether there is repression in translation initiation and whether a given stress condition induces gene-specific translation, such as that described for *ATF4*. Incorporating these experimental strategies are important for extending the analyses of the UPR beyond simple signaling events as measured by increased levels of eIF2α~P and expression of downstream target genes using immunoblot analyses. Furthermore, it is important to distinguish whether the translational control is a direct consequence of eIF2α~P, as other mechanisms can direct translation control through alternative initiation or elongation regulatory processes.

2. Measuring Activation of the eIF2 Kinase Pathway During ER Stress

2.1. Strategies for measuring induction of eIF2α~P and the ISR

When designing stress response experiments, it is important to consider the dosage and duration of stress exposure. eIF2α~P is rapid and can be detected in cultured cells within about 2 h of treatment with agents that can induce ER stress. Commonly used ER stress agents in cell culture studies include 500–1000 nM thapsigargin (stock 1 mM in DMSO), which depletes calcium stores in the ER, 500–2000 nM tunicamycin (2 mM stock in DMSO), which blocks N-glycosylation, and 2 mM DTT (stock 1 M in H$_2$O), which interferes with the oxidative environment in the ER (Fig. 19.1). Additionally, eIF2α~P can be effectively increased by 250–1000 nM MG132, which blocks proteasome degradation of proteins including those subject to the ERAD, and 20–200 μM sodium arsenite (stock 100 mM in PBS), which can elicit general oxidative stress. Finally, salubrinal is a drug that can inhibit targeting of PP1c to mediate dephosphorylation of eIF2~P, and thus enhance the levels eIF2α~P in the absence of an underlying stress (Boyce et al., 2005, 2008). More recently, a potent derivative of salubrinal, Sal003, has been used effectively at lower concentrations which seem to improve solubility (Costa-Mattioli et al., 2007). Sal003 can be added at a final concentration of 1–10 μM, by first diluting a 10 mM stock to 100 μM in warm media prior to adding to the cultured cells. These chemical stressors may be central to the experiment at hand, or can be used as positive controls if new agents are being investigated for induction of eIF2α~P. For cultured cells, usually measurements in the 0–6 h range are sufficient to establish eIF2α~P and induced expression of immediate downstream genes. Expression of *ATF4* is subject to both translational control and transcriptional regulation (Dey et al., 2010). Preferential translation of *ATF4* occurs shortly after eIF2α~P, with continued accumulation over a course of 1–4 h. There are a number of different ATF4 target genes that can be readily monitored, including CHOP, asparagine synthetase (ASNS), and GADD34 (Kilberg et al., 2005; Ma and Hendershot, 2003; Schroder and Kaufman, 2005) (Fig. 19.1). Additionally, as noted above, eIF2α~P is subject to feedback control through GADD34 targeted PP1c dephosphorylation of this translation factor. Therefore, there may be high levels of ATF4-target genes, such as CHOP or asparagine synthetase, without readily detectable eIF2α~P.

2.2. Measurement of eIF2α~P and the ISR by immunoblot analyses

We will describe the measurements of eIF2α~P and the ISR in the context of cultured cells, but this line of experimentation can be implemented for analysis of tissue samples. Culture cells to 70–80% confluence and incubate in stressed or nonstressed conditions. It is important to prevent cells from becoming confluent, as this is by nature a stress that can induce eIF2α~P. High basal eIF2α~P levels in the absence of added stress agents is not an uncommon dilemma in experiments, and this can lead to experimental misinterpretations when appropriate controls are not included in the studies. As noted above, when studying new stress conditions or cell types for induction of eIF2α~P, it is helpful to conduct a dose response at a fixed time point, such as 6 h of exposure. Once an optimum stress condition is determined, keeping in mind the physiological parameters for a given stress, this can be followed with a time course.

After the stress treatment, cells should be washed twice with cold phosphate-buffered saline (PBS) containing 137 mM NaCl, 2.7 mM KCl, 10 mM Na_2HPO_4, and 1.76 mM KH_2PO_4, and then lysed in RIPA solution containing 50 mM Tris–HCl (pH 7.9), 150 mM sodium chloride,

1% nonylphenoxypolyethoxyethanol (nonident P-40), 0.1% sodium dodecyl sulfate (SDS), 100 mM sodium fluoride, 17.5 mM β-glycerophosphate, 0.5% sodium deoxycholate, 10% glycerol supplemented with protease inhibitors (100 μM phenylmethylsulfonyl fluoride, 0.15 μM aprotinin, 1 μM leupeptin, and 1 μM pepstatin) or EDTA-free protease inhibitor cocktail tablet (Roche). Keep lysates cold on ice and continue lysis with pulse sonication for 30 s or 8–10 passages through a 23 gauge needle. Clarify lysates by centrifugation at 6000×g for 10 min at 4 °C. Determine the protein content for each preparation by a modified Bradford or Lowry assay, such as that offered by the Bio-Rad protein quantification kit for detergent lysis (cat no. 500-0114) following the manufacturer's instructions. Equal amounts of protein preparations, typically 10–30 μg, are prepared in 2× SDS sample buffer, heated at 95 °C, and then separated by electrophoresis using an SDS-PAGE. The percentage of acrylamide for these gels should be 10% or 12% to insure linearity for the size of these proteins. Polypeptide markers of known molecular weight should be included to determine the size of proteins identified in the immunoblot analysis.

The proteins separated by gel electrophoresis are then transferred to membrane filters, and the membranes are then incubated in 5% (w/v) nonfat milk powder in PBS followed by incubation with antibody that specifically recognizes the protein of interest. After washing the membrane to remove unbound antibody, add secondary antibody (dilution 1:3000 to 1:10,000) that is dye-labeled for visualization of the target protein with the Odyssey infrared imaging system (LI-COR). Alternatively, use secondary antibodies conjugated to horseradish peroxidase (HRP) with an enhanced chemiluminescent (ECL) immunoblot detection protocol. ECL protocols can use "homebrewed" solutions that can be made fresh or in bulk and kept at 4 °C for several weeks. For detection, equal volumes of solution 1 and solution 2 are mixed and applied evenly to the membrane filter for 1–2 min followed by film detection. Solution 1 contains 100 mM Tris–HCl (pH 8.5), 2.5 mM 3-aminophthalydrazide (Luminol; Sigma–Aldrich #A-8511), and 0.4 mM P-coumaric acid (Sigma–Aldrich #C-9008) and Solution 2 contains 100 mM Tris–HCl (pH 8.5) and H_2O_2 (stock 30%) at a final concentration of 0.02%. Freshly prepared solutions usually give the more intense signal, but SuperSignal West Femto from Thermo Scientific (Prod # 34094) can be used for immunoblots when more sensitivity is needed, such as some eIF2~P immunoblots. The coupled use of LI-COR and ECL immunoblot detection can also allow for detection of multiple different proteins from a single immunoblot filter. Thus this can bypass the need for multiple rounds of membrane filter stripping and also allow for detection of proteins of similar size from one blot when cutting the membrane is not an option. Measuring levels of eIF2α~P by immunoblot analysis is not the same as determining the percentage of eIF2α that is phosphorylated. It is helpful to follow up measurements of induced eIF2α~P with analyses of repressed translation initiation, as outlined in Section 3, to insure that there is consequence to the signaling changes measured by immunoblot.

There are many different antibody preparations for measuring eIF2α~P and ISR target proteins available commercially. However, these commercial antibodies can vary in their affinity and specificity. Although commercial antibodies are indicated in publications, it is important to appreciate that the efficacy can vary between commercial antibody preparation, even when procuring antibodies with the same product number and lot number. Establishing criteria for antibody specificity, in conjunction with suitable control lysate preparations, is important, especially when one is characterizing new cultured cell lines or tissue types. The following are guidelines for antibody usage based on our laboratory experience. Antibodies against eIF2α~P typically work well among different vertebrate protein preparations, as the sequences flanking the serine-51 are well conserved. We typically use antibody against eIF2α~P at a concentration of 1:250, but depending on the commercial antibody preparation, dilutions can vary in the range of 1:50 and 1:1000. Measurements of total eIF2α are also important to insure that induced eIF2α~P is not a function of changes in the levels of the translation factor. CHOP antibodies are also typically reliable and can be used at a concentration range of 1:1000 to 1:3000. In our hands, immunoblot analyses of CHOP protein can sometimes result in a doublet band that may represent CHOP phosphorylation (Ubeda and Habener, 2003). In our experience, ATF4 immunoblots offer the greatest technical hurdles, and can be a source of misinterpretation in the literature. The predicted molecular weight of ATF4 is 38 kDa, but ATF4 typically migrates with a size between 45 to 52 kDa by SDS-PAGE. Furthermore, ATF4 often migrates as a broad band in immunoblots, which upon closer inspection can represent multiple bands that appear to be a consequence of protein phosphorylation (Elefteriou *et al.*, 2006; Frank *et al.*, 2010; Wek and Cavener, 2007; Yang *et al.*, 2004). In our view, many commercial ATF4 antibody preparations can be of poor quality, and one should be cautious when assigning a specific band to be ATF4. We typically prepare our own ATF4 antibody against recombinant human ATF4, and further purify the antibody by affinity purification.

2.3. Determining whether the observed induction of eIF2α~P and the ISR results from ER stress

An observed increase in eIF2α~P and the ISR is not necessarily a demonstration of ER stress, as many different stress conditions involving other cellular compartments can activate the eIF2 kinase family. There are four strategies to ensure that an observed increase in eIF2α~P is a consequence of ER stress. Each offers certain advantages and deficiencies, which should be recognized during the design of the proposed ER stress experiments. The first strategy is indirect, and involves measuring the activity of the UPR sensors, IRE1 or ATF6, in conjunction with eIF2α~P. These UPR sensors are thought to be exclusively activated by ER stress. IRE1 facilitates cytoplasmic splicing of *XBP1* mRNA, which then leads to synthesis of the transcription activator of UPR genes (Yoshida, 2007). Splicing of *XBP1* mRNA can be measured by qPCR (Winnay *et al.*, 2010; Zhang and Kaufman, 2008), and enhanced eIF2α~P combined with spliced *XBP1* mRNA gives one a degree of confidence that the underlying stress being studied perturbs the ER. It is noted that there are some stresses that have been reported to have discordant activation of PERK and IRE1 (Lee *et al.*, 2003; Park *et al.*, 2010; Winnay *et al.*, 2010), which can complicate negative experimental findings.

Additional strategies for establishing whether a studied stress induces eIF2α~P due to ER stress include measuring eIF2α~P in $PERK^{-/-}$ mouse embryo fibroblast (MEF) cells, or specifically knocking down this eIF2 kinase using siRNA (Jiang and Wek, 2005; Jiang *et al.*, 2004; Lu *et al.*, 2009; Xue *et al.*, 2005). A collection of MEF knockout cells have been described that are deleted for one or more of the eIF2 kinase genes, and significant reductions in levels of stress-induced eIF2α~P in PERK depleted cells support the idea that there is an underlying ER stress (Jiang *et al.*, 2004). However, it is noted that longer periods of ER stress exceeding 6 h can induce eIF2α~P in $PERK^{-/-}$ cells. It is suggested that secondary eIF2 kinases are activated during these extended periods of stress. Another related strategy is to measure the phosphorylation status of PERK, which is known to occur during the activation of this eIF2 kinase. Commercial antibodies are available for measuring phosphorylation of PERK at the activation loop (threonine-980), and measuring phospho-PERK by immunoblot analysis along with total PERK levels is a strategy for determining whether a stress activates this ER stress-inducible eIF2 kinase (Harding *et al.*, 1999, 2001; Ma *et al.*, 2002). It is noted that immunoblot analysis of phospho-PERK can be problematic, as PERK is a large membrane-bound protein (~125 kDa), and the phosphorylated version of this eIF2 kinase can migrate as several tightly packed bands in SDS-PAGE. In fact the increase in migration of PERK was originally used to establish its phosphorylation and activation during ER stress (Harding *et al.*, 1999; Ma *et al.*, 2002). Frequently, measures of phospho-PERK have involved first immunoprecipitating PERK, which can improve the sensitivity and specificity of the phospho-PERK antibody in immunoblot analysis (Harding *et al.*, 2000, 2001; Zhang and Kaufman, 2008). This protocol is provided in Section 2.4. A final strategy for establishing whether eIF2α~P is a consequence of ER stress and activation of PERK is to carry out *in vitro* kinase assays using recombinant eIF2α substrate and immunoprecipitated

PERK prepared from stressed cells showing enhanced eIF2α~P (Section 2.5) (Sood et al., 2000). Increased PERK eIF2 kinase activity in this in vitro assay would provide confidence that the studied stress condition involves ER stress and the UPR.

2.4. Measuring the phosphorylation status of PERK during cellular stress

There are two strategies for measuring the phosphorylation status of PERK. The first involves a stress-induced increase in the size migration of PERK as judged by SDS-PAGE and immunoblot analysis (Harding et al., 1999, 2000; Ma et al., 2002). A second method, which is currently most frequently used, is measuring the phosphorylation status of PERK with antibodies specific to phospho-PERK (threonine-980) (Harding et al., 2000; Ma et al., 2002). Detection of phospho-PERK by immunoblot analysis can be problematic, but success can be improved by first immunoprecipitating PERK protein (Harding et al., 2001; Zhang and Kaufman, 2008). One should first establish that eIF2α~P and its target genes are significantly enhanced in response to a studied stress condition (Section 2.2). For immunoprecipitation of PERK protein followed by immunoblot analysis for phospho-PERK, culture and prepare cell lysates as described in Section 2.2. Next preclear upward of 1 mg of protein lysate in a 500 µl volume by adding a 50 µl slurry of washed protein A agarose beads (~25 µl packed bead volume) to the protein lysate. Gently mix for up to 1 h at 4 °C. Clarify the protein lysate by microcentrifugation and then combine the supernatant with 5 µl PERK antibody. Mix by rotation from 4 h to overnight at 4 °C to bind the primary antibody to PERK. In parallel, one can use preimmune serum as a control for the immunoprecipitation. Following a brief centrifugation to remove any precipitant, add washed protein A agarose beads (~25 µl packed bead volume) to the antibody/protein lysate mixture. Follow with 4 h of mixing at 4 °C to bind the PERK/antibody complex to the beads. At the conclusion of the incubation, collect the complex by microcentrifugation and then wash the resulting PERK/protein A agarose pellet three times with 1 ml of solution containing 0.1% SDS, 1% Triton X-100, 2 mM EDTA, 10 mM Tris–HCl (pH 8.0), 150 mM NaCl, 100 mM NaF, 17.5 mM β-glycerophosphate, and protease inhibitors (100 µM phenylmethylsulfonyl fluoride, 0.15 µM aprotinin, 1 µM leupeptin, and 1 µM pepstatin).

Elute PERK from the protein A agarose beads by adding 2× SDS sample buffer, and heat at 95 °C for 5 min. Addition of fresh 2-mercaptoethanol to stock buffer may be necessary to completely reduce immunoglobulin disulfide bonds. Separate the proteins eluted from the immunoprecipitation by electrophoresis in a 7.5% SDS-PAGE. The proteins are then transferred to membrane filters and PERK is detected by using antibody specific to phospho-PERK or total PERK. It is anticipated that phosphorylation of PERK will occur rapidly upon ER stress compared to the nonstressed control. No phospho-PERK should be detected from the immunoprecipitated sample using preimmune serum. When analyzing phospho-PERK in response to a stress condition not yet linked with ER stress, one should include a positive control, such as treatment of cell cultures with thapsigargin or tunicamycin for 2 h.

2.5. Detection of UPR activation through PERK *in vitro* kinase assay

PERK purified from cells cultured in the presence or absence of stress can be used for *in vitro* eIF2 kinase assays (Sood *et al.*, 2000). Conduct the immunoprecipitation of PERK following the general strategy described in Section 2.4 using RIPA solution (50 mM Tris–HCl (pH 7.9), 150 mM NaCl, 1.0% nonident P-40, 0.5% sodium deoxycholate, 17.5 mM β-glycerophosphate and 0.1% SDS) in the presence of protease inhibitors. Avoid including NaF in the PERK immunoprecipitation mixture as this will inhibit the protein kinase assay. Wash the final PERK/protein A argarose pellet twice with the RIPA solution, then twice in kinase reaction mixture containing 20 mM Tris–HCl (pH 7.9), 50 mM KCl, 10 mM MgCl$_2$, 2 mM MnCl$_2$, 5 mM 2-mercaptoethanol and protease inhibitors (100 μM PMSF, 0.15 μM, aprotinin, 1 μM leupeptin, and 1 μM pepstatin). Note that extended incubation with SDS can reduce PERK activity, so this detergent can be omitted from the immunoprecipitation procedure, if desired. However, elimination of the SDS can reduce the purification of PERK in the immunoprecipitation so it is important to establish specificity for the serine-51 phosphorylation site in the assay. For the eIF2 kinase assay, incubate the PERK pellet in 25 μl of the kinase reaction mixture supplemented 2 μM of recombinant eIF2α (either wild-type or mutant version with alanine substituted for serine-51 [S51A]), and 10 μCi [γ^{32}P]ATP in a final concentration of 10 μM ATP. Note that peptide substrates do not work well with eIF2 kinases, as residues not directly flanking the serine-51 phosphorylation site are critical for recognition. Incubate the reaction at 37 °C for between 2 and 10 min. Linearity of the assay should be determined empirically, and the control analysis of an eIF2α-S51A substrate will insure that the measured protein kinase activity is specific for this regulatory site. Terminate the kinase assay by adding 2× SDS sample buffer and heat at 95 °C for 5 min. Separate radiolabeled proteins by SDS-PAGE (12% gels), fix by Coomassie staining, dry and visualize by autoradiography. If the underlying stress disrupts the ER and activates PERK, it would be anticipated that there would be significantly enhanced phosphorylation of the eIF2α substrate as visualized by ^{32}P radiolabeling. Only a low level of eIF2 kinase activity would be anticipated in the PERK immunoprecipitated from nonstressed cells.

3. INVESTIGATING GENERAL AND GENE-SPECIFIC TRANSLATION CONTROL

This section describes methods for measuring global translation and gene-specific translation, with an emphasis on the biological consequences of eIF2α~P during ER stress. It should be noted that while *ATF4* translational control is illustrated in these examples because it is a well-characterized target of PERK phosphorylation of eIF2α, these methods can be used broadly to study other genes which are candidates for translational control as part of the UPR, or more generally the ISR.

3.1. Measuring protein synthesis by [^{35}S] methionine/cysteine incorporation

Measuring incorporation of radiolabeled amino acids into proteins is a classical approach for determining changes in global translation. A PERK-deficient strain or an eIF2α (S51A) knockin mutant MEF cell line deficient for eIF2 phosphorylation is a useful control to determine if a given stress condition reduces global protein synthesis in an eIF2α~P-dependent manner (Harding *et al.*, 2000; Jiang *et al.*, 2003; Scheuner *et al.*, 2001). Begin by culturing wild-type and mutant MEF cells, or other desired cell types, to 50–70% confluence in Dulbecco's modified Eagle's medium. Treat cells with the desired ER stress agent for a designated length of time and subject to a 60 min pulse labeling prior to harvesting. For the pulse label, wash cells with warm PBS and then incubate in 4 ml translabeling media lacking methionine and cysteine, supplement with 10% dialyzed fetal bovine serum and 500 µCi Express Protein Label Mix containing [^{35}S]methionine and [^{35}S]cysteine (PerkinElmer Life Sciences), along with the desired stress agent. After culturing the cells for 60 min, wash cells twice with cold PBS containing nonradiolabeled methionine and cysteine and then lyse cells with RIPA buffer. Measure the [^{35}S] radiolabel in the whole cell lysate by scintillation counting to insure equal amounts of cellular uptake, as this can be reduced upon exposure to certain stress conditions. Separate equal amounts of total protein from each lysate preparation by SDS-PAGE, and visualize the radiolabeled proteins by autoradiography. The imaging can be used in combination with autoradiographic image intensifiers, such as Autofluor. The incorporation of [^{35}S]methionine/cysteine into the newly synthesized peptides can also be quantified by scintillation counting. For this, precipitate proteins by adding an equal volume of 20% trichloroacetic acid (TCA) to each cell lysate, followed by a 10-min incubation on ice and a 20-min centrifugation at $10,000 \times g$ at 4 °C. Wash the protein pellet two times with 1 ml cold acetone. Remove the final acetone wash and air dry samples. At this point, scintillation counting of the dried, insoluble pellet is used to determine radioactivity of the incorporated [^{35}S]methionine/cysteine, thus enabling the quantification of *de novo* polypeptide synthesis during the 60 min pulse.

The [^{35}S]methionine/cysteine pulse-labeling method can be extended to measure the synthesis of a given protein of interest by first immunoprecipitating the protein using a specific antibody, as outlined in Section 2.4, followed by SDS-PAGE and autoradiography. Assuming a similar cellular uptake of the [^{35}S]methionine/cysteine between stressed and nonstressed cells, each of the immunoprecipitation preparations can be normalized for total protein levels. If there is increased radiolabeling of a protein of interest coincident with eIF2α~P, it is important to also measure the transcript levels encoding the target protein by qPCR to address whether the increased protein expression is a consequence of preferential translation or by elevated levels of mRNA. When studying a new protein for possible translational control, it may be useful to carry out a control immunoprecipitation of ATF4, which is anticipated to have a significant increase in expression in response to ER stress.

3.2. Polysome profiling by sucrose gradient centrifugation

An attractive alternative to radioactive pulse labeling is polysome profiling using sucrose gradient fractionation (Dey *et al.*, 2010; Law *et al.*, 2005; Qin and Sarnow, 2004; Warner *et al.*, 1963; Zhou *et al.*, 2008). In addition to replacing the hazards and waste products associated with radioactivity, this technique enables one to determine whether translation initiation is repressed in response to a given stress condition, and allows for collection and characterization of preferentially translated mRNAs during the desired stress treatment. It is important to have a reliable method of generating linear sucrose gradients, which will reproducibly separate fractions based on density. Additionally, solutions and reagents used during this process typically contain cycloheximide, an antibiotic that blocks translation elongation, which insures that polysome will be visualized in the gradient. Some researchers omit cycloheximide and rely on cold temperatures to inhibit elongation in the lysate preparation. Our laboratory prepares sucrose gradients on the day of the experiment using a tilted tube rotation method on a gradient station equipped with a Piston Gradient Fractionator™ and a Gradient Master™ from BioComp (NB, Canada). Preparations of 10% and 50% sucrose solutions are made in 50-ml volumetric flasks in a solution containing 20 mM Tris–HCl (pH 7.5), 100 mM NaCl, 5 mM MgCl$_2$, and 50 µg/ml cycloheximide. Mark the appropriate level indicated by the marker block on Seton open-top polyclear™ centrifuge tubes (14 × 89 mm). Fill each centrifuge tube up to the indicated mark with the 10% sucrose solution, taking care that sucrose is completely solubilized in both solutions. Next, with a 4″ layering cannula, slowly layer the 50% sucrose solution below the 10% solution until the tube is completely filled to the top. If done properly, a sharp interface separating two distinct sucrose layers should be visible with the less dense 10% solution on top. Once each tube has been filled, securely add the rubber caps. Carefully add these tubes to the gradient forming tube holders in the upright position. Lastly, follow the program directions on the gradient station to generate a 10–50% gradient in about 1–2 min and gently move the resulting solutions to a 4 °C refrigerator for later use. Alternatively, sucrose gradients can be generated manually using a simple gradient maker, such as those offered by Hoefer Scientific Instruments, San Francisco. This method relies on adding the two different density solutions to separate housing chambers and regulating the rate of mixture by opening the adjoining valves and using either gravity or a peristaltic pump to collect the resulting gradient.

Cells to be analyzed in the presence or absence of stress should optimally be collected about 2 h after making the sucrose gradients. Prior to harvesting, cells are incubated in culture media containing 50 µg/ml cycloheximide for 10 min at 37 °C. After 10 min, the plates are transferred to an ice tray, the media removed by aspiration, and the cells rinsed twice with ice-cold PBS containing 50 µg/ml cycloheximide. Lyse the cells with 500 µl of cold lysis buffer consisting of 20 mM Tris (pH 7.5), 100 mM NaCl, 10 mM MgCl$_2$, 0.4% nonident P-40, 50 µg/ml cycloheximide, and EDTA-free protease inhibitor cocktail tablet (Roche). Culture plates are then scraped and the lysates transferred to labeled microcentrifuge tubes. Pass each lysate through a sterile syringe with a 23 gauge needle 8–10 times and incubate the resulting samples on ice for 10 min. Whole cell lysates are then clarified at $8000 \times g$ for 10 min and the supernatant recovered.

To prepare samples for ultracentrifugation, remove approximately 300 µl of sucrose from the top of each Seton centrifuge tube. Next, gently layer 400 µl of the cell lysate supernatant to the corresponding labeled sucrose gradient, and weigh each sample to ensure proper balancing during ultracentrifugation. Gradients of similar mass should be placed opposite one another in the rotor. If two samples vary significantly in mass, make up the difference in one gradient by layering additional lysis buffer until similar masses are achieved. Arrange the gradients appropriately in a Beckman SW41Ti rotor or equivalent and centrifuge at 40,000 rpm for 2 h at 4 °C. At the end of the centrifugation period, sucrose fractions and the resulting polysome profiles for each sample are collected using a Piston Gradient Fractionator and a 254-nm UV monitor with Data Quest software, respectively. When collecting the fractions, follow the detailed instructions provided in the Biocomp operator's manual. It is also important to use milliQ water to both blank the UV monitor before operating the system, and to flush lines between samples.

Generating polysome profiles is a convenient means to figuratively illustrate a stress-induced repression in global mRNA translation (Fig. 19.2). Furthermore, by calculating the areas under the monosome (80S), or combined 40S + 60S + 80S fractions, compared with polysomes (disome and larger) peaks, a polysome/monosome (P/M) ratio can be derived. P/M ratios provide a means to make quantitative comparisons between profiles for different treatment conditions. As illustrated in Fig. 19.2, the polysome profile generated from MEF cells treated with the ER stress agent thapsigargin demonstrates a sharp increase in the monosome fraction, coincident with lowered polysomes. This is the consequence of limiting translation initiation during eIF2α~P. Upon completion of the elongation phase, ribosomes are released from the mRNA, and then accumulate as inactive 80S complexes devoid of initiation factors. Subsequent translation initiation is blocked by the inability to recruit Met-tRNA$_i^{Met}$ due to eIF2α~P and lowered eIF2-GTP levels.

Another significant benefit of this method is that it enables one to also characterize shifts in the polysome distribution of mRNAs of interest. To do so, the Piston Gradient Fractionator can be coupled to a fraction collector. Program the fraction collector so that it collects 14 fractions of

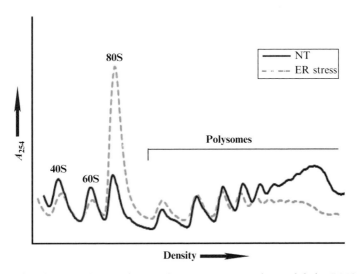

Figure 19.2 Phosphorylation of eIF2α during ER stress reduces global mRNA translation initiation as measured by sucrose gradient centrifugation. MEF cells were exposed to the ER stress agent thapsigargin for 6 h, or no stress treatment. Lysates were then analyzed by centrifugation in a 10–50% sucrose gradient, and the profiles were monitored by absorbance at 254 nm. The 40S and 60S ribosomal subunits, 80S monosomes, and polysomes are highlighted. During ER stress, the reduced polysomes, coincident with an increase in the 80S monosomes, are indicative of repressed translation initiation by eIF2α~P.

equivalent volume over the course of the entire collection process. For instance, using the above model and centrifuge tubes, a collection sampling rate of approximately 30 s yields 14 fractions of 700–750 μl of sample each. If the fraction collector and Data Quest software are appropriately coupled, the resulting polysome profile can be divided into 14 equal units to establish which peak corresponds to which sucrose fraction. These fractions can then either be frozen in an −80 °C freezer for future analysis, or immediately processed for RNA extraction, as detailed in the next section. For the latter, we suggest pooling the fractions in such a manner that the final result is seven total fractions with a portion to be analyzed for each RNA preparation (e.g., combine fractions 1 and 2, 3 and 4, etc.).

3.3. Isolation and characterization of RNA in sucrose gradients

To investigate if a specific mRNA is preferentially translated under a stress condition, it is helpful to spike the sucrose samples with an external RNA control for later normalization purposes as endogenous housekeeping mRNAs are distributed differentially throughout the sucrose gradient. This can be accomplished by spiking each pooled sucrose fraction with 10 ng/ml firefly luciferase RNA (Promega), allowing measurements of the relative amounts of the transcript of interest to be normalized to the luciferase mRNA present. For RNA isolation, add 750 μl of Trizol Reagent LS and 200 μl of chloroform to 250 μl of each sample and vortex thoroughly. Incubate the samples for 5 min at room temperature and collect by microcentrifugation at $10,000 \times g$ for 10 min at 4 °C. Collect the top aqueous layer in a new tube and add 0.5 μl Glyco-blue (Ambion) and 500 μl isopropanol, vortex thoroughly and incubate at room temperature for 5 min. At the end of the incubation, centrifuge at $10,000 \times g$ for 10 min at 4 °C and aspirate off the isopropanol. Resuspend each pellet (which should be visible as a result of the Glyco-blue) in 750 μl of 75% cold ethanol and collect by microcentrifugation at $5000 \times g$ for 7 min at 4 °C. Carefully aspirate off the ethanol and allow the pellet to air dry until all traces of ethanol are absent. Resuspend the resulting pellet in 20 μl of DEPC-treated H_2O and determine the concentration and quality of the purified RNA by measuring the absorbance at 260 and 280 nm using a Nanodrop spectrophotometer. The resulting RNA can now be used to generate cDNA for analysis by qPCR.

To identify mRNAs that are preferentially translated during a given stress, one can again rely on *ATF4* as an example, as *ATF4* is both transcriptionally and translationally regulated during ER stress (Dey et al., 2010; Siu et al., 2002). While both semiquantitative and qPCR can be used to compare levels of mRNA present in the various fractions, the benefit of the latter is in the ability to both normalize fractions to the amount of luciferase mRNA present and to analyze the data in such a manner that the distribution of the percent of total mRNA of a gene is displayed. Figure 19.3 illustrates the results of a representative experiment investigating the shift in *ATF4* mRNA during ER stress. In response to ER stress, there is a sharp change in the distribution of the *ATF4* transcripts from the free ribosome and monosome fractions to the polysome fractions. As ER treatment also causes a three- to fourfold transcriptional induction of *ATF4*, this redistribution of the *ATF4* mRNA is accompanied by an increase in the amount of *ATF4* transcript. This combination of transcriptional regulation and translational control of *ATF4* is illustrated in Fig. 19.3A, where there is about a threefold increase in the amount of *ATF4* mRNA in the polysome fractions (fractions 5–7) during ER stress as compared to the nonstressed condition.

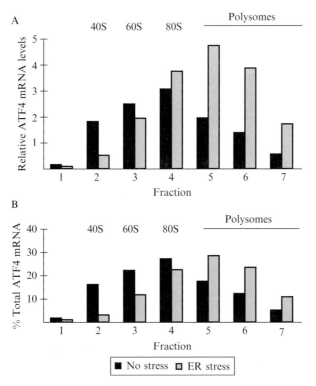

Figure 19.3 *ATF4* mRNA is preferentially translated during ER stress. Fractions were collected by sucrose gradient analyses of lysates prepared from MEF cells exposed to thapsigargin for 6 h (ER stress), or no stress treatment (No stress). Total RNA was then extracted from each of the seven fractions, and the relative total total *ATF4* mRNA levels present in each were determined by qRT-PCR (A). The percentage of total *ATF4* mRNA in each of the seven fractions was then calculated for the ER stress or nontreated cells (B).

If one wishes to focus solely on the translational control of *ATF4*, the amount of *ATF4* mRNA present in each gradient fraction can be represented as a percentage of the total *ATF4* mRNA for that treatment group (Fig. 19.3B). That is, the percentage of *ATF4* in all seven fractions should sum to 100%. Upon doing this, the amount of the *ATF4* mRNA in the polysome fractions increases about twofold in the ER stress cells compared to nonstressed. This illustrates the key features of the preferential translation of *ATF4* mRNA in response to eIF2α~P, and these strategies are applicable to new candidate genes proposed to be subject to translational control.

3.4. *ATF4* 5′UTR luciferase assay to study preferential translation

An *ATF4*-luciferase assay is available to study translational control in response to eIF2α~P (Vattem and Wek, 2004). The P_{TK}-*ATF4*-Luc plasmid contains the full-length of the 5′-leader encoded in the *ATF4* mRNA, along with the *ATF4* start codon, inserted between the thymidine kinase (TK) promoter and firefly luciferase gene in plasmid pGL3. The P_{TK}-*ATF4*-Luc plasmid can be used in a cotransfection assay to establish and characterize the extent of eIF2α~P in response to a given stress condition, and the experimental details are applicable to a broad range of cells that can be transfected. The preferential translation of *ATF4* involves two upstream ORFs (uORFs) situated in the 5′-leader of the *ATF4*-luc reporter. In nonstressed conditions when eIF2α~P is low, ribosomes loading onto the 5′-end of the *ATF4* mRNA will initiate at uORF1 (Fig. 19.4A). Following translation of this short uORF, ribosomes are thought to rapidly reinitiate translation at the next ORF, the inhibitory uORF2. Following translation of uORF2, ribosomes dissociate from the *ATF4* mRNA and, therefore, there are low levels of translation of the *ATF4* coding region. During ER stress there is increased eIF2α~P, which reduces the levels of eIF2-GTP. As a consequence, there is a delay in ribosome reinitiation following translation of uORF1. This delay allows scanning ribosomes to bypass the inhibitory uORF2, leading to enhanced translation of the *ATF4* coding region. This is illustrated by the approximately fourfold increase in ATF4-luciferase activity in response to ER stress (Fig. 19.4B). Only low levels of luciferase activity are expressed when there is a mutation in the start codon of the positive-acting uORF1 because scanning ribosomes only translate the inhibitory uORF2. Loss of the inhibitory uORF2 alone, or in combination with uORF1, leads to high levels of *ATF4*-Luc expression independent of ER stress.

The wild-type version of the P_{TK}-*ATF4*-Luc plasmid, and select mutant controls, can be transfected individually into the cultured cells of choice. Culture cells to 40–50% confluence and cotransfect, in triplicate, the P_{TK}-*ATF4*-luciferase reporter and a *Renilla* luciferase plasmid using the

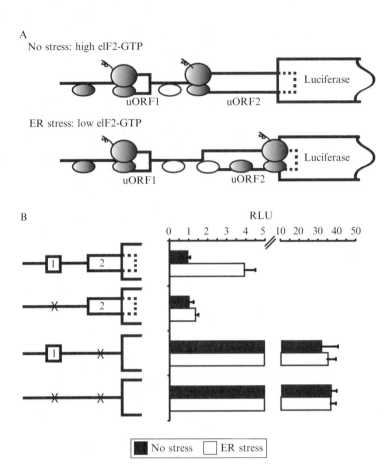

Figure 19.4 Phosphorylation of eIF2α leads to preferential translation of *ATF4* mRNA by a mechanism involving delayed translation reinitiation. (A) Model for translational control of *ATF4* in response to ER stress and induced eIF2α~P. During nonstressed conditions, when eIF2α~P is low and there are high levels of eIF2-GTP, ribosomes translate the 5′-proximal uORF1. The uORF1 is a positive-acting element in *ATF4* translation control, facilitating ribosome reinitiation at a downstream ORF. When eIF2-GTP is readily available, ribosomes rapidly reinitiate at the next available ORF, uORF2. The uORF2 is inhibitory, and following translation of this coding region, ribosomes are suggested to dissociate from the *ATF4* mRNA, resulting in low synthesis of ATF4 transcription factor. In this diagram, the ATF4 coding region is fused to the firefly luciferase reporter gene to allow for a rapid measure of translational control. During ER stress, there is increased eIF2α~P, which lowers the recycling of eIF2-GDP to eIF2-GTP. As a consequence, following translation of uORF1, ribosomes have a delayed translation reinitiation, which allows the bypass of uORF2. During the interval between the initiation codon of uORF2 and the *ATF4*-coding region, ribosomes reacquire eIF2/GTP/Met-tRNA$_i^{Met}$ and reinitiate at the *ATF4-Luc* coding region. (B) Wild-type MEF cells were cotransfected with the P$_{TK}$-*ATF4-Luc* plasmid and a control *Renilla* luciferase plasmid. The transfected cells were then treated with 1 μ*M* thapsigargin for 6 h (ER stress), or no stress agent, as indicated in the figure legend. To illustrate the importance of each uORF in the *ATF4* translational control, mutant versions of the 5′-leader of the *ATF4-Luc* mRNA were also analyzed. These include a mutation in the start codon of uORF1, indicated by the "X" designations, which abolishes the positive-acting element for translational control. Additionally, *ATF4-Luc* reporter genes were analyzed with mutations present in the start codons for uORF2, or both uORFs. For clarity, the histograms are represented in two different scales.

FuGENE Transfection Reagent (Roche), or an equivalent cationic polymer, following the manufacturer's instructions. The *Renilla* luciferase is used as an internal normalization standard and the ratio of *Renilla* luciferase to *ATF4*-luciferase plasmids is transfected in a ratio of 1:10–1:30. Too much *Renilla* luciferase plasmid will result in a relative light unit ratio (RLU) that exceeds the ability to accurately normalize firefly luciferase. At 24 h after transfection, treat the cultured cells with ER stress or vehicle control for the desired length of time (typically 6–12 h). Wash cells with cold PBS, followed by a 15-min incubation with passive lysis solution (Promega) mix by pipetting, and then clarify by centrifugation at $6000 \times g$ for 10 min at 4 °C. Remove the supernatant and keep on ice or store at -80 °C. Luciferase activity is measured using a Luminometer equipped with an automated dual injection system. A volume of 20 μl of cell lysate is transferred to a microcentrifuge tube and placed in the injection chamber. The reflective metallic dome on the bottom of the injection chamber should be kept clean and free of salt for accurate measurements. Our default run protocol automatically injects 100 μl luciferase substrate (Promega) and RLUs are measured after 2 s followed by a 10-s integration. A volume of 100 μl of the Stop'n'Glow (Promega) solution is then injected which quenches the first luciferase reaction. The Stop'n'Glow solution also contains the *Renilla* luciferase substrate, RLUs from the second reaction are measured after 2 s followed by a 10-s integration. Measured values are a ratio between the luciferase activity and *Renilla* activity; if desired these relative light values can be further normalized to obtain a fold change with respect to the untreated or repressed samples. Protocols can also be modified for manual injection using the same or related systems.

ACKNOWLEDGMENTS

The authors acknowledge support from NIH grants GM49164 and GM64350 and a predoctoral fellowship T32DK064466 to B. F. T. We would also like to thank members of the Wek laboratory, including Souvik Dey, Reddy Palam, Brad Joyce, and Donghui Zhou for helpful discussions and advice on experimental protocols.

REFERENCES

Anthony, T. G., McDaniel, B. J., Byerley, R. L., McGrath, B. C., Cavener, D. R., McNurlan, M. A., and Wek, R. C. (2004). Preservation of liver protein synthesis during dietary leucine deprivation occurs at the expense of skeletal muscle mass in mice deleted for eIF2 kinase GCN2. *J. Biol. Chem.* **279,** 36553–36561.

Bertolotti, A., Zhang, Y., Hendershot, L. M., Harding, H. P., and Ron, D. (2000). Dynamic interaction of BiP and ER stress transducers in the unfolded-protein response. *Nat. Cell Biol.* **2,** 326–332.

Boyce, M., Bryant, K. F., Jousse, C., Long, K., Harding, H. P., Scheuner, D., Kaufman, R. J., Ma, D., Coen, D. M., Ron, D., and Yuan, J. (2005). A selective inhibitor of eIF2alpha dephosphorylation protects cells from ER stress. *Science* **307**, 935–939.

Boyce, M., Py, B. F., Ryazanov, A. G., Minden, J. S., Long, K., Ma, D., and Yuan, J. (2008). A pharmacoproteomic approach implicates eukaryotic elongation factor 2 kinase in ER stress-induced cell death. *Cell Death Differ.* **15**, 589–599.

Brush, M. H., Weiser, D. C., and Shenolikar, S. (2003). Growth arrest and DNA damage-inducible protein GADD34 targets protein phosphatase 1 alpha to the endoplasmic reticulum and promotes dephosphorylation of the alpha subunit of eukaryotic translation initiation factor 2. *Mol. Cell. Biol.* **23**, 1292–1303.

Chen, J. J. (2007). Regulation of protein synthesis by the heme-regulated eIF2alpha kinase: Relevance to anemias. *Blood* **109**, 2693–2699.

Cherkasova, V. A., and Hinnebusch, A. G. (2003). Translational control by TOR and TAP42 through dephosphorylation of eIF2alpha kinase GCN2. *Genes Dev.* **17**, 859–872.

Cherkasova, V., Qiu, H., and Hinnebusch, A. G. (2010). Snf1 promotes phosphorylation of the alpha subunit of eukaryotic translation initiation factor 2 by activating Gcn2 and inhibiting phosphatases Glc7 and Sit4. *Mol. Cell. Biol.* **30**, 2862–2873.

Costa-Mattioli, M., Gobert, D., Stern, E., Gamache, K., Colina, R., Cuello, C., Sossin, W., Kaufman, R., Pelletier, J., Rosenblum, K., Krnjevic, K., Lacaille, J. C., et al. (2007). eIF2alpha phosphorylation bidirectionally regulates the switch from short- to long-term synaptic plasticity and memory. *Cell* **129**, 195–206.

Dey, S., Baird, T. D., Zhou, D., Palam, L. R., Spandau, D. F., and Wek, R. C. (2010). Both transcriptional regulation and translational control of ATF4 is central to the Integrated Stress Response. *J. Biol. Chem.* **285**, 33165–33174.

Elefteriou, F., Benson, M. D., Sowa, H., Starbuck, M., Liu, X., Ron, D., Parada, L. F., and Karsenty, G. (2006). ATF4 mediation of NF1 functions in osteoblast reveals a nutritional basis for congenital skeletal dysplasiae. *Cell Metab.* **4**, 441–451.

Frank, C. L., Ge, X., Xie, Z., Zhou, Y., and Tsai, L. H. (2010). Control of ATF4 persistence by multisite phosphorylation impacts cell cycle progression and neurogenesis. *J. Biol. Chem.* **285**, 33324–33337.

Harding, H. P., Zhang, Y., and Ron, D. (1999). Protein translation and folding are coupled by an endoplasmic-reticulum-resident kinase. *Nature* **397**, 271–274.

Harding, H. P., Novoa, I., Zhang, Y., Zeng, H., Wek, R., Schapira, M., and Ron, D. (2000). Regulated translation initiation controls stress-induced gene expression in mammalian cells. *Mol. Cell* **6**, 1099–1108.

Harding, H., Zeng, H., Zhang, Y., Jungreis, R., Chung, P., Plesken, H., Sabatini, D. D., and Ron, D. (2001). Diabetes mellitus and exocrine pancreatic dysfunction in Perk−/− mice reveals a role for translational control in secretory cell survival. *Mol. Cell* **7**, 1153–1163.

Harding, H. P., Zhang, Y., Zeng, H., Novoa, I., Lu, P. D., Calfon, M., Sadri, N., Yun, C., Popko, B., Paules, R., Stojdl, D. F., Bell, J. C., et al. (2003). An integrated stress response regulates amino acid metabolism and resistance to oxidative stress. *Mol. Cell* **11**, 619–633.

Hinnebusch, A. G. (2005). Translational regulation of GCN4 and the general amino acid control of yeast. *Annu. Rev. Microbiol.* **59**, 407–450.

Jiang, H. Y., and Wek, R. C. (2005). GCN2 phosphorylation of eIF2alpha activates NF-kappaB in response to UV irradiation. *Biochem. J.* **385**, 371–380.

Jiang, H. Y., Wek, S. A., McGrath, B. C., Scheuner, D., Kaufmann, R. J., Cavener, D. R., and Wek, R. C. (2003). Phosphorylation of the a subunit of eukaryotic initiation factor 2 is required for activation of NF-kB in response to diverse cellular stress. *Mol. Cell. Biol.* **23**, 5651–5663.

Jiang, H. Y., Wek, S. A., McGrath, B. C., Lu, D., Hai, T., Harding, H. P., Wang, X., Ron, D., Cavener, D. R., and Wek, R. C. (2004). Activating transcription factor 3 is integral to the eukaryotic initiation factor 2 kinase stress response. *Mol. Cell. Biol.* **24,** 1365–1377.

Kilberg, M. S., Pan, Y. X., Chen, H., and Leung-Pineda, V. (2005). Nutritional control of gene expression: How mammalian cells respond to amino acid limitation. *Annu. Rev. Nutr.* **25,** 59–85.

Law, G. L., Bickel, K. S., MacKay, V. L., and Morris, D. R. (2005). The undertranslated transcriptome reveals widespread translational silencing by alternative 5' transcript leaders. *Genome Biol.* **6,** R111.

Lee, A. H., Iwakoshi, N. N., Anderson, K. C., and Glimcher, L. H. (2003). Proteasome inhibitors disrupt the unfolded protein response in myeloma cells. *Proc. Natl. Acad. Sci. USA* **100,** 9946–9951.

Lu, P. D., Harding, H. P., and Ron, D. (2004). Translation reinitiation at alternative open reading frames regulates gene expression in an integrated stress response. *J. Cell Biol.* **167,** 27–33.

Lu, W., Laszlo, C. F., Miao, Z., Chen, H., and Wu, S. (2009). The role of nitric-oxide synthase in the regulation of UVB light-induced phosphorylation of the alpha subunit of eukaryotic initiation factor 2. *J. Biol. Chem.* **284,** 24281–24288.

Ma, Y., and Hendershot, L. M. (2003). Delineation of a negative feedback regulatory loop that controls protein translation during endoplasmic reticulum stress. *J. Biol. Chem.* **278,** 34864–34873.

Ma, K., Vattem, K. M., and Wek, R. C. (2002). Dimerization and release of molecular chaperone inhibition facilitate activation of eukaryotic initiation factor-2 kinase in response to endoplasmic reticulum stress. *J. Biol. Chem.* **277,** 18728–18735.

Marciniak, S. J., and Ron, D. (2006). Endoplasmic reticulum stress signaling in disease. *Physiol. Rev.* **86,** 1133–1149.

Marciniak, S. J., Yun, C. Y., Oyadomari, S., Novoa, I., Zhang, Y., Jungreis, R., Nagata, K., Harding, H. P., and Ron, D. (2004). CHOP induces death by promoting protein synthesis and oxidation in the stressed endoplasmic reticulum. *Genes Dev.* **18,** 3066–3077.

McCullough, K. D., Martindale, J. L., Klotz, L. O., Aw, T. Y., and Holbrook, N. J. (2001). Gadd153 sensitizes cells to endoplasmic reticulum stress by downregulating BcL2 and perturbing the cellular redox state. *Mol. Cell. Biol.* **21,** 1249–1259.

Novoa, I., Zeng, H., Harding, H. P., and Ron, D. (2001). Feedback inhibition of the unfolded protein response by GADD34-mediated dephosphorylation of eIF2alpha. *J. Cell Biol.* **153,** 1011–1022.

Novoa, I., Zhang, Y., Zeng, H., Jungreis, R., Harding, H. P., and Ron, D. (2003). Stress-induced gene expression requires programmed recovery from translational repression. *EMBO J.* **22,** 1180–1187.

Park, S. W., Zhou, Y., Lee, J. N., Lu, A., Sun, C., Chung, J., Ueki, K., and Ozcan, U. (2010). Regulatory subunits of PI3K, p85a and p85b, interact with XBP1 and increase its nuclear translocation. *Nat. Med.* **16,** 429–437.

Puthalakath, H., O'Reilly, L. A., Gunn, P., Lee, L., Kelly, P. N., Huntington, N. D., Hughes, P. D., Michalak, E. M., McKimm-Breschkin, J., Motoyama, N., Gotoh, T., Akira, S., *et al.* (2007). ER stress triggers apoptosis by activating BH3-only protein Bim. *Cell* **129,** 1337–1349.

Qin, X., and Sarnow, P. (2004). Preferential translation of internal ribosome entry site-containing mRNAs during the mitotic cycle in mammalian cells. *J. Biol. Chem.* **279,** 13721–13728.

Ron, D., and Walter, P. (2007). Signal integration in the endoplasmic reticulum unfolded protein response. *Nat. Rev. Mol. Cell Biol.* **8,** 519–529.

Rutkowski, D. T., Arnold, S. M., Miller, C. N., Wu, J., Li, J., Gunnison, K. M., Mori, K., Sadighi Akha, A. A., Raden, D., and Kaufman, R. J. (2006). Adaptation to ER stress is mediated by differential stabilities of pro-survival and pro-apoptotic mRNAs and proteins. *PLoS Biol.* **4,** e374.

Scheuner, D., Song, B., McEwen, E., Liu, C., Laybutt, R., Gillespie, P., Saunders, T., Bonner-Weir, S., and Kaufman, R. J. (2001). Translational control is required for the unfolded protein response and in vivo glucose homeostasis. *Mol. Cell* **7,** 1165–1176.

Schroder, M., and Kaufman, R. J. (2005). The mammalian unfolded protein response. *Annu. Rev. Biochem.* **74,** 739–789.

Siu, F., Blain, P. J., LeBlanc-Chaffin, R., Chen, H., and Kilberg, M. S. (2002). ATF4 is a mediator of the nutrient-sensing response pathway that activates the human asparagine synthetase gene. *J. Biol. Chem.* **277,** 24120–24127.

Sonenberg, N., and Hinnebusch, A. G. (2009). Regulation of translation initiation in eukaryotes: Mechanisms and biological targets. *Cell* **136,** 731–745.

Sood, R., Porter, A. C., Ma, K., Quilliam, L. A., and Wek, R. C. (2000). Pancreatic eukaryotic initiation factor-2alpha kinase (PEK) homologues in humans, *Drosophila melanogaster* and *Caenorhabditis elegans* that mediate translational control in response to endoplasmic reticulum stress. *Biochem. J.* **346**(Pt. 2), 281–293.

Staschke, K. A., Dey, S., Zaborske, J. M., Palam, L. R., McClintick, J. N., Pan, T., Edenberg, H. J., and Wek, R. C. (2010). Integration of general amino acid control and target of rapamycin (TOR) regulatory pathways in nitrogen assimilation in yeast. *J. Biol. Chem.* **285,** 16893–16911.

Ubeda, M., and Habener, J. F. (2003). CHOP transcription factor phosphorylation by casein kinase 2 inhibits transcriptional activation. *J. Biol. Chem.* **278,** 40514–40520.

Vattem, K. M., and Wek, R. C. (2004). Reinitiation involving upstream open reading frames regulates *ATF4* mRNA translation in mammalian cells. *Proc. Natl. Acad. Sci. USA* **101,** 11269–11274.

Warner, R. C., Samuels, H. H., Abbott, M. T., and Krakow, J. S. (1963). Ribonucleic acid polymerase of *Azotobacter vinelandii*, II. Formation of DNA–RNA hybrids with single-stranded DNA as primer. *Proc. Natl. Acad. Sci. USA* **49,** 533–538.

Wek, R. C., and Cavener, D. R. (2007). Translational control and the unfolded protein response. *Antioxid. Redox Signal.* **9,** 2357–2371.

Wek, R. C., Jiang, H. Y., and Anthony, T. G. (2006). Coping with stress: eIF2 kinases and translational control. *Biochem. Soc. Trans.* **34,** 7–11.

Winnay, J. N., Boucher, J., Mori, M. A., Ueki, K., and Kahn, C. R. (2010). A regulatory subunit of phosphoinositide 3-kinase increases the nuclear accumulation of X-box-binding protein-1 to modulate the unfolded protein response. *Nat. Med.* **16,** 438–445.

Xue, X., Piao, J. H., Nakajima, A., Sakon-Komazawa, S., Kojima, Y., Mori, K., Yagita, H., Okumura, K., Harding, H., and Nakano, H. (2005). Tumor necrosis factor alpha (TNFalpha) induces the unfolded protein response (UPR) in a reactive oxygen species (ROS)-dependent fashion, and the UPR counteracts ROS accumulation by TNFalpha. *J. Biol. Chem.* **280,** 33917–33925.

Yang, X., Matsuda, K., Bialek, P., Jacquot, S., Masuoka, H. C., Schinke, T., Li, L., Brancorsini, S., Sassone-Corsi, P., Townes, T. M., Hanauer, A., and Karsenty, G. (2004). ATF4 is a substrate of RSK2 and an essential regulator of osteoblast biology; implication for Coffin-Lowry Syndrome. *Cell* **117,** 387–398.

Yoshida, H. (2007). Unconventional splicing of XBP-1 mRNA in the unfolded protein response. *Antioxid. Redox Signal.* **9,** 2323–2333.

Zhang, K., and Kaufman, R. J. (2008). Identification and characterization of endoplasmic reticulum stress-induced apoptosis in vivo. *Methods Enzymol.* **442,** 395–419.

Zhou, D., Palam, L. R., Jiang, L., Narasimhan, J., Staschke, K. A., and Wek, R. C. (2008). Phosphorylation of eIF2 directs ATF5 translational control in response to diverse stress conditions. *J. Biol. Chem.* **283,** 7064–7073.

Author Index

A

Abbott, M. T., 346
Abe, H., 203, 206, 208–209
Abe, K., 2
Abiru, Y., 54
Abplanalp, W. A., 240
Ackerman, S. L., 262
Adachi, N., 54
Adachi, Y., 296
Aderem, A., 179–180
Aebi, M., 243–244
Aesif, S. W., 325
Agui, T., 240
Ahlfors, H., 179
Ahlke, E., 123
Aiba, S., 11
Aimanianda, V., 5, 14, 23
Akai, R., 83, 94, 144, 197–199
Akamatsu, H., 22
Akeroyd, M., 5
Akira, S., 78, 337
Akita, O., 2
Akiyama, M., 79
Akram, A., 179–180
Alanen, H., 244
Albang, R., 2, 11
Albermann, K., 2, 11
Alcocer, M. J., 4
Aldaz, C. M., 179–180, 184
Alessi, D. R., 88
Algenstaedt, P., 151
Allen, A. E., 176
Allen-Jennings, A. E., 180
Allen, J. R., 80
Allfrey, V. G., 160
Alloza, I., 240
Al-Sheikh, H., 4
Alston-Mills, B., 179–180
Alt, F., 76
Altmann, T., 307
Amano, H., 54, 62, 64
Amorim, F. T., 125
Anai, M., 151
Ancuta, P., 108
Andersen, M. R., 2, 11
Anderson, K. C., 77, 342
Andjelkovich, M., 88
Andrews, D. W., 55, 60

Anfinsen, C. B., 239
Ang, D., 238
Angst, C., 54
Anichini, A., 220
Anthony, T. G., 177, 336
Antonetti, D. A., 151
Appenzeller-Herzog, C., 249
Apsley, K., 239
Aragon, T., 278
Araki, E., 78
Araki, K., 77, 239, 248
Arantes, S., 179–180, 184
Arap, M. A., 219–220
Arap, W., 219–220
Arava, Y., 12–13
Archer, D. B., 1–5, 7, 9, 11
Arellano, J., 139
Arentshorst, M., 5–6, 8, 14, 20
Arenzana, N., 41, 85
Arima, T., 2
Arioka, M., 3
Arnold, S. M., 337
Arvan, P., 240
Arvas, M., 4
Asahara, T., 296, 309
Asai, K., 2
Asano, T., 77, 151
Ascani, S., 183
Asfari, M., 77
Ashworth, J. L., 240
Askew, D. S., 5, 14, 23
Austin, R. C., 78, 138, 230
Aw, T. Y., 337

B

Backes, B. J., 87, 197, 275, 278, 287
Back, S. H., 138, 150, 210, 213, 276–277
Badr, C. E., 95
Bafna, V., 11
Baig, E., 240
Baird, T. D., 333, 346, 349
Bai, X., 179–180
Bakken, K. S., 240
Bali, P., 162
Ballatore, C., 220
Bandyopadhyay, S. K., 179–180, 260
Bansal, V., 11
Barbacid, M., 54

Barber, G. N., 261
Barde, Y. A., 54
Barron, E., 218–219, 222
Bartnicki-Garcia, S., 5
Bartoli, A., 218
Bashir, A., 11
Baskin, C. R., 261
Basseri, S., 138
Batchvarova, N., 87
Batoux, M., 296–298
Battig, P., 243
Batt, J., 179–180
Baumeister, P., 165, 219, 224–225, 227, 230
Behrman, S. L., 282
Belle, A., 236
Bell, J. C., 76, 334, 336–337
Bendtsen, J. D., 2, 11
Benedetti, C., 240
Benson, M. D., 341
Bent, A. F., 305
Berger, J., 95
Bergeron, J. J. M., 240–241, 243
Bernales, S., 272, 274
Bernal-Mizrachi, E., 79, 88
Berruyer, R., 11, 16, 23
Bertoli, G., 248
Bertolotti, A., 74–76, 83, 87, 148, 206, 275, 296, 335
Beug, H., 12
Bhamidipati, A., 244
Bialek, P., 341
Bickel, K. S., 346
Bielmann, R., 244
Bies, C., 239, 296
Billen, L. P., 55, 60
Birge, R. B., 151
Bisgaard, K., 183
Bittel, P., 296–298
Bjorkman, M., 179
Blain, P. J., 349
Blakely, C. M., 262
Blond-Elguindi, S., 245
Blumberg, R. S., 118
Blum, R., 239, 296
Bode, H., 108
Bodenmiller, B., 249
Bodmer, D., 243
Bohmler, S., 245
Bolden, J. E., 160–162
Bolle, C., 303
Bolouri, H., 179–180
Bonner-Weir, S., 144, 345
Boot Handford, R. P., 272
Borasio, G. D., 54
Bordallo, J., 245
Borisova, S., 241
Borregaard, N., 55
Bortell, R., 80

Boston, R. S., 297
Bottone, F. G. Jr., 179–180
Bouchara, J. P., 11, 16, 23
Boucher, J., 151–153, 342
Boulme, F., 12
Boussuges, A., 124
Boutry, M., 312
Bouvier, M., 240
Bower, K., 236
Bowers, R. R. Jr., 323–324, 327, 329
Boyce, M., 83, 339
Brach, T., 310, 312, 314
Bradner, J. E., 161
Brancorsini, S., 341
Brandizzi, F., 310, 312
Bratton, D. L., 55
Breakefield, X. O., 95
Breakspear, A., 4
Breestraat, S., 5
Brenchley, J. M., 108
Brerro-Saby, C., 124
Brewer, J. W., 139
Brichory, F., 219
Bridges, J. P., 239
Briggs, M. D., 240, 272
Brodsky, J. L., 239, 245
Broekhuijsen, M., 20
Brooks, Y., 177, 179–180
Brown, A. J., 5
Brown, M. S., 76
Brown, S., 241
Brunsing, R., 44–45
Brush, M. H., 337
Bryant, K. F., 83, 339
Budge, S., 5
Buerger, E., 245
Bulleid, N. J., 240, 245, 248
Burda, P., 243–244
Burgess, S. M., 304
Burikhanov, R., 220, 230
Buschhorn, B. A., 244
Butler, P. C., 84
Byerley, R. L., 336
Byrne, M. C., 76, 144
Bytebier, B., 305

C

Cabibbo, A., 240
Calanca, V., 243
Calder, P. C., 125–126
Calfon, M., 35, 74, 76, 83, 150, 154, 178, 242–243, 276, 334, 336–337
Cali, T., 244
Callaghan, B., 287
Calmes, B., 11, 16, 23
Campbell, E. I., 20
Cao, C., 276, 278, 280

Author Index

Cao, S., 272, 274
Cao, X., 124
Caramelo, J. J., 241
Cardiff, R. D., 185
Cardozo, A. K., 78, 87
Carroll, A. M., 16, 18
Carvalho, N. D. S. P., 1, 5, 14
Carvalho, P., 246
Castro, O. A., 241
Cavener, D. R., 83, 144, 149, 177, 334, 336, 341–342, 345
Cawley, K., 31
Ceccarelli, C., 183
Chabner, B. A., 160
Chang, E. B., 133–134
Chang, J., 123, 132–133
Chang, S., 177, 179–180
Chang, Y. S., 177, 179, 219
Chao, C. C., 219
Chapman, R., 196–197
Chatterjee, S., 126
Chaudhuri, M., 151
Chaves, A., 180
Chen, G., 77
Chen, H., 137–139, 141–145, 339, 342, 349
Chen, J. J., 180, 336
Chen, L., 108, 245
Chen, T. C., 124, 218
Chen, W.-T., 217
Chen, X., 46, 76, 149–150
Chen, Z., 220
Cherkasova, V. A., 337
Cherry, B., 11
Chevalier, M. W., 278
Chiba, H., 54, 62
Chinchilla, D., 296–298
Chinnaiyan, P., 159
Choi, D. Y., 64
Cho, J. H., 76
Choudhary, C., 164
Chrispeels, M. J., 197, 296, 299, 304
Christianson, J. C., 244, 248
Christianson, T. W., 206
Chumley, F. G., 16, 18
Chung, C. W., 64
Chung, J., 151, 342
Chung, K. T., 296
Chung, P., 75–76, 83, 87, 144, 275, 342–343
Chung, W. I., 301
Chu, Z. H., 296–298
Cidlowski, J. A., 122
Clark, A. E., 179–180
Clark, R. L., 210, 213, 276–277
Clark, S. G., 74, 83, 150, 154, 243
Clauss, I., 76, 144
Clerc, S., 244
Cleveland, J. L., 139
Clotet, B., 108

Clough, S. J., 305
Cnop, M., 78
Coen, D. M., 83, 339
Coffey, E., 179
Cohen, H. R., 275
Cohen, P., 88
Cole, W. G., 240
Colina, R., 339
Collett, J. R., 11
Conesa, A., 3
Conti, C., 180
Contreras, R., 20
Corral, J., 121, 123, 126–131
Costa-Mattioli, M., 339
Cotterill, S. L., 240
Coughlan, S., 312
Coward, J. C., 125–126
Cox, J. S., 197, 260, 276
Craig, E. A., 262
Credle, J. J., 202, 210, 213, 276–277
Creemers, J. W., 144
Creighton, T. E., 239
Cresswell, P., 240
Cross, D. A., 88
Cuchacovich, M., 220
Cuello, C., 339
Cui, D., 139
Cullen, B. R., 95–96
Culley, D. E., 11
Cullinan, S. B., 35
Cunnea, P. M., 248
Cwirla, S. E., 245
Cygler, M., 241
Cyr, D. M., 245–246
Czechowski, T., 307

D

Dambly, S., 312
Damdimopoulos, A. E., 248
Damveld, R. A., 6, 8, 20
Dante, M., 206
Darby, N. J., 239–240
Darveau, A., 262
Das, I., 272
Datta, R., 262
Dautrevaux, M., 123
Dave, U. P., 76
Davidson, D. J., 220
Davie, S. A., 185
Davis, A., 244
Deblaere, R., 305
Deboeck, F., 305
Deegan, S., 31, 94
de Faire, U., 125
De Falco, F., 218
De Greve, H., 305
De La Cruz, F. J., 248

Delliaux, S., 124
DeMartino, G. N., 245
DenBoer, L. M., 40
Denecke, J., 312
Denic, V., 244, 246
Denzel, A., 241
Dephoure, N., 236
Depicker, A., 304
De Prat-Gay, G., 241
Deprez, P., 244
Der, S. D., 260
de Ruiter-Jacobs, Y. M., 20
Dever, T. E., 276, 278, 280
Devlin, C. M., 78
de Vries, R. P., 2, 11
DeWille, J., 179
de Winde, J. H., 2, 11
Dey, M., 276, 278, 280
Dey, S., 337, 346, 349
Dickie, P., 241
Diehl, J. A., 35
Diekman, K., 244
Dignard, D., 243
Di Ianni, M., 218
Di Jeso, B., 240
Dijkstrat, K., 239
Distelhorst, C. W., 124
Ditzel, H. J., 220
Dobson, C. M., 148
Dogusan, Z., 78
Dohi, K., 296, 309
Domingos, P. M., 55, 57, 60, 62
Dong, D., 218–219
Dong, M., 239
Dong, X., 180
Donkers, S., 5
Donovan, P. J., 54
Dorfler, S., 276
Dorweiler, B., 139
Dotto, G. P., 177, 179–180
Douek, D. C., 108
Dower, W. J., 245
Dressman, J., 108
Drinkwater, C. C., 54
Druzhinina, I. S., 11
Dubeau, L., 219
Dubois, M., 76
Dudek, J., 238–239, 296
Dunn-Coleman, N. S., 4
Du, S., 95, 101–103
Dyer, P. S., 2, 11
Dziak, E., 241
Dziunycz, P., 177, 179–180

E

Edeal, J. B., 12
Edenberg, H. J., 337

Edwards, K., 298
Egan, D. A., 220
Egea, P. F., 197, 199, 279, 284, 291
Eizirik, D. L., 78, 87
Ekiel, I., 240
Elbein, A. D., 299
Elefteriou, F., 341
Eling, T. E., 179–180
Ellgaard, L., 240, 249
Ellies, L. G., 185
Ellis, R. E., 74
El-Samad, H., 278
Emi, M., 272
Endo, T., 243, 245, 249
Enjalbert, B., 5
Enokido, Y., 54
Eri, R. D., 272
Etkin, A., 76, 144

F

Fabbri, M., 240
Fagioli, C., 248
Falzetti, F., 218
Fan, C. Y., 245
Farcasanu, I. C., 203, 206, 208–209
Farmery, M. R., 240
Farrall, L., 16, 18
Fassio, A., 240
Fatrai, S., 79, 88
Fazilev, F., 323–324
Felding-Habermann, B., 220, 230
Feldmesser, M., 5, 14, 23
Feng, B., 78
Feng, D., 83, 144
Feng, Y. M., 78
Fettucciari, K., 218
Fewell, S. W., 245
Filen, S., 179
Findlay, V. J., 323–324
Finer-Moore, J. S., 197, 199, 202, 210, 213, 276–277, 279, 284, 291
Fischer, R., 3
Fischer, W. H., 220
Fisher, E. A., 78
Fisher, T. L., 151
Fitzgerald, K. A., 139
Fitzgerald, U., 94
Flamez, D., 144
Fleischhack, G., 123
Flexner, C., 108
Flitter, S. J., 5
Flocco, M. T., 260
Florin, T. H., 272
Flynn, G. C., 260
Fonseca, S. G., 80
Formstecher, P., 123
Forster, M. L., 240

Forster, S. J., 240
Fortes, M. B., 124
Fra, A. M., 248
Frank, A., 83
Frank, C. L., 239, 261, 341
Frankel, S. K., 55
Frasch, S. C., 55
Fraser, J. P., 323–324
Fraternali-Orcioni, G., 183
Freedman, R. B., 239–240
Freeman, B. C., 261–262
Frei, P., 240
Fribley, A., 272
Frickel, E. M., 240
Friend, D. S., 76, 144
Frien, M., 239
Frostegard, J., 125
Fujikake, N., 55
Fukuda, S., 296, 309
Fukuhara, A., 139
Fukushima, Y., 151
Fukutomi, H., 87
Fuller, K. K., 5, 14, 23
Funaki, M., 151
Furudate, S. I., 240
Furuta, E., 179–180
Fuster, J. L., 126–131
Fu, Y., 219

G

Galabru, J., 261
Gale, M. J. Jr., 260–262
Galli, C., 243–244, 248
Gamache, K., 339
Gane, P. J., 239
Gangi, L., 180
Gannon, M., 83
Garcia-Sanz, J. A., 12
Gardai, S. J., 55
Gardner, K. L., 180
Garrett, T. E., 324
Garrison, J. L., 239, 260–262
Gauss, R., 246
Gaut, J. R., 148
Gawdi, G., 220
Gehring, K., 240
Geiger, R., 95
Gent, M. E., 4
Georgopoulos, C., 238
Gerstein, M., 12
Gething, M. J., 197, 199, 206, 245, 260, 278, 280, 287
Ge, X., 341
Geysens, S., 2
Ghaemmaghami, S., 236
Ghosh, R., 80, 87
Gilbert, H. F., 240

Gilchrist, M., 179–180
Gildersleeve, R. D., 138
Gillece, P., 248
Gillespie, P., 144, 345
Gill, P., 219, 228–231
Gish, G. D., 151
Givol, D., 239
Glasheen, E., 151
Glazebrook, J., 298, 300, 305, 310
Glimcher, L. H., 35, 38, 41, 74, 76–77, 144, 150, 272, 274, 342
Glockshuber, R., 240
Gobert, D., 339
Goder, V., 246
Goeckeler, J., 239
Goldberger, R. F., 239
Goldstein, J. L., 76
Gole, Y., 124
Gomi, F., 275
Gomi, K., 2
Gonzalez-Gronow, M., 219–220
Gonzalez, T. N., 276
Goodman, A. G., 239, 260–262
Goosen, T., 4–7, 9, 11
Gorman, A. M., 272
Goswami, A., 220, 230
Gotoh, T., 78, 337
Gouka, R. J., 3
Grau, H., 239
Gravallese, E. M., 76
Gray, P. C., 220
Greene, L. A., 56
Green, M., 240
Greiner, D. L., 80
Greiner, M., 238
Grimble, R. F., 125–126
Groshen, S., 219
Grusby, M. J., 76, 144
Gubbins, E. F., 220
Guerrero, J. A., 121
Guillemette, T., 1, 4–5, 7, 9, 11, 16, 23
Gunnison, K. M., 337
Gunn, P., 337
Gupta, S., 31, 38, 55, 94
Gurlo, T., 84
Gustafsson, J. A., 248
Guy, C. T., 185
Gu, Z., 323, 327

H

Haataja, L., 35, 84
Habener, J. F., 341
Hackett, A., 80
Hagen, A., 87, 197, 275, 278, 287
Hahn, M., 241
Hai, T., 175–177, 179–180, 342
Hajitou, A., 219–220

Halban, P. A., 77
Hamada, M., 55
Hamada, T., 296, 309
Hamari, Z., 16
Hammarback, J. A., 218
Hanash, S. M., 219
Hanauer, A., 341
Hanawalt, P. C., 219
Han, B., 179–180
Han, D., 87, 197, 275, 278, 287
Han, J., 138
Han, K. H., 16
Hansen, W. J., 261–262
Harada, A., 76, 144, 296
Harding, H. P., 34, 74–76, 78, 83, 86–87, 139, 144, 148–150, 154, 163, 177–178, 197, 206, 242–243, 260, 275–276, 280, 284, 287, 291, 296, 334–337, 339, 342–343, 345
Harris, S. D., 3
Hartl, L., 5, 14, 23
Hartman, M. G., 176, 180
Hartmann, E., 239
Hartmann-Petersen, R., 249
Hasegawa, K., 86, 243
Haser, W. G., 151
Hashizume, Y., 240
Haskell, C., 220
Haslam, D. B., 239
Hatanaka, H., 54, 56
Hatanaka, M., 79
Hatano, O., 55
Hatzivassiliou, G., 88
Hauber, J., 95
Hauber, R., 95
Hawkins, H. C., 239
Hayakawa, K., 95, 101–104
Hayashi, E., 80
Haynes, C. M., 280, 284, 287, 291
Haze, K., 35, 76, 84, 150, 153, 260
Hegde, R. S., 214, 239, 260–262
He, L., 323–324, 327, 329
Helenius, A., 240
Hellens, R., 305
Hell, R., 310, 312, 314
Hemmings, B. A., 88
Hendershot, L. M., 33, 139, 148–150, 163, 206, 238–240, 245, 248, 260–262, 296, 335, 337, 339
Henderson, W. R. Jr., 179–180
Henkel-Rieger, R., 77
Henkin, J., 220
Henson, P. M., 55
Heras, I., 126–131
Herchuelz, A., 78
Hermann, S., 248
Hernandez-Espinosa, D., 123, 126–131
Herrera, P. L., 79
Herrgen, H., 245

Herscovics, A., 86, 243–244
Herzog, V., 245
Hetz, C., 272, 274
Heumann, R., 54
Hewett, J. W., 95
He, Y., 138–139, 141–145
Hieter, P., 199, 206
Higashio, H., 208
High, S., 240
Hildeman, D. A., 55
Hillmer, S., 296–299, 301–304, 306–309, 312
Hinck, A. P., 267
Hinnebusch, A. G., 336–337
Hino, S., 218
Hinton, D. R., 218–219, 222
Hinz, G., 310, 312
Hiramatsu, N., 93, 95, 101–104, 109
Hirao, K., 244, 246
Hirata, A., 196, 199, 202–203, 210, 212–213, 277
Hirota, M., 11
Hirota, S., 179–180
Hirsch, C., 244, 246
Hirsh, D., 304
Hlodan, R., 239
Ho, C. Y., 262
Hofbauer, G. F., 177, 179–180
Hofman, F. M., 218
Hofmann, G., 2, 11
Hohmeier, H. E., 77
Holbrook, N. J., 337
Hollien, J., 87, 197, 275
Holmgren, A., 329–330
Holt, K. H., 151
Hondel, C. A., 4–7, 9, 11
Honjo, T., 296, 309
Hooykaas, P. J., 21
Hopkins, D. A., 262
Hori, O., 272
Hori, S., 75, 87, 275
Horton, H. F., 76, 144
Hosaka, T., 151
Hoseki, J., 239, 248
Hosoda, A., 94, 243, 248
Hosokawa, N., 74, 86, 243–244, 246, 248
Hovanessian, A. G., 261
Howson, R. W., 236
Hruska, K. A. Jr., 240
Huang, C. J., 84
Huang, T., 95
Hubbard, S. R., 74, 83, 150, 154, 243, 276
Hughes, P. D., 337
Huh, W. K., 236
Hui, D. Y., 108
Huikko, M. P., 248
Huntington, N. D., 337
Huppa, J. B., 248
Hutchens, S., 323–324, 327, 329
Huuskonen, A., 4

I

Ichijo, H., 75, 87, 275
Iida, K. I., 83, 144, 240
Iiizumi, M., 179–180
Ikegami, H., 77
Ikenaka, K., 56
Ikeuchi, T., 53–55, 62, 64–66
Ilic, T. V., 322
Imaizumi, K., 55, 275
Inazawa, J., 296, 309
Inoue, K., 75, 87, 275
Inukai, K., 151
Inze, D., 304
Iparraguirre, A., 179
Ischiropoulos, H., 322
Ishibashi, M., 179–180
Ishigaki, S., 80
Ishiguro, T., 180
Ishihara, H., 80
Ishiwata-Kimata, Y., 196, 199, 202–203, 205, 208–210, 212–213, 277
Isobe, K. I., 240
Isosaki, M., 55
Itagaki, H., 11
Itamoto, T., 296, 309
Itoh, Y., 22
Ito, T., 196, 199, 202–203, 210, 212–213, 277
Iversen, J. J., 5–6, 11
Iwai, K., 246
Iwakoshi, N. N., 35, 38, 41, 74, 76–77, 150, 342
Iwawaki, T., 44–46, 83, 94, 144, 197–200, 202–203, 205–210, 212–213, 243, 278
Izumi, T., 79

J

Jackson, G. C., 240
Jacobs, D. I., 5
Jacquot, J. P., 312, 314
Jacquot, S., 341
Jakob, C. A., 243–244
Jakobsen, C. G., 220
Jalgaonkar, S., 175
Jammes, Y., 124
Janda, K. D., 220, 230
Jang, J. H., 219
Janjic, D., 77
Jansen, G., 239–240, 248
Janssen-Heininger, Y. M., 325
Jarosch, E., 246
Jeenes, D. J., 4
Jefferson, R. A., 304
Jiang, H. Y., 34, 177, 336, 342, 345
Jiang, J., 267
Jiang, L., 346
Jin Kwon, M., 5, 14
Ji, X., 323–324
Johansson, L., 249
Johno, H., 100
Johnson, A. E., 245
Johnson, C. D., 179–180
Johnstone, C., 298
Jolicoeur, E. M., 197, 243
Jones, K., 180
Jones, S. J., 12
Jones, T. R., 248
Jorda, L., 296–298
Jorgensen, T. R., 5–6, 11
Joshi, S., 245
Joubert, A., 1, 11, 16, 23
Jousse, C., 83, 339
Jovanovic, M., 322
Jovicic, A., 322
Jung, D. Y., 219
Jung, M., 239, 296
Jungreis, R., 337, 342–343
Jungries, R., 76, 83, 144
Jun, J. Y., 219
Junker, R., 123
Jurczyk, A., 80
Justice, N. J., 220

K

Kadowaki, H., 87
Kahali, S., 159, 162, 164–167
Kahn, C. R., 147, 151–153, 342
Kakizuka, A., 75, 87, 275
Kalies, K. U., 239
Kamauchi, S., 296, 299, 301, 304
Kamigori, Y., 243
Kamiya, D., 244
Kamiya, K., 296, 309
Kamiya, Y., 244
Kampinga, H. H., 262
Kam, T. I., 64
Kanapin, A., 236
Kanda, A., 151
Kanemoto, S., 218
Kang, S. J., 16, 18, 240
Kang, S. W., 239, 260–262
Kano, F., 84
Karaveg, K., 244
Karimi, M., 304
Karin, M., 260
Karsenty, G., 341
Kasai, A., 95, 101–104
Kaser, A., 118
Kashiwagi, Y., 2
Katagiri, H., 151
Katayama, T., 55, 275
Katiyar, S., 245
Katoh-Semba, R., 56
Kato, K., 56, 244
Kato, Y., 245
Katze, M. G., 239, 260–262
Kaufmann, G. F., 220, 230

Kaufman, R. J., 34–35, 41, 74, 76, 83, 85, 111, 138–139, 144, 148, 150, 163, 177, 197, 210, 213, 218, 230, 239, 242–243, 245–246, 260–262, 272, 274, 276–277, 287, 334–337, 339, 342–343, 345
Kaung, G., 86
Kawaguchi, Y., 166
Kawahara, T., 38, 199
Kawamura, A., 207
Kayo, T., 79
Keasling, J. D., 3
Keefer, L. K., 323–324
Kelber, J. A., 220
Keller, S., 296–299, 301–304, 306–309, 312
Kelley, R., 180
Kelly, P. N., 337
Kelly, V., 240
Kemmink, J., 239
Kennedy, K., 179–180
Kharroubi, I., 78
Kherzai, A., 220
Kielty, C. M., 240
Kiessling, R., 125
Kiguchi, K., 179–180, 184
Kikkert, M., 245
Kikuchi, M., 151
Kilberg, M. S., 339, 349
Kimata, Y. I., 35, 150, 195–197, 199, 202–203, 205–210, 212–213, 243, 248, 277–278
Kim, I., 36, 39, 72, 138, 145
Kim, J. K., 219
Kim, J. S., 179–180
Kim, P. S., 240
Kim, R., 272
Kim, S. J., 54
Kim, T.-Y., 40
Kim, W., 244
Kim, Y., 220, 230
King, F., 151
Kinghorn, J. R., 20
Kinoshita, E., 139
Kinoshita-Kikuta, E., 139
Kirchhoff, S. R., 123, 132–133
Kishi, S., 54–55, 62, 64–65
Kiss-Papp, M., 296–298
Kitagaki, M., 11
Kitamoto, K., 3
Kitamura, M., 93–95, 100–104
Kitao, Y., 272
Klaasmeyer, J. G., 12
Klappa, P., 240
Klee, H., 305
Kloetzel, P. M., 245
Klomkleaw, W., 180
Klotz, L. O., 337
Knoblaugh, S. E., 261
Knowlton, A. A., 123, 132–133
Kobayashi, S., 86

Kociba, G. J., 180
Kodama, M., 22
Koda, Y., 54
Ko, H. J., 219
Kohmoto, K., 22
Kohno, K., 4, 83, 94, 144, 196–197, 199, 202–203, 205–210, 212–213, 243, 248, 277–278
Koike, N., 296, 309
Koike, T., 139
Koivu, J., 240
Koivunen, P., 240
Koizumi, A., 78–79
Koizumi, N., 197, 301
Kojima, Y., 342
Kokame, K., 40–41
Komuro, R., 76, 139
Kondo, S., 218
Kong, Y. Y., 64
Koning, F., 245
Kooistra, R., 5
Kopito, R. R., 86, 244, 248
Kopsch, K., 238
Korb, M., 179–180
Korennykh, A. V., 197, 199, 278–279, 282, 284, 291
Kornfeld, R., 241
Kornfeld, S., 241
Korostelev, A. A., 197, 199, 279, 284, 291
Korth, M. J., 239, 260–262
Koster, J. C., 108
Kostka, S., 239
Kostova, Z., 244–245
Kovacs, J. J., 162
Kozaki, K. I., 240
Kozlov, G., 240
Kozutsumi, Y., 260
Kraft, R., 239
Krakow, J. S., 346
Krause, K. H., 241
Kreye, S., 312, 314
Krnjevic, K., 339
Kubelka, L., 179
Kubo, S. K., 79
Kubo, T., 54
Kubota, T., 95
Kudo, M., 54–55, 62, 64, 66
Kudo, T., 55
Kulseth, M. A., 240
Kumagai, T., 2
Kumar, K., 260
Kunugi, H., 54
Kuriakose, G., 78
Kurnit, D. M., 74
Kuroda, M., 87, 197, 243
Kusumoto, K., 2
Kwon, C. S., 301
Kwon, M. J., 5

L

Lacaille, J. C., 339
Lacey, G. A., 4
Lacombe, S., 296–298
Ladiges, W. C., 261
Ladriere, L., 78
Laenkholm, A. V., 220
Lafer, E. M., 267
Lahdenranta, J., 219–220
Lahesmaa, R., 179
Lai, Y. K., 219
Lam, E., 179–180, 299
Lane, A. A., 160–161
Lanthaler, K., 4, 7, 9, 11
Lapointe, T., 5
Laszlo, C. F., 342
Latge, J. P., 5, 14, 23
Lavail, M. M., 275
Law, G. L., 346
Laxell, M., 4
Laybutt, R., 144, 345
Leber, J. H., 4
LeBlanc-Chaffin, R., 349
LeBoeuf, R. C., 261
Lechleider, R. J., 151
Le Crom, S., 11
Leder, A., 160
Lee, A.-H., 35, 38–39, 41, 74, 77, 150, 342
Lee, A. S., 148, 217–220, 222, 224–225, 227–231, 260
Lee, B., 218–219, 222
Lee, D. Y., 160
Lee, H. K., 64, 163
Lee, J. N., 151, 240, 342
Lee, J. Y., 64
Lee, K. P. K., 74, 150, 260, 271, 276, 278, 280, 284, 287, 291
Lee, L., 337
Leemans, J., 305
Lee, T. G., 261
Lefort, K., 177, 179–180
Lehmann, L., 11
Leighton, M. P., 240
Leinonen, S., 248
Lemne, C., 125
Le Naour, F., 219
Lencer, W. I., 240
Lennarz, W. J., 241, 245, 248, 323
Leong, L. E., 316
Leren, T. P., 240
Lerner, A. G., 87, 197, 275, 278, 287
Lesniewski, R., 220
Leung-Pineda, V., 339
Lhotak, S., 138
Liao, C. P., 219
Li, B., 179–180
Li, C. W., 88, 219

Li, D., 180
Li, E., 55, 62–64
Li, G., 77
Li, H., 87, 197, 275, 282
Li, J., 78, 218–219, 222, 296–298, 337
Li, L., 341
Li, Y., 78, 83, 144, 241
Liberek, K., 238
Libert, F., 87
Lichtenthaler, H. K., 303
Lightfoot, R. T., 87
Lilley, B. N., 245–246, 248
Lillo, A. M., 220
Lindsay, R. M., 54
Lindsten, T., 88
Lin, J. H., 36, 87, 118, 163, 197, 272, 275
Lipshutz, R. J., 245
Lipson, K. L., 80, 87
Lipton, S. A., 323, 327
Li, S., 83
Litvak, M., 179–180
Liu, C. Y., 74, 144, 148, 210, 213, 276–277, 345
Liu, R., 219, 228–231
Liu, W., 179–180
Liu, X., 341
Liu, Y., 220, 230
Liu, Z., 88, 138, 145
Lizak, B., 238
Llanos, C., 220
Llewellyn, D. H., 240
Lockhart, D. J., 3, 79, 144, 260, 277
Lock, L. F., 54
Logue, S. E., 272
Long, K., 83, 339
Lourie, R., 272
Lovell, R. J., 125
Lu, A., 151, 342
Lu, D., 79, 177, 179, 342
Lu, P. D., 76, 178, 334, 336–337
Lu, S., 80
Lu, W., 342
Lubertozzi, D., 3
Lucca, P., 243
Lumb, R. A., 248
Lundstrom, J., 329–330
Luo, B., 219
Luo, S., 219
Luz, J. M., 248

M

MacAuley, A., 261
Macdonald, D. C., 55
MacGregor, G. R., 88
Machida, M., 2
MacKay, V. L., 346
MacKenzie, D. A., 4
MacLennan, D. H., 241

MacLeod, C. L., 185
MacLeod, M. C., 179–180, 184
Ma, D., 83, 339
Madden, J., 125–126
Madden, L. A., 125
Madonna, M. B., 133–134
Madrid, S. M., 10–11
Maeda, S., 101
Maes, E. G., 267
Maggioni, C., 248
Maglione, J. E., 185
Magnuson, J. K., 11
Magnusson, N. E., 87
Mahaingam, D., 162
Maillet, I., 5
Majest, S., 220
Ma, K., 83, 335, 342–344
Mäkitie, O., 240
Malhotra, J. D., 111, 138, 272, 274
Malim, M. H., 95–96
Malone, M. H., 124
Mancini, R., 248
Mandrup-Poulsen, T., 78
Manevich, Y., 323–324, 327, 329
Manis, J., 76
Manner, C. K., 185
Mao, C., 219, 230
Marchetti, G., 108
Marcil, A., 243
Marciniak, S. J., 55, 334–337
Marconi, P., 218
Marks, A. R., 78
Markus, A., 54
Marrack, P., 55
Marsee, D. K., 176
Marszalek, J., 238
Martens, E., 240
Martina, N., 296, 309
Martindale, J. L., 337
Martinez, C., 126–131
Martinez, I. M., 197, 296, 299, 304
Martinez-Martinez, I., 123, 126–131
Martin, J. E., 11, 241
Martinon, F., 36
Marty, L., 312, 314
Maruyama, T., 87
Masaki, T., 144
Masliah, E., 323, 327
Masoom, H., 179–180
Mast, S. W., 244
Masuda, Y., 296, 309
Masuoka, H. C., 341
Mathis, H., 11
Matlack, K. E., 245
Matsuda, K., 341
Matsuda, M., 139
Matsui, T., 55, 65, 74, 76, 144, 243
Matsumoto, A., 219

Matsumoto, T., 54
Matsuoka, N., 54
Matsuzawa, A., 75, 87, 275
Mattern, I. E., 7
Maudoux, O., 312
Ma, W., 197, 260, 278, 280, 287
Ma, Y., 33, 163, 323, 327, 337, 339
McArthur, M. J., 179–180, 184
McCarthy, J. E., 12
McCarthy, J. W., 5, 14, 23
McClintick, J. N., 337
McConoughey, S. J., 179–180
McCourt, D. W., 240
McCullough, K. D., 337
McDaniel, B. J., 336
McEwen, E., 144, 345
McGrath, B. C., 83, 144, 177, 336, 342, 345
McGuckin, M. A., 272
McIntosh, C. H., 54
McKimm-Breschkin, J., 337
McLaughlin, S. H., 239
McNaughton, K., 83
McNaughton, L., 125
McNurlan, M. A., 336
Meda, P., 77
Medicherla, B., 244
Melville, M. W., 260–262
Meng, Y., 95, 101
Menke, H., 5
Menon, S., 240
Merksamer, P. I., 79
Mesaeli, N., 241
Meulenberg, R., 5
Meunier, L., 239–240, 296
Meyer, A. J., 310, 312, 314
Meyer, H. A., 239
Meyer, V., 5–6, 8, 14, 20, 23
Miao, H., 138
Miao, Z., 342
Michael, C. W., 219
Michalak, E. M., 337
Michalak, M., 150, 241, 276
Michielse, C. B., 21
Mikoshiba, K., 56
Miley, M. D., 5, 14, 23
Miller, C. N., 83, 144, 337
Miller, J., 261
Miller, M. A., 183
Minano, A., 123, 126–131
Minden, J. S., 339
Mintz, P. J., 219–220
Miralpeix, M., 151
Miranda-Vizuete, A., 248
Mirmira, R. G., 180
Misek, D. E., 219
Miura, M., 83, 94, 144, 199
Miyaishi, O., 240
Miyata, Y., 139

Miyazaki, J., 77
Miyazaki, M., 79
Miyazono, K., 75
Miyoshi, K., 55
Mller, A., 238
Mochida, Y., 75
Mohinta, S., 179–180
Molinari, M., 240–241, 243–244, 248
Momany, M., 4
Moncada, S., 322
Monot, F., 11
Montminy, M. R., 180
Moon, J. S., 301
Moon, Y., 179–180
Moremen, K. W., 244
Mori, K., 55, 65, 74, 76, 84, 144, 150, 153, 196–197, 199, 206, 242–243, 245–246, 260, 272, 274, 276, 278, 280, 287, 296, 337, 342
Morikawa, K., 218
Mori, M. A., 78, 151–153, 342
Morimoto, R. I., 74, 260, 262
Morishima, N., 55, 62–64
Morito, D., 244, 246
Moriyama, T., 244, 248
Morris, D. R., 346
Morris, T. E., 324
Morsomme, P., 312
Mortarini, R., 220
Morton, J. F., 261
Moseley, P., 125
Mota, R., 123, 126–131
Mothes, W., 245
Motoyama, N., 337
Movafeghi, A., 312
Moye-Rowley, W. S., 151
Mulder, H. J., 5, 10–11, 14, 19, 23, 77
Muller, C., 310, 312
Muller, W. J., 185
Mullineaux, P., 305
Mullner, E. W., 12
Murakami, S., 272
Murakami, T., 218
Murata, M., 84
Murphy, D., 316
Murray, J., 79
Musch, M. W., 133–134
Muto, T., 180
Myers, M. G., 151
Myllyharju, J., 240
Myllyla, R., 240

N

Naamane, N., 87
Nadanaka, S., 34, 46, 84
Nagai, A., 87
Nagai, K., 55, 65, 95
Nagai, Y., 55
Nagasawa, K., 244, 248

Nagata, K., 74, 86, 235, 239, 243–246, 248, 337
Naidoo, N., 272
Naito, M., 180
Nakagawa, T., 55, 62–64
Nakahira, K., 56
Nakajima, A., 342
Nakajima, M., 180
Nakajima, S., 95
Nakamura, K., 241
Nakamura, T., 296, 309, 323, 327
Nakano, C., 296, 299, 301, 304
Nakano, H., 75, 342
Nakano, T., 296, 309
Nakatani, H., 296, 299, 301, 304
Nakatani, Y., 55, 65
Nakatsukasa, K., 243
Nakayama, H., 55
Narasimhan, J., 346
Nastainczyk, W., 238–239, 296
Natsuka, S., 244
Natsuka, Y., 244, 248
Neculai, D., 276, 278, 280
Neel, B. G., 151
Negishi, M., 76, 84, 150, 153
Negredo, E., 108
Nekrasov, V., 296–298
Newgard, C. B., 77
Ng, D. T. W., 244
Ng, P. W., 316
Nguyen, B. C., 177, 179–180
Nguyen, L. X., 80
Nian, C., 54
Ni Chonghaile, T., 55
Nielsen, J., 11
Nikolaev, I., 5, 14, 19, 23
Nilges, M., 239
Nilsson, A., 260
Ni, M., 148, 218–219, 222, 224–225, 227–231
Nishikawa, S. I., 243, 245, 249
Nishimura, M., 54, 62, 64
Nishitoh, H., 75, 87, 275
Nita-Lazar, M., 244
Nitsche, B. M., 5
Noguchi, S., 144
Noguchi, T., 87
Noh, J. Y., 64
Noh, S. J., 301
Noiva, R., 323
Nomura-Furuwatari, C., 243
Nomura, Y., 323, 327
Nonomura, T., 54
Nookaew, I., 11
Normington, K., 199, 206, 260
Novoa, I., 76, 86, 178, 239, 242, 261, 334, 336–337, 342–343, 345
Nowak-Gottl, U., 123
Nowroozalizadeh, S., 108
Numakawa, T., 54

Numakawa, Y., 54
Nystrom, J., 179

O

Oakes, S. A., 87, 197, 275, 278, 287
Oberhauser, S., 239, 296
Oda, Y., 243, 245–246
Ogata, M., 218
Ogata, R., 95
Ogawa, S., 272
Oggier, D. M., 244
Ogihara, T., 151
Ogle, M., 323–324
Oh, D. H., 301
Ohsugi, M., 79, 88
Ohta, Y., 79
Oikawa, D., 34, 195–197, 199–200, 202–203, 205–210, 212–213, 277–278
Oikawa, T., 54, 64
Okada, T., 55, 65, 74, 76, 84, 144, 150, 153, 243, 245–246, 276, 296
Okamoto, Y., 180
Okamura, K., 208
Okamura, M., 95, 101–103
Oka, T., 54
Okawa, K., 244, 248
Oka, Y., 77, 80, 151
Okuda-Shimizu, Y., 245, 248
Okuda, T., 243
Okumura, K., 342
Okumura, N., 55, 65
Olivari, S., 244
Oliver, J. D., 240
Oliver, S. G., 4
Olson, L., 151
Olsthoorn, M. M., 5
Oono, K., 275
Opas, M., 241
Ordonez, A., 123, 126–131
O'Reilly, L. A., 337
Orkin, S. H., 76, 144
Orntoft, T. F., 87
Orrenius, S., 55, 60
Ortis, F., 78
O'Shea, E. K., 236
Oslowski, C. M., 71, 73, 80
Otani, H., 22
Otero, J. H., 238
Outinen, P. A., 78
Owen, M. J., 241
Oyadomari, S., 78, 239, 260–262, 337
Ozcan, U., 151, 342

P

Paganetti, P., 243–244, 248
Pagani, M., 240
Pai, S. K., 179–180

Pakula, T. M., 4
Palam, L. R., 337, 346, 349
Pamidi, S., 78
Panagiotou, G., 11
Panning, B., 275
Pan, T., 337
Pan, Y. X., 339
Papa, F. R., 79, 87, 197, 202, 210, 213, 272, 274–278, 283, 287
Papalas, J., 219
Parada, L. F., 341
Park, M. A., 36
Park, S. W., 151, 342
Park, Y. N., 240
Parodi, A. J., 241
Parry, J. W. L., 239
Pasqualini, R., 219–220
Patel, R., 260
Patil, C. K., 3, 79, 144, 242, 260, 277
Paton, A. W., 95
Paton, J. C., 95
Patterson, C., 245
Patterson, J. B., 218
Paules, R., 76, 178, 334, 336–337
Pawson, T., 151
Pazoles, C. J., 324
Peisach, D., 210, 213, 276–277
Pel, H. J., 2, 11
Pelletier, J., 339
Pelletier, M. F., 240
Peng, C., 179–180
Pen, L., 218–219
Pennacchio, L., 11
Pennathur, S., 138
Penttila, M., 4, 10–11, 18
Perez-Ceballos, E., 126–131
Perkins, A., 76, 144
Permutt, M. A., 79–80, 88
Pessin, J. E., 151
Petrova, K., 239, 260–262
Pfeifer, S. I., 78
Phillips, D. M., 160
Piao, J. H., 342
Piccaluga, V., 248
Piccioli, M., 183
Pieren, M., 248
Pieri, F., 183
Pihet, M., 11, 16, 23
Pilati, S., 240
Pileri, S. A., 183
Pilon, M., 245
Pimpl, P., 312
Pincus, D., 278
Pipe, S. W., 138
Pirot, P., 87
Pizzo, S. V., 219–220
Plemper, R. K., 245
Plesken, H., 76, 83, 144, 342–343

Ploegh, H. L., 245–246, 248
Plumridge, A., 5
Pockley, A. G., 125
Podor, T. J., 78
Poduval, T. B., 126
Poggi, S., 183
Pohl, J., 260
Pollock, S., 240
Polyak, S. J., 261
Pons, S., 151
Ponting, C. P., 297
Popko, B., 76, 178, 334, 336–337
Porter, A. C., 343–344
Porter, A. G., 316
Pouwels, P. H., 7, 20
Pradet-Balade, B., 12
Prehn, S., 239
Premachandran, S., 126
Prentki, M., 77
Price, J., 280, 284, 287, 291
Price, R. D., 54
Pryme, I. F., 310
Prywes, R., 41, 76, 85, 149–150, 276
Przedborski, S., 322
Punt, P. J., 3–4, 20, 23
Puravs, E., 219
Puthalakath, H., 337
Py, B. F., 339
Pyle, A. D., 54
Pyrko, P., 218

Q

Qi, L., 137–139, 141–145, 180
Qin, X., 346
Qiu, H., 337
Qiu, S., 220, 230
Quan, E. M., 244, 246
Quaroni, A., 109
Quilliam, L. A., 343–344

R

Raden, D., 337
Radzimanowski, J., 296–299, 301–304, 306–309, 312–313
Ram, A. F. J., 1, 5–6, 8, 11, 14, 20–21, 23
Ramos-Vara, J. A., 181–183, 185
Rampino, G., 218
Ramsey, S. A., 179–180
Rangnekar, V. M., 220, 230
Ranheim, T., 240
Rao, K. V., 179
Rao, R., 162, 164–165
Rapoport, T. A., 245–246, 248
Rasheva, V. I., 55, 57, 60, 62
Rasmussen, N., 220
Rasool, O., 179
Rasschaert, J., 78

Ratnofsky, S., 151
Rauchman, M., 262
Ravaud, S., 296–299, 301–304, 306–309, 312–313
Rawson, R. B., 76
Reddy, R. K., 230
Reed, J. C., 72, 138, 145, 272
Reimold, A. M., 76, 144
Reinert, J., 83
Remotti, H., 87
Ren, H. Y., 245
Retsky, J., 133–134
Reyes-Dominguez, Y., 16
Reynaert, N. L., 325
Reynaga-Pena, C. G., 5
Rhen, T., 122
Rhodes, J. C., 5, 14, 23
Ribick, M., 138, 144
Richard-Mereau, C., 123
Richie, D. L., 5, 14, 23
Richon, V. M., 160
Riemer, J., 249
Riggs, A. C., 79
Riggs, M. G., 160
Robergs, R., 125
Roberts, T., 151
Robinson, D. G., 312
Robson, G. D., 4, 7, 9, 11
Rocchi, M., 240
Roderick, H. L., 240
Roebuck, Q. P., 240
Romano, P. R., 262
Romero, P., 244
Romisch, K., 245, 248
Ron, D. P., 33–35, 55, 72, 74–76, 78, 83, 86–87, 138–139, 144, 148–150, 154, 177, 197, 206, 213, 239, 242–243, 245–246, 248, 260–262, 272, 274–276, 280, 284, 287, 291, 296, 334–337, 339, 341–343, 345
Rong, J. X., 78
Rosales, C., 262
Rosati, E., 218
Rose, M. D., 203, 206, 208–209
Rosenberger, C. M., 179–180
Rosenblum, K., 339
Ross, A. J., 88
Rosser, M. F. N., 245
Rothe, M., 75
Roth, J., 243
Rothman, J. E., 260
Roth, R. A., 240
Rougon, A., 296–298
Rouhier, N., 312, 314
Rouschop, K. M., 35
Roux, M., 296–298
Roy, B., 39, 41
Roy-Burman, P., 219
Rubbelke, T. S., 262

Rubio, C. A., 278
Ruddock, L. W., 240, 244
Ruegsegger, U., 4
Russell, S. J., 240
Rust, A. G., 179–180
Rutkowski, D. T., 76, 214, 218, 239, 260–262, 337
Ryazanov, A. G., 339

S

Saavedra, J. E., 323–324
Sabatini, D. D., 76, 83, 144, 342–343
Sabatini, R., 218
Sabattini, E., 183
Sadighi Akha, A. A., 337
Sadri, N., 76, 178, 334, 336–337
Saegusa, K., 75, 87, 275
Saeki, S., 79
Saga, S., 240
Sahin, A., 179–180, 184
Saito, A., 218
Saitoh, M., 75
Saito, T., 79
Sakoh-Nakatogawa, M., 249
Sakon-Komazawa, S., 342
Salazar, M. P., 11
Saloheimo, M., 4, 10–11, 18
Salo, K. E. H., 240, 244
Samali, A., 31, 36–37, 43, 45, 48, 55, 94, 272
Sambrook, J. F., 197, 199, 206, 245, 260, 278, 280, 287
Samuels, H. H., 346
Sanderson, T. H., 36
Sandstrom, M. E., 125
Sano, H., 197
Sano, M., 2, 56
Sant, A., 199
Santer, M., 7
Sarcar, B., 159
Sargent, K. E. G., 80
Sarkis, A. S., 219–220
Sarnow, P., 346
Sasaya, H., 53–54, 62, 64
Sassone-Corsi, P., 341
Sato, M., 76, 144
Sato, T., 76, 144
Saunders, T., 76, 144, 345
Scazzocchio, C., 16
Schaap, P. J., 2, 11
Schackwitz, W., 11
Schaiff, W. T., 240
Schalkwijk, C. G., 272
Schapira, M., 76, 336, 342–343, 345
Scheible, W. R., 307
Schekman, R., 239, 245
Schell, J., 305
Scheuner, D., 83, 138, 144, 260, 272, 339, 345

Schindelin, H., 323
Schinke, T., 341
Schlessinger, K., 76
Schmidt, R. E., 79
Schmitt, A., 239
Schmitz, A., 245
Schneider, A., 220
Schneider, S. M., 125
Schobess, R., 123
Schonthal, A. H., 218
Schott, A., 295–299, 301–304, 306–309, 312–313
Schrag, J. D., 241
Schroder, M., 34–35, 148, 150, 163, 177, 242, 276, 334–336, 339
Schuit, F. C., 144
Schulze, A., 245
Schumann, C., 123
Schwabe, D., 123
Schwaller, M., 240
Schwartz, B. D., 240
Schwarze, S. R., 220, 230
Scott, A., 76, 144
Screpanti, I., 218
Scroggins, B. T., 162
Seburn, K., 262
Seeger, M., 245
Segal, M., 260
Seiboth, B., 11
Selim, M. A., 219
Seo, J. A., 16
Seo, S. J., 64
Serafimova, I. M., 278
Sha, B., 259–262
Sha, H., 138–139, 142, 145
Shah, G. N., 262
Shaler, T. A., 244, 248
Shamas-Din, A., 55, 60
Shamu, C. E., 139, 197, 260, 276, 278
Shani, G., 220
Shan, Y., 179–180
Sharma, A. M., 138
Shearman, C. P., 125–126
Shearstone, J. R., 80
Shen, J., 34, 41, 85, 149–150
Shenolikar, S., 337
Shen, X., 74, 150, 276
Shero, J. H., 206
Shibahara, K., 296, 309
Shibasaki, Y., 77
Shibata, Y., 245–246, 248
Shi, H., 139, 142
Shimada, T., 55, 65
Shimamura, Y., 54, 64
Shimizu, Y., 199, 202–203, 205–206, 208–209, 213
Shimoke, K., 53–55, 62, 64–66
Shimomura, I., 139

Shim, S. M., 64
Shin, B. C., 151
Shin, B. K., 219
Shinoda, K., 79
Shin-Ya, K., 55, 65
Shiosaka, S., 218
Shirozu, M., 296, 309
Shi, Z. Q., 323, 327
Shoelson, S. E., 151
Shoji, J. Y., 3
Shokat, K. M., 197, 199, 275, 278–279, 283–284, 286–287, 290–291
Shooter, E. M., 54
Shukla, J., 126
Shuttleworth, C. A., 240
Sicheri, F., 271, 276, 278, 280, 284, 287, 291
Sickmann, A., 239
Sidrauski, C., 196–197, 276
Siegler, J. C., 125
Sifers, R. N., 244
Sikorski, R. S., 199, 206
Simmen, T., 248
Simmons, R. M., 179–180
Simoneau, P., 1, 11, 16, 23
Sims, A. H., 4
Sinning, I., 296–299, 301–304, 306–309, 312–313
Sitia, R., 240, 248
Sitnikov, D., 240
Siu, F., 349
Sivaprasad, U., 176
Sivick, K., 240
Skach, W. R., 245
Sleno, B., 244
Sly, W. S., 262
Smith, C. W., 12
Smith, K. D., 179–180
Snyder, M., 12
Solomon, A., 74
Solomon, L., 220
Sommer, T., 244–246
Sonenberg, N., 336
Song, B., 76, 83, 138, 144, 345
Song, J., 260, 262
Song, S., 64
Songyang, Z., 151
Sood, R., 343–344
Sood, S. K., 78
Sopher, B. L., 239, 261
Sorensen, S., 240
Sossin, W., 339
Sousa, R., 267
Sowa, H., 341
Soyk, S., 310, 312
Spandau, D. F., 346, 349
Spear, E. D., 244
Spirig, U., 243
Sprang, S. R., 245

Spyrou, G., 248
Srinivasan, S., 88
Stafford, W. F., 240
Stahl, U., 6, 8, 20
Stam, H., 4, 7, 9, 11
Stamp, G., 241
Standera, S., 245
Starbuck, M., 341
Staschke, K. A., 337, 346
Stehouwer, C. D., 272
Steinberg, J. G., 124
Steiniger, S. C., 220, 230
Steitz, J. A., 38
Stern, E., 339
Stevens, J. L., 87
Stevens, N., 87, 197, 275
Stiles, C., 218
Stirling, C. J., 260
Stitt, M., 307
Stojdl, D. F., 76, 334, 336–337
Storling, J., 78
Story, C. M., 248
Strahl, S., 295–299, 301–304, 306–309, 312–313
Straus, D., 133–134
Stroud, R. M., 197, 199, 202, 210, 213, 276–277, 279, 284, 291
Suh, E., 109
Sumii, M., 296, 309
Sun, C., 151, 342
Sun, F. C., 219
Sun, L., 123, 132–133
Sun, S., 138–139, 141–145
Sun, X. J., 151
Suortti, P., 4
Suter, U., 54
Suzuki, M., 151
Suzuki, T., 196, 199, 202–203, 210, 212–213, 277
Swathirajan, J., 76
Sweeney, M., 78
Sweigard, J. A., 16, 18
Szathmary, R., 244
Szegezdi, E., 33–34, 36, 39, 55, 272
Szomolanyi-Tsuda, E., 76

T

Tabas, I., 139
Tada, H., 296, 309
Tae Chung, K., 239–240
Tagaw, Y., 95
Taguchi, T., 54
Tai, M. L., 64
Takahashi, M., 296, 309
Takahashi, S., 100
Takano, Y., 102
Takata, K., 79, 151

Takayama, C., 56
Takeda, K., 75, 78, 87, 275
Takeda, M., 95, 101–103
Takei, N., 54
Takeshita, N., 3
Takeuchi, M., 196, 199, 202–203, 206, 208–210, 212–213, 277
Takeuchi, T., 79
Takiyama, K., 139
Tamura, T., 244
Tanabe, K., 79, 272
Tanaka, K., 246
Tanaka, S., 79
Tanaka, T., 2
Tang, N. M., 260–262
Taniguchi, M., 55, 218
Tanii, I., 218
Tannous, B. A., 95
Tan, S. H., 316
Tan, S. L., 260–262
Tao, J., 259–262
Tarn, W.-Y., 38
Tashiro, K., 296, 309
Taunton, J., 239, 260–262
Taylor, A. B., 267
Tbarka, N., 123
Teishima, J., 296, 309
Terai, G., 2
Teske, B. F., 333
Tew, K. D., 322–324, 327, 329
Thames, H. D., 179–180, 184
Thomas, D. Y., 239–241, 243, 248
Thomenius, M. J., 124
Thompson, C. B., 88, 298
Thompson, S., 261
Thorsson, V., 179–180
Tian, G., 323
Tien, C. L., 54
Tieu, K., 322
Till, J. H., 74, 83, 150, 154, 243, 276
Tirasophon, W., 41, 85, 150, 197, 243, 276, 287
Tischler, A. S., 56
Tobiume, K., 75, 87, 275
Tohyama, M., 55, 275
Tokuda, M., 94, 197, 200
Tokunaga, F., 246
Tokunaga, M., 207
Tomita, J., 261
Tomovic, M., 322
Tonnesen, M., 78
Tortorella, D., 248
Totsuka, T., 56
Townes, T. M., 341
Townsend, D. M., 321–324, 327, 329
Traber, P. G., 109
Tra, J., 219
Travers, K. J., 3, 79, 144, 245, 260, 277
Treglia, A. S., 240

Tremblay, L. O., 86, 243–244
Trempe, J. F., 240
Treuting, P. M., 179–180
Trigueros, C., 241
Tripathi, S., 179
Trusina, A., 79
Tsai, B., 240, 248
Tsai, L. H., 341
Tsukamoto, K., 144
Tsuneoka, M., 54
Tsuru, A., 208, 243, 248
Tsuru, M., 79
Tsuruo, T., 180
Tu, B. P., 33, 79
Turnbull, E. L., 245
Turner, G., 2–3, 11
Tyler, R. E., 244, 248
Tyson, J. R., 260

U

Ubeda, M., 341
Uchida, H., 54, 62, 64
Udvardi, M. K., 307
Ueda, K., 79
Uehara, T., 323, 327
Uehara, Y., 79
Ueki, K., 151–153, 342
Uemura, A., 38
Uesato, S., 54, 64
Ulianich, L., 240
Unkles, S. E., 20
Upton, J. P., 87, 197, 275, 278, 287
Urade, R., 296–297, 299, 301, 304
Urano, F., 71, 73–76, 80, 83, 87, 150, 154, 242–243, 275–276
Urayama, S., 133–134
Urbanas, M. L., 240
Urzua, C., 220
Usherwood, Y. K., 239–240, 296
Ushioda, R., 235, 239, 248
Utsumi, T., 54, 62, 64
Uusitalo, J., 4
Uys, J. D., 321

V

Valent, B., 16, 18
Vale, W., 220
Valkonen, M., 4, 18
Vallabhajosyula, P., 76
van Anken, E., 278
van Breda, E., 151
Vandenbroeck, K., 240
van den Hondel, C. A., 3, 5–8, 19–21
van der Hoeven, R. A., 5
van der Kallen, C. J., 272
Vander Mierde, D., 144
Van Der Wal, F. J., 240

Author Index

Vande Walle, L., 87, 197, 275, 278, 287
van Esse, H. P., 296–298
Van Eylen, F., 78
van Gorcom, R. F., 19
van Greevenbroek, M. M., 272
van Hartingsveldt, W., 7
van Luijk, N., 3
Van Montagu, M., 305
van Peij, N. N., 4, 7, 9, 11
Van Voorden, S., 245
van Waes, M. A., 245
van Zeijl, C. M., 7
Vattem, K. M., 335–336, 342–343, 351
Velmurgan, S., 241
Vicente, V., 121, 123, 126–131
Vidal, S. E., 278
Viotti, C., 296–299, 301–304, 306–309, 312
Virrey, J. J., 218–219
Vishniac, W., 7
Vitale, A., 297
Vlzing, C., 239
Vongsangnak, W., 11

W

Wada, I., 86, 243–244, 248
Wada, K., 56
Wagener, R., 240
Waheed, A., 262
Wahlman, J., 245
Walker, K. W., 240
Walker, P. A., 316
Waller, S., 316
Walter, K. A., 220
Walter, P., 3–4, 33–35, 72, 79, 87, 138–139, 144, 148, 177, 196–197, 199, 202, 210, 213, 242, 260, 272, 274–279, 282–284, 287, 291, 334, 336–337
Wambach, M., 260–262
Wang, A., 179–180, 184
Wang, C., 139, 142
Wang, F., 260
Wang, H., 18, 219
Wang, J., 76, 79, 138
Wang, L., 267
Wang, M., 218–219
Wang, X. Z., 35, 75–76, 83, 87, 177, 197, 243, 275, 342
Wang, Y., 41, 85, 179–180
Wang, Z., 12, 124
Ward, M., 18
Warner, R. C., 346
Wasson, J., 79
Watabe, M., 179–180
Watanabe, N., 299
Watson, A. J., 4
Weaver, T. E., 239
Weed, H. G., 180
Weibezahn, J., 244
Weigel, D., 298, 300, 305, 310
Wei, J., 148
Wei, S., 219
Weiser, D. C., 337
Weissman, J. S., 3, 33, 79, 87, 144, 197, 236, 244–246, 260, 275, 277
Weitz, J. I., 78
Wek, R. C., 34, 36, 76, 149, 177, 333–337, 341–346, 349, 351
Wek, S. A., 177, 342, 345
Welch, W. J., 261–262
Welihinda, A. A., 139, 197, 243, 287
Wellburn, A. R., 303
Welling, C., 79
Weninger, W., 179
West, A. B., 76, 179
Wey, S., 218–219, 224–225, 227
White, M. F., 151, 180
White, S., 5, 14, 23
Whitham, M., 124
Whitlock, B. B., 55
Whitmore, M. M., 179
Whyteside, G., 2
Wiertz, E. J. H. J., 245, 248
Wilkinson, B., 240
Williams, B. R., 179, 260
Williams, D. B., 240
Wilson, R., 240
Wimalasena, T. T., 5
Winkeler, A., 245
Winnay, J. N., 147, 151–153, 342
Winter, K., 54
Winters, M. S., 5, 14, 23
Wiseman, R. L., 280, 284, 287, 291
Wittinghofer, A., 54
Wodicka, L., 3, 79, 144, 260, 277
Wolf, D. H., 244–245
Wolfgang, C. D., 176, 179
Wolford, C. C., 175, 177, 179–180
Wolfson, A., 138
Wollheim, C. B., 77
Wong, H. N., 83, 144
Woo, C. W., 139
Woods, T. L., 12
Wortelkamp, S., 239
Wu, J., 76, 260, 337
Wu, R., 125
Wu, S., 342
Wu, X., 108, 115, 177, 179–180
Wu, Y., 260

X

Xiang, S., 323
Xie, Z., 341
Xiong, S., 219
Xiong, Y., 321, 323–324, 327, 329

Xu, C., 111
Xu, F., 124
Xu, J., 55, 62–64
Xu, W. S., 72, 87, 138, 145, 160–161, 197, 272, 275, 278, 287
Xu, Y., 239
Xu, Z., 210, 213, 276–277
Xue, X., 342
Xue, Z., 138–139, 141–145

Y

Yagita, H., 342
Yamada, M., 54
Yamada, P. M., 125
Yamagishi, S., 54
Yamaguchi, M., 180
Yamaji, T., 54
Yamamoto, A., 55, 65, 74, 243
Yamamoto, K., 33, 35, 39–40, 76, 144, 296
Yamamura, K., 77
Yamanaka, S., 197, 199
Yamato, E., 77
Yamazaki, H., 95
Yan, A., 241
Yanagi, H., 76, 84, 150, 153, 199, 260
Yanai, A., 79
Yang, K., 74
Yang, L., 137–139, 141–145
Yang, X., 139, 142, 341
Yankner, B. A., 55, 62–64
Yan, L., 179–180, 184
Yan, W., 239, 260–261
Yao, D., 323, 327
Yao, J., 95, 101–104
Yao, P. M., 78
Yasumoto, H., 296, 309
Yasumura, D., 275
Yau, G. D., 76
Yazaki, Y., 151
Ye, J., 76, 163
Ye, R., 218–219
Ye, W., 219
Ye, X., 179–180
Ye, Y., 245–246, 248
Yen, T. S., 272, 275
Yim, A. M., 219
Yin, X., 175, 179–180
Yin, Z., 5
Ylikoski, E., 179
Yokomaku, D., 54
Yokota, T., 87
Yokouchi, M., 102
Yoneda, T., 55, 275
Yoo, S. E., 240
Yorihuzi, T., 86, 243
Yoshida, H., 55, 65, 74, 76, 84, 144, 148, 150, 153, 163, 243, 245–246, 260, 276, 296, 342
Yoshida, M., 144, 160
Yoshinaga, K., 218
Yoshizumi, M., 55
Young, D., 185
Younger, J. M., 245
You, Z., 86
Yuan, J., 55, 62–64, 83, 88, 339
Yu, H., 86
Yu, J. H., 16
Yu, M., 239
Yu, Q. C., 88
Yu, X., 162
Yui, D., 275
Yun, C. Y., 76, 178, 245–246, 248, 334, 336–337
Yura, T., 76, 84, 150, 153, 199, 260

Z

Zaborske, J. M., 337
Zahedi, R. P., 239
Zambito, F., 83
Zekert, N., 3
Zeng, H., 74, 76, 83, 86, 144, 150, 154, 178, 243, 276, 334, 336–337, 342–343, 345
Zenno, A., 139, 142
Zhang, C., 197, 199, 275, 278–279, 283–284, 287, 291
Zhang, D. D., 35, 78
Zhang, J., 76, 144
Zhang, K., 83, 138, 144, 150, 260, 272, 342–343
Zhang, P., 83
Zhang, W., 83, 144
Zhang, X., 139, 142
Zhang, Y., 75–76, 83, 86–87, 144, 148–149, 151, 178, 197, 206, 218–219, 228–231, 240, 243, 260, 275, 280, 284, 287, 291, 296, 334–337, 342–343, 345
Zhan, R., 179–180
Zhao, J., 55
Zhao, L., 262
Zhao, R., 219
Zhao, Y., 220, 230
Zhong, F., 124
Zhou, B., 220
Zhou, D., 346, 349
Zhou, H., 107–108, 112, 219, 224–225, 227
Zhou, J., 34, 210, 213, 276–277
Zhou, Y., 151, 341–342
Zhu, H., 55, 62–64
Zhu, L. J., 80
Zhu, M. X., 180
Zimmermann, R., 238–239, 296
Zinszner, H., 87
Zmuda, E. J., 180
Zong, W. X., 88
Zvaritch, E., 241
Zylicz, M., 238

Subject Index

A

Activating transcription factor 3 (ATF3)
  cellular adaptive–response network
    level, mRNA, 176
    stress signals, 177
  disease models
    double-edge sword nature, 179
    loss/gain-of-function approaches, 179
    mouse, 179–180
    roles, 179
  IHC protocol
    factors, pilot experiments, 181–186
    procedures, 187–192
    required materials, 186–187
  UPR
    ER-membrane proteins, 177
    signaling pathways, 178
    stress signals, 178
Activating transcription factor 4 (ATF4)
  bZIP family, 163
  CHOP in U251 cells, 164
  mRNA encoding, 163
Activating transcription factor 6 (ATF6)
  activation, ATF6-GFP fusion constructs
    limitation, 46
    protocol, 47–48
  ERSE II activation, 39–40
  nonstressed cells, 34
  reporter assays
    activity, 41, 44
    protocol, 44
  transfection, PC12 cells, 58–59

B

Batch cultivations
  controlled, 6
  prolonged, 8–9
  short, 8

C

Calciumbinding protein 1 (CaBP1), 298
Cardiomyocytes protection
  adult rat cardiac myocytes, 132
  animal models, 132
  dexamethasone treatment, 132
  heat-shock factors analysis, 133
  Western blot analysis, 133
Caspase-3 activity measurement
  ER stress-mediated apoptosis, 62
  fluorogenic substrates
    lysates preparation and required equipments, 63
    peptide, 62
    plating and treatment, 63
Caspase-12 fragmentation
  activation mechanism, 63–64
  detection
    ER stress-mediated apoptosis, 64
    modified Western blotting method, 64–65
    plating and treatment, 64
    required reagents and equipments, 65
CD measurements. See Circular dichroism measurements
C/EBP homologous protein (CHOP), 84
Cell surface proteins biotinylation
  avidin, 228
  description, 230
  materials and reagents, 228–229
  SDS-PAGE sample, 230
Chaperone protein and signaling regulator
  cell surface GRP78 detection
    biotinylation, 228–230
    FACS analysis, 230–231
  cytosolic GRP78 isoform detection
    codon downstream, 224–225
    mRNA and protein levels, 224
    quantitative real-time PCR, 226
    RT-PCR, 225
    single-copy gene, 224
    small interfering RNA, 227–228
    Western blot, 227
  ER stress, 218
  mouse models, 219
  TGF-β signaling, 220
  total GRP78/BiP detection
    quantitative real-time PCR, 220–222
    RT-PCR, 220
    Western blot, 222–223
Chemiluminescent and formazan assay
  luciferase and formazan, 100
  SEAP, 99–100
CHOP. See C/EBP homologous protein
Circular dichroism (CD) measurements, 330
Core stress-sensing region (CSSR)
  anti-aggregation activity, 213
  BiP, 209–210

Core stress-sensing region (CSSR) (cont.)
  IRE1α, 202
  maltose binding protein (MBP), 210
  X-ray structure, 210
CSSR. See Core stress-sensing region

D

Dithiothreitol (DTT)
  cDNA subtraction, 4
  disulfide-bond formation, 77
  effectiveness, 301
  ER-stress reagents, 11
  protein folding, 302
  thiol-reducing reagent, 197
Dulbecco's modified eagle's medium (DMEM)
  IEC-18 cells, 133
  PC12 cells, 56, 60
  siRNA transfection solution, 61

E

Enhanced chemiluminescent (ECL)
  immunoblot detection protocol, 348
  membrane-bound antibodies, 80
  protein signal detection, 223
Effector activation mechanism
  back-to-back dimers, 280–281
  crystal structure, 278
  dimerization interfaces, 282
  "effector oligomer/polymer" model, 280
  face-to-face and side-to-side arrangements, 281
  IRE1 molecules, 278
  novel alpha-helical domain, 279
  nuclease active site and dimer interfaces, 279
  oligomerization/polymerization, 280
eIF2α. See Eukaryotic initiation factor 2α
eIF2 kinase pathway measurement and ER stress
  eIF2αP and ISR
    antibody preparation, 341
    ATF4 target gene, 339
    culture cells, 339
    ER stress agents, 338
    LI-COR and ECL immunoblot detection protocol, 340
    MEF cells, 342
    membrane and membrane filters, 340
    phospho-PERK immunoblot analysis, 342
    stress condition, 341–342
    washed cells, 339–340
  in vitro kinase assay
    cells culture, 344
    protein A argarose, 344
    serine-51 phosphorylation site, 344
  PERK phosphorylation status
    immunoblot analysis, 343
    membrane filters, 343–344
    protein A agarose, 343

EM. See ER dilation
Endoplasmic reticulum (ER) stress
  alternative splicing, 219
  associated stress induction
    analysis, hacA transcript splicing, 9–10
    DTT and tunicamycin, 8–9
    reagents concentration and Aspergillus niger expression analyses, 11
    RT-PCR method, 10–11
  cell culture system
    cell lines and primary cells, 76–77
    mammalian cells, 76
  chaperones, 323
  defined, 196
  description, 72
  detecting levels
    CHOP and XBP1 proteins, 143
    PERK downstream targets, 142–143
  eIF2αP, 338
  ESTRAP assay
    in culture cells, 96–101
    in vivo, 101–104
  expression analysis, 11–14
  folding capacity, 218
  induced expression pattern
    GUS reporter fusion, 304–306
    qRT-PCR, 307
    SDF2-specific antibodies, 307–309
  inducers
    pharmaceutical, 77
    physiological, 78
  inducing and monitoring
    disulfide bond formation, 299–300
    DTT, 299
    phenotypes, 302–304
    plant growth and application, 300–301
    UPR induction monitoring, 301
  ISR, 337
  measurement
    EM, 79
    real-time redox, 79
  mediated apoptosis, 87–88
  MEF cells, 348
  methods, 94
  quantitating, UPR sensors
    Tg-treated MEFs, 142
    visualization and quantitation, 141–142
  role
    folding and maturation, 73
    yeast, 3–4
  SEAP reporter system
    advantages, 95–96
    GRP78 expression, 96
  secreted luciferase and SEAP, 95
  secretory proteins, 94–95
  tumor cells, 218
  UPR detection
    cell lysates, 139–140

Subject Index

gel running and transfer, 140
phos-tag gels, 139
Western blot, 140–141
ER. *See* Endoplasmic reticulum
ER-associated protein degradation (ERAD)
  composing steps, 236–237
  ERQC mechanism
    degradation signals and recognition, 243–244
    disulfide bonds reduction, 248–249
    ER stress–UPR protection system, 242–243
    nonglycosylated proteins, 244–245
    retrotranslocation, dislocon channels, 245–246
    ubiquitination, 246–248
  folded and disulfide-bonded ternary structure, 237
  misfolded proteins, 237
  newly synthesized proteins
    BiP and Hsp40 family proteins, 238–239
    CNX and CRT, 241
    oxidoreductases, 239–240
  and protein stability, 86
  secretory and membrane proteins, 236
ER dilation (EM), 79
ER quality control (ERQC) mechanism
  degradation signals and recognition
    EDEM, 244
    ERAD pathway, 243
    N-glycosylated substrates, 243
  disulfide bonds reduction, 248–249
  ER stress–UPR protection system
    ATF6, 242
    cytosolic RNase domain, 243
    glycosylation inhibitors, 242
    sensors, 242–243
  nonglycosylated proteins
    misfolded substrates, 245
    N-glycan, 244–245
  pathways, 236
  retrotranslocation, dislocon channels
    Derlin family proteins, 245
    Sec61 translocon channel, 245
  ubiquitination, 246–248
ERSE. *See* ER stress–response element
ER stress-activated indicator (ERAI)
  IRE1α–expression, 200–201
  reporter, 198
  splicing-susceptible fragment, 199–200
  strong and deregulated induction, 203
  transient co-transfection, 205
ER stress detection approach
  ATF6 activation, ATF6–GFP fusion constructs
    limitation, 46
    protocol, 47–48
  IRE1 activation, XBP1–*venus* reporter
    design, 44–45

expression and tunicamycin treatment, 44–45
protocol, 45–46
qRT-PCR, UPR target genes
  HERP and Northern blotting, 40
  mammalian ERSE, 39–40
  protocol, 40–41
  Taqman assays, 41–42
sXBP-1 mRNA
  conventional RT-PCR protocol, 38–39
  mammals and yeast, 36–38
  semiquantitative RT-PCR use, 37–38
  spliceosome *vs.* unconventional splicing, 38
Western blotting, UPR target genes, 41
XBP-1 and ATF-6 activity, 41, 44
ER stress expression, filamentous fungi
  polysome analysis
    required materials, 12–13
    RNA fractionation, 14–15
    sample preparation, 13
    sucrose gradient preparation, 13–14
    translation efficiency determination, 12
  transcriptomic analysis
    affymetrix format arrays, 11
    sequencing method advantages, 12
ER stress induction, PC12 cells
  culture and treatment
    alamarBlue® and Nunc easy flasks, 56–57
    cell viability and NGF, 56
    DMEM and collagen coated plates, 56
    PC12h cell line use, 56
  GRP78 detection, 55–56
  upregulation, BH3-only protein expression
    Bcl-2 family, 60
    plating and treatment, 60
    siRNA, 60–61
    TUNEL and PI staining, 61–62
  XBP1 mRNA and protein use
    p50ATF6 isolation and detection, 59
    PERK, Ire1 and ATF6 effects, 57–58
    required equipments, 59
    total RNA isolation, 57
    transfection, human ATF6 expression plasmid, 58–59
ER stress-response element (ERSE)
  ATF6α, 150
  cis-acting, 41
  mammalian, 39
  response elements, 85
ER stress-responsive alkaline phosphatase (ESTRAP)
  culture cells
    chemiluminescent and formazan assays, 99–100
    data analysis, 101
    reporter cells formation, 96–98
    treatment, reporter cells, 98–99

ER stress-responsive alkaline phosphatase (ESTRAP) (cont.)
mice
characterization, 101–102
endotoxemia application, 102–103
heavy metal intoxication, 103–104
in vivo detection, ER stress, 102
reporter cell-implanted, 103–104
ESTRAP. See ER stress-responsive alkaline phosphatase
Eukaryotic initiation factor 2α (eIF2α), 34, 36

F

FACS analysis. See Fluorescence-activated cell sorting analysis
Filamentous fungi culture conditions
cultivation methods, 6
inoculation
conidia, 7
short and prolonged batch cultivations, 8–9
mycelium growth, 5–6
required materials
Aspergillus niger strains, 6–7
bioreactor apparatus devices, 6
MM, reagents and disposables, 7
Filamentous fungi, UPR investigation methods
Aspergillus nidulans and *Trichoderma reesei*, 4
commercial exploitation, 3
culture conditions
cultivation methods, 6
inoculation, 7–8
mycelium growth, 5–6
required materials, 6–7
ER-associated stress induction
analysis, hacA transcript splicing, 9–10
DTT and tunicamycin, 8–9
reagents concentration and *Aspergillus niger* expression analyses, 11
RT-PCR method, 10–11
ER role, yeast, 3–4
ER stress expression analysis
polysome, 12–14
transcriptomic, 11–12
hacA gene role, 5
Hac1p synthesis and RESS, 4
hyphae, 2
secretion stress responses, *Aspergillus niger*, 4–5
secretory pathway, 2–3
signaling modification, gene replacement
constitutive activation, 18–20
deficient strains construction, 14–18
mutant verification, 23
PEG-mediated transformation, protoplasts, 21–22
phenotypic analysis, 23–26
Fluorescence-activated cell sorting (FACS) analysis

GRP78
domain determination, 231
plasma membrane, 230
instrument, 231
isotype-matched control, 231
materials and reagents, 230–231
Formalin-fixed paraffin embedded (FFPE), 186
Formazan assay, 100

G

Gene-specific translational control
ATF4 5′UTR luciferase assay
eIF2αP stress and nonstressed conditions, 351
Renilla luciferase plasmid and RLU, 353
ribosomes and uORF, 351
polysome profiling
cell lysates, 347
fraction collector, 348–349
global mRNA translation, 348
sucrose gradient fractionation, 346
tube rotation method, 346–347
ultracentrifugation, 347
protein synthesis measurement
lysate preparation, 345
methionine and cysteine, 345–346
radiolabeled amino acids, 345
stressed and nonstressed cells, 346
RNA isolation and characterization
ATF4 transcripts and translational control, 350–351
incubation, 349
normalization purposes, 349
Glucose-regulated protein 78 (GRP78)
dissociation, 34, 36
expression upregulation
p50ATF6 isolation and detection, 59
PERK, Ire1 and ATF6 effects, 57–58
required equipments, 59
total RNA isolation, 57
transfection, human ATF6 expression plasmid, 58–59
hyperinduction
detection by Northern blotting, 66–68
mechanism, 65–66
Glucose-regulated protein 94 (GRP94), 296
GRP78. See Glucose-regulated protein 78
GRP78 acetylation
ATF4 and CHOP, 165–166
cell lines and treatment, 167–168
clonogenic survival
experimental procedure, 171–172
materials, 171
HDAC inhibition, 164
immunoblot analysis
anti-acetylated lysine residues, 164
experimental procedure, 169–170

materials, 169
immunoprecipitation
  experimental procedure, 168–169
  materials, 168
loading control, 164
PERK downstream signaling, 164, 166
protein regulation, 163
small interfering RNA
  experimental procedure, 170–171
  materials, 170
UPR activation, 166
GRP78 detection
  BiP
    quantitative real-time PCR, 220–222
    RT-PCR, 220
    Western blot, 222–223
  cell surface
    biotinylation, 228–230
    fluorescence-activated cell sorting (FACS) analysis, 230–231
  isoform
    codon downstream, 224–225
    mRNA and protein levels, 224
    quantitative real-time PCR, 226
    RT-PCR, 225
    single-copy gene, 224
    small interfering RNA, 227–228
    Western blot, 227
  Northern blotting, 66–67
  Western blotting, 67–68

H

Heat stress response, dexamethasone
  cardiomyocytes protection
    adult rat cardiac myocytes, 132
    animal models, 132
    dexamethasone treatment, 132
    heat-shock factors analysis, 133
    Western blot analysis, 133
  epithelial cells protection
    oxidant-induced stress, 133
    Sprague-Dawley rats, 133–134
  peripheral blood mononuclear cells, 125–126
  plasma HSPs, 124–125
  UPR in mice and cellular models
    L-ASP-induced conformational disease, 126–131
    LPS-induced septic shock, 126
HERP. See Homocysteine-induced ER protein
Highly active antiretroviral therapy (HAART), 108
Histone acetyltransferases (HATs), 161
Histone deacetylases (HDACs)
  enzymes and cancer
    chromatin structure, 159
    histones, amino-terminal lysine residue, 160
  inhibitors
    anti-cancer agents, 161
    chaperone protein function, 162
    chromatin structure, 161
    nonhistone substrates, 163
    stages, clinical evaluation, 160–161
    tumorigenesis role, 160
  UPR and inhibition
    clonogenic survival, 171
    ER stress sensors, 163
    Grp78 acetylation, 163–169
    immunoblot analysis, 169–170
    novel nonhistone target, 162
    small interfering RNA, 170–171
HIV PI-induced ER stress
  barrier integrity disruption
    histological examination, 116–117
    intestinal permeability in vivo measurement, 115
    Western blot analysis, 115–116
  paracellular permeability measurement, 113–114
  real-time RT-PCR and Western blot analysis, 110–113
  SEAP in IECs, 109–110
  UPR activation in cultured IECs
    detection, 110
    epithelial stem cells, 109
    real-time RT-PCR and Western blot analysis, 110–113
    SEAP, 109
Homocysteine-induced ER protein (HERP)
  ER membrane, 40
  misfolded nonglycosylated proteins, 248
  nonglycosylated proteins, 247
  promoter, 39

I

IHC protocols. See Immunohistochemistry protocols
Immunoglobin binding protein (BiP), 84
Immunohistochemistry (IHC) protocols
  factors, pilot experiments
    antigen retrieval, 182
    primary antibodies, 181–182
    secondary antibody and detection methods, 182–184
    signal specificity, 184–186
  procedures
    counterstain and coverslipping, 191
    deparaffinization and rehydration, 187
    peroxidase reaction, 190–191
    pressure cooker method, 189
    primary antibody, blocking and incubation, 189–190
    secondary antibody, 190
    semi-quantification, 192
  required materials

Immunohistochemistry (IHC) protocols (*cont.*)
  devices, 186–187
  disposables, 187
  reagents and buffers, 187
  samples, 186
Inositol-requiring enzyme 1 (IRE1)
  activation
    antibodies, ER stress markers detection, 80–81
    design, 44–45
    expression and tunicamycin treatment, 44–45
    knockout MEFs, 83
    phosophorylation levels, 79–80
    primers, ER stress markers detection, 80, 82
    protocol, 46
    XBP-1 mRNA, 83
  dimerization, 34
  endoribonuclease activity, 35
  immunoblot analysis, 153
  XBP-1 splicing assay
    DNA fragments, 154–155
    ER stress, 154
    required materials, 155
  XBP-1s translocation
    immunoblot analysis, 156
    required materials, 156
Integrated stress response (ISR)
  cultured cells, 339
  eIF2α phosphorylation, 177–178
  ER stress, 337
  transcriptome, 338
Intestinal epithelial cells (IECs)
  barrier integrity disruption, 114–117
    histological examination, 116–117
    intestinal permeability *in vivo* measurement, 115
    Western blot analysis, 115–116
  HIV PI-induced ER stress and UPR activation analysis, 109–114
  pathophysiological relevenace, 108
IRE1. *See* Inositol-requiring enzyme 1
IRE1 family proteins
  arabidopsis, 197
  BiP
    calcium-phosphate method, 207–208
    ER-stress sensors, 206
    immunoprecipitation, 206–207
    results and insights, 208–210
  dithiothreitol (DTT), 197
  ER-stress sensors., 197–198
  in vivo activity monitoring
    activation in mammalian cells, 203–204
    cellular luminescence activity, 200
    CSSR deletions, 205
    deletion-scanning mutagenesis, 202
    downstream molecular events, 199
    ERAI, 199–201

  expression plasmids, mammalian, 200
  high-order oligomer, 199
  loosely folded structure, 205
  mammalian cells, 199
  RT-PCR amplification, 201–202
  stress-dependent splicing, 203
  structure and mutations, 202–203
  unfolded proteins with yeast
    E. coli production and purification, 210–211
    in vitro anti-aggregation assay, 211–212
  results and insights, 213
  XBP1u and ERAI-reporter, 198
IRE1 modulation principles
  ATP-competitive compounds, 288
  back-to-back dimerization, 288
  dimer-spanning groove, 291
  drug
    discovery, 283
    interaction, 290
  exert indirect control, 290
  focal points
    back-to-back dimer arrangement, 287
    dimerization, 286
    "effector dimer" model, 287
    fungal and metazoan systems, 286
    genetic modulation using 1NM-PP1, 283
    kinase-inhibitors mimick ADP, 283–284
    nucleotide-binding site, 287–288
  proapoptotic
    kinase activity, 292
    signaling, 282
  prosurvival and proapoptotic signals, 287
  protein–protein interactions, 287
  Q-site binding compounds, 290
  quercetin
    action mechanism, 286
    in vitro and *in vivo* mutational analysis, 285–286
    small-molecule modulators, 284
  RNA cleavage, 289
  UPR
    ATF6, 273
    biosynthetic capacity, 272
    cytoplasmic effector region, 275
    ER stress, 274
    intracellular signaling, 275
    signaling pathways, 272–273
    using chemical tools, 288–289
  XBP1 signaling structural biology
    effector activation, 278–282
    hydrophobic membrane-spanning, 276
    intracellular signaling, 276
    misfolded protein detection, 277–278
    resultant exons, 276
    XBP1 and HAC1, 276–277
ISR. *See* Integrated stress response

Subject Index

K

Knockout (KO) mice
  ATF3-negative controls, 184
  murine mammary tumors virus (MMTV), 185
  tumors, 186

L

L-Asparaginase (L-ASP)-induced conformational disease
  conformationally sensitive proteins
    α1-antitrypsin polyclonal antibodies, 128–129
    antithrombin, 128
    dexamethasone effects, 127
    Dpex mounting medium, 127
    immunofluorescence evaluation, 127–128
    polymerization, 128
  heat stress response and UPR
    dexamethasone effect, 129–130
    GRP78 levels, 130–131
    immunohistochemistry, 130
    proteins intracellular expression, 129
    PVDF membranes, 131
    Western blot analysis, 129
  in vivo and *in vitro* studies, 126–127
Luciferase assay, 100

M

Maltose binding protein (MBP)
  CSSR, 212
  yeast Ire1, 210
MEF cells. *See* Mouse embryo fibroblast cells
Misfolded protein detection mechanism
  dimer-spanning groove, 277
  hLD and yLD structures, 277–278
  IRE1 structural biology, 277
Mouse embryo fibroblast (MEF) cells
  ER stress agent, 348
  Renilla luciferase plasmid, 352
  sucrose gradient, 350
Murine mammary tumors virus (MMTV), 185–186

N

Nerve growth factor (NGF)
  Caspase-3 activity measurement
    ER stress-mediated apoptosis, 62
    fluorogenic substrates, 62–63
  ER stress-mediated apoptosis, 55
  fragmentation, Caspase-12
    activation mechanism, 63–64
    detection, 64–65
  GRP78 hyperinduction
    detection by Northern blotting, 66–68
    mechanism, 65–66
  induction, ER stress
    culture and treatment, 56–57
    GRP78 detection, 55–56
    upregulation, BH3-only protein expression, 60–62
    XBP1 mRNA and protein use, 57–59
  NTs binding, 54
  PI3-K and signaling pathways, 54–55
Neurotrophins (NTs), 54

O

Oxidoreductases, ER
  electron transfer, 240
  intermolecular disulfides, 239
  major histocompatibility complex (MHC), 240
  oxidative reaction, 240
  thioredoxin like domains, 239–240

P

PDI. *See* Protein disulfide isomerase
PERK. *See* PKR-like ER kinase
Pheochromocytoma 12 (PC 12) cells
  culture and treatment, 56–57
  GRP78 detection, 55–56
  upregulation, BH3-only protein expression, 60–62
  XBP1 mRNA and protein use, 57–59
Phosphatidylinositol 3-kinase (PI3-K), 54
PI 3-K. *See* Phosphatidylinositol 3-kinase
PI 3-kinase regulatory subunits, UPR
  ER transmembrane transcription factor, 149
  luminal domain, 150
  modulators
    lipid kinases, 150
    p85α and p85β structure, 150–151
    p85α expression, 151
  nascent polypeptides, 148
  pathway activation
    ATF6α, 153
    IRE1α, 153–156
    PERK, 152–153
PKR-like ER kinase (PERK)
  cell extracts preparation, 152–153
  described, 34
  phosphorylation, 35
  phospho-specific antibodies, 152
  proximal sensors, 148–149
  signaling in cytosol, 260
  trans-autophosphorylation, 41
  type I transmembrane protein, 148
Polyvinylidene difluoride (PVDF) membranes, 169

Protein disulfide isomerase (PDI)
　chaperones, 323
　ER and UPR, 322–323
　GSH/GSSG ratio, 322
　nitrosative stress, 323
　redox reactions, 322
　ROS/RNS, 322
　S-glutathionylated proteins
　　identification and confirmation, 324–326
　　structural and functional consequences, 329–331
　target cysteine residues
　　biotin switch assay, 327
　　materials, 327
　　methods, 328–329
P58(IPK), UPR
　bind unfolded proteins using subdomain I
　　hydrophobic residues, 264–265
　　ideal binding site, 265
　　mapping, 265–266
　crystal structure, TPR domain, 263–264
　dual-function protein
　ER
　　resident Hsp40, 262
　　stress, 260
　J domain–BiP interaction, 267
　knocking out, mouse models, 261–262
　working model, 267

R

Reactive oxygen and nitrogen species (ROS/RNS), 322
Reporter cells
　formation
　　dual and singular, 97
　　lack, ER stress, 97–98
　　LLC-PK1 cells, 96
　　transient transfection use, 98
　treatment with ER stress inducers
　　continuous monitoring, 99
　　quantitative assessment, 98–99
Repression under secretion stress (RESS), 4
Resident molecular chaperones
　ERADof, 239
　ERdj1–7, 238
　eukaryotes, 238
　Hsp70 family proteins, 238
　Shiga toxin, 239
ROS/RNS. *See* Reactive oxygen and nitrogen species

S

Secreted alkaline phosphatase (SEAP)
　assay, 99–100
　expression levels, 97–98
　reporter system, 95–96
　vs. secreted luciferase, 95

S-glutathionylated proteins
　identification and confirmation
　　anti-SSG, 325
　　dose-curve, 325–326
　　2-D SDS-PAGE and MALDI-TOF, 326
　　in vitro and *in vivo*, 324
　　materials, 325
　　protein A/G plus agarose, 326
　　target cysteine residues, 324
　　wash cells and PBS, 325
　structural and functional consequences
　　CD measurements, 330
　　isomerase activity determination, 330–331
　　materials, 329
　　spectroscopic analysis, 329
　　UPR activation, 329
Spliced X box-binding protein-1 (sXBP1)
　mRNA detection
　　conventional RT-PCR protocol, 38–39
　　mammals and yeast, 36–38
　　semiquantitative RT-PCR use, 37–38
　　spliceosome *vs.* unconventional splicing, 38
　role, ER stress, 35
Src-homology 2 (SH2) domains, 150
Stromal-cell derived factor 2 (SDF2), *Arabidopsis*
　chaperone systems, ER, 296
　developmental and morphological defects, 297
　E. coli, GST recombinant purification
　　affinity, 315
　　fusion construct, 313–314
　　tag removal, 316
　ER stress-induced expression pattern
　　GUS reporter fusion, 304–306
　　qRT-PCR, 307
　　SDF2-specific antibodies, 307–309
　inducing and monitoring ER stress
　　disulfide bond formation, 299–300
　　DTT, 299
　　phenotypes, 302–304
　　plant growth and application, 300–301
　　UPR induction monitoring, 301
　principle mechanisms, 297
　subcellular localization
　　confocal microscopy, 312–313
　　sucrose density gradient centrifugation, 310–312
　　transient expression, 309–310
　T-DNA insertion lines
　　description, 297–298
　　genotyping mutants, 298–299
sXBP1. *See* Spliced X box-binding protein-1

T

Target cysteine residues
　biotin switch assay, 327
　cells, 328

Subject Index

in vitro, 328
  mass spectrometry identification, 328–329
  materials, 327
Tissue microarrays (TMAs), 181
Toll-like receptors (TLR), 35–36
Translational control, UPR
  eIF2α phosphorylation
    mammalian cells, 336
    mTOR signaling pathway, 337
    stress-related genes, 336
  eIF2 kinase PERK
    ER stress, 334
    gene expression, 334
    GRP78 repression, 336
    type I ER transmembrane protein, 335
  experimental strategies and methods
    eIF2αP and downstream target genes, 337
    ER stress, 338
    global and genespecific translation, 338
    ISR transcriptome, 337–338
  gene-specific
    ATF4 5′UTR luciferase assay, 351–353
    polysome profiling, 346–349
    protein synthesis measurement, 345–346
    RNA isolation and characterization, 349–351
Triphosphate nick end-labeling (TUNEL)
  method, 61–62

U

Unfolded protein response (UPR)
  activation and signaling, 138
  defined, 236
  detection
    cell lysates, 139–140
    gel running and transfer, 140
    phos-tag gels, 139
    Western blot, 140–141
  downstream markers and responses
    ERAD and protein stability, 86
    ER stress-mediated apoptosis, 87–88
    immunostaining and immunofluorscence, 84–85
    mRNA degradation, 87
    transcriptional activation, 85–86
    translational attenuation measurement, 86
  eIF2 kinase pathway measurement and ER Stress
    eIF2αP and ISR, 338–343
    PERK, 343–344
  hyperphosphorylation, 139
  IECs activation
    HAART, 108
    inflammatory cytokine production, 108
    using SEAP, 109–110
  inhibition
    clonogenic survival, 171
    ER stress sensors, 163
    Grp78 acetylation, 163–169
    immunoblot analysis, 169–170
    novel nonhistone target, 162
    small interfering RNA, 170–171
  measuring methods, activation
    ATF6α, 83–84
    IRE1α, 79–83
    PERK, 83
  modulators
    lipid kinases, 150
    p85α and p85β structure, 150–151
    p85α expression, 151
  pathway activation
    ATF6α, 153
    IRE1α, 153–156
    PERK, 152–153
  primary functions, 322–323
  protection system
    ATF6, 242
    cytosolic RNase domain, 243
    glycosylation inhibitors, 242
    sensors, 242–243
  quantitating, sensors
    Tg-treated MEFs, 142
    visualization and quantitation, 141–142
  regulators
    ATF6, 76
    IRE1, 74–75
    PERK, 76
  response categories, 73–74
  signaling modification
    constitutive activation, 18–20
    deficient strains construction, 14, 16–18
    mutant verification, 23
    PEG-mediated transformation, protoplasts, 21–22
    phenotypic analysis, 23–26
  translational control
    eIF2α phosphorylation, 336–337
    eIF2 kinase PERK, 334–336
    experimental strategies and methods, 337–338
    gene-specific investigation, 345–353
UPR. *See* Unfolded protein response
UPR detection assays
  activation, ER transmembrane receptors, 33–34
  ATF6 and GRP78, 34
  eIF2α phosphorylation and OSU-03012, 36
  ER lumen *vs.* cytosol, 33
  ER stress detection approaches
    IRE1 activation and ATF-6 translocation, 44–48
    mRNA and protein levels, 39–43
    sXBP-1 mRNA, 36–39
    XBP-1 and ATF-6 activity, 41, 44
  human diseases and ER stress, 36
  IRE1 activation and sXBP1, 35

UPR detection assays (cont.)
  PERK role, 34–35
  TLR role, 35–36
UPR downstream markers
  ERAD and protein stability, 86
  ER stress-mediated apoptosis
    BAX and BAK, 87–88
    components, 87
    measuring methods, 88
  immunostaining and immunofluorscence
    CHOP, BiP and PDI, 84
    Lab-Tek chambers, 84–85
    ProLong Gold antifade mounting, 85
  mRNA degradation, 87
  transcriptional activation
    ER stress inducers activation, 85–86
    IRE1α, PERK, and ATF6α, 85
  translational attenuation, 86

X

X-box binding protein 1 (XBP1)
  endoribonuclease activity, 35
  ERAD-associated proteins, 38
  mRNA and protein use
    p50ATF6 isolation and detection, 59
    PERK, Ire1 and ATF6 effects, 57–58
    required equipments, 59
    total RNA isolation, 57
    transfection, human ATF6 expression plasmid, 58–59
  rat, human and mouse, 39
  splicing, 36
  venus reporter
    design, 44–45
    expression and tunicamycin treatment, 44–45
    protocol, 45–46

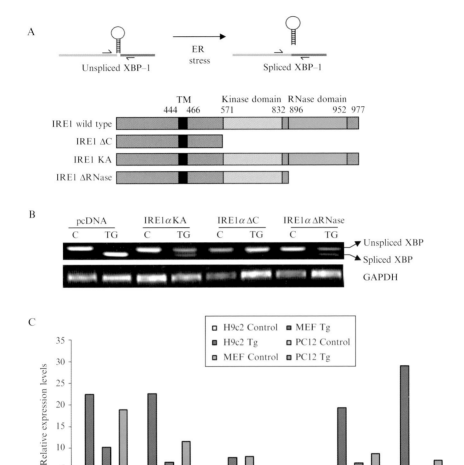

Karen Cawley et al., Figure 2.1 Detection of transcript levels of UPR target genes by RT-PCR. (A) Upper panel, cartoon of XBP1 splicing during ER stress. Lower panel, schematic representation of various mutant constructs of IRE1. (B) Modulation of XBP1 splicing by mutant IRE1. Total RNA was isolated from HEK 293 cells were transfected with IRE1 mutants, either untreated or treated with thapsigargin (0.5 μM) 6 h, and RT-PCR analysis of total RNA was performed to simultaneously detect both spliced and unspliced XBP1 mRNA and GADPH. (C) Induction of UPR target genes upon exposure to thapsigargin. Total RNA was isolated from indicated cells after treatment thapsigargin (Tg) and the expression levels of the indicated genes were determined by real-time RT-PCR, normalizing against GAPDH expression. Adapted from Samali et al., 2010.

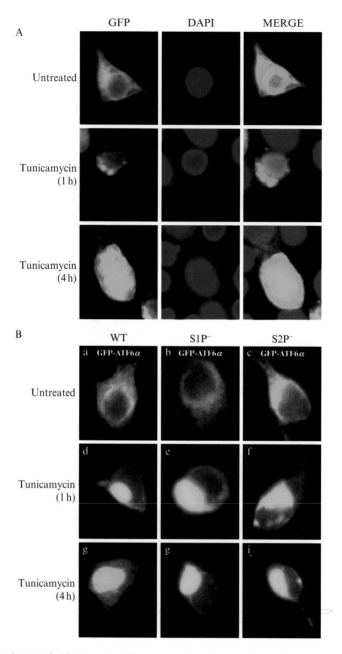

Karen Cawley et al., Figure 2.4 ER stress-induced processing and nuclear translocation of GFP–ATF6. (A) Twenty-four hours after transfection with pCMVshort-EGFP–ATF6 (WT), 293T cells were left untreated or treated with 1-μg/ml tunicamycin for the indicated periods. Cells were fixed in 4% paraformaldehyde, stained with DAPI, and then analyzed by fluorescence microscopy. (B) Twenty-four hours after transfection with pCMVshort-EGFP–ATF6 (WT), pCMVshort-EGFP–ATF6(S1P−), or pCMVshort-EGFP–ATF6(S2P−), 293T cells were left untreated or treated with 1-μg/ml tunicamycin for the indicated periods and then analyzed by fluorescence microscopy. Adapted from Samali et al., 2010.

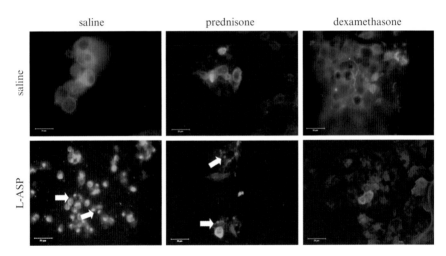

José A. Guerrero *et al.*, Figure 7.1 Effect of dexamethasone and prednisone on the intracellular accumulation of antithrombin (arrows) caused by L-asparaginase (L-ASP) in HepG2 cells and evaluated by immunofluorescence.

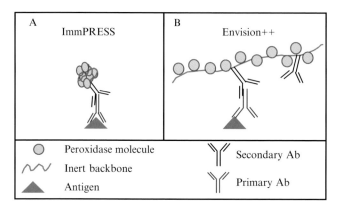

Tsonwin Hai et al., Figure 11.3 A schematic of two commercially available polymer-based systems. A flexible micropolymer backbone (ImmPRESS reagent; A) or dextran (Envision++ reagent; B) bridges a large number of peroxidase molecules to each secondary antibody. Ab, antibody.

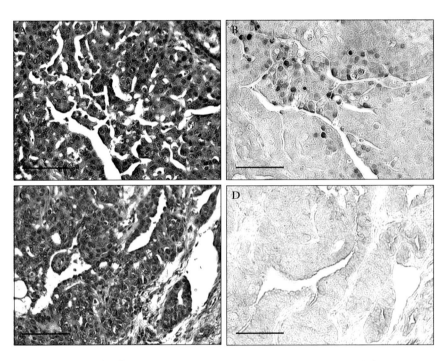

Tsonwin Hai et al., Figure 11.6 Analyses of ATF3 in tumors derived from WT or ATF3 KO mice. Serial sections of tumors from MMTV-PyMT transgenic mice in either WT (A, B) or ATF3 KO (C, D) background were stained with hematoxylin and eosin (A, C) or Atlas anti-ATF3 antibody (B, D). The slower developing DAB substrate was used in this experiment. No counterstain was applied, as it reduced the ability to discern different signal intensities. Scale bars: 100 μm.

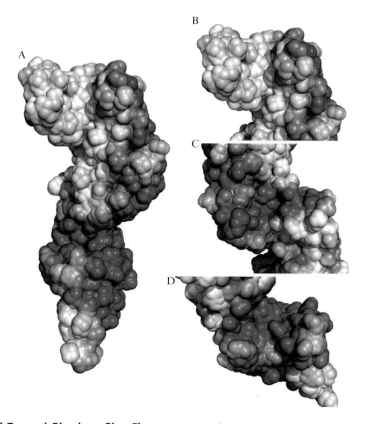

Jiahui Tao and Bingdong Sha, Figure 15.2 Surface potential drawing of P58(IPK) TPR domain. Solvent-accessible surface of P58(IPK) TPR domain is colored according to their potential, red for negative, white for neutral, and blue for positive (Fig. 15.2A). The groove of subdomain I is neutral (Fig. 15.2B), while grooves of subdomain II and III are negatively charged (Fig. 15.2C,D).

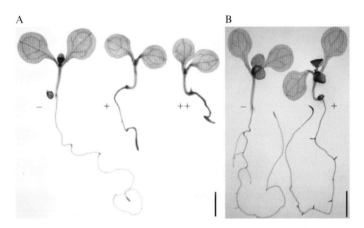

Andrea Schott and Sabine Strahl, Figure 17.2 Analysis of *SDF2* expression. (A, B) *SDF2* promoter activity detected in GUS reporter plants. Transgenic plants harboring the *ProSDF2:GUS* reporter were stained for GUS activity to monitor sites of *SDF2* expression. (A) Seven-day-old seedlings either grown on MS plates (−) or MS medium supplemented with 2 mM (+) or 2.5 mM DTT (++). Bar: 1 mm. (B) Seven-day-old seedlings grown overnight in liquid MS medium were treated for 24 h with 1 mM DTT (+). Mock-treated seedlings were grown in liquid MS medium (−). Bar: 2 mm. (This research was originally published in Schott *et al.* 2010. Arabidopsis stromal-derived factor 2 (SDF2) is a crucial target of the unfolded protein response in the endoplasmic reticulum. © The American society of Biochemistry and Molecular Biology.)

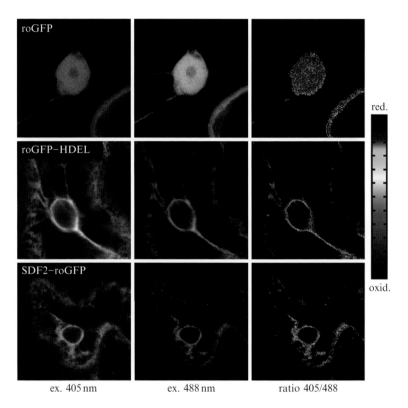

Andrea Schott and Sabine Strahl, Figure 17.3 Ratiometric imaging of SDF2–roGFP. SDF2–roGFP fusion indicates ER localization of SDF2. SDF2–roGFP, the ER-localized roGFP–HDEL (Meyer *et al.*, 2007), and cytosolic roGFP (Meyer *et al.*, 2007) were transiently expressed in *N. benthamiana* leaf epidermal cells under control of the CaMV 35S promoter. Images were taken by confocal laser scanning microscopy with excitation at 405 and 488 nm as indicated. Ratio images 405/488 nm were calculated and converted into pseudocolors. The color scale shows the pseudocolor coding for reduced (blue) and oxidized roGFP (red).